This book addresses both fundamental and applied aspects of ocean waves including the use of wave observations made from satellite. More specifically it describes the WAM model, its scientific basis, its actual implementation, and its many applications. This model has been developed by an international group (the WAve Modelling group), and is based on a detailed physical description of air/sea interactions. It is widely used for wave forecasting for meteorological and oceanographic purposes.

The volume describes the basic statistical theory and the relevant physical processes; the numerical model and its global and regional applications; and satellite observations, their interpretation and use in data assimilation.

Written by leading experts, it is a comprehensive guide and reference for researchers and advanced students in physical oceanography, meteorology, fluid dynamics, coastal engineering and physics.

T0182160

DYNAMICS AND MODELLING OF OCEAN WAVES

DYNAMICS AND MODELLING OF OCEAN WAVES

G.J. KOMEN, L. CAVALERI, M. DONELAN,
K. HASSELMANN, S. HASSELMANN
AND P.A.E.M. JANSSEN

CAMBRIDGE
UNIVERSITY PRESS

CAMBRIDGE UNIVERSITY PRESS
Cambridge, New York, Melbourne, Madrid, Cape Town, Singapore, São Paulo

Cambridge University Press
The Edinburgh Building, Cambridge CB2 2RU, UK

Published in the United States of America by Cambridge University Press, New York

www.cambridge.org
Information on this title: www.cambridge.org/9780521470476

First published 1994
First paperback edition (with corrections) 1996

A catalogue record for this publication is available from the British Library

Library of Congress Cataloguing in Publication data

Dynamics and modelling of ocean waves / G.J. Komen . . . [*et al.*].
p. cm. –
Includes bibliographical references and index.
ISBN 0-521-47047-1
1. Ocean waves. I. Komen, G.J.
GC211.2.D96 1994
551.47′02–dc20 94-65395 CIP

ISBN-13 978-0-521-47047-6 hardback
ISBN-10 0-521-47047-1 hardback

ISBN-13 978-0-521-57781-6 paperback
ISBN-10 0-521-57781-0 paperback

Transferred to digital printing 2006

Final report of the WAM group/SCOR wg 83

Contents

Contributors

G. Barzel

E. Bauer

L. Bertotti

C. Brüning

G. J. H. Burgers

C. J. Calkoen

V. Cardone

L. Cavaleri

M. Donelan

R. Flather

H. C. Graber

H. Günther

B. Hansen

K. Hasselmann

S. Hasselmann

M. M. de las Heras

M. W. Holt

L. Holthuijsen

P. A. E. M. Janssen

R. E. Jensen

K. Kahma

G. J. Komen

P. Lionello

R. B. Long

V. K. Makin

D. Masson

C. Mastenbroek

J. Monbaliu

H. L. Tolman

C. F. De Valk

S. L. Weber

X. M. Wu

I. R. Young

Y. Yuan

L. Zambresky

Affiliations and addresses are given on page 488.

Preface

De uiterlijke eigenheden der zee, de kleuren en de verschillende golfvormen en de wijze waarop de grote baren omslaan, kokend aanbruisen, schuim-betijgerd uitzwalpen en zich terugslepen van 't vlakke zand – die herdacht zij en minde zij zoals een minnaar de trekken en gebaren van zijn geliefde herdenkt en mint†.

Frederik van Eeden, Van de koele meren des doods, 1900

Those were the waves my friend ...

The beginning

In the spring of 1984 Klaus Hasselmann invited wave modellers to Hamburg to discuss possible joint work. The WAM (Wave Modelling) group emerged. I dutifully reported on progress (Komen, 1985a,b, 1986, 1987a,b, 1990, 1991a,b). Now, ten years later, the WAM group has achieved what it wanted to achieve: a third generation computer model has been developed, which is able to predict the wave conditions at sea; it is used for global and regional applications; the model is a useful tool for interpreting satellite observations of the ocean and increased our understanding of the role of waves in air/sea interaction and the coupling between the atmosphere and the ocean. The results obtained are presented in this book.

WAM was one of a number of international collaborations: JONSWAP, SWAMP, SWIM, WAM, ... Observations during the Joint North Sea Wave Project (Hasselmann *et al*, 1973) and also by Mitsuyasu and collaborators (Mitsuyasu, 1966, 1969, Mitsuyasu *et al*, 1971) had established the importance of the nonlinear transfer in governing the shape and evolution of a wind sea spectrum. Based on this finding 'parametric' wave models had been developed based on the approximation of a self-similar shape of the wind-sea spectrum. But then the Sea Wave Modeling Project revealed the weak point of this approach (SWAMP group, 1985). The weakness, the wind-sea–swell transition, had been known from the beginning. In real situations (whimsical coast lines, capricious bottoms and winds that change continuously, not only from time to time but also from place to place) the two-dimensional spectra which describe the waves can take many different shapes. Typically, several

---

† The exterior peculiarities of the sea, the colours and the different wave forms and the way in which the large billows turn over, boil and sprinkle, surge capped with foam and withdraw from the flat sand – she remembered them and loved them as a lover remembers and loves the traits and gestures of his beloved one.

hundred numbers are needed to characterize the sea state at one single (grid) point and therefore an adequate description with a few parameters is simply impossible in general.

The establishment of WAM followed intensive work at the Max-Planck-Institut für Meteorologie on numerical taming of the Boltzmann integral describing the energy transfer due to nonlinear interactions. This work was an essential building block for a wave model that could treat wave evolution in a physical way. I contributed to this in 1983, the year preceding the start of WAM, when I spent a summer in Hamburg analysing fetch- and duration-limited wave evolution in the first third generation wave model ever: EXACT-NL, which had been presented already in 1981 in Miami.

The WAM group/SCOR wg 83

EXACT-NL was one-dimensional and time consuming. The WAM group had the task of making a useful wave model out of this. All I remember from the now historic meeting in Hamburg, is how Klaus Hasselmann proudly wrote the new acronym on the blackboard and how he suggested that I be the chairman of the group. After a day's hesitation I accepted and work started. To bundle forces we dovetailed interests. Some of us were interested in physics, others in the quality of operational wave forecasts, still others were mainly concerned with climate applications and with the use and interpretation of satellite observations of the ocean surface. As a result the aims for WAM came out as follows:

- to develop jointly a third generation wave model, based on a full description of the physical processes governing wave evolution,
- to implement a global version of the model and to test medium-range forecasting,
- to develop regional versions of the third generation model to be nested with the global model,
- to perform physical studies of wave dynamical processes in order to extend our understanding of wave evolution, where needed, and
- to develop data assimilation techniques which would make it possible to make full use of satellite observations of the sea state.

In 1986 the Scientific Committee on Oceanic Research (SCOR) established working group 83 (Wave Modelling) which accommodated the WAM group. The terms of reference were simply taken over but SCOR provided an international mechanism to broaden the involvement. As a result, the geographical scope of the group was enlarged.

In some respects the organization of the WAM group was loose and easy, in other respects it was firm and full of complex structure. This contributed to the vitality of a voluntary collaboration, in which each participant could contribute his specific expertise. There was the WAM community, consisting of those wave modellers attending the annual meetings, but within this community many more intensive nuclei formed, with more specific aims. A core project was the 'special project' carried out at the European Centre for Medium-Range Weather Forecasts (ECMWF), which aimed specifically at the implementation of the model in a global version. Other active groups were the model implementation group (responsible for developing and maintaining the integrity of the model), the Mediterranean subgroup, the growth curve reanalysis subgroup, the data-assimilation subgroup and several others. Often these subgroups were funded through separate contracts which provided their own structure. *Ad-hoc* meetings were frequently necessary. In the beginning the 'steering committee' played an important role. This role was taken over later by SCOR and the annual meetings of the SCOR working group members. The final operationalisation of the model at ECMWF was a joint venture of a number of European Met Services.

The WAM model

The WAM model (WAMDI, 1988) is now one of the best tested wave models in the world. The code is well documented; it can run on many different machines; it has been distributed among over 40 research groups which have used the model for forecasting on global and regional scales (ECMWF, FNOC, KNMI and, with modified numerics, NMC), for case studies (Mediterranean, North Sea, hurricanes, ...) and in measurement campaigns (LEWEX, SWADE, ERS-1 calibration/validation campaign). The model is continuously updated. Cycle 4 is the most recent version. It is an improvement over earlier cycles in that it contains current refraction, nesting and it has improved wind input and dissipation. It has a shell for both real-time and non real-time applications. The code is highly optimized. Post- and preprocessing programmes (to set up the grid, interface with wind input, etcetera) and standard test cases come with the main programme for only US $50 to anyone willing to pay and to accept conditions that are only mildly restrictive. (Originally, model ownership was restricted to a happy few. Others could get a copy of the model on the condition that they helped improve the model. They earned ownership with their contribution to the development. This was a successful policy for a while, but then it lost its meaning when the circle had widened. In the end we decided to relax the conditions.)

The physics of the model has been improved. The *wind input* was originally described by the Bight of Abaco expression, rescaled in terms of friction velocity. Now there is an improved version which takes into account the feedback of the waves on the atmospheric flow. The *nonlinear transfer* is computed in the discrete interaction approximation. This approximation is fast, but it has sometimes been criticized because of its limited quantitative agreement with the exact transfer. This, however, does not seem to be too critical. One of the marvels of third generation wave modelling is the interplay between the source terms and the detailed shape of the wave spectrum. The nonlinear transfer has the effect of adjusting the details of the spectrum on a fast time scale. It then works on the adjusted spectrum on a slower time scale for the dynamical evolution. The discrete interaction approximation has the required qualitative behaviour, and it gives the correct temporal evolution of the spectrum, even if this spectrum deviates in small details from the one you would get with EXACT-NL. *Dissipation due to whitecapping* is a relatively weak point of the model. It is clear now that there is some freedom in redefining the wind input and retuning the dissipation constants. Therefore, we are not completely certain about the absolute magnitude of wind input and dissipation. An independent experimental or theoretical estimate of this dissipation term would be most welcome. In shallow water *dissipation due to bottom friction* and other effects such as refraction are taken into account.

The global implementation

The WAM model has been coupled to the operational atmospheric forecast model of ECMWF. It runs on their CRAY YMP on a 3° x 3° global grid. The resolution will be improved in the near future. The spectrum is discretized in 25 frequency and 12 directional bands. For several years now it has been run daily to compute the sea state from analysed surface winds. The results of the wave model have been compared with ship, buoy and satellite observations. Typically, a bias of about 10 cm is obtained and a scatter index of 20%. Five day forecasts have also been made routinely, and their quality has been documented. Comparing the forecasts with the corresponding analysis, it was found that the percentage of correct forecasts (having the significant wave height correct within 0.5 m) decreases slowly from 86% on day one to 65% on day five. The model has been 'operational' since the beginning of 1992. The results can be used for many applications, such as ship routing, which has become very attractive through reliable medium-range weather forecasts, global satellite data and the global WAM model.

The regional versions

Many regional implementations have been made, in fact too many to be mentioned here. Among the earliest studies were the WHIST hindcasts for the North Sea and the north west North Atlantic and three hurricane hindcasts for the Gulf of Mexico. Later work ranged from Mediterranean and Adriatic hindcasts to typhoon hindcasts in the South China Sea. The Koninklijk Nederlands Meteorologisch Instituut (KNMI) is using the model for real-time forecasting of North Sea waves. At ECMWF WAM has been implemented on a 0.5° x 0.5° grid in the routine daily procedure to provide a five-day forecast for the Mediterranean Sea. A Mediterranean version with the same resolution has been operational at the Fleet Numerical Oceanography Center

(FNOC) since 1990. Recent hindcasting applications have been given for such places as the Adriatic Sea and Bristol Channel. Much attention was given to hindcasting for major experiments, such as the Labrador Extreme Wave Height Experiment (LEWEX) and the Surface Wave Dynamics Experiment (SWADE). One of the objectives of this latter study was to improve modelling of wave–current interaction in the Gulf Stream. To this end the WAM model was run on nested grids at three different levels.

Physics

One of the aims of WAM was to carry out the recommendations of the SWAMP study. The first was to reanalyse existing fetch-limited observations rather than to make new observations. A second recommendation was to study the directional behaviour of the spectrum under changes of wind direction, as this was one of the key features of model behaviour. An extension of SWAMP to shallow water (SWIM, 1985) had revealed distinct differences between different wave models in their treatment of shallow water effects. Therefore WAM considered shallow water aspects also and increased our understanding of the dependence of bottom dissipation on the detailed structure of the bottom.

Satellite observations and data assimilation

The advent of ocean observing satellites has had an enormous impact on wave modelling. SEASAT was the first satellite to illustrate the usefulness of sea surface observations from space. Not only winds, but also waves were measured. Later the GEOSAT altimeter gave high quality wave height observations. Since 1991 ERS-1 has provided observations of winds over sea with the scatterometer, observations of wave height and wind with the radar altimeter and observations of two-dimensional wave spectra with the Synthetic Aperture Radar. Several groups within WAM spent years of preparation and are now successfully using the large stream of satellite observations of the ocean surface.

Satellite observations and wave models can interact at different levels. First of all, *retrieval* of geophysical parameters from raw data can be improved when first-guess wind and wave parameters are available. This is particularly relevant for the SAR observations, where impressive progress has been made, but it should also apply to the scatterometer. Secondly, these first-guess model results provide a means of assessing the quality of the observations (*quality control*). Accepted observations can then be used for model validation. Finally, the data can be *assimilated* by merging first-guess model fields with all available observations. This not only leads to the best possible estimate of global wind and wave fields, but it also improves the quality of (medium-range) forecasts.

The general data assimilation problem is a challenging one. It involves the minimization of the difference between model guesses and all observations, taking into account dynamical constraints and observational errors. There are several general techniques which are still being explored. Simpler methods are already applied operationally.

The role of waves in the coupled ocean–atmosphere system

As a byproduct of the investigations it was found that waves have an effect on the momentum flux from the atmosphere to the ocean. For a given wind speed young waves support a larger momentum flux than old waves. As such waves play a role in the interaction between the atmosphere and the ocean.

This book

This book is the final report of the WAM group/SCOR WG 83. Its scope and contents were first discussed during the 1990 SCOR WG 83 meeting in Canada. The concept grew during bilateral *ad-hoc* discussions in a number of places, and was finally consolidated during the 1991 meeting on Sylt. From the outset two principles were taken as guidelines: the book should reflect the multi-disciplinary nature of WAM (basic physics, modelling, applications and satellites) and it should be suitable as an introduction to the field. As a result we chose a three level approach: contributors, chapter coordinators and a final editor. Drafts were sent to anonymous reviewers, where possible from outside the WAM group. A unique result emerged. Each chapter coordinator followed the way that best suited the subject and his or her personality, but the whole is fairly consistent because it is the result of many years of intense collaboration and of extensive discussions between contributors and chapter coordinators.

The first chapter gives the general background. With the exception of § I.4 it was drafted by myself, but I interacted intensively with many colleagues, in particular with Peter Janssen and Klaus Hasselmann. The result is a fairly thorough treatment of the basic statistical theory of ocean wave forecasting. MARK DONELAN, in the physics chapter, collected contributions from many specialists, while concentrating himself with Yeli Yuan on the gap in our understanding formed by dissipation theory. PETER JANSSEN and LUIGI CAVALERI wrote chapters III and IV using their original work and material published elsewhere or provided by other members of the WAM group. KLAUS HASSELMANN wrote chapter V, with different co-workers in the different sections. SUSANNE HASSELMANN provided the general parts of the data assimilation chapter and she collected and edited contributions from several groups which give an exciting and colourful picture of the state of the art in this new research area. The outlook (chapter VII) is a joint product of all chapter coordinators. In March 1993 a first draft of the book was completed. On the basis of this draft we had many intense discussions, which led to much rewriting and certainly contributed to the homogeneity of the report.

Acknowledgements

I can remember many little events which were essential in creating the right atmosphere for our collaboration. The efficiency of secretarial support during meetings, for instance. Or the mediating role of Danielle Merelle, who had to share a small flat in Reading with WAM members from different countries; the Dutch wanted dinner at six, the French preferred to have it between seven and eight and the Italian occupant insisted on having dinner after eight. Or Jane Frankenfield, conscientiously correcting the English of the programme for the happening in the *Basilica dei Frari* during the fourth WAM meeting. Or – just as another example – Dorothee von Berg, who cheered all of us up when progress was slow and who knew how to get higher priority from the computer operators. Or *Kaffee mit Kuchen* in the sun at the *Blaue Hütte* on Sylt. It would be impossible to memorize all of these events and the people involved. Even a skilled novelist might have a hard time. Yet, let me mention a few names.

First, and most of all I would like to thank Klaus and Susanne Hasselmann for their inspiration, their hospitality and their time, which they made available despite their many other commitments. The discussions I had with them were held in many places, over lunch, in the car, in the sauna, during the intermission of concerts, ... Many of the good ideas came from Klaus, who intervened time and again during the project and during the writing of this book. Susanne worked with success and dedication, throughout the years. Without her the WAM model would not have existed.

Thanks go to Peter Janssen as well. Together we learned about waves, sharing our love for basic physics *and* useful applications. Peter gave momentum to the project by moving to England for a while when this was needed. He helped unravel many problems, organized meetings and surely was one of the decisive forces behind the present book.

Then I would like to thank the steering committee members Luigi Cavaleri, Peter Francis, Leo Holthuijsen, Wolfgang Rosenthal and Willem de Voogt for their loyality and their efforts, which were essential in shaping WAM and in keeping it going. Unfortunately, Willem did not live to see the completion of this book. He died in January 1994. In a later stage of the project the steering work was shared with the other SCOR wg 83 members, in particular with Vince Cardone, Mark Donelan, Hans Graber, Anne Guillaume, Yeli Yuan and Vladimir Zakharov.

Special thanks go to those members who actively contributed to the final report. Everyone was already very committed, so much was done outside regular working hours. I want to acknowledge the kind hospitality extended to me in Kayhude, Reading, Burlington and Venice, while we worked on the book. I remember with pleasure the picnic lunches Luigi Cavaleri and I ate from his desk on Saturdays and Sundays in an otherwise empty

Palazzo Papadopoli, with the gondoliers on the Canal Grande singing in the background.

The special project at ECMWF was one of the core projects of WAM. Major contributions to this project were made by Peter Janssen, Liana Zambresky and Heinz Günther, who worked on the model for many years, and all left their unique imprints (Peter introduced an integer starting with the letter B in the Fortran code, and both Liana and Heinz beautified and streamlined the code according to their personal insight, so that for a while their model versions deviated from the canonical one; nevertheless we managed to end up with a well documented standardized code, blending the best from different versions). Besides carrying out their own work, they assisted the many visitors that came to use the model or to contribute to its development: Eva Bauer, Luciana Bertotti, Claus Brüning, Gerrit Burgers, Luigi Cavaleri, Vince Cardone, Juan Carlos Carretero, Spiros Christopoulos, John Ewing, Arthur Greenwood, Anne Guillaume, Bjorn Hansen, Miriam de las Heras, Jean-Michel Lefèvre, Piero Lionello, Henrique Oliveira-Pires, Renate Portz, Magnar Reistad, Claire Ross, Rachel Stratton, Bechara Toulany and Gerbrant van Vledder. The excellent computer support at ECMWF by Norbert Kreitz, Brian Norris, Dave Dent and John Greenaway is acknowledged, as is the moral support from ECMWF staff and directorate: Lennart Bengtsson, Dave Burridge and Tony Hollingsworth. We thank Bert Kamp for his diplomatic activities in getting European Met Services to agree on the operationalisation of the global model. Peter Groenewoud, Bill Perrie, Roger Flather and others worked in Hamburg and elsewhere on the model.

The annual meetings were essential. Their organization involved a good many people, whose dedication I gratefully acknowledge. By far the poorest meeting was the one at KNMI (De Bilt, The Netherlands, 1984), so let me not mention the local organizer. Others did better: Johannes Guddal (Bergen, Norway, 1985), Luigi Cavaleri (Venice, Italy, 1986), Hans Graber (Woods Hole, USA, 1987), Anne Guillaume (Paris, France, 1988), Jose Enrique de Luis Guillen (Valencia, Spain, 1989), Mark Donelan (Lake Couchiching, Canada, 1990), Susanne Hasselmann (Sylt, Germany, 1991), Henrique Oliveira-Pires (Sintra, Portugal, 1992). The secretarial support during all of these meetings was of great help. The repeated efficient assistance of Mrs Gross of SCOR deserves special mention.

WAM related activities were funded by many agencies: the World Meteorological Organization (WMO), the Scientific Committee on Oceanic Research (SCOR), the European Space Agency (ESA), the European Community, the NATO Science panel, the Office of Naval Research (ONR), the US Army Corps of Engineers and the National Aeronautic and Space Administration (NASA). This help was essential.

At KNMI I received much support in scientific and organizational matters. First of all I would like to thank Greet de Graaf, Brigitta Kamphuis and

Martin Visser for their secretarial support in communication (of course we had a newsletter, the last issue is still due), organization and preparation of this final report. Thanks are also due to Jeroen Teeuwissen who patiently worked over the figures. Scientific support and suggestions came from Evert Bouws, Gerrit Burgers, Bert Holtslag, Peter Janssen, Han Janssen, Arie Kattenberg, Vladimir Makin, Kees Mastenbroek, Wiebe Oost, Aad van Ulden, Aart Voorrips and Nanne Weber. In Hamburg Marion Grunert provided support for the book, in Burlington Kay Goodwill was of great help. John Ewing read through the manuscript and made valuable suggestions for improvement. I would like to thank Patrick Heimbach for his help with the preparation of the subject index.

I would like to end this preface by thanking the anonymous reviewers of draft texts of this book, as well as all contributors for giving time and energy and I apologize for what may have gone wrong.

De Bilt,   Gerbrand Komen
April, 1994

Chapter I
Basics
G. J. Komen *et al*

I.1 Wave modelling in historical perspective

The third generation ocean wave models considered in this book are numerical models which integrate the dynamical equations that describe the evolution of a wave field. Their development followed progress in understanding ocean wave dynamics and experience with practical forecasting methods.

The study of ocean wave dynamics has a very long history indeed. Khandekar (1989) quotes Aristotle, Pliny the Elder, Leonardo da Vinci and Benjamin Franklin. In Phillips (1977) it is recalled how Lagrange, Airy, Stokes and Rayleigh, the 'nineteenth century pioneers of modern theoretical fluid dynamics', sought to account for the properties of surface waves. Subsequent progress in the twentieth century has been enormous, as any reader of, for instance, the *Journal of Fluid Mechanics* will realize. The subject has grown so large that it is nearly impossible for one person to have a comprehensive knowledge of all aspects. There are a number of useful textbooks, devoted totally or in part to water waves (Krauss, 1973, Whitham, 1974, Phillips, 1977 and LeBlond and Mysak, 1978, for example). Subjects that are of special relevance to wave modelling are the propagation of waves, their generation by wind, their nonlinear properties, their dissipation and their statistical description. We will discuss all these aspects in later sections where we will also indicate the historical perspective. Here it is enough to recall the milestones formed by the theoretical work of Phillips and Miles on wave generation and by Hasselmann's theory of four wave interactions. An important role has also been played by a number of large experiments, such as the Pacific swell attenuation study (Snodgrass *et al*, 1966), JONSWAP (Hasselmann *et al*, 1973), Mitsuyasu *et al* (1971) and the Bight of Abaco measurements of Snyder *et al* (1981).

Interest in wave prediction grew during the Second World War because of the practical need for knowledge of the sea state during landing operations. The first operational predictions were based on the work of Sverdrup and Munk (1947), who introduced a parametrical description of the sea state and who used empirical wind sea and swell laws. An important advance was the introduction of the concept of a wave spectrum (Pierson *et al*, 1955). This was not yet accompanied, however, by a corresponding dynamical equation describing the evolution of the spectrum. This step was made by Gelci *et al* (1956, 1957) and others who introduced the concept of the spectral transport equation, which will be discussed at some length in the following sections. Because of the lack of adequate theories, Gelci was forced to use a purely empirical expression for the net source function governing the rate of change of the wave spectrum. It was only after the new theories of wave generation by Phillips (1957) and Miles (1957) had been published and the source

3

function for the nonlinear transfer had been derived (Hasselmann, 1962) that it was possible to write down the general expression for the source function (Hasselmann, 1962), consisting of three terms representing the input from the wind, the nonlinear transfer and the dissipation by whitecapping (or bottom friction) – in a form which is still used today. The subsequent developments can be traced in the book by the SWAMP group (SWAMP, 1985). SWAMP made a distinction between first generation models, in which nonlinear interactions are neglected, and second generation models which do describe them, but in a simplified parametrized form. Two types of second generation model were distinguished: 'coupled hybrid', and 'coupled discrete'. In 'coupled hybrid' models the wind sea spectrum, which is strongly controlled by the nonlinear interactions, is assumed to adjust rapidly to a universal quasi-equilibrium form in which only a single scale parameter – normally the wind sea energy – or at the most a second frequency scale parameter need be predicted as slowly varying parameters. The swell, which is not affected by nonlinear interactions, is then treated as a superposition of independent components in the same way as in a first generation model. 'Coupled discrete' models retain the traditional discrete spectral representation, but have a parametrization of the nonlinear transfer with limited validity, so that the potential advantage of a more flexible representation of the wind sea spectrum and a uniform representation of the swell–wind-sea transition regime cannot be properly exploited. The SWAMP study compared the results of nine different first and second generation models in simple, hypothetical conditions, such as fetch- and duration-limited growth in a constant wind field, growth in a slanting fetch situation, swell propagation out of a half plane wind field and a diagonal front. Also, the response to a sudden change in wind direction was studied. A final test considered the response of the models to a hurricane wind field. In this last test especially it was found that the models behaved quite differently. For example, the extreme significant wave height ranged between 8 and 25 metres, demonstrating our lack of knowledge. A careful analysis of the tests revealed that much of the discrepancy could be traced to details of parametrizations affecting the spectral behaviour of the models and that differences in predicted wave height could be partly traced back to differences in spectral shape. In reality, wave spectra showed much more variability than was originally assumed in parametric models, and two-dimensional aspects were found to be more important than expected. Therefore a 'third generation' model was developed: a full spectral model with an explicit representation for the physical processes relevant for wave evolution and which gives a full two-dimensional description of the sea state.

Application of third generation models requires considerable computing power, which has become available in recent years. This has coincided with the development of remote sensing techniques for measurements of the sea surface with the help of microwave instruments (altimeter, scatterometer and synthetic aperture radar or SAR). The relationship between wave model development is much closer than one might expect at first sight. Satellite observations can be used to validate the model and the model also gives a first check on the accuracy of the observations. Furthermore, a detailed description of the dynamics of the sea surface is important for a correct interpretation of the radar signals. At present, the WAM model is used operationally in global and regional implementations to make forecasts of the sea state, which can be used for many applications, such as ship routing and offshore activities, and for the validation and interpretation of satellite observations.

Finally, much progress has been made and is being made on the assimilation of satellite observations into wave models. This will lead to a more accurate description of the waves of the world oceans. Deviations between predicted and observed waves are normally indicators of errors in the predicted wind fields, so that effective wave data assimilation procedures correct both the wave and wind fields. This is not only important for applications, but also because it will give us a better understanding of the role of waves in the broader context of the coupled ocean/atmosphere system. The assimilation of wave observations is potentially very useful in obtaining better estimates of the fluxes involved in air/sea interaction, which play a crucial role both in weather prediction and in climate modelling.

I.2 The action balance equation and the statistical description of wave evolution

G. J. Komen and K. Hasselmann

The action balance equation gives a statistical description of the time evolution of sea waves.

Suppose one *knows* the sea state at a given moment. How then can one use the *general laws of physics* to compute the sea state at a later time? The answer to this question is provided by the action balance equation (1.207), which describes the evolution of interacting weakly nonlinear waves. The action balance equation plays a central role in modern wave modelling. In this section we discuss its derivation which is quite lengthy.

In general, an instantaneous picture of the sea surface will not indicate in which direction the waves are going. Therefore, to specify fully the sea state, velocity information should be given in addition to the height of the sea surface everywhere. To determine the time evolution one must also specify the external conditions, such as wind and tides. Because it is impossible, in practice, to specify the initial state in complete detail, one normally uses a statistical description, in which the probability of finding a particular sea state is considered. In general, this involves knowledge of the joint probability function $P(\eta_1, \eta_2, \ldots \eta_n)$, where

$$P(\eta_1, \eta_2, \ldots \eta_n) \mathrm{d}\eta_1 \mathrm{d}\eta_2 \ldots \mathrm{d}\eta_n \qquad (1.1)$$

is the probability that the surface displacements η_i, $i = 1, \cdots n$ (or their derivatives) at the points $(\mathbf{x}_i, t_i)$ have values between η_i and $\eta_i + \mathrm{d}\eta_i$. Fortunately, the probability distribution of the sea surface is nearly Gaussian, so a good approximate description is provided by the covariance function

$$< \eta(\mathbf{x}_1, t_1) \eta(\mathbf{x}_2, t_2) >, \qquad (1.2)$$

the Fourier transform of which is known as the *wave spectrum*. The wave spectrum also has a physical meaning, because it can be shown to be the density function specifying the distribution of energy over wave components with different wavenumber vectors and frequencies. Its integral over all wave components is proportional to the total wave energy per unit area.

To *derive* the action balance equation, one has to start from more fundamental equations. Different choices are possible, depending on what aspects one wants to study. We will therefore begin with a review of the different forms of these basic equations. Then we will introduce the normal modes of the system, which are elementary waves with a well-defined relation between wavenumber k and frequency ω. Following this, we will describe more general wave trains with slowly varying amplitude and wavenumber. We will also discuss 'wave packets', which are wave trains of finite extent. Next we will extend our discussion to the case of many waves and will show how the normal modes can be used to solve the initial value problem. This will be followed by a description of the statistical treatment of sea waves. The remaining part of the section will deal with the wave spectrum in a two scale approach. On the fast scale the motion of the sea surface is thought to be a Gaussian, statistically stationary and homogeneous stochastic process, characterized locally by the wave spectrum. On the slow space and time scale this spectrum is allowed to vary in interaction with the environment. The action balance equation is the equation describing the variation of the wave spectrum on the slow space and time scale. It is conveniently expressed in terms of the wave action density, which is the wave spectrum divided by the intrinsic frequency. This approach was pioneered by Gelci *et al* (1956, 1957), Hasselmann (1960) and Gelci and Cazalé (1962). The action balance equation is also known as the radiative transfer equation or the transport equation. The first name derives from an analogy with the radiative balance in radiation theory; the latter name originates from an analogy between the sea and a molecular (kinetic) gas, which explains the name kinetic equation used by Russian investigators.

I.2.1 Dynamics

We discuss the basic equations of motion, give an expression for the total energy and indicate how the equations of motion can be derived from a Lagrangian or a Hamiltonian function.

The dynamics of classical physics is governed by Newton's law, which, for a fluid, takes the form of the Navier-Stokes equation (Batchelor, 1967). The atmosphere/ocean system consists of fluids of two densities, air and water. So, in order to determine its time evolution we have to solve the Navier-Stokes equation for a two-layer fluid. This is so complicated, however, that one cannot solve it directly for cases of practical interest, either analytically or numerically. Fortunately, the density of air is much smaller than the density of water. As a result, waves can be described in good approximation by the linearized Navier-Stokes equations for a one-layer fluid in a gravitational field.

The Coriolis force is normally negligible, because the inverse Coriolis parameter is much larger than a typical wave period. For certain higher order effects, such as the Stokes drift with zero frequency, this is no longer correct. As such effects do not play a role in wave modelling they will not be discussed here.

To first order ocean waves are free, that is to say dissipation and forcing by the environment can be neglected. For the longer waves (with wavelength $\lambda > 1$ m) viscosity and surface tension can also be neglected, so that, in a first approximation, it is sufficient to consider the Euler equation for a one-layer fluid.

A further simplification can be made, because waves happen to be irrotational. This has been explained in a number of ways. It is well known (Batchelor, 1967) that a free fluid with zero vorticity initially, will have zero vorticity at all times. If vorticity is confined to a finite region it can spread only slowly by diffusion. In the case of swell, sea waves spread rapidly from the generation region to a region that was initially at rest. Therefore, swell waves cannot have vorticity. During active generation waves are not free, so that the above argument does not apply to wind sea. However, another theorem asserts that pressure forces acting on a fluid cannot generate vorticity inside the fluid. One then argues that waves are actually generated by pressure forces and not by tangential stresses, so that the wave motion must be vorticity free. But this argument – true as it may be – is a rather poor one as it invokes a nontrivial theorem on the way in which waves are being generated. A third approach, which we prefer, starts off by constructing a complete set of solutions of the linearized Euler equation. One then simply *finds* that the wave modes which are identified with the surface wave solutions are irrotational. In addition, one finds other modes which can have vorticity. However, these are time-independent, 'zero-frequency modes', representing constant currents and are less interesting from a dynamical point of view. Nevertheless, even in the framework of Euler's equation, the total water motion can have vorticity. In the linear approximation this vorticity is constant, but in the nonlinear case it may vary in time. Of course, Euler's equation is only an approximation to the real system. In reality forcing and shear flow instability contribute to the generation of vorticity.

Once one has established the irrotationality of the surface wave motion, the flow can be described by a velocity potential (Batchelor, 1967). In addition, water is nearly incompressible, so that the flow is also divergence free, which implies that the velocity potential satisfies Laplace's equation.

To formalize all of these statements we will discuss more explicitly a hierarchy of equations, starting from the two-layer Navier-Stokes (Reynolds) equation and ending with the Laplace equation. But before we embark on this, a number of remarks on the notation must be made. In this section, three-dimensional space coordinates will be denoted by (x_1, x_2, x_3), the velocity three-vector by (u_1, u_2, u_3), pressure by $p(x_1, x_2, x_3, t)$, time by t and density by $\rho(x_1, x_2, x_3, t)$. When it is convenient the vertical coordinate x_3 and the vertical velocity component u_3 will also be denoted by z and w, respectively. Differentiation will be indicated in a number of ways:

$$\frac{\partial}{\partial x_i} = \nabla_i, \quad \frac{\partial}{\partial t} = \partial_t, \quad \frac{\partial}{\partial z} = \partial_z. \tag{1.3}$$

If convenient the subscript notation will also be used. For example,

$$\phi_{x_1} = \frac{\partial}{\partial x_1}\phi. \tag{1.4}$$

The introduction of different notations for differentiation may seem strange, but it can be quite useful in bringing out the structure of equations more clearly. Bold face will be used to indicate two-dimensional vectors. For example, $\mathbf{x} = (x_1, x_2)$, $\mathbf{u} = (u_1, u_2)$ and also

$$\nabla\phi = (\nabla_1, \nabla_2)\phi = (\phi_{x_1}, \phi_{x_2}). \tag{1.5}$$

Later, when confusion could arise, we will give the vector ∇ a subscript x or k to distinguish between differentiation with respect to x and k. When convenient, we will use the Einstein convention implying summation over repeated indices. For the two-dimensional inner product we use the dot notation

$$\mathbf{k} \cdot \mathbf{x} = k_1 x_1 + k_2 x_2. \tag{1.6}$$

Introducing g for the acceleration due to gravity and denoting the interface between air and water by $\eta(x_1, x_2, t)$ we then have

$$\nabla_i u_i = 0,$$

$$\partial u_i / \partial t + u_j \nabla_j u_i = -(1/\rho)\nabla_i p - g\delta_{3i} + \nabla_j \tau_{ij}, \quad i = 1, 2, 3, \qquad (1.7)$$

$$\rho = \begin{cases} \rho_a, & z > \eta, \\ \rho_w, & z < \eta. \end{cases}$$

Here ρ_a (ρ_w) is the density of air (water) and δ_{3i} is zero except for $i = 3$, when it equals one. Velocities and forces (pressure and tangential stress) are continuous at the interface. A particle on either side of the surface will move in time Δt from

$$(\mathbf{x}, \ z = \eta(\mathbf{x}, t)) \qquad (1.8)$$

to

$$(\mathbf{x} + \Delta \mathbf{x}, \ z + \Delta z = \eta(\mathbf{x} + \Delta \mathbf{x}, t + \Delta t)), \qquad (1.9)$$

with $\Delta \mathbf{x} = \mathbf{u}\Delta t$ and $\Delta z = w\Delta t$, so one obtains in addition the kinematic boundary condition

$$\eta(\mathbf{x} + \Delta \mathbf{x}, t + \Delta t) = \eta(\mathbf{x}, t) + w\Delta t, \qquad (1.10)$$

or equivalently

$$w_+ = w_- = \partial \eta / \partial t + \mathbf{u} \cdot \nabla \eta = D\eta/Dt. \qquad (1.11)$$

The subscripts $+$ and $-$ indicate that the argument has to be taken just above or below the interface. To complete the set of equations one has to express the stress tensor τ in (1.7) in terms of the mean flow. The stress contains the viscous stress and it may or may not contain the additional stress resulting from fluctuating motion (the Reynolds stress). The interpretation of the velocity field depends on the exact definition of τ. If the Reynolds stress is included the equation is known as the Reynolds equation and it describes the mean flow; otherwise it refers to the actual flow. (For a discussion we refer the reader to Monin and Yaglom (1971, 1975) and Phillips (1977).) For $z \to \pm\infty$ one has to specify boundary conditions. For a wave solution the oscillatory motion should vanish in these limits. However, instead, in water of finite depth the orthogonal velocity component at the bottom should vanish. Also, for a finite basin one should, in principle, specify the lateral boundaries. In practice this is often ignored because typical oceanic dimensions are very much larger than the typical wavelength, so effectively the basin is of infinite size. For very short waves the continuity of pressure is not correct, because the effect of surface tension becomes important (see Phillips, 1977). At high wind speeds, when multiple whitecaps, bubbles and spray appear, the assumption on the existence of a well-defined sea surface η breaks down and the equations lose their meaning. An attempt to describe this situation was made by Newell and Zakharov (1992).

As has already been discussed in general terms, one may simplify the equations in a number of ways. One way is to neglect viscosity. In that case one obtains the Euler equations in which the stress terms have been dropped. Continuity of stress at the interface is no longer required, which allows the parallel velocity at the interface to be discontinuous. This approximation can be made in air and water. A further approximation of the air motion is to neglect it altogether. As noted above, yet another approximation takes the velocity in water as irrotational, so that the motion can be described by potential flow. In this case one introduces the velocity potential $\phi(x_1, x_2, z, t)$ with the property

$$u_i = \phi_{x_i}, \quad i = 1, 2, 3, \tag{1.12}$$

where we use the short hand notation for differentiation. The full nonlinear equations then read

$$\phi_{x_1 x_1} + \phi_{x_2 x_2} + \phi_{zz} = 0, \quad z < \eta(x_1, x_2, t), \tag{1.13}$$

$$\left. \begin{array}{l} \eta_t + \phi_{x_1}\eta_{x_1} + \phi_{x_2}\eta_{x_2} = \phi_z, \\[4pt] \phi_t + \frac{1}{2}[\phi_{x_1}^2 + \phi_{x_2}^2 + \phi_z^2] + g\eta = 0, \end{array} \right\} \quad z = \eta(x_1, x_2, t). \tag{1.14}$$

The second boundary condition makes use of Bernoulli's relation to express continuity of pressure at the surface. Since pressure is zero in the vacuum just above the water, it should be zero just below the sea surface.

In all of these approaches the general nonlinear equations can be simplified by linearization, i.e. by neglecting quadratic and higher order terms and by expanding arguments which depend on η retaining only linear terms. In the next subsection we will illustrate the linearization procedure for the case in which the water motion is described by Euler's equations and in which the space above the sea surface is taken to be vacuum.

For later use, it is important to write down the expression for the energy of the fluid in motion. As usual this is the sum of potential energy and kinetic energy. For $\mathcal{E}$, the energy density (J/m^2), one obtains in the potential flow description

$$\mathcal{E} = \frac{1}{2}g\rho\eta^2 + \frac{1}{2}\rho \int_{-\infty}^{\eta} \left[\phi_{x_1}^2 + \phi_{x_2}^2 + \phi_z^2\right] \, dz, \tag{1.15}$$

where $\rho = \rho_w$, the density of water. (We will drop the subscript when confusion with the density of air is unlikely.) The total energy is given by

$$E_{tot} = \int\int \mathcal{E} \, dx_1 dx_2. \tag{1.16}$$

The x-integrals extend over the total basin considered. If one considers an infinite basin, the integral is infinite, unless the wave motion is localized within a finite region. In any case one can also introduce $\bar{\mathcal{E}}$, the energy per unit area, by dividing (1.16) by the total surface. For a basin of dimension $L \times L$

$$\bar{\mathscr{E}} = \frac{1}{L^2} \int \int \mathscr{E} \, dx_1 dx_2. \tag{1.17}$$

The expression for the total momentum can be written down in a similar way (see Phillips, 1977).

We have formulated fluid dynamics in terms of Eulerian coordinates, by specifying the velocity of the fluid in fixed space points **x** as a function of time. Since the fluid flows, this velocity refers to different fluid elements when time progresses. The fundamental law of mechanics, Newton's law, refers to the acceleration of a physical body (fluid element) when it is subjected to forces. The more fundamental formulation of fluid mechanics is therefore in terms of Lagrangian coordinates. This description connects directly to classical mechanics, which considers forces between material particles with a fixed 'identity'. Therefore, Lagrangian coordinates specify the velocity of a fluid element while it flows. One normally does not use Lagrangian coordinates because the boundary conditions at a fixed wall become complicated and time-dependent. However, for free surfaces, Lagrangian coordinates have certain advantages and are therefore often used in wave studies. Eulerian coordinates on the other hand are basically noncanonical. One advantage of Lagrangian coordinates is that they are canonical variables and satisfy Hamilton's equations. They are also quite useful for describing phenomena such as the Stokes drift. In a linear approximation, Lagrangian coordinates are identical with Eulerian ones, so it does not matter which are used. However, when one considers nonlinear systems one knows from the outset from basic mechanics, that a Lagrangian exists, which is a definite advantage over the Eulerian case. The transformation from Lagrangian to Eulerian coordinates is noncanonical, so it is not at all obvious that a Lagrangian in terms of Eulerian coordinates exists. When working with Lagrangian coordinates one can simply work with the Lagrangian from basic mechanics and carry out a (canonical) coordinate transformation to normal mode. This approach, which applies to an arbitrary system consisting of all sorts of different waves, currents, bottom topography, is discussed in Hasselmann (1968).

In view of the above discussion it is remarkable that $\mathscr{E}$ may be used as a Hamiltonian density by choosing appropriate variables. This was found independently by Zakharov (1968), Broer (1974) and Miles (1977), who showed that with

$$\eta(x_1, x_2, t) \quad \text{and} \quad \psi(x_1, x_2, t) = \phi(x_1, x_2, z = \eta, t) \tag{1.18}$$

the boundary conditions at the surface are equivalent to Hamilton's equations:

$$\frac{\partial \eta}{\partial t} = \frac{\delta E_{tot}}{\delta \psi}, \quad \frac{\partial \psi}{\partial t} = -\frac{\delta E_{tot}}{\delta \eta}. \tag{1.19}$$

Here $\delta E_{tot}/\delta \psi$ is the functional derivative of E_{tot} with respect to ψ. The variables η and ψ are canonical conjugates. Later, in § II.3, this formulation will be used for the derivation of the resonant four-wave interaction.

There is also a Lagrangian formulation of the water wave problem (not in Lagrangian coordinates, but in terms of a Lagrangian function, which has a minimum for the true solution). Luke (1967) found that the variational principle

$$\delta \int \int \int \mathscr{L} \, dx_1 dx_2 dt = 0, \tag{1.20}$$

with

$$\mathscr{L} = -\rho \int^{\eta} [\phi_t + \frac{1}{2}(\phi_{x_1}^2 + \phi_{x_2}^2 + \phi_z^2) + gz] \, dz, \tag{1.21}$$

gives the Laplace equation as well as the boundary conditions. This is discussed in some detail in Whitham (1965, 1974). The relation with Hamilton's formalism is given in Seliger and Whitham (1968).

As usual in mechanics one can derive constants of the motion from the symmetries of the equations of motion. For water waves this is discussed by Brooke Benjamin and Olver (1982). Investigating the symmetries of the full water wave equations they obtain a number of conserved quantities, 8 in the one-dimensional case and 12 in two dimensions. Most of these quantities, for example energy (1.16), momentum and angular momentum have an obvious physical interpretation. Some are less easily interpreted.

Even in the presence of currents, the potential flow approximation can be used, provided (1.12) is generalized to

$$(u_1, u_2, w) = (U_1, U_2, 0) + (\phi_{x_1}, \phi_{x_2}, \phi_z), \tag{1.22}$$

with $\mathbf{U}(\mathbf{x}, z, t)$ a current. The governing Lagrangian becomes in that case

$$\mathscr{L} = -\rho \int_{-h(\mathbf{x})}^{H(\mathbf{x},t)+\eta} [\phi_t + \frac{1}{2}(\phi_{x_1}^2 + \phi_{x_2}^2 + \phi_z^2) + \mathbf{U} \cdot \nabla \phi + gz] \, dz, \tag{1.23}$$

where $H(\mathbf{x}, t)$ is the mean sea level, obtained after averaging over the wave motion. We will use the Lagrangian approach later to describe the behaviour of waves when they encounter currents and variable depth.

I.2.2 Linearization and normal modes

We find the normal mode solutions of the free, linearized equations and obtain the dispersion relation. Conserved quantities, such as the energy, are given in the linearized approximation.

The general nonlinear equations discussed in the previous subsection can be simplified by linearization, that is by expanding arguments depending on η and by retaining only terms linear in the wave steepness. This wave steepness, considered to be a small parameter, can be defined in a general way by assuming that η is characterized by both a vertical and a horizontal length scale, in such a way that their ratio is small. This ratio is then taken as the expansion parameter ϵ. To obtain the linearized equations one makes a formal expansion around $\epsilon = 0$.

We first consider the linearized Euler equations, which are obtained from (1.7) by putting $\tau = \rho_a = 0$:

$$\nabla \cdot \mathbf{u} + \partial_z w = 0,$$

$$\partial_t \mathbf{u} = -\nabla p/\rho,$$

$$\partial_t w = -\partial_z p/\rho - g, \tag{1.24}$$

$$\partial_t \eta = w, \quad z = 0,$$

$$p = p_{atm} = 0, \quad z = 0.$$

The last equation expresses continuity of pressure at the surface. The atmospheric pressure is taken to be equal to zero. To find elementary wave-like solutions of these equations we write

$$\begin{pmatrix} \eta \\ \mathbf{u} \\ w \\ p + \rho g z \end{pmatrix} = \begin{pmatrix} \tilde{\eta} \\ \tilde{\mathbf{u}} \\ \tilde{w} \\ \tilde{p} \end{pmatrix} e^{i\mathbf{k}\cdot\mathbf{x}} + \text{complex conjugate (c.c.).} \tag{1.25}$$

Substitution in (1.24) gives

$$\partial_z \tilde{w} + i\mathbf{k} \cdot \tilde{\mathbf{u}} = 0, \tag{1.26}$$

$$\partial_t \tilde{\mathbf{u}} = -i\mathbf{k}\tilde{p}/\rho, \tag{1.27}$$

$$\partial_t \tilde{w} = -\partial_z \tilde{p}/\rho - g, \tag{1.28}$$

$$\partial_t \tilde{\eta} = \tilde{w}, \quad z = 0, \tag{1.29}$$

$$\tilde{p} = g\rho\tilde{\eta}, \quad z = 0. \tag{1.30}$$

It is now useful to distinguish between prognostic fields, which provide a complete description of the instantaneous state of the system and whose time evolution is determined explicitly by the equations of motion, and diagnostic fields which can be determined in terms of other fields at the same time. There are two diagnostic variables. The first is obtained from (1.26) as

$$\tilde{w} = -i \int_{-\infty}^{z} \mathbf{k} \cdot \tilde{\mathbf{u}} dz'. \tag{1.31}$$

The second can be obtained from (1.26), (1.27) and (1.28) as

$$(\partial_z^2 - k^2)\tilde{p} = 0 \quad \rightarrow \quad \tilde{p} = \rho g \tilde{\eta} e^{kz}, \tag{1.32}$$

with $k = \sqrt{k_1^2 + k_2^2}$. We assume $\tilde{p} \to 0$ for $z \to -\infty$. The prognostic fields are $\tilde{\eta}$ and $\tilde{\mathbf{u}}$. Writing $\Psi = (\tilde{\eta}, \tilde{\mathbf{u}})$ the general form of the prognostic equations reads

$$\partial_t \Psi = \Lambda \Psi \tag{1.33}$$

with Λ some kind of operator. Now, first we split Ψ into a longitudinal component and a transverse component by decomposing the horizontal velocity vector into components parallel and perpendicular to the wavenumber:

$$\tilde{\mathbf{u}} = \tilde{\mathbf{u}}_{/\!/} + \tilde{\mathbf{u}}_{\perp}. \tag{1.34}$$

Then the parallel component may be further split into a barotropic (bt) and a baroclinic (bc) part by introducing

$$\bar{\mathbf{u}}_{/\!/} = k \int_{-\infty}^{0} \tilde{\mathbf{u}}_{/\!/}(z) dz, \tag{1.35}$$

so that one may define $\tilde{\mathbf{u}}_{/\!/}^{bc}$ with the property $\int \tilde{\mathbf{u}}_{/\!/}^{bc}(z) dz = 0$ by

$$\tilde{\mathbf{u}}_{/\!/}^{bc} = \tilde{\mathbf{u}}_{/\!/}(z) - \bar{\mathbf{u}}_{/\!/} e^{kz}, \tag{1.36}$$

and

$$\tilde{\mathbf{u}}_{/\!/}^{bt} = \bar{\mathbf{u}}_{/\!/} e^{kz}. \tag{1.37}$$

Finally, collecting the results we obtain

$$\Psi = \Psi_{\perp} + \Psi_{/\!/}^{bt} + \Psi_{/\!/}^{bc}, \tag{1.38}$$

with

$$\Psi_{\perp} = \begin{pmatrix} 0 \\ \tilde{\mathbf{u}}_{\perp} \end{pmatrix}, \quad \Psi_{/\!/}^{bt} = \begin{pmatrix} \tilde{\eta} \\ \tilde{\mathbf{u}}_{/\!/}^{bt} \end{pmatrix}, \quad \Psi_{/\!/}^{bc} = \begin{pmatrix} 0 \\ \tilde{\mathbf{u}}_{/\!/}^{bc}(z) \end{pmatrix}. \tag{1.39}$$

Working out Λ, one explicitly has

$$\partial_t \Psi_{\perp} = \partial_t \Psi_{/\!/}^{bc} = 0 \tag{1.40}$$

for the transverse and the baroclinic modes, so they are both independent of time. They are the zero frequency modes representing constant currents and carrying the vorticity prescribed by the initial state. We have included them in our analysis for completeness. Dynamically they are less interesting and therefore from now on we will concentrate on the parallel barotropic mode, so that we can drop the subscripts on **u**. We obtain

$$\partial_t \tilde{\eta} = -i \int_{-\infty}^{0} \mathbf{k} \cdot \tilde{\mathbf{u}} \, dz \qquad (1.41)$$

and

$$\partial_t \tilde{\mathbf{u}} = -ig\tilde{\eta}e^{kz} \qquad (1.42)$$

for the barotropic mode. From these equations one may eliminate either function to obtain, for example

$$(\partial_t^2 + gk)\tilde{\eta} = 0. \qquad (1.43)$$

This has the following two solutions:

$$\tilde{\eta}(t) = ae^{i\sqrt{gk}t} \qquad (1.44)$$

and

$$\tilde{\eta}(t) = ae^{-i\sqrt{gk}t}, \qquad (1.45)$$

with a an arbitrary constant. For each wavenumber vector **k** there are two 'normal mode' solutions:

$$\omega_\pm = \pm\sigma(k), \qquad (1.46)$$

with the dispersion relation

$$\sigma(k) = \sqrt{gk}, \qquad (1.47)$$

relating the wavenumber k to the frequency ω. It is now straightforward to write down the general solution for the barotropic mode:

$$\begin{pmatrix} \eta \\ \mathbf{u} \\ w \\ p + \rho g z \end{pmatrix} = a \begin{pmatrix} 1 \\ \omega_\pm \mathbf{k}/ke^{kz} \\ -i\omega_\pm e^{kz} \\ \rho g e^{kz} \end{pmatrix} e^{i(\mathbf{k}\cdot\mathbf{x} - \omega_\pm t)} + \text{c.c.} \qquad (1.48)$$

As can be verified explicitly, this solution is irrotational. So one does not have to assume irrotational flow, but one can actually derive that the linear, periodic, barotropic modes are irrotational. The other modes can have nonvanishing vorticity, but in the framework of the linear Euler equation they cannot vary with time. The Gerstner waves, discussed, for example, in Lamb (1932), are time varying and they do have vorticity, but their vorticity vanishes in the limit of infinitesimal amplitude, so that they do not violate the present result.

   Once the irrotational nature of the elementary solutions has been established, one may use a potential flow description without lack of generality.

We will do this from now on. For completeness we also give the linearized potential flow equations, which read

$$\phi_{x_1x_1} + \phi_{x_2x_2} + \phi_{zz} = 0, \; z < 0, \tag{1.49}$$

$$\left.\begin{array}{l} \eta_t = \phi_z, \\ \phi_t + g\eta = 0, \end{array}\right\} z = 0. \tag{1.50}$$

Their normal mode solutions are

$$\begin{pmatrix} \eta(\mathbf{x}, t) \\ \phi(\mathbf{x}, z, t) \end{pmatrix} = a \begin{pmatrix} 1 \\ -i\omega_{\pm}e^{kz}/k \end{pmatrix} e^{i(\mathbf{k}\cdot\mathbf{x}-\omega_{\pm}t)} + \text{c.c.} \tag{1.51}$$

Note that the average energy per unit area can be obtained by substituting (1.51) in (1.17). The result is

$$\bar{\mathscr{E}} = 2\rho g|a|^2. \tag{1.52}$$

So far we have considered deep water waves by assuming $\phi \to 0$ as $z \to -\infty$. In shallow water, for a flat bottom at depth $z = -h$ one has instead

$$\phi_z = 0, \quad z = -h. \tag{1.53}$$

The linearized equations can be analysed as before but instead of (1.51) one obtains

$$\phi = -i\omega_{\pm}a\frac{\cosh k(z+h)}{k \sinh kh}e^{i(\mathbf{k}\cdot\mathbf{x}-\omega_{\pm}t)} + \text{c.c.}, \tag{1.54}$$

and the dispersion relation becomes

$$\omega_{\pm} = \pm\sigma(k, h) \quad \sigma^2 = gk \tanh kh. \tag{1.55}$$

Implicitly, this is an expression for the depth dependence of the relation between wavelength and wave period.

I.2.3 Slowly varying waves

We consider wave trains and wave packets in the 'geometrical optics' approximation. For propagation through an inhomogeneous medium (variable depth, currents) wave action is conserved, but other processes not conserving wave action may also be described in this approximation.

In the previous subsection we analysed the free, linearized, water wave equations, in a basin with a constant depth. We now want to extend the analysis to more general situations. To this end we will consider a generalization of (1.51) in which amplitude, frequency and wavelength are allowed to vary slowly with respect to the intrinsic space scale (the wavelength) and time scale (the wave period). To be specific, one may think of swell on currents varying slowly in space and time in a shallow sea in which the bottom depth also varies slowly. However, the idea is quite general and applicable to other situations as well. In fact, it is also useful in case of free, linear waves with an inhomogeneous initial condition. One writes

$$\eta(\mathbf{x}, t) = a(\mathbf{x}, t) \exp is(\mathbf{x}, t) + \text{c.c.} \tag{1.56}$$

The phase function s is also known as the eikonal. The approach is known as the geometrical optics approximation (or the WKB or WKBJ method, in quantum mechanics). Locally the eikonal may be expanded as

$$s(\mathbf{x}, t) = s(0, 0) + \mathbf{x} \cdot \nabla_x s + t \partial_t s, \tag{1.57}$$

so that a local frequency and wavenumber can be defined by

$$\mathbf{k}(\mathbf{x}, t) = \nabla_x s(\mathbf{x}, t), \quad \omega(\mathbf{x}, t) = -\partial_t s(\mathbf{x}, t), \tag{1.58}$$

where the subscript x on the nabla denotes differentiation with respect to space coordinates, to distinguish it from differentiation with respect to wavenumber which will be introduced shortly. Equation (1.58) implies the following consistency relation:

$$\partial_t \mathbf{k} + \nabla_x \omega = 0. \tag{1.59}$$

For dispersive waves the dispersion relation dictates

$$\omega(\mathbf{x}, t) = \Omega[\mathbf{k}(\mathbf{x}, t), \mathbf{U}] = \mathbf{k} \cdot \mathbf{U} + \sigma[\mathbf{k}(\mathbf{x}, t), h(\mathbf{x})], \tag{1.60}$$

with the σ defined in (1.55) and $\mathbf{U}(\mathbf{x}, t)$ the current, which we have taken to be independent of depth for simplicity. The first term on the right in (1.60) is the Doppler shift due to this current.

Written explicitly in terms of s, the dispersion relation yields the eikonal equation

$$(\partial_t s + \mathbf{U} \cdot \nabla_x s)^2 - g|\nabla_x s| \; \tanh(|\nabla_x s|h) = 0. \tag{1.61}$$

It is straightforward to solve the time evolution of the eikonal numerically. Given $s(\mathbf{x}, 0)$, one determines $\mathbf{k}(\mathbf{x}, 0) = \nabla_x s(\mathbf{x}, 0)$ and then one solves (1.59) written as

$$\partial_t \mathbf{k} + \nabla_x \Omega[\mathbf{k}, \mathbf{U}] = 0 \tag{1.62}$$

to determine $\mathbf{k}(\mathbf{x}, t + \Delta t)$ and so on. Alternatively, equation (1.61) can be solved by the method of characteristics, which yields the well-known Hamilton equations (see below).

Free waves in an inhomogeneous medium

In the case of an inhomogeneous medium (varying depth and varying currents) it is possible to derive a relatively simple expression for the time evolution of a:

$$\partial_t N + \nabla_x[\mathbf{v}_D N] = 0, \tag{1.63}$$

with

$$N(\mathbf{x}, t) = 2|a(\mathbf{x}, t)|^2/\sigma \qquad (1.64)$$

(the factor 2 has been introduced for later convenience) and

$$\mathbf{v}_D = \nabla_k \Omega = \mathbf{U} + \mathbf{c}_g = \mathbf{U} + \nabla_k \sigma(\mathbf{k}(\mathbf{x}, t)), \qquad (1.65)$$

where the subscript k now denotes differentiation with respect to wavenumber. Note that equation (1.63) has the general form of a conservation law, in which the local rate of change of a density is determined by a flux of that density. It implies that the integral over all space of N is conserved in time. Denoting this integral by

$$N_{tot} = \int d\mathbf{x} N(\mathbf{x}, t), \qquad (1.66)$$

its constancy follows immediately by integrating (1.63)

$$\dot{N}_{tot} = \frac{d}{dt} \int_{\mathscr{S}} d\mathbf{x} N(\mathbf{x}, t) = -\oint d\mathbf{r} \cdot [\mathbf{v}_D(\mathbf{r}) N(\mathbf{r}, t)] = 0. \qquad (1.67)$$

The last equality follows from the fact that one has zero flux at the boundaries of the basin.

The characteristic propagation velocity in (1.63) is $\mathbf{v}_D$. It is the sum of the group velocity $\mathbf{c}_g = \nabla_k \sigma$ and the current velocity $\mathbf{U}$.

The detailed derivation of (1.63) (and (1.60)) will not be given here. A discussion can be found in Bretherton and Garrett (1968) and Whitham (1974). The essence is that expression (1.56) is substituted in the Lagrangian, which then is averaged over the rapid variations. The average Lagrangian is a function of frequency, wavenumber and amplitude, for which one postulates the average variational principle (Hamilton's principle or the principle of least action)

$$\delta W = 0, \quad W = \int\int \bar{\mathscr{L}}(\omega, \mathbf{k}, a) d\mathbf{x} dt. \qquad (1.68)$$

The resulting Euler-Lagrange equations are the equations given above. Multiplied by a factor ρg, the quantity N_{tot} has the dimension of action (energy multiplied by time). For reasons to be explained below

$$A_{tot} = \rho g N_{tot} \qquad (1.69)$$

is called the wave action and $\rho g N(\mathbf{x}, t)$ is the corresponding wave action density.

For the special case of no currents and a flat bottom equation (1.63) reduces to

$$\partial_t F + \nabla_x \cdot [\mathbf{c}_g F] = 0, \qquad (1.70)$$

with

$$F(\mathbf{x}, t) = 2|a(\mathbf{x}, t)|^2, \quad \mathbf{c}_g = \nabla_k \sigma(\mathbf{k}). \qquad (1.71)$$

The quantity F has a physical interpretation which may be obtained by substituting (1.56) in (1.15) and by averaging over a wavelength. One then obtains

$$\bar{\mathscr{E}}(\mathbf{x}, t) = \rho g F(\mathbf{x}, t), \tag{1.72}$$

so F is proportional to $\bar{\mathscr{E}}$, the mean energy density per unit surface. Note that equation (1.72) is more general than equation (1.52), because now the mean energy is allowed to vary slowly with $\mathbf{x}$ and t. Equation (1.70) expresses conservation of energy. The group velocity $\mathbf{c}_g$ can be interpreted as the propagation velocity of energy. The total energy in a certain area only changes because of energy flowing in and out through the boundaries. However, one should realize that energy conservation only holds in the absence of currents and energy changing processes. In general, the energy may change, because the waves may lose or gain energy through interaction with the currents. The more general conservation law therefore is equation (1.63) expressing conservation of wave action.

The action conservation law is different from the other conservation laws. Energy and momentum, for example, can be obtained by considering the symmetry of the Lagrangian under time and space translations, but for wave action we do not, in general, have such a symmetry. (Later we shall see that a truncated form of the Lagrangian is phase-shift invariant and therefore conserves wave action.) In classical mechanics the action is given by (1.68). In addition the '(abbreviated) action' (Landau and Lifshitz, 1960) plays a role. It is defined by

$$W_{abbr} = \int p dq, \tag{1.73}$$

where p and q are the canonical conjugate momentum and coordinate. It can be shown that W_{abbr} is conserved for a system in which the energy varies slowly with an external parameter and for solutions that are near-periodic. In that case one can rewrite the above equation as

$$A = \oint p dq. \tag{1.74}$$

Whitham (1974) argues that energy and frequency change when the external parameter changes, but that

$$A(E, \omega) = \text{constant} \quad \rightarrow \quad \frac{dA(E, \omega)}{dt} = 0. \tag{1.75}$$

A is an 'adiabatic invariant'. For a harmonic oscillator (consider a pendulum in which the length is varied slowly) one obtains

$$A = E/\omega \quad \rightarrow \quad \frac{d}{dt} E/\omega = 0. \tag{1.76}$$

For a slowly varying wave train it is equation (1.63) that expresses the adiabatic invariance. In this case A_{tot} as given in (1.69) is the adiabatic invariant corresponding to the reduced action (1.73). It is from this property that it derives the name wave action.

Wave packets

A wave packet is a function of the form (1.56) with a finite extent. It has a mean position $\mathbf{X}$ given by

$$\mathbf{X}(t) = \int \mathbf{x} a(\mathbf{x}, t) d\mathbf{x} \Big/ \int a(\mathbf{x}, t) d\mathbf{x}. \qquad (1.77)$$

Typically a wave packet may contain 10 to 100 wave crests. The mean position of a wave packet can be shown to propagate with the group velocity

$$\dot{\mathbf{X}} = \nabla_k \Omega(\mathbf{K}, \mathbf{U}(\mathbf{X}, t), h(\mathbf{X})) \qquad (1.78)$$

in which the mean wavenumber $\mathbf{K}$ has been defined by

$$\mathbf{K} = \mathbf{k}(\mathbf{X}, t). \qquad (1.79)$$

The time evolution of $\mathbf{K}$ follows from (1.59). The path defined by equations (1.78) and (1.79) is called a *ray*. Wave packets propagate along rays. One may summarize this by writing

$$\dot{\mathbf{X}} = \nabla_k \Omega, \quad \dot{\mathbf{K}} = -\nabla_x \Omega. \qquad (1.80)$$

These are the characteristic equations of the eikonal equation (1.61) (cf. Courant and Hilbert, 1962). They correspond to Hamilton's equations in classical mechanics (cf. Landau and Lifshitz, 1960), from which they are obtained by the replacement $(p, q, H) \rightarrow (K, X, \Omega)$.

For a wave packet the total wave action is conserved. It can be obtained by integrating the action density over its (finite) extension. The total energy of the packet is, in general, not conserved, because the wave amplitude may be modified by the interaction with the currents.

The simplest example of a wave packet is obtained by considering linear waves in deep water without currents. In that case $\dot{\mathbf{K}} = 0$, and $\dot{\mathbf{X}}$ is simply given by the constant group velocity. The shape of the packet is completely determined by the initial condition. In this case the Fourier approach to the time evolution of linear waves, to be discussed in the next subsection, is an alternative to the present WKB treatment.

Other perturbations

So far we have considered free waves on varying currents in basins with variable depth. However, the WKB approach can also be used in other situations. For example, one may study the coupled problem which takes into account the effect of wind on waves. Or, one may consider nonlinear effects or dissipation. As long as these perturbations are small they can be added

$$\dot{a} = \sum_i \dot{a}_i, \tag{1.81}$$

where the sum is over different perturbations. This is an important observation because it allows one to study the perturbations separately.

I.2.4 The superposition of linear waves and the initial value problem

We discuss the superposition of normal mode solutions, which is nothing but the Fourier representation of the sea surface. We show how this representation can be used to solve the Cauchy problem (initial value problem) in the linear case. The energy is the sum of the energies of the separate modes.

In the previous subsection we discussed the time evolution of special solutions. We will now consider the general solution of the initial value problem.

It can be shown that the normal mode solutions of § I.2.2 form a complete set for the irrotational motion of the sea, that is for irrotational initial conditions, the function $\eta(\mathbf{x}, t)$ can be written as a superposition of plane waves with wavenumber $\mathbf{k}$, provided one includes the two modes for each $\mathbf{k}$ involved. For a finite basin one would expect to obtain a discrete set of $\mathbf{k}$ values, as one has to satisfy lateral boundary conditions. For an infinite basin a continuum of modes would be allowed. As in other branches of physics a problem arises because of the large dimension of the ocean basin in comparison with a typical wavelength. Although formally the basin is finite and one has a discrete set of modes, in fact the system can be treated as infinite with a continuum of modes. (Even so – as we shall see later – it is often convenient to use a discrete Fourier sum representation for a small area of the ocean.) In the continuum case one would write

$$\eta(\mathbf{x}, t) = \int_{-\infty}^{+\infty} d\mathbf{k} \, \hat{\eta}_+(\mathbf{k}) e^{i(\mathbf{k} \cdot \mathbf{x} - \omega_+ t)} + \int_{-\infty}^{+\infty} d\mathbf{k} \, \hat{\eta}_-(\mathbf{k}) e^{i(\mathbf{k} \cdot \mathbf{x} - \omega_- t)}. \tag{1.82}$$

There are two terms because there are two different modes for each $\mathbf{k}$. However, since η has to be real it is straightforward to show that

$$\hat{\eta}_-(\mathbf{k}) = \hat{\eta}_+^*(-\mathbf{k}) \tag{1.83}$$

and this allows one to simplify (1.82) to

$$\eta(\mathbf{x}, t) = \int_{-\infty}^{+\infty} d\mathbf{k} \, \hat{\eta}(\mathbf{k}) e^{i(\mathbf{k} \cdot \mathbf{x} - \omega t)} + \text{c.c.}, \tag{1.84}$$

where we can omit the subscript $+$ on $\hat{\eta}$. It is easy to work out the inverse

$$\hat{\eta}(\mathbf{k}) = \frac{1}{8\pi^2} \int d\mathbf{x} \; e^{-i\mathbf{k}\cdot\mathbf{x}} \left[\eta(\mathbf{x},0) + \frac{i}{\omega(k)} \eta_t(\mathbf{x},0) \right]. \qquad (1.85)$$

For an infinite ocean, the integral on the right hand side does not exist in the conventional sense. We shall regard the representation in (1.82) as defined first over a finite region and consider the limit of an infinite region later in a more general statistical context.

It is often convenient to represent η by a discrete Fourier representation. One way of obtaining such a representation is to subdivide the k_{x_1}- and k_{x_2}-axes into equidistant intervals with the help of a bandwidth $\triangle k$ and to approximate (1.84) by

$$\eta(\mathbf{x},t) = \sum_{\mathbf{k}} \eta_{\mathbf{k}} + \text{c.c.} = \sum_{\mathbf{k}} \tilde{a}_{\mathbf{k}} e^{i\mathbf{k}\cdot\mathbf{x}} + \text{c.c.} = \sum_{\mathbf{k}} a_{\mathbf{k}} e^{i(\mathbf{k}\cdot\mathbf{x}-\omega t)} + \text{c.c.}, \qquad (1.86)$$

with

$$\mathbf{k} = (\pm n\triangle k, \pm m\triangle k) \quad n,m = 1, \cdots N_{max} \qquad (1.87)$$

and

$$a_{\mathbf{k}} = \int_{\mathbf{k}-\frac{1}{2}\triangle k}^{\mathbf{k}+\frac{1}{2}\triangle k} \hat{\eta}(\mathbf{k}) d\mathbf{k}. \qquad (1.88)$$

Formally, (1.86) is equivalent to the expression that would be obtained for a finite box with dimension $L \times L$, provided one identifies $\triangle k = 2\pi/L$. It should be noted that $\mathbf{k}$ has a different meaning now, because it is a discrete label rather than a continuous variable. This is sometimes confusing, but since it is commonly used we will not introduce another notation.

In the discrete approach the inverse of (1.86) becomes:

$$a_{\mathbf{k}} = \frac{1}{2L^2} \int_{-L/2}^{+L/2} dx_1 \int_{-L/2}^{+L/2} dx_2 \; e^{-i\mathbf{k}\cdot\mathbf{x}} \left[\eta(\mathbf{x},0) + \frac{i}{\omega_{\mathbf{k}}} \eta_t(\mathbf{x},0) \right]. \qquad (1.89)$$

This integral, which is the finite domain counterpart of (1.85), exists in the usual Riemann sense.

The problem of determining wave propagation from an initial disturbance is known as the Cauchy-Poisson problem, after the first analysis of the problem by Poisson (1816) and Cauchy (1827). It is discussed in more detail elsewhere (Courant and Hilbert, 1953, Morse and Feshbach, 1953, LeBlond and Mysak, 1978 and Whitham, 1974). Once the normal modes have been derived its solution is straightforward. It can be schematized by:

$$\{\eta(\mathbf{x},0), \eta_t(\mathbf{x},0)\} \xrightarrow{(1.85)} \hat{\eta}(\mathbf{k}) \qquad (1.90)$$

followed by

$$\hat{\eta}(\mathbf{k}) \xrightarrow{(1.84)} \eta(\mathbf{x},t). \qquad (1.91)$$

In words: from the surface elevation and its time derivative at time $t = 0$, one can compute the normal mode amplitudes. From this one may obtain the surface elevation at a later time. The essence is, of course, that we know, through the dispersion relation, for each normal mode, what its time evolution is.

One important application is the study of the evolution of free wave packets. By substituting (1.56) and its time derivative for $t = 0$ in (1.90) and by analysing the result of (1.91) one obtains the results discussed before, namely, that the energy and the mean position of a linear wave packet propagate with the group velocity. It is also found that dispersion leads to an increase of the effective size of a packet with time. Quantitatively, one finds that the extension grows proportionally with the square root of time.

We have applied the Fourier approach to free waves in a homogeneous medium. The general discussion of propagation in nonhomogeneous media is more easily performed within the frame of a WKB approach. In this approach one considers a superposition of WKB modes of the form

$$\eta(\mathbf{x}, t) = \sum_n a_n(\mathbf{x}, t)e^{i(\mathbf{k}_n \cdot \mathbf{x} - \omega_n t)} + \text{c.c.}, \tag{1.92}$$

which may be considered as a generalization of (1.86). We will postpone the discussion of (1.92) for a while, as we need to discuss statistical aspects first.

The energy distribution and the wave spectrum

Substitution of (1.82) in the energy expression gives an infinite answer unless the waves are confined to a finite area. This is as it should be, because an infinite ocean has infinite energy. For the discrete representation (1.86), which is on a finite box, one does not have this problem. In this case it is straightforward to calculate the total energy and the average energy density, defined by (1.17). Substituting (1.86) in (1.17) and (1.16), the average energy per unit area can be obtained. To this end one must express ϕ in terms of η, so that each term in (1.15) can be computed separately. In doing this one finds that the potential energy density and the kinetic energy density are not, in general, equal at a given time. This disparity occurs because there may be standing waves, for which there is a periodic exchange of energy between both terms of (1.15). With and without standing waves, the following result is obtained for the desired energy density

$$\bar{\mathscr{E}} = \rho g \sum_{\mathbf{k}} F_{\mathbf{k}}, \tag{1.93}$$

where we have introduced

$$F_{\mathbf{k}} = 2|a_{\mathbf{k}}|^2, \tag{1.94}$$

which, apart from the factor ρg, is the mean energy per Fourier mode (in the statistical description introduced in the following subsection, standing waves are excluded by the requirement of quasi-homogeneity and stationarity).

It is also interesting to introduce the (continuous) energy density per wavenumber band, i.e. the energy per unit area contained in modes with $\mathbf{k}$ between $\mathbf{k} - \frac{1}{2}\Delta\mathbf{k}$ and $\mathbf{k} + \frac{1}{2}\Delta\mathbf{k}$. This is most conveniently done by introducing a function $\varepsilon(\mathbf{k})$ with the definition

$$\varepsilon(\mathbf{k}) = \begin{cases} 1 & \text{if } -\frac{1}{2}\Delta\mathbf{k} < \mathbf{k} < \frac{1}{2}\Delta\mathbf{k} \\ 0 & \text{otherwise.} \end{cases} \tag{1.95}$$

With this definition we can write

$$F(\mathbf{k})\Delta\mathbf{k} = \sum_{\mathbf{k}'} F_{\mathbf{k}'}\varepsilon(\mathbf{k} - \mathbf{k}'), \tag{1.96}$$

so that the sum is over all the modes with wavenumbers contained in the above mentioned interval. The factor $\Delta\mathbf{k}$ has been introduced to make sure that the integral of $F(\mathbf{k})$ over a particular $\mathbf{k}$ interval is equal to the energy contained in the discrete modes having wavenumber values in that interval. In particular, one may re-express (1.93) as:

$$\bar{\mathscr{E}} = \rho g \int d\mathbf{k} \, F(\mathbf{k}). \tag{1.97}$$

The quantity $F(\mathbf{k})$ as well as its discrete counterpart $F_{\mathbf{k}}$ is known as the wave (energy) spectrum. It is of crucial and central importance in the description of ocean waves. It should be noted how much simpler equation (1.93) is than the original equation (1.15). This is because we linearized the equations and because we expanded in normal modes. The energy as given by (1.93) does not depend on time, simply because the $a_{\mathbf{k}}$ do not. Once one has made this observation, an infinite number of constants of the motion can be constructed, because to each function $\Gamma(\mathbf{k})$ there corresponds an invariant

$$\mathscr{C}_\Gamma = \int d\mathbf{k}\Gamma(\mathbf{k})F(\mathbf{k}). \tag{1.98}$$

For $\Gamma = \mathbf{k}/\sigma$ it can be shown that one obtains the mean momentum per unit area

$$\bar{\mathscr{M}} = \rho g \int d\mathbf{k}\frac{\mathbf{k}}{\sigma}F(\mathbf{k}). \tag{1.99}$$

Another invariant, obtained for $\Gamma = 1/\sigma$,

$$\rho g \int d\mathbf{k}F(\mathbf{k})/\sigma, \tag{1.100}$$

can be interpreted as the wave action per unit area. To see this, it is convenient to introduce a quantity $N_{\mathbf{k}}$ which, apart from a factor ρg, is the wave action density per Fourier mode

$$N_{\mathbf{k}} = 2|a_{\mathbf{k}}|^2/\sigma_{\mathbf{k}}, \qquad (1.101)$$

and its continuous counterpart

$$N(\mathbf{k})\triangle\mathbf{k} = \sum_{\mathbf{k}'} N_{\mathbf{k}'}\varepsilon(\mathbf{k} - \mathbf{k}'). \qquad (1.102)$$

One can then write for the mean action per unit area

$$\bar{\mathscr{A}} = \rho g \sum_{\mathbf{k}} N_{\mathbf{k}} = \rho g \int \mathrm{d}\mathbf{k} N(\mathbf{k}). \qquad (1.103)$$

Comparing (1.94) and (1.101) one has

$$N_{\mathbf{k}} = F_{\mathbf{k}}/\sigma \qquad (1.104)$$

and also

$$N(\mathbf{k}) = F(\mathbf{k})/\sigma, \qquad (1.105)$$

so the right hand side of (1.103) is indeed identical with (1.100). For free, linear waves, $\bar{\mathscr{E}}$, $\bar{\mathscr{M}}$, $\bar{\mathscr{A}}$ and an infinity of other quantities are conserved, i.e. they do not vary with time. However, only the energy, the momentum and ten other invariants (not the wave action) are also invariants of the full nonlinear system (Brooke Benjamin and Olver, 1982).

I.2.5 Statistical description in the linear theory

We introduce the homogeneous and stationary statistical theory of linear random waves. In such a theory wave components are necessarily independent. Its Gaussian statistics are determined by the two-point covariance function, which is the Fourier transform of the wavenumber spectrum. We also introduce the frequency spectrum. The significant wave height, defined as four times the root-mean-square surface displacement, is the characteristic length scale in the probability distribution of wave maxima.

In practice one never considers the deterministic initial value problem for a realistic sea basin. The reason is that it is virtually impossible to determine the Fourier modes with the correct phases. To overcome this problem one resorts to a statistical description.

We will now summarize the statistical theory of a linear random wave field (see, for instance, Monin and Yaglom, 1975). In the next section we will consider the generalization to a quasi-homogeneous and quasi-stationary description of interacting sea waves.

Knowledge of the probability density P of equation (1.1) allows one to compute all statistical properties. Equivalently one may ask for the set of all moments:

$$< \eta(\mathbf{x}_1, t_1) >= \int_{-\infty}^{+\infty} \eta_1 P(\eta_1) d\eta_1,$$

$$< \eta(\mathbf{x}_1, t_1)\eta(\mathbf{x}_2, t_2) >= \int_{-\infty}^{+\infty} \eta_1 \eta_2 P(\eta_1, \eta_2) d\eta_1 d\eta_2,$$

$$< \eta(\mathbf{x}_1, t_1)\eta(\mathbf{x}_2, t_2)\eta(\mathbf{x}_3, t_3) >= \int_{-\infty}^{+\infty} \eta_1 \eta_2 \eta_3 P(\eta_1, \eta_2, \eta_3) d\eta_1 d\eta_2 d\eta_3, \qquad (1.106)$$

$$\vdots$$

$$< \eta(\mathbf{x}_1, t_1)\eta(\mathbf{x}_2, t_2) \ldots \eta(\mathbf{x}_n, t_n) >=$$
$$\int_{-\infty}^{+\infty} \eta_1 \eta_2 \ldots \eta_n P(\eta_1, \eta_2, \ldots, \eta_n) d\eta_1 d\eta_2 \ldots d\eta_n.$$

The indices 1, 2, ... in the integrals refer to the points $(\mathbf{x}_1, t_1), (\mathbf{x}_2, t_2), \ldots$.

There are two approaches for defining the ensemble average $< \cdots >$. In either approach it is essential that the wave field is statistically stationary and homogeneous.

The first approach is a practical method which is used when one actually works with data. The region of the ocean in which one is interested is imagined as being divided up into a large number N_x of boxes. For each box the time history of the wave field is similarly divided into a large number N_t of time periods (chunks). The size L of the box and the period T of the chunks are assumed to be large compared with the characteristic wavelength and periods of the waves, while at the same time the overall space-time dimensions $N_x L$ and $N_t T$ of the region of analysis are assumed to be still sufficiently small that the wave field can be regarded as statistically homogeneous and stationary within this region. The average of a quantity q is then defined as the mean over the set of $N = N_x N_t$ box-chunks ℓ,

$$< q >= \frac{1}{N} \sum_{\ell} q_\ell. \qquad (1.107)$$

Homogeneity and stationarity are defined here as the property that essentially the same value for the mean is obtained if the average is taken over any subset of box-chunks, as long as the subset is large. Definition (1.107) of the ensemble average is clearly possible and meaningful only if the wave field is indeed (quasi-)homogeneous and stationary.

Although it is useful to keep this picture in mind, it has proved difficult to develop a rigorous statistical theory founded on this practice-oriented ensemble construction (see also von Mises' (1952) elegant but ultimately unsuccessful attempt). Accordingly, modern stochastic theories are based on an alternative approach in which it is simply postulated that an ensemble of realizations λ exists in some abstract space Λ endowed with some probability measure $dP(\lambda)$, and the ensemble average of a quantity q is defined as

$$< q > = \int q(\lambda)dP(\lambda). \tag{1.108}$$

To establish a relation between this abstract formalism and the real, single ocean (that is to translate theoretical results into numerical procedures) it is again necessary that the field is statistically either homogeneous or stationary (in practice both simultaneously apply). Statistical stationarity and homogeneity is now defined most conveniently as the invariance of all ensemble averages under space and time translations $\mathbf{x} \to \mathbf{x}' = \mathbf{x} + \Delta\mathbf{x}, t \to t' = t+\Delta t$, a property that can be inferred also from the previous definition for the first approach. As we will see in a moment, homogeneity and stationarity are extremely important properties because they make it possible to work out the theory without determining the probability measure explicitly.

To show this we begin by considering the two-point covariance function

$$\mathscr{F}(\boldsymbol{\xi},\tau) = < \eta(\mathbf{x} + \boldsymbol{\xi}, t + \tau)\eta(\mathbf{x}, t) > . \tag{1.109}$$

Inserting (1.86) we obtain

$$\begin{aligned}\mathscr{F} &= \sum_{\mathbf{k},\mathbf{k}'}\{[< a_{\mathbf{k}}a_{\mathbf{k}'} > e^{i(\mathbf{k}+\mathbf{k}')\cdot\mathbf{x}-i(\omega+\omega')t} \\ &+ < a_{\mathbf{k}}a_{\mathbf{k}'}^* > e^{i(\mathbf{k}-\mathbf{k}')\cdot\mathbf{x}-i(\omega-\omega')t}]e^{i(\mathbf{k}\cdot\boldsymbol{\xi}-\omega\tau)}\} + \text{c.c.} \tag{1.110}\end{aligned}$$

Clearly, the right hand side can represent a function which is independent of $\mathbf{x}$ and t only if we have

$$< a_{\mathbf{k}}a_{\mathbf{k}'} > = 0, \tag{1.111}$$

$$< a_{\mathbf{k}}a_{\mathbf{k}'}^* > = |a_{\mathbf{k}}|^2\delta_{\mathbf{k}\mathbf{k}'} = \frac{1}{2}F_{\mathbf{k}}\delta_{\mathbf{k}\mathbf{k}'}. \tag{1.112}$$

The Kronecker δ is defined by

$$\delta_{\mathbf{k}\mathbf{k}'} = \begin{cases} 1 & \text{if } \mathbf{k} = \mathbf{k}', \\ 0 & \text{otherwise.} \end{cases} \tag{1.113}$$

Equations (1.111) – (1.112) are quite general and do not depend on a specific form of the probability measure (1.108). They may be substituted into (1.110) with the result

$$\mathscr{F}(\boldsymbol{\xi},\tau) = \frac{1}{2}\sum_{\mathbf{k}} F_{\mathbf{k}}e^{i(\mathbf{k}\cdot\boldsymbol{\xi}-\sigma\tau)} + \text{c.c.}, \tag{1.114}$$

expressing the two-point covariance function in terms of the wavenumber spectrum introduced in equation (1.94). Note that equation (1.114) is similar to (1.86). It can be inverted, with a result similar to (1.89):

$$F_{\mathbf{k}} = \frac{1}{2L^2}\int_{-L/2}^{+L/2} d\boldsymbol{\xi}\, e^{-i\mathbf{k}\cdot\boldsymbol{\xi}}\left[\mathscr{F}(\boldsymbol{\xi},0) + \frac{i}{\omega_{\mathbf{k}}}\mathscr{F}_{\tau}(\boldsymbol{\xi},0)\right]. \tag{1.115}$$

This dynamical result may be compared with the standard way of defining the frequency–wavenumber-dependent variance spectrum of a random space-time-dependent surface, without reference to the dynamics of the surface, as the three-dimensional Fourier transform of $\mathscr{F}$ both with respect to ξ and τ:

$$F_{\mathbf{k}}(\omega) = \frac{1}{L^2} \int_{-L/2}^{+L/2} d\xi \int_{-\infty}^{+\infty} d\tau \mathscr{F}(\xi, \tau) e^{i(\mathbf{k}\cdot\xi - \omega\tau)}. \qquad (1.116)$$

From this the wavenumber spectrum is then defined as

$$F_{\mathbf{k}} = \int_0^\infty F_{\mathbf{k}}(\omega) d\omega. \qquad (1.117)$$

In this approach one can reintroduce the dynamics by substitution of (1.114) in (1.116) which leads to:

$$F_{\mathbf{k}}(\omega) = F_{\mathbf{k}}\delta(\omega - \sigma(k)) + F_{-\mathbf{k}}\delta(\omega + \sigma(k)), \qquad (1.118)$$

where $\delta(\ )$ is the Dirac delta-function. Equation (1.118) tells you that linear waves have to satisfy the dispersion relation. It can be shown that substitution of (1.118) in (1.117) recovers (1.115), so both expressions for $F_{\mathbf{k}}$ in terms of $\mathscr{F}$ are equivalent.

Note that the simpler transform (the 'frozen surface spectrum')

$$\hat{F}_{\mathbf{k}} = \frac{1}{L^2} \int d\xi e^{i\mathbf{k}\cdot\xi} \mathscr{F}(\xi, 0) = \frac{1}{L^2} \int d\xi e^{i\mathbf{k}\cdot\xi} < \eta(\mathbf{x} + \xi, t)\eta(\mathbf{x}, t) > \qquad (1.119)$$

gives an answer that differs from (1.117), because it adds the energy from waves going in positive and negative $\mathbf{k}$ directions:

$$\hat{F}_{\mathbf{k}} = \frac{1}{2}(F_{\mathbf{k}} + F_{-\mathbf{k}}). \qquad (1.120)$$

This is not surprising: from a frozen picture of the sea surface one cannot see in what direction the waves are moving. The spectrum $\hat{F}$ consists of the mean of the directional spectrum F in the two propagation directions $+\mathbf{k}$ and $-\mathbf{k}$ and is thus an even function of the wavenumber. The directional wave spectrum completely determines the other two spectra and thus provides a complete description of a Gaussian sea surface, while the spectrum $\hat{F}$ (which is measured by imaging devices such as a synthetic aperture radar; see also §§ I.4, V.3 and V.4) provides an incomplete description of the sea surface due to the ambiguity in the propagation direction of a measured component $\mathbf{k}$. This is a problem when one attempts to convert image spectra into wave spectra.

As already stated equations (1.111) and (1.112) are quite general and do not depend on a specific form of the probability measure (1.108). Nevertheless, it may be useful to consider the construction of an ensemble in an explicit example. Such an example is the well-known, random phase case in which it is assumed that the magnitude of the amplitudes of the Fourier modes are deterministic and that the phases are distributed randomly. In this case it is convenient to write

$$a_{\mathbf{k}} = |a_{\mathbf{k}}|e^{i\mu}\mathbf{k}, \tag{1.121}$$

where $\mu_{\mathbf{k}}$ is the random phase of wave component $\mathbf{k}$. The moments can now be computed as

$$< (\ldots) > = \frac{1}{(2\pi)^N} \int_0^{2\pi} \int_0^{2\pi} \cdots \int_0^{2\pi} d\mu_1 d\mu_2 \ldots d\mu_N(\ldots). \tag{1.122}$$

This gives, for example,

$$< a_{\mathbf{k}} > = \frac{1}{2\pi} \int d\mu_{\mathbf{k}} |a_{\mathbf{k}}|e^{i\mu}\mathbf{k} = 0, \tag{1.123}$$

which implies

$$< \eta(\mathbf{x}, t) > = 0, \tag{1.124}$$

as it should for a steady mean sea level. Averaging products of $a_{\mathbf{k}}$s in a similar way, we recover (1.111) and (1.112). However, it must be stressed that this example is a simplification, because, in general, both the amplitude and the phases are random variables.

So far we have discussed the consequences of homogeneity and stationarity for the second moments only. For the higher moments it can be shown that these conditions require the ensemble-mean Fourier-component products to satisfy simultaneously $s_1\mathbf{k}_1 + \cdots + s_n\mathbf{k}_n = 0$ and $s_1\omega_1 + \cdots + s_n\omega_n = 0$ (where s_n is a sign index). This implies that nonzero contributions to the higher order moments can arise only from terms in the mean products which are decomposed into pairs of mean products, in accordance with the higher moment decomposition relations for a Gaussian field, or in other words, the cumulant contributions must vanish. (There exists actually a subspace outside the subspace of paired wavenumbers for which both wavenumber and frequency sums vanish, but this is of measure zero.) Thus a linear wave field which is both homogeneous and stationary must necessarily be Gaussian. This result is at first sight surprising. It appears plausible that a wave field should be both homogeneous and stationary, but why should it then automatically also be Gaussian? Surely it is possible to specify an initial value problem for a statistical ensemble of wave components which initially has non-Gaussian statistics. The answer is that this is indeed true, but that the wave field then rapidly loses its non-Gaussian properties. We note first that, as pointed out above, in order to talk meaningfully about statistics, the initial wave ensemble must at least be statistically homogeneous. Formally it is possible to construct statistical averages from an ensemble of 'boxes' without also invoking statistical stationarity (although in practice one normally averages over time 'chunks'). Homogeneity implies that different wave components with different wavenumbers – to within a sign – must be uncorrelated. However, wave components with opposite wavenumbers propagating in opposite directions can be correlated. These yield statistically nonstationary standing waves. If the initial distribution is non-Gaussian there will then also generally exist higher order mean products of wave amplitudes whose wavenumber sums vanish, $s_1\mathbf{k}_1 + \ldots + s_n\mathbf{k}_n = 0$, as required for homogeneity, but whose frequency sums do not necessarily vanish, $s_1\omega_1 + \ldots + s_n\omega_n \neq 0$ and which are therefore also nonstationary. Because the frequency side condition does not apply, these products cannot be collapsed into products of spectra, as in the case of a stationary (and therefore Gaussian) wave field. However, it can be shown that all nonstationary, non-Gaussian properties of the wave field are oscillatory with respect to the wavenumbers, the rate of oscillation increasing with time. They degrade therefore into a fine structure which, in practice, cannot be detected by measurements (Hasselmann, 1968). Thus both the nonstationarity and non-Gaussian properties of the initial wave field are rapidly destroyed by phase mixing.

Rather than deriving this in detail mathematically, one may also invoke a simple physical argument based on representation (1.92) which described the wave field as a linear superposition of a large number of dynamically independent wave packets. In this picture, the wave field is thought as built up by many finite size wave packets. For linear waves the slow $\mathbf{x}$ and t dependences of a simply describe propagation and dispersion. To justify the use of (1.92) one argues that all representations of that form will lead to the same statistical theories as long as the a_n are independent and yield the same mean values. The wave packet picture is not very useful for solving the Cauchy problem, but it has proved extremely helpful in visualizing statistical aspects of the theory. It represents the wave field in the vicinity of any point in time and space as the superposition of many different wave packets which converged into that particular region by propagating from far distant, widely separated regions. They can therefore be regarded as statistically independent, and it follows then from the Central Limit Theorem that the field must be Gaussian (details are given in Hasselmann, 1968). For such a Gaussian distribution, the probability function P can be expressed in terms of the two-point or covariance function, now considered as a matrix with elements

$$\mathscr{F}_{ij} = <\eta_i\eta_j> = <\eta(\mathbf{x}_i, t_i), \eta(\mathbf{x}_j, t_j)> . \qquad (1.125)$$

In fact, one has

$$P(\eta_1, \eta_2, \ldots, \eta_n) = \frac{1}{(2\pi)^{n/2} \mid \mathscr{F} \mid^{1/2}} e^{-\frac{1}{2}\boldsymbol{\eta}^T \mathscr{F}^{-1} \boldsymbol{\eta}}, \qquad (1.126)$$

where $\boldsymbol{\eta}$ is the vector $(\eta_1, \eta_2, \ldots, \eta_n)$ and $\boldsymbol{\eta}^T$ is its transpose. It is quite natural therefore that the two-point function $\mathscr{F}$ plays a central role in the statistical theory. In (1.114) we have already seen that $\mathscr{F}$ is the Fourier transform of the wavenumber spectrum $F_{\mathbf{k}}$. Therefore, once one knows the spectrum, one can compute $\mathscr{F}$ and from this the probability distribution (1.126). For this reason it is understandable that wave research and wave modelling are strongly focussed on the determination of the wave spectrum.

Although this section is not on wave measurements, it may be conceptually helpful to imagine how the spectrum is actually measured. What is basically needed is a determination of $\eta(\mathbf{x}, t_0)$ and its time derivative over a sufficiently large area at time t_0. Such a determination is not trivial. Stereophotography, for example, would not be sufficient because it would not give the time derivatives. But let us assume we have measured both $\eta(\mathbf{x}, t_0)$ and $\partial_t \eta(\mathbf{x}, t_0)$. We can then perform a two-dimensional Fourier analysis to obtain the $a_{\mathbf{k}}$, which, upon quadrature, gives $F_{\mathbf{k}}$. Alternatively, one could determine the correlation function $\mathscr{F}(\boldsymbol{\xi}, 0)$ and its derivative $\partial_\tau \mathscr{F}(\boldsymbol{\xi}, 0)$ to obtain the wave spectrum with the help of (1.115). To get a reliable estimate of the spectrum we have to average over neighbouring boxes and time chunks (see the discussion above).

The average energy

It is instructive to calculate the mean wave energy $< \mathscr{E} >$, from the expression given for $\mathscr{E}$ given in (1.15), as an ensemble average. In this case the average potential and kinetic energy are equal, so we have the simple relation

$$< \mathscr{E} >= \rho g < \eta^2 > . \tag{1.127}$$

Putting $\boldsymbol{\xi}$ and τ equal to zero in equation (1.114) we obtain

$$< \eta^2 >= \sum_{\mathbf{k}} F_{\mathbf{k}}, \tag{1.128}$$

which leads to

$$< \mathscr{E} >= \rho g \sum_{\mathbf{k}} F_{\mathbf{k}}. \tag{1.129}$$

Comparison with (1.93) shows that the ensemble mean energy is identical with the spatially averaged energy, as indeed it must be.

The continuous case

For certain applications it is useful to have an expression for the moments of the continuous Fourier transforms $\hat{\eta}(\mathbf{k})$. This is most conveniently obtained by substituting (1.84) in the covariance function (1.109). One obtains

$$\mathscr{F} = \int d\mathbf{k} d\mathbf{k}' < \hat{\eta}(\mathbf{k})\hat{\eta}(\mathbf{k}') > e^{i[\mathbf{k}\cdot(\mathbf{x}+\boldsymbol{\xi})-\sigma(k)(t+\tau)+\mathbf{k}'\cdot\mathbf{x}-\sigma(k')t]}$$

$$< \hat{\eta}^*(\mathbf{k})\hat{\eta}(\mathbf{k}') > e^{i[-\mathbf{k}\cdot(\mathbf{x}+\boldsymbol{\xi})+\sigma(k)(t+\tau)+\mathbf{k}'\cdot\mathbf{x}-\sigma(k')t]} + \text{c.c.,} \tag{1.130}$$

which is the continuous-spectrum form of (1.110). For this to be independent of $\mathbf{x}$ and t we must have

$$< \hat{\eta}(\mathbf{k})\hat{\eta}(\mathbf{k}') >= 0, \tag{1.131}$$

$$< \hat{\eta}(\mathbf{k})\hat{\eta}^*(\mathbf{k}') >= \frac{1}{2}F(\mathbf{k})\delta(\mathbf{k} - \mathbf{k}'). \tag{1.132}$$

These equations are equivalent to (1.111) and (1.112) and express the independence of wave components in the continuous representation. It is easily confirmed that the function appearing on the right hand side of (1.132) is indeed equal to the wave spectrum $F(\mathbf{k})$ by substituting (1.132) in (1.130), which gives

$$\mathscr{F}(\boldsymbol{\xi},\tau) = \frac{1}{2} \int d\mathbf{k} F(\mathbf{k}) e^{i(\mathbf{k}\cdot\boldsymbol{\xi} - \omega\tau)} + \text{c.c.} \tag{1.133}$$

Setting $\boldsymbol{\xi}$ and τ equal to zero gives

$$< \eta^2 >= \int d\mathbf{k} F(\mathbf{k}) \tag{1.134}$$

or using (1.127)

$$< \mathscr{E} >= \rho g \int d\mathbf{k} F(\mathbf{k}), \tag{1.135}$$

which, indeed, may be identified with equation (1.129).

It should be noted that neither the continuous Fourier integral (1.84) nor the discrete Fourier series (1.86) is a mathematically rigorous representation of an infinite, statistically homogeneous wave field. Both apply only for a finite region of the ocean. In the limit of an infinite region neither representation converges to a Fourier integral in the normal Riemannian sense, since the wave field does not vanish at infinity. In fact, as the spectral resolution of the Fourier sum representation, for example, is increased, the discrete sequence of Fourier components becomes a more and more dense sequence of uncorrelated values, randomly scattered about zero. In the limit of an infinite analysis area, the representations (1.84) and (1.86) must be replaced by Fourier-Stieltjes (Lebesgue) integrals. However, if one considers the means of the modulus-squared Fourier amplitudes, averaged within individual narrow, but fixed, finite spectral bands, these quantities converge to a well-defined, smooth function of the spectral band index as the analysis area tends to infinity. After the limit of an infinite analysis area has been taken (keeping the averaging bandwidth constant), the spectral bandwidth can then also be allowed to approach zero, yielding in the limit a continuous, Riemann-integrable function, the wave spectrum. Thus the Fourier transform relation (1.133) between the covariance function and the wave spectrum represents a normal Riemann integral. To avoid the rather cumbersome Fourier-Stieltjes notation we shall consider the limit of an infinite analysis region in the following only in second moment expressions involving the wave spectrum, not in the representation of individual realizations $\eta(\mathbf{x}, t)$ of the wave field. For individual wave field realizations we shall use either the sum representation or the Fourier integral representation, noting that both only apply formally to a finite box.

The frequency spectrum

In practice one often works with the frequency-directional spectrum. This is defined by

$$F(f, \theta)\mathrm{d}f\mathrm{d}\theta = F(\mathbf{k})\mathrm{d}\mathbf{k} \quad 2\pi f = \sigma(k), \quad \theta = \mathrm{atan2}(k_{x_1}, k_{x_2}), \tag{1.136}$$

(the angle $\theta = \mathrm{atan2}(k_{x_1}, k_{x_2})$ is defined by $k_{x_1} = \sin\theta$ and $k_{x_2} = \cos\theta$) or equivalently

$$F(f, \theta) = 2\pi k F(\mathbf{k})/(\partial\sigma/\partial k). \tag{1.137}$$

Often, directional information is lacking. In such cases it is convenient to work with a frequency spectrum

$$F(f) = \int_0^{2\pi} \mathrm{d}\theta F(f, \theta). \tag{1.138}$$

It can be shown that F is the Fourier transform of the temporal correlation function

$$F(f) = \int_0^\infty \mathrm{d}\tau e^{-2\pi i f\tau} < \eta(\mathbf{x}, t + \tau)\eta(\mathbf{x}, t) > . \tag{1.139}$$

As a result $F(f)$ can be determined from a time series measurement in a single point.

It is often convenient to introduce the moments of the spectrum:

$$m_n = \int_0^\infty f^n F(f) \, \mathrm{d}f. \tag{1.140}$$

The zeroth moment is equal to the variance:

$$m_0 = \int_0^\infty F(f) \, \mathrm{d}f = < \eta^2 >, \tag{1.141}$$

as follows directly from (1.134), (1.136) and (1.138).

The probability distribution of the wave maxima

As was pointed out above, knowledge of the wave spectrum – for linear waves – allows the reconstruction of the full probability structure of the sea surface. The probability that the sea level is between η and $\eta + d\eta$ can be simply obtained from (1.126) as

$$P_\eta(\eta) = \frac{1}{(2\pi m_0)^{\frac{1}{2}}} \exp(-\frac{1}{2}\eta^2/m_0). \qquad (1.142)$$

Note that $m_0^{1/2}$ plays the role of characteristic length scale. This is normally expressed in terms of the significant wave height, defined as

$$H_S = 4m_0^{1/2}. \qquad (1.143)$$

Since the spectrum can be measured, the significant wave height can be determined by integration of the observed wave spectrum.

Less simple, but still relatively straightforward, is the computation of the distribution of *wave maxima*, as carried out by Cartwright and Longuet-Higgins (1956). The general result is somewhat complicated, but for a narrow spectra it simplifies to a Rayleigh distribution:

$$P_H(\eta) = \begin{cases} (\eta/m_0)\exp(-\frac{1}{2}\eta^2/m_0) & \eta \geq 0, \\ 0 & \eta \leq 0, \end{cases} \qquad (1.144)$$

where $P_H(\eta)d\eta$ denotes the probability that a sea level maximum occurs with height between η and $\eta + d\eta$. Again H_S is the effective length scale and for this reason the significant wave height is commonly used as a measure for the average height of the waves. A short derivation of (1.144) can be found in Phillips (1977). There it is also shown how one can use the probability distribution of maxima to calculate $H_{1/3}$, defined as the average height of the highest 1/3 maxima, which is the relevant – or 'significant' – quantity estimated traditionally by means of visual observation of the sea state. It can be shown that $H_{1/3}$ is approximately equal to the significant wave height:

$$H_S \sim H_{1/3}. \qquad (1.145)$$

This explains the historical origin of the term significant. For a narrow spectrum $H_{1/3}$ and H_S are exactly equal.

The initial value problem

In the deterministic case we found how the initial sea state and its time derivative could determine the sea state at a later time. It is natural to wonder then, how this is treated in a stochastic theory of free linear waves.

To compute forward correlations, for instance, one can use a prediction sequence similar to the deterministic one:

$$(\mathscr{F}(\xi,0), \mathscr{F}_\tau(\xi,0)) \overset{(1.115)}{\longrightarrow} F_\mathbf{k}, \qquad (1.146)$$

followed by

$$F_{\mathbf{k}} \xrightarrow{(1.114)} \mathscr{F}(\boldsymbol{\xi}, \tau). \tag{1.147}$$

However, if the objective is the prediction of wave heights, the answer is extremely simple, because stationarity tells you that the two-dimensional wave spectrum and its corresponding significant wave height will remain as they are:

$$\frac{\partial H_S}{\partial t} = 0 \tag{1.148}$$

(one also has $\nabla_x H_S = 0$ from homogeneity). A constant wave height does not describe nature very realistically, but this only shows that homogeneous, free, linear waves are not very realistic. Equation (1.148) does make sense in a two-scale approach, when it simply means that the wave height does not vary on the fast time scale. We will discuss this two-scale approach in the next subsection.

I.2.6 The superposition of random WKB trains

We extend the statistical theory to allow for quasi-homogeneity and quasi-stationarity. To this end two scales are distinguished. We consider the random superposition of general WKB trains and wave packets. It is shown that the statistical properties of linear theory can be extended to this more general situation. We elaborate on the 'particle picture' by following the trajectories of wave packets in phase space and we define the density of packets in this space.

So far we have mainly considered free, linearized water waves in homogeneous environmental conditions. This is a poor approximation to reality which is much more complicated. Therefore, in § I.2.3 we started to consider slowly varying wave trains to account for perturbing effects. However, one such train still does not realistically describe the sea surface. Therefore, we will now generalize (1.56) by considering the superposition of many wave trains:

$$\eta(\mathbf{x}, t) = \sum_{n}^{N} a_n(\mathbf{x}, t) e^{i s_n(\mathbf{x}, t)} + \text{c.c.} + \text{small corrections.} \tag{1.149}$$

This equation is of the form (1.92), but the phase is now expressed through the eikonal functions s_n. The summation extends over a discrete index n which is not directly associated with a fixed wavenumber value. The small corrections refer to terms that one finds, for example, when considering nonlinear corrections, such as the Stokes correction. Their dynamical significance appears to be small, so we will neglect them.

As in § I.2.3 the amplitudes a_n and the phase angles (or eikonals) s_n are assumed to be slowly varying functions of $\mathbf{x}$ and t. Locally the wave trains are like plane waves with wavenumber and frequency given by

$$\omega_n = -\partial_t s_n, \quad \mathbf{k}_n = \nabla_x s_n, \tag{1.150}$$

so one may rewrite (1.149) as

$$\eta(\mathbf{x}, t) = \sum_n a'_n(\mathbf{x}, t) e^{i[\mathbf{k}_n(\mathbf{x},t)\cdot\mathbf{x} - \omega_n(\mathbf{x},t)t]} + \text{c.c.}, \tag{1.151}$$

where the slowly varying mean phase has been included in a'. This allows for a more direct comparison with our earlier expansion (1.86), because one can now see that not only the amplitude and phase are slowly varying with x and t, but also the wavenumber corresponding with a particular label n. If one sets out with equidistant $\mathbf{k}_n(\mathbf{x}, t_0) = \pm(n_1, n_2)\Delta k$, one may find that at a later time the distribution of $\mathbf{k}_n$ values over the k-axis is no longer uniform.

We have discussed at some length the problems involved in the transition from a discrete set of wave components to a continuous set, and we have pointed out that, in general, it is best to work with a discrete set and to take the continuum limit at the very end. Yet, for some applications it is convenient to work with a continuous expression

$$\eta(\mathbf{x}, t) = \int_{-\infty}^{+\infty} d\mathbf{k} \, \hat{\eta}(\mathbf{k}, t) e^{i(\mathbf{k}\cdot\mathbf{x} - \omega t)} + \text{c.c.} + \text{small corrections}, \tag{1.152}$$

which is a straightforward generalization of (1.82). In this representation the $\hat{\eta}$s slowly vary with time as in (1.151), but there is no need to introduce an explicit x dependence in the amplitudes. In the discrete representation we follow many wave components, each of which may slowly change its wavenumber; in the continuous case we have, at any instant, wave components for all $\mathbf{k}$. The slow change of $\mathbf{k}$ in the discrete case is now reflected in an extra change of amplitude at a particular wavenumber.

After this side remark let us now consider the generalization of equations (1.93) – (1.105) for the case that there are variations on the slow space and time scale. To lowest order the mean energy density of the wave train is still given by

$$\bar{\mathscr{E}}(\mathbf{x}, t) = \rho g \sum_n F_n(\mathbf{x}, t) = \rho g \int d\mathbf{k} F(\mathbf{k}, \mathbf{x}, t), \tag{1.153}$$

with

$$F_n(\mathbf{x}, t) = 2|a_n(\mathbf{x}, t)|^2 \tag{1.154}$$

and

$$F(\mathbf{k}, \mathbf{x}, t)\Delta k = \sum_n F_n(\mathbf{x}, t)\varepsilon(\mathbf{k} - \mathbf{k}_n). \tag{1.155}$$

Similarly one has for the wave action density

$$\bar{\mathscr{A}}(\mathbf{x}, t) = \rho g \sum_n N_n(\mathbf{x}, t) = \rho g \int d\mathbf{k} N(\mathbf{k}, \mathbf{x}, t), \qquad (1.156)$$

with

$$N_n(\mathbf{x}, t) = F_n(\mathbf{x}, t)/\sigma_n \qquad (1.157)$$

and

$$N(\mathbf{k}, \mathbf{x}, t) = F(\mathbf{k}, \mathbf{x}, t)/\sigma(k). \qquad (1.158)$$

Now the spectral densities are functions of space and time on the slow scale. One may integrate over all wavenumbers and over the whole basin on the slow scale to obtain

$$E_{tot} = \rho g \int \int d\mathbf{x} d\mathbf{k} F(\mathbf{k}, \mathbf{x}, t), \qquad (1.159)$$

$$A_{tot} = \rho g \int \int d\mathbf{x} d\mathbf{k} N(\mathbf{k}, \mathbf{x}, t) \qquad (1.160)$$

for the total energy and the total wave action of the basin.

The particle picture

Before we can discuss the statistical aspects of the WKB approach we need to elaborate on the particle interpretation of the superposition described by equation (1.149).

In a statistical theory, the particular representation chosen for η is not unique and is, in fact, irrelevant, as long as one obtains the same statistical averages. We want to make use of this by considering a representation, in which the sea is made up of a large number of *wave packets*, with a length which is intermediate between the wave length and the scale on which the amplitudes and phases change.

As discussed in § I.2.3 each packet can be characterized by a mean position $\mathbf{X}_n$ and a corresponding wavenumber $\mathbf{K}_n$. The motion of the wave packets follows directly from a generalization of (1.80)

$$\dot{\mathbf{X}}_n = \nabla_k \Omega(\mathbf{K}_n, \mathbf{X}_n), \quad \dot{\mathbf{K}}_n = -\nabla_x \Omega(\mathbf{K}_n, \mathbf{X}_n), \qquad (1.161)$$

with Ω given as in (1.78) and (1.60). All the packets move along their own rays.

These packets behave as the particles in a collisionless gas. One can define the number of wave packets with $\mathbf{X}$ and $\mathbf{K}$ in grid boxes around $\mathbf{x}$ and $\mathbf{k}$ as

$$\mathscr{N}(\mathbf{k}, \mathbf{x}, t)\Delta\mathbf{k}\Delta\mathbf{x} = \sum_n \varepsilon(\mathbf{K}_n - \mathbf{k})\varepsilon(\mathbf{X}_n - \mathbf{x}), \qquad (1.162)$$

where the function ε was defined in (1.95). The conservation law

$$\dot{\mathcal{N}}_{tot} = \frac{\mathrm{d}}{\mathrm{dt}} \int \int \mathrm{dxdk} \mathcal{N}(\mathbf{k}, \mathbf{x}, t) = 0 \qquad (1.163)$$

follows directly from the interpretation of $\mathcal{N}$ and the fact that the total number of modes remains constant in time. This, of course, is only true as long as (1.149), the basic WKB expansion, holds and the amplitudes of the wave trains are not changed by external forcing (for example by the wind) or by internal nonlinear interactions.

Statistical aspects

One may wonder what is the correct generalization of the statistically homogeneous theory discussed in § I.2.5 to the case of slowly varying modes. Intuitively the answer is rather simple. The process of phase mixing through linear wave dispersion can be expected to dominate over the weak processes causing the slow changes, such as the wind forcing and the nonlinear coupling between the waves. The weak nonlinear coupling does induce weak non-Gaussian features but these can be computed by a perturbation expansion and expressed in the Gaussian properties of the linear field. This will be discussed in § II.3.

The general problem of formulating a statistical theory of nonlinear fields is not at all a trivial matter. This can be understood in the simple random phase picture discussed before. In that picture one starts off with wave modes having independent phases. So one has

$$< a_n(x, 0) a_m^*(x, 0) > = \frac{1}{2} F_n(x, 0) \delta_{nm} \qquad (1.164)$$

at $t = 0$, but then the evolution of the (nonlinear) system will lead, in general, to dependencies in phases, such that (1.164) does not hold at a later time.

However, in the case of dispersive ocean waves we may use the intuitive arguments given above to show that (1.164) holds, in fact, at all times. We recall the argument which uses the wave packet picture for a quasi-homogeneous and quasi-stationary ocean. In this picture we may consider a 'region of interest' which is large with respect to the wavelength but small with respect to the ocean as a whole. Whatever the initial state, after the elapse of sufficient time the wave packets in the region of interest will have dispersed and the wave field in the region of interest will consist of contributions which have propagated in from far-away, well-separated regions of the ocean. That is, the wave field is now the sum of many contributions, which are independent for all practical purposes. Consequently, (1.164) can be extended to read (at least approximately)

$$< a_n(x, t) a_m^*(x, t) > = \frac{1}{2} F_n(x, t) \delta_{nm}. \qquad (1.165)$$

As a result all the statistical relations discussed before generalize. For example, one has

$$P(\eta_1, \eta_2, \ldots, \eta_n; \mathbf{x}, t) = \frac{1}{(2\pi)^{n/2} |\mathscr{F}|^{1/2}} e^{-\frac{1}{2}\eta^T \mathscr{F}^{-1} \eta}, \qquad (1.166)$$

with

$$\mathscr{F}(\xi, \tau; \mathbf{x}, t) = \frac{1}{2} \sum_n F_n(\mathbf{x}, t) e^{i(\mathbf{k}_n \cdot \xi - \sigma_n \tau)} + \text{c.c.,} \qquad (1.167)$$

as the generalization of (1.126) and (1.114). In fact, all the equations of § I.2.5 will continue to hold, provided one includes the slow variations in the appropriate way.

The statistical description outlined so far has also become known as the weak turbulence description of surface waves, because it belongs to the part of statistical fluid mechanics that is concerned with weakly nonlinear waves. This theory will be discussed in more detail in § II.3, where we will also relate the Gaussian assumption to the irreversibility of the theory.

I.2.7 The slow evolution of the spectrum due to weak nonlinearity and weak interaction with the environment

We relate the variation of the wave spectrum to the evolution of the WKB amplitudes on the slow scale and we obtain a generic expression for the slow time evolution of the spectrum due to weak nonlinearity and weak interaction with the environment.

It is possible to distinguish between two different types of perturbation. The first leads to wave amplitudes whose variations on the slow time scale are homogeneous, that is they do not depend on spatial gradients. In this case one may derive an ordinary differential equation for the time evolution of the amplitude. Examples are air/sea interaction, whitecapping and nonlinear interactions. The other perturbations are more complex because the rate of change does depend on spatial gradients, so one obtains a partial differential equation for the evolution of $a(\mathbf{x}, t)$. This type of behaviour occurs when one considers waves on variable currents in a sea of variable depth. We will discuss the simple perturbations first.

In this case one may write in general to lowest order

$$\partial_t a_n = \sum_i \varepsilon_i \chi_n^i(\vec{a}, \cdots), \quad \vec{a} = \{a_n\}, \qquad (1.168)$$

where the summation is over different perturbations, ε is the small parameter characterizing that perturbation and the forcing functions ε; χ_n^i, representing the net rate of change of the component a_n due to interactions with all other

wave components and with external fields, are functionals of the wave amplitudes (and possibly the eikonals) and environmental parameters (indicated
by the dots), such as the surface wind vector. The actual computation of
the χs is the subject of chapter II, so we will not discuss specific forms at
this stage. Instead we will make use of the general form (1.168) to obtain
an expression for the evolution of the spectrum. In a deterministic approach
this is easily done (multiply by a_n^* and add the complex conjugate), but then,
in general, the resulting right hand side will not close to an equation for the
spectrum. This is quite different in a weakly nonlinear, first order, Gaussian
process, because on the one hand one has

$$2\partial_t < a_n a_n^* >= \partial_t F_n, \qquad (1.169)$$

and, on the other hand, one may write

$$
\begin{aligned}
2\partial_t < a_n a_n^* > \ &= \ 2\{< \dot{a}_n a_n^* > + < a_n \dot{a}_n^* >\} \\
&= 2\sum_i \varepsilon_i(< \chi_n^i(\vec{a})a_n^* > + < a_n \chi_n^{i*}(\vec{a}) >) \\
&\equiv \sum_i S_n^i(\vec{F}, \cdots), \quad \vec{F}= \{F_n\}.
\end{aligned}
\qquad (1.170)
$$

The last step, in which we introduced the *source terms* $S_n^i(\vec{F})$, can be made,
because the wave distribution is Gaussian. However complex the functions
χ may be and whatever powers of a they may contain, it is always possible
to reduce $< \chi a^* >$ to products of two-point functions. Therefore, combining
equations (1.169) and (1.170) one obtains an equation for F_n that closes on
itself:

$$\partial_t F_n = \sum_i S_n^i(\{F_n\}, \mathbf{u}). \qquad (1.171)$$

We have introduced $\mathbf{u}$ as a generic symbol for environmental parameters
such as the surface wind.

It is straightforward to extend the result to the continuous case. Defining

$$S_i(\mathbf{k}; F, \cdots)\triangle\mathbf{k} = \sum_n S_n^i(F, \cdots)\varepsilon(\mathbf{k}_n - \mathbf{k}), \qquad (1.172)$$

one obtains

$$\frac{\partial}{\partial t}F(\mathbf{k}, \mathbf{x}, t) = \sum_i S_i(\mathbf{k}; F, \mathbf{u}), \qquad (1.173)$$

a relation giving nearly, but not quite, the evolution equation we are seeking
to derive. One important aspect is still missing. This will be discussed in
§ I.2.9. First, however, we will explore a general property of the source terms.

I.2.8 Quasi-linearity of nonlinear source functions

Under fairly general conditions any process which is weak-in-the-mean, even if it is highly nonlinear locally, will yield a quasi-linear source function.

In general, the source terms on the right hand side of (1.170) or (1.173) will be nonlinear functionals of the wave spectrum. However, it can be shown that any process which is weak-in-the-mean, even if it is highly nonlinear locally, will yield a source function $S(\mathbf{k})$ at wavenumber $\mathbf{k}$ which to lowest order is proportional to the spectrum $F(\mathbf{k})$ at the same wavenumber with a proportionality constant that still may depend nonlinearly on integral properties of the spectrum. This applies to many different processes, such as whitecapping, dissipation by bottom friction, percolation, inelastic bottom movement and the generation of waves by wind in accordance with Miles- or Jeffreys-type mechanisms. However, for some processes, such as the Phillips wind input term or the nonlinear transfer expression, the lowest order quasi-linear term vanishes and the source function has a different structure. All of these source terms will be discussed in chapter II. In this subsection we will consider properties that are independent of the details of the process. In particular, we will derive the general result that the source functions of processes which are weak-in-the-mean are quasi-linear to lowest interaction order and discuss the conditions under which this result applies. It will be found that source functions which do not have the general quasi-linear structure are produced by higher order interactions.

Before deriving the general result from the dynamical wave equations, we present first a simple derivation of the quasi-linearity of a purely dissipative process given by Snyder *et al* (1993). Consider an expansion of the source function of a general purely dissipative process in a Taylor series with respect to the wave spectrum of the first order, linearized, homogeneous, stationary wave field,

$$
\begin{aligned}
S(\mathbf{k}) = \ & -[K_1 F(\mathbf{k}) + \int\int K_2(\mathbf{k}_1,\mathbf{k}_2)\delta(\mathbf{k}-\mathbf{k}_1-\mathbf{k}_2)F(\mathbf{k}_1)F(\mathbf{k}_2)d\mathbf{k}_1 d\mathbf{k}_2 \\
& + \cdots + \int \cdots \int K_n(\mathbf{k}_1,\ldots,\mathbf{k}_n)\delta(\mathbf{k}-\mathbf{k}_1\cdots-\mathbf{k}_n) \\
& \times F(\mathbf{k}_1)\cdots F(\mathbf{k}_n)\,d\mathbf{k}_1\cdots d\mathbf{k}_n + \cdots].
\end{aligned}
\tag{1.174}
$$

If the process is purely dissipative the source function is always negative or zero, for all conceivable spectra. In that case the (symmetrized) kernels K_n must be everywhere nonnegative. But this implies that the kernels must contain a factor $\delta(\mathbf{k}-\mathbf{k}_1)$ such that the spectrum at the wavenumber $\mathbf{k}$ can be factored out of the integral. If this were not the case one could consider a spectrum which vanished at some particular wavenumber $\mathbf{k}_0$ but was non-zero at other wavenumbers. For a suitable choice of the spectrum the integral would then yield a finite negative contribution at $\mathbf{k}_0$, which would generate a negative spectrum on time integration. This conclusion can be avoided only when the integral contains the spectrum $F(\mathbf{k}_0)$ as a factor, or, equivalently, when the source function is quasi-linear.

To derive the general result for arbitrary processes we need to consider (1.170) in more detail. For simplicity we will single out one process (and drop the index i). We then can rewrite (1.170) as

$$S_n = 2\epsilon\{< \chi_n(\ldots a_n, a_n^*, \ldots)a_n^* > + < a_n\chi_n^*(\ldots, a_n, a_n^*, \ldots) >\}. \tag{1.175}$$

If the rate of change of the spectrum due to the interactions χ_n is weak-in-the-mean (which is, in fact, a necessary condition for the meaningful description of a quasi-homogeneous, quasi-stationary wave field in terms of a wave spectrum), the source term S_n can be computed to lowest interaction order by substituting the unperturbed, linear, Gaussian wave field into the right hand side of (1.175). The relevant correlation between χ_n and the Fourier component a_n^* (and the complex conjugate expression) can then be evaluated rather simply by noting that the wave component a_n^* is infinitesmal. We may therefore write

$$\chi_n = \chi_n^0 + a_n^*\partial\chi_n^0/\partial a_n^* + a_n\partial\chi_n^0/\partial a_n, \tag{1.176}$$

where the superscript ^{0} denotes that χ and its partial derivative are taken at the value $a_n = a_n^* = 0$. Since the functions depend then only on the remaining wave components, which are statistically independent of the wave component a_n (and its complex conjugate a_n^*), it follows that (1.175) reduces to

$$\begin{aligned} S_n &= 2\epsilon \{< \chi_n^0 >< a_n^* > + < \chi_n^{0*} >< a_n >\} \\ &\quad + \epsilon \{< \partial\chi_n^0/\partial a_n > + < \partial\chi_n^{0*}/\partial a_n^* >\}F_n. \end{aligned} \tag{1.177}$$

Noting that the expectation value of a_n vanishes, (1.177) yields then the quasi-linear form

$$S_n = \gamma F_n, \tag{1.178}$$

where

$$\gamma = \epsilon(< \partial\chi_n^0/\partial a_n > + < \partial\chi_n^{0*}/\partial a_n^* >). \tag{1.179}$$

The continuous equivalent of (1.178) reads

$$S(\mathbf{k}) = \gamma F(\mathbf{k}). \tag{1.180}$$

The proportionality constant γ may depend on integrals over products of spectra, so that (1.180) is only *quasi*-linear in $F(\mathbf{k})$. The computation of γ is not always straightforward, but can be accomplished in a suprising number of cases, including situations, such as whitecapping or the nonlinear turbulent bottom boundary layer, in which the interactions are locally strongly nonlinear.

There exist, however, interaction processes for which γ is zero. In this case one needs to go to the next order in ϵ to obtain a finite source term, and the source function is, in general, not quasi-linear. An example is Phillips' theory of wave generation by turbulent pressure fluctuations. Since the turbulent pressure fluctuations are assumed to be independent of the wave field, the partial derivative of χ with respect to the wave component a_n and therefore γ vanishes. Phillips' mechanism is thus of second order in the perturbation parameter ρ_a/ρ_w as compared with the first order quasi-linear Miles-Jeffreys-type process. Similarly, conservative nonlinear interactions yield a zero gamma term to lowest order: the relevant lowest order cubic products of three wave components which excite the fourth component a_n of a resonant quartet do not contain the component a_n itself, so that the partial derivative of χ_n again vanishes. One has to go to higher order in the expansion to obtain a mean source term (§ II.3).

Despite these exceptions, in most applications the expectation value of the partial derivative of the forcing functional χ_n with respect to the wave component a_n is found to be nonzero, and the source function is consequently a quasi-linear function of the wave spectrum.

I.2.9 Propagation in an inhomogeneous medium

We derive an expression for the variation of the wave spectrum on the slow time and space scale when the waves are propagating in an inhomogeneous medium. This takes care of slow depth variations and interaction with slowly varying currents. It is shown that the total wave action is thus conserved in this situation. In a wave packet representation this corresponds to the conservation of the number of packets.

We have given a general discussion of perturbations not involving spatial gradients. Let us now also discuss situations in which waves propagate over variable depth and encounter variable currents. At first sight it may look strange that the currents and winds are treated differently, since from the geophysical point they both represent flowing fluids. However, there are also large differences. As will be discussed in the next chapter, the wind speed as observed at anemometer height is often larger than the wave propagation velocity (the phase speed). In that case there is a height at which the wind speed equals the phase velocity. The existence of such a critical height allows an intense wind–wave interaction, which does not occur in the case of wave–current interactions, because then, in general, there is no critical layer. In the case of currents, it is the sensitivity of waves to current variations that matters. Consequently the interaction between waves and currents primarily involves the current gradients.

Let us consider the effect of variable currents and variable depth. To derive the evolution equation for this case we can follow Willebrand (1975).

In essence, Willebrand noted that the conservation of wave action, (1.63), holds for every wave component separately:

$$\frac{\partial}{\partial t}N_n + \nabla_x \cdot [\nabla_k \omega_n N_n] = 0, \quad N_n = 2\frac{|a_n|^2}{\sigma_n} \qquad (1.181)$$

(N_n was defined in (1.157)) and, in addition,

$$\omega_n(\mathbf{x}, t) = \Omega(\mathbf{k}_n), \quad \Omega(\mathbf{k}) = \sigma(k) + \mathbf{k} \cdot \mathbf{U}, \qquad (1.182)$$

$$\frac{\partial}{\partial t}\mathbf{k}_n + \nabla_x \omega_n = 0. \qquad (1.183)$$

Willebrand (1975) shows how this equation can be written in a continuous form as well. The density of k-modes is a function of time and this leads to an extra term involving $\nabla_x\Omega$ in the equation for $N(\mathbf{k}, \mathbf{x}, t)$ (defined in (1.102)):

$$\frac{\partial N(\mathbf{k}, \mathbf{x}, t)}{\partial t} + (\nabla_k\Omega) \cdot \nabla_x N(\mathbf{k}, \mathbf{x}, t) - (\nabla_x\Omega) \cdot \nabla_k N(\mathbf{k}, \mathbf{x}, t) = 0. \qquad (1.184)$$

Explicitly, in terms of the wave spectrum F (using (1.158)), one has

$$\left\{ \frac{\partial}{\partial t} + (\nabla_k\Omega) \cdot \nabla_x - (\nabla_x\Omega) \cdot \nabla_k \right\} \left(\frac{F(\mathbf{k}, \mathbf{x}, t)}{\sigma} \right) = 0. \qquad (1.185)$$

Another useful form for this equation is obtained by using the trivial identity

$$\nabla_x \cdot (\nabla_k\Omega) - \nabla_k \cdot (\nabla_x\Omega) = 0. \qquad (1.186)$$

This leads to the 'flux form'

$$\frac{\partial}{\partial t}\left(\frac{F}{\sigma} \right) + \nabla_x \cdot \left[(\mathbf{c}_g + \mathbf{U})\left(\frac{F}{\sigma} \right) \right] - \nabla_k \cdot \left[\nabla_x\Omega\left(\frac{F}{\sigma} \right) \right] = 0. \qquad (1.187)$$

This equation has the form of a conservation law. It also holds in other coordinates, for example in terms of ω, θ. In particular, it implies conservation of the total wave action defined in (1.160). Provided fluxes vanish at the boundary of the integration one has

$$\dot{A}_{tot} = 0. \qquad (1.188)$$

The total wave action is thus conserved.

The particle interpretation of the action balance equation

In the particle interpretation (equation (1.162) and the discussion following it) one may assign a particular (total) energy and (total) wave action to each packet. In fact, for each packet one can write down the expression for the total energy and the total wave action *per wave packet* as

$$E_{tot,n}(t) = \rho g \int d\mathbf{x} F_n(\mathbf{x}, t), \tag{1.189}$$

$$A_{tot,n}(t) = \rho g \int d\mathbf{x} N_n(\mathbf{x}, t). \tag{1.190}$$

As in the single mode case $E_{tot,n}(t)$ will be time dependent in general, while Whitham's analysis shows that $A_{tot,n}(t) = A_{tot,n}(0)$ is conserved. One may compute now the energy and action in packets with their centres of mass in a grid box around the point $(\mathbf{x}, \mathbf{k})$ in phase space

$$f_E(\mathbf{k}, \mathbf{x}, t)\Delta k \Delta x = \sum_n \varepsilon(\mathbf{K}_n - \mathbf{k})\varepsilon(\mathbf{X}_n - \mathbf{x})E_{tot,n}, \tag{1.191}$$

$$f_A(\mathbf{k}, \mathbf{x}, t)\Delta k \Delta x = \sum_n \varepsilon(\mathbf{K}_n - \mathbf{k})\varepsilon(\mathbf{X}_n - \mathbf{x})A_{tot,n}. \tag{1.192}$$

As in (1.158) we find

$$f_A = f_E/\sigma. \tag{1.193}$$

The total energy and action of the basin are

$$E_{tot} = \int \int d\mathbf{x} d\mathbf{k} f_E(\mathbf{k}, \mathbf{x}, t) = \sum_n E_{tot,n}, \tag{1.194}$$

$$A_{tot} = \int \int d\mathbf{x} d\mathbf{k} f_A(\mathbf{k}, \mathbf{x}, t) = \sum_n A_{tot,n}. \tag{1.195}$$

At this point one may make an essential assumption, namely

$$\rho g F(\mathbf{k}, \mathbf{x}, t) d\mathbf{k} d\mathbf{x} \simeq f_E(\mathbf{k}, \mathbf{x}, t) d\mathbf{k} d\mathbf{x}. \tag{1.196}$$

The left hand side gives the energy contained in a grid box around $(\mathbf{x}, \mathbf{k})$, the right hand side gives the energy in wave packets that have $\mathbf{X}_n$ and $\mathbf{K}_n$ in that box. The assumption is clearly plausible and requires only that the packets have not dispersed to a length scale comparable to the slow scale of variation of the amplitudes, which is implicit in the two-scale representation. Combining (1.193) and (1.196) one obtains

$$f_A(\mathbf{k}, \mathbf{x}, t) = \rho g F(\mathbf{k}, \mathbf{x}, t)/\sigma. \tag{1.197}$$

An elegant picture emerges if one constructs the packets in such a way that they all have equal (total) action ϕ:

$$A_{tot,n} = \phi. \tag{1.198}$$

In general, for a given grid box, the action in waves with a particular wavenumber will have a value that is different from ϕ, but then one can split this action over as many elementary packets as needed. This is a good approximation provided one chooses the elementary action to be sufficiently small. Equation (1.192) then reduces to

$$f_A(\mathbf{k}, \mathbf{x}, t) = d\mathcal{N}(\mathbf{k}, \mathbf{x}, t);$$ (1.199)

in other words *the action density function is just proportional to the density of modes in* $\mathbf{x} - \mathbf{k}$ *phase space.* For the total action one obtains

$$A_{tot} = \mathcal{N}_{tot}\, d.$$ (1.200)

Using (1.197) we also have

$$\mathcal{N}(\mathbf{k}, \mathbf{x}, t) = \frac{\rho g}{d}\left(\frac{F(\mathbf{k}, \mathbf{x}, t)}{\sigma}\right),$$ (1.201)

so that *conservation of wave action is equivalent to the conservation of the number of wave packets.*

It should be pointed out that the number of waves is an artifact of the description. Indeed, for a given spectrum $F(\mathbf{k}, \mathbf{x}, t)$ the total action is fixed so that the number of waves computed as A_{tot}/d depends on the particular value of d taken. Useful as $\mathcal{N}_{tot}$ is in establishing the particle analogy, its actual value is an arbitrary number.

In analogy with the derivation of the transport equation for a Boltzmann gas one may now derive the 'transport equation' for the 'wave packet gas'. One finds

$$\frac{\partial}{\partial t}\mathcal{N} + \nabla_x \cdot [\dot{\mathbf{X}}\mathcal{N}] + \nabla_k \cdot [\dot{\mathbf{K}}\mathcal{N}] = 0,$$ (1.202)

with $\dot{\mathbf{X}}$ the velocity and $\dot{\mathbf{K}}$ the wavenumber change rate of the wave packets in a phase space grid box around $(\mathbf{x}, \mathbf{k})$. This equation can be derived in the standard way by noting that $\mathcal{N}(t + \Delta t)$ can be related to $\mathcal{N}(t)$ by adding the number of particles that enter the relevant phase space grid box and subtracting the number of particles leaving it.

Equation (1.202) is in the flux form. From (1.161) it follows that

$$\dot{\mathbf{X}} = \nabla_k \Omega,$$ (1.203)

$$\dot{\mathbf{K}} = -\nabla_x \Omega,$$ (1.204)

and one obtains

$$\nabla_x \dot{\mathbf{X}} + \nabla_k \dot{\mathbf{K}} = 0,$$ (1.205)

so that (1.202) can be rewritten in the Liouville form (compare Dorrestein, 1960)

$$\frac{\partial}{\partial t}\mathcal{N} + \dot{\mathbf{X}} \cdot \nabla_x \mathcal{N} + \dot{\mathbf{K}} \cdot \nabla_k \mathcal{N} = 0.$$ (1.206)

Using (1.201) we recover equations (1.185) and (1.187) respectively for the evolution of the wave action density F/σ.

I.2.10 The action balance equation

The results of the previous two subsections are combined into an equation for the variation of the wave spectrum on the slow space and time scale.

The action balance equation is now obtained by combining results from the previous two subsections (equations (1.173) and (1.185)). The general result for the evolution of the wave action density $F(\mathbf{k}, \mathbf{x}, t)/\sigma(\mathbf{k}, \mathbf{x})$ reads

$$\left\{ \frac{\partial}{\partial t} + (\mathbf{c}_g + \mathbf{U}) \cdot \frac{\partial}{\partial \mathbf{x}} - (\nabla_x \Omega) \cdot \frac{\partial}{\partial \mathbf{k}} \right\} \left(\frac{F}{\sigma} \right) = \sum_{\ell} S'_{\ell}(F; \mathbf{u}), \qquad (1.207)$$

where the summation is over the different perturbations, we define $S' = S/\sigma$ (the prime will be omitted when confusion is unlikely) and $\mathbf{u}$ refers to 'environmental parameters', such as the surface wind speed and direction. The group velocity $\mathbf{c}_g$ was defined in (1.71), $\mathbf{U}$ is the surface current and Ω was given in (1.182). The equation describes the evolution of the wave variance spectrum $F(\mathbf{k}, \mathbf{x}, t)$, which characterizes the statistical properties of the sea surface. Rapid phase oscillations are not followed. Instead the slow evolution of the energy in each wave component is given. The left hand side describes propagation through a nonhomogeneous medium (currents, variable depth), which conserves total wave action as an adiabatic invariant. The right hand side consists of source terms describing wind input (S_{in}), sinks (whitecapping dissipation S_{ds} + bottom terms) and a nonlinear interaction term S_{nl}. These source terms will be discussed in more detail in the following chapters.

Equation (1.207) is also known as the radiative transfer equation, the transport equation or the kinetic equation. The name Boltzmann equation is used when one considers the sole effect of four wave interactions (see § II.3). In that case the equation describes conservative interactions ('elastic collisions') between wave packets and is analogous to the equation describing the distribution of momentum over the molecules in a gas.

In deep water, without currents, (1.207) reduces to the simpler form

$$\left\{ \frac{\partial}{\partial t} + \mathbf{c}_g \cdot \frac{\partial}{\partial \mathbf{x}} \right\} F(\mathbf{k}, \mathbf{x}, t) = \sum_{\ell} S_{\ell}(F; \mathbf{u}) = S_{in} + S_{nl} + S_{ds} \qquad (1.208)$$

or equivalently

$$\frac{\mathrm{D}F}{\mathrm{D}t} = S_{in} + S_{nl} + S_{ds}, \qquad (1.209)$$

where $\mathrm{D}/\mathrm{D}t$ denotes differentiation when moving with the group velocity. In this form the equation is also known as the energy balance equation, since F is directly proportional to the energy spectrum.

This book focusses on the solution of equations (1.207) and (1.208). Prediction of significant wave height requires prediction of the wave spectrum,

because all wave components have their own propagation and interaction
characteristics. One usually begins by specifying the (action) spectrum everywhere at a particular time. This can be done on the basis of earlier
forecasts, measurements or a mixture of both. One must also specify surface
winds (and currents) at all times and then one can solve (1.207) to obtain
the spectra (and the significant wave heights) at a later time. Schematically:

$$F(\mathbf{k}, \mathbf{x}, 0) \stackrel{u(x,t'),\ 0 < t' < t}{\Longrightarrow} F(\mathbf{k}, \mathbf{x}, t), \qquad (1.210)$$

where $\mathbf{x}$ extends over the area or basin considered.

This completes our discussion of how one can predict the sea state, once
it is known at a given time, using the general laws of physics.

I.3 Sea surface winds and atmospheric circulation

G. J. Komen

I.3.1 General

The art of predicting ocean waves depends to a large extent on our understanding of the wind field over sea. Early wave models extracted wind simply
from synoptic weather charts of surface pressure. The surface winds were
then calculated as the geostrophic wind plus a rule of thumb correction.
With the present sophistication of wave modelling and with the advent of
numerical weather prediction with GCMs (= general circulation models) this
approach has lost its prominence.

The importance of the wind will show up time and time again in this
book. For example, for a calculation of the wave growth it is crucial to know
the shape of the wind profile in the lowest 30 metres of the atmosphere.
A related topic of great and recurrent interest is the relation between wind
speed and wind stress (or momentum flux) near the sea surface. In this
relation the effective sea surface roughness plays an important role, and it is
not surprising that this roughness depends on the wave spectrum. Therefore,
a third topic is the effect of waves on the atmospheric flow through the way
in which they modify the effective sea surface roughness. Other relevant
topics are the effect of atmospheric stratification and subgrid scale wind
variability.

It is the purpose of this section to give the generalities of the description
of winds over sea and to provide a framework for understanding material
covered in later chapters. Basically, winds are driven by unequal solar
heating of the atmosphere and the resulting density differences and pressure

gradients. Therefore, we can only understand winds over sea if we understand the thermo-hydrodynamics of the atmosphere.

I.3.2 Atmospheric flow

The physical state of the atmosphere is described by seven fields: density, pressure, temperature, moisture and the three components of velocity. Each of these varies with time and with the three space coordinates. To describe the atmosphere one must know their time evolution. The relevant equations are well known. There are six prognostic (time-dependent) equations: conservation of mass and moisture; Newton's law relating the change of the three components of momentum to forces; and the heat equation which gives the change in internal energy and temperature in terms of sources and sinks. The seventh, diagnostic, equation is the equation of state, giving density in terms of pressure, temperature and moisture. A good description of the full set of equations can be found in the textbooks by Holton (1992) or Gill (1982). Here we will focus on the equations for horizontal momentum. They read

$$\mathrm{D}u/\mathrm{D}t - fv = -\rho_a^{-1}\partial p/\partial x + \partial\tau_x/\partial z, \qquad (1.211)$$

$$\mathrm{D}v/\mathrm{D}t + fu = -\rho_a^{-1}\partial p/\partial y + \partial\tau_y/\partial z, \qquad (1.212)$$

and they express the balance between the rate of change of momentum (the accelerations), the effect of earth rotation (the Coriolis force), pressure gradient forces and vertical momentum exchange. Fluctuations have already been averaged out: u and v are the x- and y-components of the *mean* wind speed. $\mathrm{D}/\mathrm{D}t$ is the total derivative (moving with the fluid), f, the Coriolis parameter, is related to the rotation of the earth, ρ_a is the density of air, p is the pressure and τ_x and τ_y are the x- and y-components of the vertical momentum flux or stress. This flux is given by

$$\tau = v_a\partial\mathbf{u}/\partial z + \tau_{fluc}, \qquad \tau_{fluc} = - <\mathbf{u}'w'> . \qquad (1.213)$$

The sign is such that τ is in the $\mathbf{u}$ direction for a downward flux. Strictly speaking $\rho_a\tau$ is the stress, but for convenience we have dropped the factor ρ_a from the definition. The τ so obtained is also known as the kinematic stress. In (1.213) v_a is the kinematic, molecular viscosity in air, and $\mathbf{u}' = (u', v')$ and w' are the fluctuations of the wind velocity around their mean value. The first term gives molecular effects; the second term describes momentum transfer due to wind speed fluctuations.

I.3.3 Two extreme limits

As is well known (Holton, 1992), the free atmosphere has a dynamic balance that is quite different from the one in the boundary layer. In free flow the

Coriolis term and pressure gradient dominate all other terms. Therefore one can write

$$fv = \rho_a^{-1}\partial p/\partial x + \text{corrections}, \tag{1.214}$$

$$fu = -\rho_a^{-1}\partial p/\partial y + \text{corrections}. \tag{1.215}$$

From this one may infer that $\mathbf{u} = \mathbf{u}_g + \mathbf{u}_a$, where the geostrophic part is defined by

$$u_g = -(1/f)\rho_a^{-1}\partial p/\partial y, \tag{1.216}$$

$$v_g = (1/f)\rho_a^{-1}\partial p/\partial x, \tag{1.217}$$

and where the ageostrophic components $\mathbf{u}_a$ are small corrections.

In the lowest part of the boundary layer the friction term dominates, and (1.211) and (1.212) simply reduce to

$$\partial \tau/\partial z = 0. \tag{1.218}$$

This implies that the total downward momentum flux near the sea surface does not depend on height. This is not surprising (forces are negligible there, so momentum is conserved), but it is quite nice to have something constant when the wind speed itself is – by definition – varying rapidly with height.

The approximation (1.218) works only near the surface. For sufficiently strong winds it can describe the lowest 10 metres. For greater heights, or at low wind speed, the pressure gradient term cannot be neglected. In between the two extreme limits, free flow and geostrophy on the one hand, and the constant stress boundary layer on the other, one also distinguishes the 'Ekman layer' or planetary boundary layer in which stress, Coriolis force and pressure gradients are of comparable magnitude. For an analysis of the relevant importance of the various terms as a function of height and their scaling with boundary layer parameters we refer the reader to Tennekes (1973, 1981).

I.3.4 Numerical weather prediction

Together with the other prognostic equations, (1.211) and (1.212) form a complete set of equations. When complemented with boundary conditions they can be solved numerically and this is exactly what is done in numerical weather prediction. There is, however, a complication which is best made clear by making the customary distinction between 'physical' and 'dynamical' aspects.

Physical aspects concern the parametrizations. The essence is that (1.211) and (1.212) contain unknown terms, such as τ_{fluc}, which are neither prognostic nor diagnostic variables. Before the prognostic equations can be solved these unknown terms have to be expressed in terms of model variables and this is what is called parametrization. A wide range of processes need to be parametrized, such as turbulent and convective transfer, condensation and radiation. An example, the parametrization of the momentum flux, will be discussed in more detail below.

The dynamical aspects mainly concern the advection terms (including the choice of coordinates) and the numerical questions related to their treatment. One usually considers horizontal and vertical discretization, time integration aspects and horizontal diffusion.

Another essential part of weather prediction is the correct initialization, that is the determination of initial values for the prognostic fields. To obtain these, use is made of observations. But generally the number of observations and the errors in the observations do not allow a sufficiently accurate determination of the initial state. Therefore, the observations are usually combined with first-guess information in a process called data assimilation. The resulting fields are also known as analysed fields. Weather prediction only works when the initial fields are sufficiently accurate. Forecasts always start from an analysis and, in general, forecast winds are of poorer quality than analysed winds. We refer the reader to specialized textbooks for details on initialization and other aspects of numerical modelling such as the optimal choice of coordinates and specific numerical techniques. However, one topic will be discussed in more detail, because it is of special interest to wave modelling. This is the representation of the lower boundary condition for the momentum equation, with special attention to the situation of air flowing over the ocean surface. The actual computation of wind fields over sea will be addressed in § IV.2.

I.3.5 Parametrization of the momentum flux

Before we can discuss the lower boundary condition on the horizontal velocity, we need to consider the problem of parametrizing the momentum flux. In turbulence theory this is also known as the closure problem. In practice it is equivalent to finding an explicit expression for τ_{fluc} in terms of mean quantities. We first consider the situation in which density differences can be neglected. Later, in § I.3.7 we will include density effects.

In the atmosphere it is often assumed that one may write

$$\tau \simeq - <\mathbf{u}'w'> = v_e \partial \mathbf{u}/\partial z \qquad (1.219)$$

in analogy with the expression for molecular transport. This is based on the idea that turbulent mixing occurs because of the instability of shear flow: the stronger the shear the stronger the turbulence. The molecular transport represented by the first term in (1.213) (the viscous contribution) is usually neglected, because it is much smaller than the turbulent term.

In equation (1.219) the momentum flux due to fluctuating motion is expressed in terms of the vertical gradient (the shear) of the mean flow. General ideas from turbulence theory suggest that the 'eddy viscosity' v_e depends on height in the following way

$$v_e = \ell^2 \mid \partial \mathbf{u}/\partial z \mid, \qquad (1.220)$$

with ℓ the 'mixing length'. This theory was first put forward by Prandtl in 1925. An account can be found in Stull (1988). In practice, near the surface ℓ is usually taken as

$$\ell = \kappa z, \qquad (1.221)$$

with $\kappa \sim 0.4$, the von Kármán constant. There are many other ways of closing the turbulence equations, but we shall not explore them here.

It should be realized that (1.219) – (1.221) constitute an approximation which may become quite poor under specific conditions. One such condition is realized very close to the sea surface. This is because the orbital wave motion gives rise to additional fluctuations (see, for example, Phillips, 1977) which contribute to the momentum transport. In the case of air moving over water waves it is customary to distinguish between mean motion, wave induced fluctuations (correlated with the wave phase) and pure turbulent fluctuations (not correlated with the wave phase). As a result the momentum flux can be written as the sum of two contributions:

$$\tau \simeq \tau_{wave} + \tau_{turb}. \qquad (1.222)$$

Several authors (Janssen, 1982, 1989b, 1991a, Chalikov and Makin, 1991) have shown how τ_{wave} can be calculated by solving the momentum equations over waves. Their results will be discussed in § II.2. Here we can already anticipate one of the main findings, namely that τ_{wave} is a function of the wave spectrum and varies rapidly with height. Close to the water surface the wave induced motion of the air will be significant and τ_{wave} will be a significant fraction of τ; with increasing z the wave stress τ_{wave} rapidly decreases, while the sum of τ_{wave} and τ_{turb} remains constant by virtue of (1.218).

Equation (1.222) tacitly assumes that the stress can be determined right at the sea surface. This is not obvious in general, because stresses at a fixed height are not defined at a level lower than the wave crests. Conceptually, a simple way out is to introduce wave following coordinates. Alternatively one may assume that the value of τ_{wave} determined slightly above the crests can be extrapolated to $z = \eta$. But this is dangerous, because the extrapolation may be incorrect, for example when air flow separation occurs.

Organized motion in the atmosphere, not described by (1.219) – (1.221), may also affect the picture. It is well known that momentum may be transferred downwards from the planetary boundary layer in bursts, triggered by the largest eddies. Holtslag and Moeng (1991) have shown that these bursts can be parametrized with the help of a nonlocal turbulent diffusion scheme. However, in the lowest 10 metres (but outside the region with wave supported stress) (1.219) – (1.221) appear to be sufficiently accurate.

I.3.6 Boundary conditions and the wind profile over waves

A discussion of boundary conditions can be found in Gill (1982). In general, two cases are distinguished. In the case of inviscid flows, one can have discontinuities in velocity at boundaries such as the air/sea interface and one merely has to impose continuity of pressure. This approximation is an accurate first approximation to large scale atmospheric flow as the viscosity of air is relatively small. Even so its effect cannot be ignored altogether. The large scale approximations allow a discontinuity in speed at the air/sea interface, but viscosity requires continuity. This leads to a strong vertical shear near the interface. Typical velocities in the atmosphere are of the order of 10 m/s. Ocean currents rarely exceed 1 m/s. The layer with strong shear is described to a good approximation by (1.218).

Let us first consider the flow in the constant stress layer described by (1.218). In their excellent introduction to such flow problems Monin and Yaglom (1971) distinguish between flow over a smooth and a rough 'wall'.

Flow over a smooth surface

Since it is important to understand the difference between smooth and rough flow, let us first discuss the simple two-dimensional case of steady flow in the *x*-direction over an infinite, rigid, smooth, flat, nonmoving plate. Equation (1.218), together with (1.213) and (1.219) – (1.221), describes this situation. They constitute a second order differential equation which may be integrated once to yield

$$v_a \partial u / \partial z + \kappa^2 z^2 (\partial u / \partial z)^2 = \text{constant.} \qquad (1.223)$$

The constant is just the total momentum flux, which may be thought of as being specified, either from measurements at the top of the layer or from a larger scale model. Alternatively, one may also assume that the wind at the top of the layer is given. In that case the magnitude of the momentum flux will follow from the solution. At the lower boundary one has

$$u = 0, \quad z = 0. \qquad (1.224)$$

Although it is fairly straightforward to integrate this equation numerically, it is more instructive to consider two limiting cases, easily constructed with the help of dimensional arguments. One limiting case then corresponds to having small values of *z*, in which case the second term is negligible. The other is just the opposite: for large *z* the viscous term becomes negligible.

Introducing the *friction velocity* as the usual alternative measure for stress (remember that we have dropped ρ_a from the definition of τ)

$$u_* = \tau^{1/2},$$ (1.225)

we have

$$u(z) = \frac{u_*^2 z}{v_a}, \qquad z \ll z_v = v_a/u_*,$$ (1.226)

and

$$u(z) = \frac{u_*}{\kappa} \ln(z/z_0), \qquad z \gg z_v.$$ (1.227)

The region of validity of the first solution is the 'viscous sublayer'. It is mainly of theoretical interest, because, in practice, it is extremely thin ($v_a \sim 1.5 \times 10^{-5}$ m^2/s, $u_* \sim 0.1$ m/s $\rightarrow z_v \sim 0.15$ mm). The integration constant z_0 in (1.227) does not follow from our limited analysis, but it might either be inferred from observations or it could be computed by matching (1.226) and (1.227) in the transition region where $z \simeq z_v$, making use of the full equation (1.223). An estimate using asymptotic matching would give $z_0 = \text{const } v_a/u_*$, but the constant is much larger than inferred from observations which give $z_0 \sim 0.1 \, v_a/u_*$. A careful analysis by van Driest (1951) revealed that close to the surface the mixing length ℓ is reduced because of an interaction between viscosity and turbulence. This modified mixing length still yields the same 'law of the wall' (1.227), but the roughness length is in better agreement with observations.

Flow over a rough surface

It is easy now to indicate the difference between a smooth wall and a rough wall. If h denotes the mean height of the variations of the wall, then the wall is called rough if $h \gg z_v$ and smooth if $h \ll z_v$. In the case of rough flow (think of wind over a forest or a corn field) the viscous solution loses its meaning, but the logarithmic solution still applies, provided $z \gg h$. So for the wind profile we still obtain

$$u(z) = \frac{u_*}{\kappa} \ln(z/z_0), \qquad z \gg h,$$ (1.228)

but the difference with (1.227) is the meaning of the integration constant z_0, which is now called the roughness length. Matching with the solution for $z \ll h$ is no longer possible, since this solution is not available, a detailed calculation of the flow inside the forest or the corn field being beyond our capabilities. The roughness in rough flow, therefore, cannot be determined in an elementary way. Instead, it is usually determined experimentally from profile measurements.

In practice, the drag coefficient C_D is frequently used to relate the surface stress to the wind speed at a given height (the height of observation, for example). It is defined by

$$\tau = C_D(z)u^2(z). \tag{1.229}$$

For a logarithmic profile of the form (1.227) it follows immediately that

$$C_D = \frac{\kappa^2}{\ln^2(z/z_0)}. \tag{1.230}$$

The value clearly depends on the choice of z_0 and on the height of observation.

So far we have considered flow over a flat surface. In practice flow over undulating terrain (wooded hills, for example) is also of interest. In such cases, when the mean wind vector follows the slow variations in surface elevation, a two-scale approach is viable in which the effect of the slow variations is calculated explicitly and only the small scale roughness needs to be parametrized.

Flow over waves

From the geometrical point of view, air flow over waves is similar to the flow over wooded hills. Yet, over the sea, things are even more complicated (see, for example, Phillips, 1977, Donelan, 1990, Geernaert *et al*, 1986 and Geernaert and Plant, 1990), because the waves move and are sensitive to pressure variations. The boundary condition equivalent to (1.224) reads

$$u = u_{orb} + u_{drift}, \qquad z = \eta, \tag{1.231}$$

where u_{orb} denotes the wave related orbital velocity and u_{drift} the surface drift current. However, wave components from the full spectral range – from capillary ripples to long swell waves – act as roughness elements. Even the shortest (and lowest) waves stick out of the viscous sublayer (except at extremely low wind speeds), so the flow is definitely rough and (1.231) is not very useful in fixing the roughness height.

Sufficiently far away from the waves, when only the last term in (1.222) contributes, one would still find a logarithmic profile. As in the flow over a rough plate the integration constant z_0 still has to be specified. There has been considerable debate on the correct expression for this roughness length. Monin and Yaglom (1971) treat the sea as a rough wall, but they admit that the question of the roughness of the sea is far from being completely clear. Charnock (1955) introduced the following length scale on dimensional arguments:

$$z_{ch} = \alpha_{ch}u_*^2/g. \tag{1.232}$$

The Charnock constant α_{ch} was found to be of the order 0.01 (but see also the discussion in § II.2). Many applications have taken $z_0 = z_{ch}$ as the roughness height, resulting in the following profile:

$$u(z) = \bar{u}_{drift} + \frac{u_*}{\kappa} \ln(z/z_{ch}), \qquad z \gg z_{ch}, \qquad (1.233)$$

where $\bar{u}_{drift}$ is the mean drift current, which is normally ignored.

Further progress was made by several authors (Janssen, 1989b, 1991a, Chalikov and Makin, 1991), who used the continuity of stress at the surface to relate the drag to the wave spectrum. The momentum flux just above the surface is equal to the momentum flux into the surface. This, in turn, can be written as an integral over the wind input source term, which sums the momentum flowing into different wave components:

$$\tau = \tau_{wave}(z = \eta) = \rho_w \int_0^\infty \omega(\mathbf{k}/k) S_{in}(\mathbf{k}) d\mathbf{k}. \qquad (1.234)$$

If one knew the wave spectrum up into the capillary range one could simply integrate (1.234) to obtain τ. However, F is normally known only up to a certain frequency and, in fact, it is not at all obvious that the small scale roughness elements can be treated as waves. Therefore, the effect of the remaining roughness still has to be parametrized.

The drag coefficient (1.230) over sea will depend on the roughness of the sea surface. Taking $z_0 = z_{ch}$, as in (1.233), it will be a function of the friction velocity only. However, alternatively, one may start from (1.234) to compute τ. From there C_D and z_0 may be easily determined. The result takes the form

$$z_0 = z_w(u_*, F). \qquad (1.235)$$

Now the drag coefficient explicitly depends on the full (long) wave spectrum F. This latter approach seems to be in better agreement with observations than the former (see Janssen, 1992) and was confirmed by the experimental work of Maat *et al* (1991) and Smith *et al* (1992) (see also §§ II.2 and III.3).

Coarser models

So far we have considered microscale modelling of the marine boundary layer. In coarser models, which do not resolve the behaviour in the lowest metre over sea, one basically solves equations (1.211) and (1.212) with the lower boundary condition

$$u(z_0) = u_{drift}, \quad z = z_0, \qquad (1.236)$$

where u_{drift} is the surface drift current, which, again, is usually neglected. It should be noted that z_0 depends on the wind speed (and the sea state), so (1.236) is, in fact, a complicated implicit expression for u. Fortunately, this does not hamper the numerical solution. For sufficiently small time steps, one can reliably estimate the roughness at a particular time by using the wind and waves from the previous time step. As before the expression taken for z_0 is crucial. One may either take results from calculations that resolve the physics of the lowest metre, such as the ones sketched above, or one may use expressions for z_0 based on experiments. This approach works well when one wants to model the lower part of the planetary boundary layer in order to compute stresses or 10 metre winds starting from wind speed estimates at a greater height, which are obtained from the geostrophic approximation applied to a pressure analysis or from the lowest level of a GCM. One then simply has $u(100)$ (say) given as an upper boundary condition and $u(z_0)$ as a lower one.

The large GCMs take a slightly different approach. Typically they have of the order of 20 layers in the vertical throughout the 10 km or so of the troposphere. Therefore, they have rather poor resolution in the lower atmosphere and they accept a discontinuity in their description of the horizontal velocity. They still need a lower boundary condition which is taken to be of the form

$$\tau = C_D u^2, \quad z = z_{ll}, \tag{1.237}$$

where ll stands for lowest level and C_D is related to z_0 through (1.230). In general, over sea, C_D will be a function of wind speed, sea state and stability (see below) but present models differ in the degree of sophistication of their C_D parametrization.

It is important to check that surface winds and wind stresses obtained in a GCM are consistent with the parametrization used. Janssen *et al* (1992) have shown that the usual treatment of the model numerics may lead to incorrect results.

I.3.7 Density effects and atmospheric stability

As outlined above the atmospheric stratification shows up at three levels.

On the large scale, it is an important element of the full set of equations governing atmospheric flow. The corresponding physical picture of the global weather system is well known. Excessive heat in the tropics is carried away through the Hadley circulation. At the surface this leads to the trade winds. The pressure difference between the subtropics and the polar regions leads to a westerly jet. Extreme winds arise either from cyclogenesis in the tropics with the resulting tropical cyclones, or from the instability of the westerly jet which results in the typical midlatitude storms. On the larger regional scale, unequal heating of the atmosphere over sea and over land gives rise to the monsoon. All of these effects are described by numerical weather prediction models.

On mesoscales, there are other significant effects. One such effect is the sea breeze, which also arises from unequal heating of the atmosphere over sea and over land. In principle, this can be modelled numerically, but coarse models may fail to resolve the effect. Other important mesoscale effects are convective activity in the case of cold air over a relatively warm sea. This may lead to cloud streets or other organized structures which enhance the variability in wind speed. The same applies to squall lines. Although there exists some knowledge of these phenomena, large-scale-weather modellers have not yet found a way of quantifying these effects. This may have a negative effect on the quality of wave predictions.

On the smallest scale, stratification is important because density differences in combination with gravity may act as a source or sink of turbulent mixing (see, for example, Geernaert and Plant, 1990). Therefore, vertical temperature stratification can enhance or reduce the value of v_e in equation (1.219). Measurements over sea (Large and Pond, 1982) have confirmed that the effect can be parametrized in terms of the Obukhov length

$$L = -\frac{u_*^3 T_v}{\kappa g < w'T_v' >},\qquad(1.238)$$

which defines the height at which shear production and buoyant production of turbulence are equal. In (1.238) T_v is the mean virtual temperature at 10 m and the primed quantities denote fluctuations. For finite L the wind profile deviates from a logarithmic profile; also there are deviations from the neutrally stable relation between the wind speed at 10 m and the corresponding wind stress. In fact, the following mean profiles hold for velocity u, potential temperature T and specific humidity q (= density of water vapour / total density of the moist air)

$$u(z) = \frac{u_*}{\kappa}[\ln(z/z_0) - \psi_m(z/L)],\qquad(1.239)$$

$$T(z) = T_0 + \frac{t_*}{\kappa}[\ln(z/z_t) - \psi_H(z/L)],\qquad(1.240)$$

$$q(z) = q_0 + \frac{q_*}{\kappa}[\ln(z/z_q) - \psi_E(z/L)].\qquad(1.241)$$

Here T_0 and q_0 are the temperature and humidity just above the sea surface, for which the sea surface temperature and the saturation humidity at that temperature are commonly used. The quantities u_*, t_* and q_* are defined as

$$u_* = < -u'w' >^{1/2},\qquad(1.242)$$

$$t_* = -\frac{1}{u_*} < w'T' >,\qquad(1.243)$$

$$q_* = -\frac{1}{u_*} < w'q' >,\qquad(1.244)$$

where the primed quantities indicate fluctuating parts. In the steady state the starred quantities are independent of height in the constant flux layer. This is immediately clear for u_*, because under those circumstances (1.218) is valid, so the friction velocity $u_* = \tau^{1/2}$ must be independent of the height.

The ψ-functions are based on the Monin-Obukhov formalism. Their shape has been determined empirically by Businger *et al* (1971) from the Kansas experiment and by Dyer and Hicks (1970), with some differences between the two in the values of the coefficients, due to the finite accuracy of measured data. Similarity between heat and moisture leads to $\psi_H = \psi_E$, which we will further indicate as ψ_H. In the stable case ($L > 0$) the ψ-functions can be given as

$$\psi_m = \psi_H = -5z/L. \tag{1.245}$$

In the unstable case ($L < 0$) they take the form

$$\psi_m = 2\ln\left(\frac{1+x}{2}\right) + \ln\left(\frac{1+x^2}{2}\right) - 2\arctan x + \pi/2, \tag{1.246}$$

$$\psi_H = 2\ln\left(\frac{1+x^2}{2}\right), \tag{1.247}$$

with $x = (1 - 16z/L)^{1/4}$. These functions have been determined over land, but they are assumed to be valid over sea as well. At sea, Large and Pond (1982) determined the roughness lengths for temperature and humidity as

$$L > 0 \text{ (stable)}, \qquad z_t = z_q = 2.2 \ 10^{-9} \text{ m}, \tag{1.248}$$

$$L < 0 \text{ (unstable)}, \qquad \begin{cases} z_t = 4.9 \ 10^{-5} \text{ m}, \\ z_q = 9.5 \ 10^{-5} \text{ m}. \end{cases} \tag{1.249}$$

Finally, to be able to determine the Obukhov length L the relation between virtual temperature T_v and temperature T is needed:

$$T_v = T(1 + 0.608q). \tag{1.250}$$

The same relation holds for the potential temperature and virtual potential temperature. In the constant stress layer the difference between temperatures and potential temperatures can usually be neglected. Also $T_v \cong T$, because q is at most 5%.

For given values of the mean quantities u_{10}, T_{10}, q_{10} and q_{sea} we are able to determine u_*, t_*, q_* and L using (1.238) and (1.239) – (1.241) at a height of 10 m. In this way the wind profile is found in terms of parameters that can be easily observed.

I.3.8 Consequences for wave modelling

Chapter II will be concerned with the physics of wave evolution. In particular, in § II.2 we will give a discussion of the growth of waves under the influence of wind. There, the shape of the wind profile in the lowest 30 m or so will play a crucial role and the effect of waves on the parametrization of the momentum flux and the resulting wind profiles will be discussed in some detail. The relation between wind speed and wind stress (or momentum flux) near the sea surface will also be derived. This relation gives explicitly the role of the effective sea surface roughness on the drag and its dependence on the wave spectrum. This relation between U_{10} and u_* will be used time and again. It plays an important role in nondimensionalizing observations (§§ II.8 and II.9) and in parametrizing the wind input source term (§ III.3).

In chapter IV, applications will be discussed. The determination of the wind fields (§ IV.2) and their effect on the wave model results (§ IV.3) will be seen to be important. Other relevant topics, such as the effect of atmospheric stratification and subgrid scale wind variability, will be discussed in § IV.5. In § IV.10 we will finally show how waves play a role in air/sea interaction, and as such are part of the coupled atmosphere/ocean system.

I.4 Measurements of waves and winds

S. Hasselmann, K. Hasselmann, C. Brüning, R. B. Long, H. C. Graber, E. Bauer and B. Hansen

I.4.1 The role of wind and wave measurements

Wind and wave measurements are essential for the operation and validation of wave models. Wave models are driven by surface wind fields, which are computed by atmospheric models using meteorological data as input. In the past, surface wind data from ships have been the most important, but not the only, source of input for computing surface wind fields. However, more extensive global wind data have become available through the altimeter and scatterometer flown on the ERS-1 satellite, and more data from similar instruments flown on other satellites are expected in the future. We can thus expect significant improvements in the surface wind fields computed over the ocean in the future.

If the wind field is given, the wave field can, in principle, be computed by a wave model without additional wave measurements (it can be assumed, at least in an operational setting, that the initial wave field lies so far back in time that it no longer influences the current wave field). For most of the ocean – the North Sea being a typical exception – wave observations have been sparse in the past. This has precluded the application of wave measurements to improve wave model forecasts. Wave data have played an important role mainly as a means of validating wave models at a few measurement stations or to study the physics of wave models in specially designed, highly instrumented measurement campaigns like JONSWAP (Hasselmann *et al*, 1973), MARSEN (Hasselmann and Shemdin, 1982) and SWADE (Weller *et al*, 1991). These experiments were typically restricted to relatively small areas of the ocean and carried out over only limited time periods.

However, the situation has radically changed with the advent of ocean satellites providing continuous global measurements of important wave properties such as the significant wave height or even the full two-dimensional wave spectrum. It has now become meaningful to assimilate wave data in wave models both to update the current state of the wave field and to improve the wind field analysis, where errors between predicted and observed wave measurements indicate deficiencies in the wind field analysis.

The implications of these developments are considered in more detail in chapter V on global satellite wave measurements and chapter VI on the assimilation of wave data in wave models. In this section we mention only briefly the various types of instruments which are available for wind and wave measurements.

Wind and wave measuring systems can be divided into two categories:

- operational systems designed to provide continuous, global or regional observations for forecasting, data assimilation and model validation;
- wind and wave measurement systems operated with higher temporal and spatial resolution during limited measurement campaigns in order to study the physical processes of wave generation and air/sea interaction.

I.4.2 Operational wind measurements

The bulk of operational wind data over the ocean still comes from visual estimates provided by commercial ships. Fortunately, the number of anemometer observations from these ships and from ocean platforms, weather ships and ocean buoys is growing. These data nevertheless suffer from various sources of errors related to differing anemometer heights, wind gustiness, effects of stability and other boundary-layer processes (compare Cardone *et al*, 1990 and Pierson, 1990). However, when assimilated as an ensemble with mutual quality control schemes in atmospheric forecasting models, the resulting wind fields are normally fairly reliable when the density of measurements is adequate. The main problem with conventional measurements is the sparsity of measurements outside the regular shipping lanes, in particular over large areas of the Southern Ocean.

More globally distributed wind data are available from the winds inferred from cloud motions recorded with geostationary satellites, but these are also not provided with uniform spatial density and need, moreover, to be corrected from cloud levels to the surface.

Global satellite altimeter and scatterometer wind data have now become routinely available, but will not be classed here as operational, as the assimilation of these data in atmospheric models is still in a test phase.

I.4.3 Wind measurements for research applications

More sophisticated wind measurements are normally carried out in measurement campaigns designed to study particular air/sea interaction or wave generation processes. High temporal resolution can be achieved with specially designed lightweight cup, propeller, pressure or thrust anemometers, or with sonic anemometers. The application of some of these instruments in the open ocean is limited. They are used to study not only the mean wind, but also the turbulent properties of the wind or the interaction between wind fluctuations and wave motions. Accurately calibrated cup anemometers can also resolve the vertical mean-wind structure of the atmospheric boundary layer. Although essential for a detailed study of the dynamics of the moving interface and the atmospheric boundary layer to which the wave field is coupled, a discussion of these measurement techniques is beyond the scope of this book. We refer the reader to Donelan (1990) and other reviews for more detailed presentations.

I.4.4 Operational wave measurements

Operational measurements of ocean waves are currently carried out mainly from ocean platforms or with ocean buoys. Platform wave measurements are usually carried out with some form of fast responding sea level gauge (a resistance or capacitance wire, for example), which yields the one-dimensional spectrum, or, alternatively, with the aid of a transmitting heave or heave-pitch-roll buoy anchored nearby. In the latter case the measurement principle is identical to that of an autonomous ocean buoy.

Heave buoys (commonly known as waveriders after a popular type) measure the vertical acceleration of the surface, or, equivalently, after a double integration, the surface displacement, from which the one-dimensional frequency wave spectrum can be determined. Although these are useful instruments, they have been largely replaced in recent years by heave-pitch-roll buoys, which provide an estimate of the two-dimensional frequency–direction spectrum. A number of these buoys are deployed in the eastern Atlantic and the western Pacific by the US National Oceanic and Atmospheric Administration (NOAA) and by other countries in coastal seas. Their data are normally transmitted to shore via satellite, but essentially the same buoys are also deployed from platforms operated by oil companies, using simple radio or cable data links (in this case the data are not always available for operational wave forecasting, however).

A heave-pitch-roll buoy typically consists of a disc-shaped float supporting an accelerometer, mounted on gyro-stabilized gimbals, and a compass. It measures the vertical acceleration, two gimbal angles, and the orientation of the buoy relative to north. The vertical acceleration yields the time series of the surface displacement, as in a heave buoy, and the gimbal angles can be resolved, using the compass signal, into the two components of sea surface slope, slope-east and slope-north. The two-dimensional wave spectrum is then estimated from the spectral auto- and co-variances of the resulting three time series. The method by which the two-dimensional spectrum is extracted from this information (see next paragraph) applies equally well to other instruments which measure time series of several different but correlated properties of the local wave field (Long and Hasselmann, 1979).

In the linear approximation, the auto- and cross-spectra between pairs of such time series are known integral properties of the surface wave directional spectrum. The task is then to invert these integral relationships to estimate the two-dimensional wave spectrum. A number of analysis methods have been published. Older techniques involve explicitly fitting a model for the two-dimensional spectrum with a number of degrees of freedom less than or equal to the number of data provided by the observed auto- and cross-spectra. These methods differ essentially in the nature of the fitted model and what is minimized in the fitting process (see, for example, Long, 1980). Newer maximum likelihood methods do not constrain the model form *a priori* but find an optimal solution which minimizes a cost function in which the data are introduced together with other information (for example, a first-guess spectrum) as constraints (Long and Hasselmann, 1979).

In the absence of significant currents, the cross-spectra in the case of heave-pitch-roll buoys can be interpreted directly in terms of the first four coefficients of a Fourier expansion of the directional dependence of the two-dimensional spectrum at each frequency (Longuet-Higgins *et al*, 1963). This truncated representation reproduces the data exactly, but has the disadvantage of producing negative side lobes. A number of alternative models have been used, but without additional information, the choice of the model is a matter of taste. The buoy provides unambiguously only the wave frequency spectrum and the four Fourier coefficients of the directional dependence at each frequency. However, the maximum likelihood techniques are able to exploit also the side condition that the spectrum must be nonnegative, and are thereby generally able to yield higher resolution spectra than direct modelling techniques (Lawson and Long, 1983).

It is easily shown that the same information as that from a heave-pitch-roll buoy is provided by an array of colocated instruments measuring surface elevation (with a surface-penetrating wave wire or a pressure sensor, for example) and two orthogonal components of velocity in the water column. Such an instrument has been successfully deployed and its results analysed using the identical maximum likelihood technique as for the case of a heave-pitch-roll buoy (Long and Hasselmann, 1979, Long, 1980, Lawson and Long, 1983 and Cavaleri *et al*, 1989).

I.4.5 Wave measurements for research applications

Traditional single channel devices, such as resistance or capacitance wires, are routinely used in measurement campaigns to measure the surface displacement of the wave field at a fixed location, yielding the one-dimensional frequency spectrum. In addition, arrays of such instruments with spacings smaller than or comparable to a typical wavelength scale are often constructed to estimate the two-dimensional spectrum, using essentially the same cross-correlation techniques as described for the heave-pitch-roll buoy. These can consist, for example, of multiple-wire wave gauge arrays, arrays of bottom mounted pressure sensors, combinations of current meters and wire or pressure gauges, or laser instruments which monitor or scan the surface wave slope.

To measure the very small relative displacements of short waves riding on longer waves, wave-follower instruments have been developed, in which the short-wave measurement probe (for example, a small wave gauge array or a laser instrument) is mounted on a support which follows the long-wave surface (compare Shemdin and Hwang, 1988 and Hwang and Shemdin, 1988).

In addition, a variety of remote sensing techniques (in addition to the satellite measurements which will be considered in more detail in chapter V) have been developed for measuring properties of the wave field from shore-based stations or aircraft.

High-frequency methods have been applied to measure the Bragg backscatter of radio waves in the 10 – 30 m wavelength range, using either surface wave propagation, with a range of typically 50 km, (compare Crombie, 1955, Barrick, 1968 and Essen *et al*, 1989a,b) or reflections from the ionosphere, with a single-bounce range of the order of a few thousand kilometres (compare Crombie, 1955 and Shearman, 1986). The first order Bragg return yields information only on the wave spectral densities for the two Bragg wave components of half the radio wavelength propagating towards or away from the radar. However, the second order side bands yield useful information on the one-dimensional frequency spectrum (Hasselmann, 1971a, Barrick, 1977), although the technique for the extraction of this information is rather complex and has not been routinely applied.

More information on the two-dimensional wave spectrum can be ob-

Table 1.1. *Satellites deploying altimeter, scatterometer or SAR.*

satellite	from	until	altimeter	scatterometer	SAR
SEASAT	27/6/1978	9/10/1978	+	+	+
GEOSAT	1985	1990	+	-	-
GEOSAT FOLLOW ON	1994	2000	+	-	-
ERS-1	1991	1994	+	+	+
ERS-2	1995	1998	+	+	+
JERS	1992	1994	+	+	+
ENVISAT	1998	2003	+	+	+
ADEOS/N-SCAT	1995	1998	-	+	-
TOPEX-POSEIDON	1992	1995	+	-	-
RADARSAT	1994	1999	-	-	+
TRMM	1996	1999	-	-	+
EPOP/POEM	1998	–	-	-	+
NPOP/EOS	1999	–	-	-	+
JPOP/JEOS	1999	–	-	-	+

tained with various microwave sensor systems developed for deployment from aircraft. Kenney *et al* (1979) and Walsh *et al* (1985) use a laterally scanning surface contouring radar (SCR) to map the two-dimensional surface topography along a swath centred on the aircraft track, from which the two-dimensional spectrum can be extracted after compensation for the aircraft motion. In the remote ocean wave scanner (ROWS) developed by Jackson (1987), the specular backscatter from a rotating near-nadir radar is used to estimate the two-dimensional spectrum. Finally, synthetic aperture radars (SARs) flown from aircraft yield the same information on the two-dimensional wave spectrum as satellite SARs, but normally with higher spatial resolution and with significantly smaller nonlinear distortions (see also chapter V).

I.4.6 Satellite measurements of winds and waves

We have classed satellite measurements as neither operational nor research oriented: at the time of writing they must still be viewed as preoperational, although their potential for operational applications has already been convincingly demonstrated and their operational application appears imminent.

The satellite instruments of principal interest for wave modelling are flown on polar orbiting satellites. The three standard active microwave instruments used to obtain information on sea state and surface wind, which were deployed on the first (unfortunately only short-lived) ocean satellite SEASAT, are the altimeter, scatterometer and the synthetic aperture radar (SAR). These instruments can provide global, all weather, day and night data coverage. An overview of satellites deploying these microwave systems is given in table 1.1.

The vertically downward looking radar altimeter obtains information on the sea surface from the travel time, intensity and shape of short pulses reflected through specular reflection at the sea surface. The scatterometer, operating at incidence angles between 20° and 50°, measures the backscattered energy (specific cross-section), while the SAR, operating typically at incidence angles between 20° and 25°, measures both the amplitude and the phase of the backscattered signal.

The dominant backscattering mechanism for the scatterometer and SAR, which operate at incidence angles greater than the typical rms slope of the sea surface, is Bragg scattering. Wind and sea state information can be extracted from the altimeter measurements directly without any prior knowledge of the local sea state or wind field, whereas a reliable interpretation of scatterometer and SAR data requires first-guess wave spectra and/or wind data as input for inverse modelling techniques. (compare chapter V.)

A typical polar orbiting satellite orbits the earth once every 100 min, yielding a separation distance of adjacent orbits at the equator of 2800 km and a mean equatorial separation of all ascending and descending orbits within one day of 1400 km. Although this data coverage greatly exceeds the coverage available with conventional instruments in most oceans, it is still not sufficiently dense for the computation of wind and wave fields on the typical synoptic scale of weather variability by straightforward interpolation. The satellite data must therefore be assimilated with all available data from conventional observing systems in atmospheric circulation and global wave models. This enables the construction of optimally interpolated wind and wave fields from all data consistent with the dynamical constraints of the models (compare chapter VI).

In the following we discuss only wind measurements with the altimeter and the scatterometer; wave measurements with the altimeter and the SAR will be discussed in more detail in chapter V.

I.4.7 Altimeter wind measurements

The radar altimeter is a nadir looking instrument. To a good approximation the backscattered return can therefore be described by specular reflection. The measurement of the travel time of the reflected microwave pulse yields the position of the sea surface relative to the orbit of the satellite. After correction for atmospheric and ionospheric disturbances, this can be determined to within a few centimetres. After further (rather complex) computations of the satellite orbit these data can then be used to infer the earth's geoid and the geostrophic component of the near-surface ocean current.

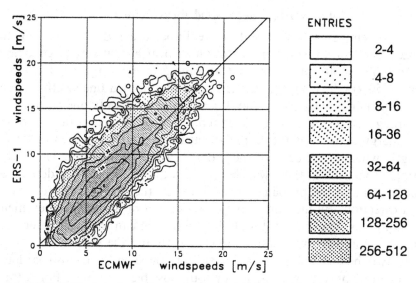

Fig. 1.1. ECMWF global wind speeds against ERS-1 altimeter wind speeds for the period July 1992.

The radar altimeter also provides useful information on the wave height, which can be deduced from the return pulse shape (compare chapter V) and on the wind speed, which can be determined from the intensity of the backscattered return pulse. For specular reflection, the backscattered energy at the nadir incidence angle (that is the specific cross-section) is proportional to the inverse of the mean slope of the surface wind waves. This, in turn, depends on the local wind speed. The relation between the wind speed and cross-section has been determined empirically and tested in numerous investigations using SEASAT, GEOSAT and, more recently, ERS-1 data. The altimeters on all three satellites had comparable performances (compare Hasselmann *et al*, 1988, Bauer *et al*, 1992, Günther *et al*, 1993 and Romeiser, 1993). For moderate winds (3 – 12 m/s) the wind speed can be determined from altimeter measurements with an accuracy of about ±2 m/s. However, the sensitivity of the measurement decreases for higher wind speeds, the instrument becoming unreliable for wind speeds above 25 m/s.

In figure 1.1 the altimeter wind speeds of July 1992 are compared with wind speeds obtained from the global operational analysis of the European Centre for Medium-Range Weather Forecasts (ECMWF). The agreement is satisfactory. Similarly good agreement has been found for SEASAT and GEOSAT data (Fedor and Brown, 1982, Dobson *et al*, 1987 and Bauer *et al*, 1992).

I.4.8 Scatterometer wind measurements

The scatterometer is a low-resolution, real-aperture radar, measuring the normalized backscatter radar cross-section σ_o over a relatively large area of the ocean (400 – 600 km swath) and averaging over a large resolution cell ~ 50 km $\times$ 50 km. The SEASAT scatterometer illuminated one swath on each side of the satellite, whereas the ERS-1 scatterometer is only one-sided.

For Bragg scattering, the backscattered return is proportional to the spectral energy density of the ripple spectrum at the Bragg backscattering wavelength, which is typically of the order of a few centimetres. The energy of the ripple spectrum at these wavelengths is found empirically to depend on the wind speed and the propagation direction of the short waves relative to the wind direction. By measuring the cross-section of a given scatterometer pixel from at least two different look directions one can therefore infer both the wind speed and the wind direction (Donelan and Pierson, 1987).

If only two beams are used, as in SEASAT, the wind direction is determined, however, only to within two or four possible solutions. For a three beam scatterometer, as in ERS-1, the solution is, in principle, unique, but in the presence of unavoidable noise, there normally remains a residual 180° ambiguity. Thus to extract reliable wind fields from the scatterometer the inversion algorithms (based on the minimization of an error cost function formulated in terms of an empirical backscatter model) must make use of first-guess wind fields or apply other meteorological information to remove the ambiguities. Present algorithms yield wind velocities with a high level of ambiguity removal (of order 90%) and accuracies of the order of 10% or ± 2 m/s for the wind speed and $\pm 20°$ for the wind direction (Anderson *et al*, 1987, Long, 1986 and Offiler, 1987).

It should be noted, however, that the purely empirical backscatter models currently used in these algorithms assume that the backscatter cross-section is a function of only the (stability corrected) local 10 m wind speed. They therefore ignore the direct impact of stability of the atmospheric boundary layer on the generation of the short backscattering ripples, and they also neglect the interaction of the ripple waves with the longer waves. Because of the long memory time of the long waves, the long wave–short wave interactions cannot be regarded as a function of the local wind speed only but must be expressed explicitly in terms of the long wave spectrum. Improved backscatter models including these effects would reduce the systematic errors which the present scatterometer models presumably exhibit in rapidly changing meteorological conditions, where the scatterometer is potentially most useful (Hasselmann *et al*, 1990). A significant improvement of the analysed surface wind fields may be expected as more experience is gained in the assimilation of this new form of global surface data in atmospheric models (compare Woiceshyn and Janssen, 1992 and Janssen and Woiceshyn, 1992).

Chapter II

The physical description of wave evolution

M. Donelan *et al*

II.1 Introduction

M. Donelan

Any numerical model of a physical process necessarily represents that process by a set of mathematical relations that approximate the underlying physical laws. The level of approximation is determined principally by two factors: (1) knowledge of the physical processes; (2) computational limitations. The latter forces one to view the problem of the evolution of waves on an oceanic scale as a statistical one. The process of global wave evolution encompasses scales as large as the basin (10^7 metres) involving ocean currents, topography and wind variations, and as small as the smallest waves (10^{-3} metres) that modify the wind stress – it could be argued that even smaller scales associated with turbulent interfacial couplings play a role in the process. The statistical approach essentially treats the bottom 4/5 of this physical range of ten orders of magnitude as though those scales respond to well-defined physical laws imposed at the nodes of a numerical grid of typical size of one degree of latitude. The task of defining appropriate mathematical expressions that reflect the essential physics of the process is one of synthesizing the results of theoretical calculations and observational programs. As discussed in chapter I, the mathematical framework is based on a statistical description of waves having a range of scales of about one metre to one kilometre. These waves evolve in response to an action balance equation in which the 'physics' is embodied in a set of source functions. In this chapter, we first discuss the source functions individually and then examine the observational evidence of spectral characteristics and wave growth in fetch-limited and in shallow water situations and directional adjustment to turning winds.

II.2 Wave growth by wind

P. A. E. M. Janssen

Perhaps one of the most intriguing problems in the theory of surface gravity waves is their generation by wind. Jeffreys (1924, 1925) assumed that air flowing over sea was sheltered by the waves on their lee side. This would give a pressure difference, so that work could be done by the wind. Subsequent laboratory experiments on solid waves showed that the pressure difference was much too small to account for the observed growth rates. As a consequence the sheltering hypothesis was abandoned, and one's every day experience of the amplification of water waves by wind remained poorly understood. This changed in the the mid-1950s, when Phillips (1957) and

Miles (1957) published their contributions to the theory of surface wave generation by wind. Both theories had in common that waves were generated by a resonance phenomenon: Phillips considered the resonant forcing of free surface waves by turbulent pressure fluctuations, while Miles considered the resonant interaction between the wave-induced pressure fluctuations and the free surface waves. Phillips' mechanism gives rise to a (linear) growth of the wave spectrum in time, but it turned out to be ineffective, because the effect is proportional to the variance spectrum of the pressure fluctuations at the resonant frequency, independent of the wave spectrum, and this is of the order of the square of the density ratio of air and water (see also § I.2.8). Miles' mechanism looked more promising. It is proportional to the wave spectrum itself, which implies exponential growth, and it is of the order of the density ratio of air and water.

Although Miles' work aroused a great, renewed interest in the problem, there was also considerable controversy. One of the main reasons for this controversy was that Miles' theory oversimplifies the problem by assuming that the air flow is inviscid and that turbulence does not play a role except in maintaining the shear flow (this is the so-called quasi-laminar approach). Another reason is that Miles neglected nonlinear effects such as wave–mean-flow interaction, which are expected to be important at the layer where the phase speed of the surface waves matches the wind speed (the so-called critical layer). Also, early field experiments, in particular by Dobson (1971), gave rates of energy transfer from wind to waves that were an order of magnitude larger than predicted by Miles (1957). The measurement of the energy transfer from wind to waves is, however, a very difficult task as it involves the determination of the phase difference between the wave-induced pressure fluctuation and the surface elevation signal (as we will see later). More recent field experiments (Snyder 1974, Snyder *et al*, 1981, D.E. Hasselmann and Bösenberg, 1991) show order of magnitude agreement with Miles' theory, although the theory still predicts energy transfer rates that are smaller than measured values, especially for relatively low-frequency waves with a phase speed that is about the same as the wind speed at 10 m height.

Nevertheless the state of affairs regarding the theory of wind-wave generation remained unsatisfactory. The quasi-laminar approach was criticized because, turbulence was not properly modelled, it ignored severe nonlinearities of the order of the ratio of the wave height to the critical layer height – which can become much larger than unity – and wave–mean-flow interaction was not considered.

There have been several attempts to overcome these shortcomings by numerical modelling of the turbulent boundary layer flow over a moving (usually sinusoidal) sea surface. The interaction of the wave-induced flow with the mean flow and boundary-layer turbulence can then be simulated explicitly (with suitable turbulent closure assumptions). This automatically takes care of the nonlinearities associated with a critical layer which is close to the surface relative to the wave height. (Stewart (1967) pointed out that this makes it geometrically impossible to accomodate the streamline phase shifts predicted by Miles within the thin critical layer.)

One such approach (see, for example, Gent and Taylor, 1976, Makin and Chalikov, 1979, Riley *et al*, 1982, Al-Zanaidi and Hui 1984, Jacobs 1987, Chalikov and Makin, 1991) considered the direct effects of small-scale turbulence on wave growth. Mixing length modelling or second order closure is then assumed to calculate the turbulent Reynolds stresses. Surprisingly, the results are not very different from the ones obtained in quasi-laminar theory, so small-scale eddies and nonlinearities in the wave steepness have a small direct effect on wave growth.

Van Duin and Janssen (1992) noted a flaw in these approaches, in which turbulence modelling relies on the analogy with molecular processes. They pointed out that this fails for low-frequency waves. Mixing length modelling of the turbulent stress assumes that the momentum transport owing to turbulence is the fastest process in the system, i.e. turbulence is assumed to be fast in comparison with the relatively slow motion of the surface gravity waves. However, this is not justified for the low-frequency waves, which have the largest phase speeds. These may interact with large eddies whose 'eddy turnover' time may become larger than the period of the waves. For those large eddies (which are identified here with gustiness) another approach is needed. Nikolayeva and Tsimring (1986) considered the effect of large-scale turbulence (gustiness) on wave growth. These authors applied a so-called kinetic model, proposed by Lundgren (1967), to the problem of wind-generated waves and a substantial enhancement of energy transfer due to gustiness was found, especially for those waves with a phase speed that is comparable to the wind speed at 10 m height. Unfortunately, a theory covering effects of both small-scale turbulence and gusts, including the case that the eddy turnover time is comparable to the wave period, is not available. Fortunately, for high-frequency waves there are no dramatic differences between Miles' theory of wind-wave generation and a theory that includes effects of turbulence.

In view of these problems and because the results of theories with turbulence are not very different from results of Miles' quasi-laminar model, we will concentrate on this latter aproach, which allows, at least partly, for an analytical treatment. It has the additional advantage that it can be easily extended to include gustiness, which is known to be important for low-frequency waves.

For a given wind profile, quasi-laminar theory is fairly successful in predicting growth rates. It ignores, however, a possible change of wind profile with the evolution of the waves. The momentum transfer from wind to waves might be considerable so that the associated wave-induced stress may be a substantial fraction of the turbulent stress (Snyder 1974, Snyder *et al*, 1981). The velocity profile over the sea waves results from a balance between turbulent and wave-induced momentum flux. One may thus expect deviations from the profile of turbulent air flow over a flat plate. In turn, the energy transfer from the air to the waves would be affected by the sea state, so that one expects a strong coupling between the turbulent boundary layer and the surface gravity waves.

Observations confirm this expectation. The most direct evidence for the dependence of the air flow on the sea waves comes from the observed dependence of the drag coefficient on the so-called wave age. The wave age parameter (to be defined more precisely later) measures the stage of development of wind sea. 'Young' wind sea, related to a wave spectrum with a relatively high peak frequency, refers to a sea state where waves have just been generated by wind, while 'old' wind sea refers to a saturated sea state, the energy of which hardly changes in time. Measurements by, for example, Donelan (1982) and Maat *et al* (1991) have confirmed the dependence of the drag coefficient on wave age. For a fixed wind speed at 10 m height Donelan found that the drag coefficient of air flow over young wind sea is some 50% larger than the drag coefficient over old wind sea.

Parallel to these experimental developments, the theory of the interaction of wind and waves was elaborated by Fabrikant (1976) and Janssen (1982). The so-called quasi-linear theory of wind-wave generation emerged, which was based on an analogy that exists between resonant wave–mean-flow interaction and the interaction of plasma waves and particles. In essence, it keeps track of the slow evolution of the sea state and its effect on the wind profile. At each particular time the wave growth follows from Miles' theory. The results are striking and have been confirmed by available observations.

The approach of this section is as follows. First of all, we discuss the linear, quasi-laminar theory of wind-wave generation, based on Miles' shear flow mechanism. Basically, this mechanism of the generation of waves by wind is a resonant interaction of the gravity waves with a plane parallel flow. Resonance occurs at a critical height z_c if $U(z_c) = c(\omega)$, where U is the air velocity and $c(\omega)$ the phase velocity of a wave with frequency ω. Only those waves grow for which the curvature of the velocity profile at the critical height is negative. The treatment, which will be given below, is fairly general so that effects of atmospheric stratification (such as gradients in the air temperature) are also taken into account (Janssen and Komen, 1985). Results of this theory are compared with available field and laboratory measurements, and a favourable agreement is found.

Next, we will give a treatment of the effect of gusts on wave growth. This treatment assumes that the time scale of the waves is shorter than the time scale of the gusts. Then, if the probability distribution of the fluctuations in stress due to gusts is known, an appropriate averaging of the Miles' result will give the effect of gusts on wave growth immediately. The growth of the low-frequency waves will be affected considerably.

After the discussion of the quasi-laminar approach we proceed with a brief account of the effects of small-scale turbulence, and we discuss at some length why mixing length modelling fails for low-frequency waves.

We conclude with a discussion of quasi-linear theory, which culminates in the derivation of the relation between the drag coefficient and the sea state. This relation is then compared with existing field data. A favourable agreement is found.

II.2.1 Linear theory of wind-wave generation

Miles (1957) was the first to give a rational theory of wind-wave generation. Although, as already stated, much controversy was aroused because of an oversimplification of the problem, his approach is still nowadays the most appealing one.

We give here a modified treatment of Miles' theory in which it is emphasized that wind-wave generation is closely related to the instability of a plane parallel shear flow. The principal difference with shear flow over a flat plate (Drazin and Reid, 1981) is that the lower boundary (the air/sea interface) is allowed to evolve in time. At the same time this gives the opportunity to generalize the analysis by including effects of stratification.

Our starting point is the set of equations for an adiabatic fluid:

$$\frac{\partial}{\partial t}\rho + \nabla \cdot \rho \mathbf{u} = 0,$$

$$\frac{\mathrm{d}}{\mathrm{d}t}\mathbf{u} = -\frac{1}{\rho}\nabla p + \mathbf{g}, \quad \frac{\mathrm{d}}{\mathrm{d}t} = \frac{\partial}{\partial t} + \mathbf{u} \cdot \nabla, \tag{2.1}$$

$$\frac{\mathrm{d}}{\mathrm{d}t}p = c_s^2\frac{\mathrm{d}}{\mathrm{d}t}\rho, \quad c_s^2 = \frac{c_p}{c_v}RT,$$

where all symbols have their usual meaning. Confining our interest to phenomena with a speed much smaller than the sound speed c_s, we obtain from equation (2.1)

$$\nabla \cdot \mathbf{u} = 0,$$

$$\frac{\mathrm{d}}{\mathrm{d}t}\mathbf{u} = -\frac{1}{\rho}\nabla p + \mathbf{g}, \tag{2.2}$$

$$\frac{\mathrm{d}}{\mathrm{d}t}\rho = 0.$$

We would like to study the stability of the equilibrium solution (see figure 2.1)

$$\mathbf{u}_0 = U_0(z)\hat{\mathbf{e}}_x, \quad \mathbf{g} = -g\hat{\mathbf{e}}_z,$$

$$\rho_0 = \rho_0(z), \quad p_0 = p_0(z) = g\int \mathrm{d}z\,\rho_0(z) \tag{2.3}$$

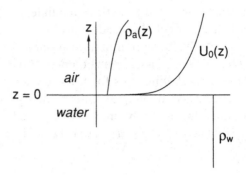

Fig. 2.1. The height dependence of velocity and density.

of equation (2.2). In other words, we have a plane parallel flow whose speed and density depend on height. The present theory, however, does not explain the height dependence of wind speed and density because effects of small-scale turbulence have not been taken into account. This is clearly a weak point in the theory as effects of turbulence on waves in the air flow are not considered as well. As discussed already, this does not matter much regarding the final results.

In order to investigate stability, we perturb the steady state (2.2) by taking $\rho = \rho_0 + \rho_1$ etc, where $\rho_1 \ll \rho_0$ and we take normal mode solutions of the form $\rho_1 \sim \exp i(kx - \omega t)$. Therefore, only wave propagation in one direction is considered. By an appeal to Squire's theorem (Drazin and Reid, 1981) results may be generalized immediately to the two-dimensional case. For waves propagating at an angle with respect to the wind not the wind itself but $\mathbf{k} \cdot \mathbf{U}_0/k$ is then seen by the waves.

As a result, one obtains from (2.2) in the linear approximation

$$
\begin{aligned}
iku + \partial w/\partial z &= 0, \\
ikWu + wW' &= -ikp_1/\rho_0, \\
ikWw &= \rho_1 p_0'/\rho_0^2 - (\partial p_1/\partial z)/\rho_0, \\
ikW\rho_1 + w\rho_0' &= 0,
\end{aligned}
\tag{2.4}
$$

where $\mathbf{u}_1 = (u, 0, w)$, $W = U_0 - c, c = \omega/k$ and the prime denotes differentiation of an equilibrium quantity with respect to z. After some tedious but straightforward algebra one then arrives at the following Sturm-Liouville type of differential equation for the displacement $\psi \sim w/W$ of the streamlines

$$
\frac{\mathrm{d}}{\mathrm{d}z}\left(\rho_0 W^2 \frac{\mathrm{d}}{\mathrm{d}z}\psi\right) - \psi\left(k^2\rho_0 W^2 + g\rho_0'\right) = 0,
\tag{2.5}
$$

which is subject to the boundary condition of vanishing displacement at infinite height or depth,

$$\psi \to 0, \quad |z| \to \infty. \tag{2.6}$$

The boundary value problem (2.5) – (2.6), in principle, determines the real and imaginary parts of the complex phase speed c, giving the growth rate $\gamma_a = \text{Im}(\omega)$ of the waves. It should be realized that without further specification of the density profile, the boundary value problem is quite general.

The (unstable) waves resulting from (2.5) – (2.6) are usually called internal gravity waves. By choosing an appropriate density profile the special case of surface gravity waves will be found. In order to obtain the case of surface gravity waves we consider the equilibrium sketched in figure 2.1.

Here, the air density is much smaller than the water density, $\epsilon = \rho_a/\rho_w \sim 10^{-3}$. By application of the Miles' theorem (Miles, 1961) one immediately infers the possibility of instability of the equilibrium given in figure 2.1. This theorem states that a sufficient condition for stability is that $dU_0/dz \neq 0$ and that the Richardson number $Ri = g\rho'/(\rho W'^2)$ is smaller than $-1/4$ everywhere. The latter condition is clearly violated in air so there is a possibility for instability, that is there is possibly an energy transfer from the shear flow $U_0(z)$ to the surface gravity waves. This transfer may be affected by buoyancy effects.

In water (no current, a constant density), the eigenvalue problem simplifies considerably. We have (for $z < 0$)

$$\frac{d^2}{dz^2}\psi_w = k^2\psi_w, \tag{2.7}$$

which gives using the boundary condition at infinity, the solution

$$\psi_w = ae^{kz}. \tag{2.8}$$

The boundary condition at the interface between air and water is derived from an integration of equation (2.5) from -0 to $+0$. Note that at $z = 0$ the density profile shows a jump so that near $z = 0$ $\rho_0' = (\rho_a(0) - \rho_w)\delta(z)$, where $\delta(z)$ is the Dirac delta-function. Requiring now that ψ be continuous across the interface we obtain from (2.5)

$$\rho_0 W^2 \frac{d}{dz}\psi \bigg|_{-0}^{+0} = \int_{-0}^{+0} dz \, \psi[\rho_0 k^2 W^2 + g\rho_0']. \tag{2.9}$$

Since in the limit only the integral involving ρ_0' gives a contribution we obtain, using (2.8), the following dispersion relation for the phase speed of the waves:

$$c^2 = \frac{g(1-\epsilon)}{k - \epsilon\psi_a'(0)}, \quad \epsilon = \frac{\rho_a(0)}{\rho_w(0)}, \tag{2.10}$$

where without loss of generality we have taken the amplitude $a = 1$ as we deal with a linear problem. Hence, the eigenvalue problem (2.5) – (2.6) may be reduced to the simplified problem

$$\frac{\mathrm{d}}{\mathrm{d}z}\left(\rho_a W^2 \frac{\mathrm{d}}{\mathrm{d}z}\psi_a\right) = \psi_a[k^2\rho_a W^2 + g\rho_a'], \quad z > 0,$$

$$\psi_a(0) = 1, \qquad\qquad\qquad\qquad\qquad\qquad\qquad (2.11)$$

$$c^2 = g(1 - \epsilon)/(k - \epsilon\psi_a'(0)),$$

$$\psi_a \to 0, \quad z \to \infty.$$

First of all, we note that in the limit $\epsilon \to 0$ (no air), the well-known dispersion relation for surface gravity waves is obtained, that is $c^2 = g/k \to \omega = \sqrt{gk}$. In practice, the effect of the air on the wave is small as $\epsilon \simeq 10^{-3}$. We therefore solve the dispersion relation in an approximate manner with the result:

$$c = c_0 + \epsilon c_1 \ldots, \qquad\qquad\qquad\qquad\qquad\qquad (2.12)$$

where $c_0 = \sqrt{g/k}$ and $c_1 = \frac{1}{2}c_0(\psi_a'/k - 1)$. As a result the problem (2.11) reduces to:

$$\frac{\mathrm{d}}{\mathrm{d}z}\left(\rho_a W_0^2 \frac{\mathrm{d}}{\mathrm{d}z}\psi_a\right) = \psi_a[k^2\rho_a W_0^2 + g\rho_a'], \quad z > 0,$$

$$\psi_a(0) = 1, \qquad\qquad\qquad\qquad\qquad\qquad\qquad (2.13)$$

$$\psi_a \to 0, \quad z \to 0,$$

where $W_0 = U_0 - c_0$ is now known. As c_0 is known already, the solution of the differential equation for ψ_a is simplified considerably, especially if one wishes to do this numerically. In addition, we now have an explicit expression for the growth rate γ_a of the waves,

$$\frac{\gamma_a}{\omega_0} = \epsilon \operatorname{Im}\left(\frac{c_1}{c_0}\right) = \frac{\epsilon}{2k}\operatorname{Im}(\psi_a') = \frac{\epsilon}{4k}\mathcal{W}(\psi_a, \psi_a^*)\bigg|_{z=0}, \qquad (2.14)$$

where the Wronskian $\mathcal{W}$ is given as

$$\mathcal{W}(\psi_a, \psi_a^*) = -i(\psi_a'\psi_a^* - \psi_a\psi_a'^*). \qquad\qquad\qquad (2.15)$$

Physically, it may be shown that the Wronskian $\mathcal{W}$ is related to the wave-induced stress $\tau_w = - <u_1w_1>$. The result (2.14) is a very elegant one as it relates the growth of the waves to the wave-induced stress.

The determination of γ_a, and hence of ψ_a', requires the solution of the boundary value problem (2.13) where c is given by the phase speed of the free surface gravity waves. Note that c_0 is real so that (2.13) is a singular problem.

It is common to work in terms of the vertical component of the wave-induced velocity instead of the displacement of the streamline $\psi \sim w/W$. Also, one normally makes the Boussinesq approximation, in which effects of density gradients are ignored, except in combination with the acceleration of gravity (see Miles 1961). One then finally arrives at the well-known Taylor-Goldstein equation for $\chi = w/w(0)$. In terms of χ (2.13) – (2.14) become:

$$\frac{d^2}{dz^2}\chi = \left[k^2 + \frac{W_0''}{W_0} + \frac{g\rho_a'/\rho_a}{W_0^2}\right]\chi,$$

$$\chi(0) = 1, \tag{2.16}$$

$$\chi \to 0, \quad z \to \infty,$$

where $W_0 = U_0(z) - c_0$, $c_0 = \sqrt{g/k}$ and the growth rate of the waves is given by:

$$\frac{\gamma_a}{\epsilon\omega_0} = \frac{1}{4k}\mathscr{W}(\chi, \chi^*)\Big|_{z=0}, \tag{2.17}$$

where the Wronskian $\mathscr{W}$ is now given by $\mathscr{W} = -i(\chi'\chi^* - \chi\chi'^*)$. In the next subsection we shall discuss some of the properties of the problem (2.16) – (2.17). Special attention will be given to the effect of stratification. The reason is that although the Richardson number is quite small for the relatively large wind speeds of interest, at the critical height (where $W_0 = 0$) the solution of (2.16) is still dominated by the density gradient. We shall, therefore, attempt to study the behaviour of the solution of (2.16) for a small critical Richardson number

$$Ri_c = g\,\frac{\rho_{a,c}'}{\rho_{a,c}}\,\frac{1}{W_c'^2}, \tag{2.18}$$

where the index c refers to evaluation at the critical height.

II.2.2 General properties of the Taylor-Goldstein equation

As mentioned above, we wish to study the solution of (2.16) – (2.17) in the limit of a small Richardson number. We give therefore the results for the growth rate first in the case of no density gradient at all in the air. Then (2.16) becomes:

$$\frac{d^2}{dz^2}\chi = \left[k^2 + \frac{W_0''}{W_0}\right]\chi,$$

$$\chi(0) = 1, \tag{2.19}$$

$$\chi \to 0, \quad z \to \infty$$

and the growth rate still follows from equation (2.17). This is the problem that Miles (1957) studied in some detail. The equation for χ is now just the Rayleigh equation, which has two independent solutions around the critical height, namely,

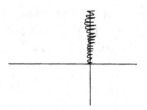

Fig. 2.2. Complex y plane; branch cut for $U'_{0c} > 0$.

$$\chi_1 = C_0 y \left(1 + \frac{1}{2} C_0 y + \ldots \right),$$

$$\chi_2 = 1 + C_0 y \ln y + \ldots, \tag{2.20}$$

where $C_0 = W''_{0c}/W'_{0c}$ and $y = z - z_c$. Clearly χ_1 is an entire function of y whereas χ_2 is a multi-valued function of y. Hence, a branch cut has to be introduced in order to make χ_2 single-valued. It should be emphasized that it is important to make a proper choice of the branch cut since the final result regarding stability or instability of the flow will depend on it. A similar problem may be found in the theory of (unstable) plasma waves. Here, there is energy transfer possible between plasma waves and electrons because of a resonant interaction. Just as in the problem of wind-wave generation the normal modes of the system are singular. A special care regarding the treatment of the poles is needed. In fact, the first attempts to solve this problem failed because effects of the singularity were disregarded. Landau (1946) was the first to give a proper solution of the problem of the interaction of particles and plasma waves. He solved an initial value problem by means of Laplace transformation and applied asymptotic techniques to find the large time behaviour of the solution. For large times the solution was found, depending on whether there was stability or instability, to be determined by the least damped or most unstable normal mode. Also a prescription of how to treat the singularity could be inferred: one simply treats the singularity in the limit of the positive imaginary part of the phase speed c.

Let us apply this rule to the problem of the choice of the proper branch cut. To that end we study the root of

$$W = U_0 - c_0 - i c_1, \quad (\lim c_1 \downarrow 0). \tag{2.21}$$

Expanding the wind speed around the critical height z_c, $U_0 = U_{0c} + y U'_{0c}$, where $U_{0c} = c_0$, we obtain $y = i c_1 / U'_{0c}$, hence in the limit $c_1 \downarrow 0$ y_c approaches the real axis from above (below) for positive (negative) U'_{0c}. The branch cut is, therefore, as shown in figure 2.2.

Therefore, for positive y the phase of y is zero whereas for negative y the phase of y is $-\pi \text{sign}(U'_{0c})$. This convention determines χ_2 uniquely.

Now, in order to determine the growth rate γ_a we need to know $\mathscr{W}(z = 0)$. By means of the Rayleigh equation (2.19) it can be shown that the Wronskian $\mathscr{W}$ is a constant except at the critical height $z = z_c$, where it may show a jump. It is therefore sufficient to determine $\mathscr{W}(z = z_c)$ at the critical height. This can be done by means of the Frobenius series (2.20). Noting that the general solution of (2.19) is given by

$$\chi = A(\chi_2 + B\chi_1), \tag{2.22}$$

we obtain for $\mathscr{W}(z = z_c)$

$$\mathscr{W}(z = z_c) = -2\pi \frac{W_{0c}''}{|W_{0c}'|} |A|^2, \tag{2.23}$$

hence, since $\chi_c = A$, we obtain for the growth rate γ_a

$$\frac{\gamma_a}{\epsilon\omega_0} = -\frac{\pi}{2k} \frac{W_{0c}''}{|W_{0c}'|} |\chi_c|^2. \tag{2.24}$$

This is Miles' classical result for the growth of surface gravity waves due to the shear flow. From equation (2.24) we obtain the well-known result that only those waves are unstable for which the curvature U_0'' of the wind profile at the critical height is negative (as is, for example, the case for a logarithmic profile). A physical explanation of this instability has been given by Lighthill (1962) and relies on the role of the vortex force. In fact, the vortex force gives the slowing down of the mean air flow by the wave. It is given by the gradient of the wave-induced stress τ_w. Since the wave-induced stress is proportional to the Wronskian $\mathscr{W}$, which has a step function discontinuity at the critical height, the vortex force is a delta-function. This suggests an important limitation of linear theory, because a considerable wave–mean-flow interaction may therefore occur, giving rise to a modified mean flow. Wave–mean-flow interaction for a spectrum of waves will be further discussed in § II.2.6.

Having discussed the case of vanishing Richardson number, let us now proceed with the finite Richardson number case. The finite density gradient case is more difficult to deal with because the singularity at the critical height is worse. Nevertheless Janssen and Komen (1985) were able to treat this singular problem and they found for the growth rate

$$\frac{\gamma_a}{\epsilon\omega_0} = -\frac{\nu}{2k} C|A|^2 \left[\frac{\sin \pi(1-\nu)S}{1-\nu} + B \sin \pi(1+\nu)S \right], \tag{2.25}$$

where

$$\nu = \sqrt{1 + 4Ri_c} \quad \text{and} \quad C = (1 - Ri_c) \frac{W_{0c}''}{W_{0c}'} + Ri_c \left(\frac{\rho_{ac}''}{\rho_{ac}'} - \frac{\rho_{ac}'}{\rho_{ac}} \right) \tag{2.26}$$

and

$$S = \begin{cases} 0, & y > 0 \\ \text{sign}(W_c'), & y < 0 \end{cases},$$

(2.27)

which, in the limit of neutral stability ($v \to 1$), reduces to the classical Miles' result (2.24).

After exploring some of the properties of the Taylor-Goldstein equation we shall, in the next subsection, study the numerical solution of the boundary value problem (2.25). Results for the growth rate γ_a in the absence of stratification will be compared with available observations, while effects of stratification will also be discussed.

II.2.3 Numerical solution and comparison with observations

The numerical solution of (2.16) is complicated somewhat by the presence of a singularity at the critical height. For the Rayleigh equation Conte and Miles (1959) and Miles (1959a) have shown how this singularity may be treated. Here we follow a similar approach for the Taylor-Goldstein equation. In this approach one writes

$$\chi = \rho e^{i\theta},$$

(2.28)

and finds that ρ and θ' are related to the Wronskian $\mathscr{W}$ in the following fashion

$$\mathscr{W} = 2\rho^2\theta'.$$

(2.29)

Using (2.29) to determine θ' one obtains the amplitude ρ from the real part of the Taylor-Goldstein equation (2.16)

$$\rho'' - \rho\theta'^2 = \rho\left[k^2 + W_0''\,\mathscr{P}\frac{1}{W_0} + g\frac{\rho_0'}{\rho_0}\,\mathscr{P}\frac{1}{W_0^2}\right],$$

(2.30)

where $\mathscr{P}$ denotes the principal value. The integration of (2.30) is performed by means of a fourth order Runge-Kutta method, except for the region in the neighbourhood of the critical height where the solution is continued analytically by means of the relevant Frobenius solutions.

The numerical programme was tested extensively by applying it to a case for which the exact solution is known. To that end an exponential profile for the density and the so-called asymptotic suction profile for the velocity was chosen (Drazin and Reid, 1981). The solution of the boundary value problem (2.16) may then be obtained in terms of hypergeometric functions. A reasonable agreement between numerical and exact growth rate was thus found.

Next, the numerical procedure was applied to the velocity and density profiles as found in the boundary layer above sea (Large and Pond, 1982).

As discussed in § I.3, in neutrally stable conditions (no density stratification by heat and moisture) the wind profile has a logarithmic height dependence

$$U_0(z) = \frac{u_*}{\kappa} \ln(1 + z/z_0). \tag{2.31}$$

This profile depends on three parameters. The von Kármán constant κ is supposed to be universal (its value is still under discussion; here we take $\kappa = 0.41$), the friction velocity u_* is a measure of the (turbulent) momentum flux and z_0, the so-called roughness length, is a parameter which reflects the loss of momentum to the sea surface. In the past the roughness length has been determined empirically. Over sea a good choice is the one proposed by Charnock (1955)

$$z_0 = \alpha u_*^2/g, \tag{2.32}$$

with g the acceleration due to gravity, and α the so-called Charnock parameter. In this subsection we will consider a constant $\alpha = \alpha_{ch} = 0.0144$. In § II.2.7 we will present arguments why α is not constant but depends on the sea state. The usefulness of (2.32) was confirmed by contributions of Garratt (1977) and Wu (1982) who compared wind-stress data from many sources. Wu showed that (2.32) leads to a wind-speed dependence of the drag coefficient which is in reasonable agreement with observations for wind speeds between 2 and 52 m/s and for long fetches.

This simple empirical picture is modified in the presence of heat or moisture fluxes, which arise when temperature or moisture stratification is present. The magnitude of this effect is commonly parametrized in terms of the so-called Obukhov length, a measure of the height at which shear production and buoyant production of turbulence are of the same magnitude (see also the discussion following (1.229)):

$$L = -u_*^3 T_v/(\kappa g < w' T_v' >). \tag{2.33}$$

Here T_v is the mean virtual temperature at 10 m height, and $< w' T_v' >$ is the vertical turbulent flux of virtual temperature. The complete parametrization of the boundary layer was given in § I.3.7.

Having established the appropriate mean profiles, we are now in a position to solve the boundary value problem for the Taylor-Goldstein equation (2.16). Before we do this we introduce dimensionless quantities in order to see which dimensionless parameters determine the problem. To that end we scale velocities with u_* and lengths with u_*^2/g, thus

$$
\begin{aligned}
z_* &= gz/u_*^2, & z_{0*} &= gz_0/u_*^2, & z_{t*} &= gz_t/u_*^2, \\
c_* &= c/u_*, & U_{0*} &= U_0/u_*, & k_* &= u_*^2 k/g
\end{aligned}
\tag{2.34}
$$

and the dimensionless Obukhov length becomes

$$L_* = gL/u_*^2. \tag{2.35}$$

The boundary value problem (2.16) has, in terms of these dimensionless quantities, the same shape except that the wind profile is given as (cf. § I.3.7)

$$U_{0*} = \frac{1}{\kappa}[\ln(1 + z_*/z_{0*}) - \psi_m(z_*/L_*)], \tag{2.36}$$

while the temperature profile is

$$T = T_{sea} + t_*[\ln(1 + z_*/z_{t*}) - \psi_H(z_*/L_*)], \tag{2.37}$$

where $t_* = - <w'T_v'> /\kappa u_*$. For a given wave characterized by its dimensionless phase speed c_*, we can solve for the dimensionless growth rate β of the energy of the waves, which is twice the growth rate of the amplitude of the waves

$$\beta = 2\gamma_a/\omega_0. \tag{2.38}$$

The quantities z_{0*} and L_* can be seen as parameters characterizing the mean wind speed, while z_{t*} and L_* characterize the effect of buoyancy $\rho_a'/\rho_a W_*^2$. Remarkably, with Charnock's relation (2.32), we have $z_{0*} = \alpha_{ch}$ which, for the moment, we regard as a constant independent of u_*. Therefore, the dimensionless growth β depends on the dimensionless phase speed c_* and two parameters L_* and z_{t*} which measure deviations from a logarithmic wind profile. For a neutrally stable atmosphere, $L_* \to \infty$, so that in that case the dimensionless growth is a function of c_* only, independent of u_*. In the presence of stratification, β then depends in addition on L_* and z_{t*}. However, it can be shown numerically that the dependence on z_{t*} is weak, so that there is effectively only one stability parameter L_*.

For neutrally stable flow the numerical results of the dimensionless growth rate $2\pi\beta = \gamma/f$ are plotted as a function of $c_*^{-1} = u_*/c$ in figure 2.3.

For comparison, we have also shown measurements of wave growth from the field (Snyder *et al*, 1981) and the laboratory (Plant and Wright, 1977). It is concluded that there is a fair agreement between Miles' theory and the observations. However, it should be remarked in this context that these results are sensitive to the choice of the Charnock constant α_{ch} since the dimensionless critical height is directly proportional to α_{ch}. This may be seen from the definition of critical height, $W_*(z_{*c}) = 0$, which gives, upon using the logarithmic wind profile,

$$z_{*c} \simeq z_{0*} \exp \kappa c_*. \tag{2.39}$$

Therefore the dimensionless growth rate β is inversely proportional to the Charnock constant α_{ch} because of the factor $W_c''/|W_c'|$. Estimating the Charnock constant from observations should be done with great care, as this constant depends exponentially on the drag coefficient C_D. Using again the

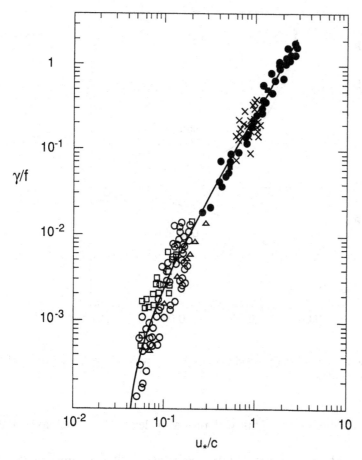

Fig. 2.3. Dimensionless wave growth vs dimensionless frequency; a comparison of quasi-laminar Miles' theory with observations.

logarithmic wind profile one has

$$\alpha_{ch} = \frac{g z_{obs}}{u_*^2} \exp\left(-\kappa/C_D^{\frac{1}{2}}\right). \qquad (2.40)$$

Here, z_{obs} is the height above the sea surface where the drag coefficient was measured. Now from the field data of Snyder *et al* (1981) one finds that on the average $C_D = 1.1 \times 10^{-3}$ and $U_5 = 7$ m/s. This gives a Charnock constant $\alpha_{ch} \simeq 0.004$. However, an error in the drag coefficient of 20% is certainly feasible. Increasing C_D by 20% gives an increase in α_{ch} of a factor of 3.5, giving a Charnock constant which is close to the one we have used here.

This sensitive dependence of the Charnock constant on the drag coefficient suggests that not only the mean value of C_D is needed in estimating α_{ch} but

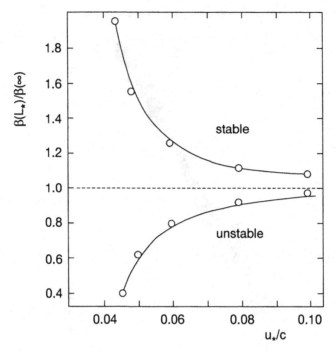

Fig. 2.4. Numerical results for the modification of the wave growth due to the effect of atmospheric stability.

also the standard deviation. This will probably lead to an increase in the estimate of α_{ch}.

Having discussed the neutrally stable case, we shall now investigate the effects of stratification. Intuitively, one would expect that an unstable stratification which, for fixed wind speed, corresponds to an increase in the friction velocity, would result in enhanced wave growth. However, things are a bit more complicated than that because in the case of stratification, the wind profile changes its shape and the equation for the wave-induced velocity is different. Instead of the Rayleigh equation one now has the Taylor-Goldstein equation.

The effects of stratification are displayed in figure 2.4, where we have plotted the ratio $\beta(L_*)/\beta(\infty)$ as a function of $c_*^{-1} = u_*/c$. Here, the curve labelled 'stable' has $L_* = 4193$, and the curve labelled 'unstable' has $L_* = -1770$. These values correspond to a wind speed of $U_{10} = 10$ m/s, and an air/sea temperature difference of $\pm 10\,°C$. The most dramatic effects are found for small values of u_*/c. The reason for this is simply that the critical Richardson number (see also equation (2.18) is increasing with increasing c/u_*, so that effects of stratification are expected to be noticeable for large dimensionless phase speed c_*.

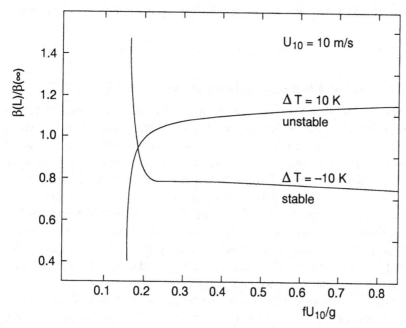

Fig. 2.5. As the previous figure, but now the growth ratio is given as a function of fU_{10}/g.

It came as a surprise that stable stratification gave, as a function of u_*/c, the highest growth rate. A simple explanation for this is hard to give. First of all we found that the expression for the growth rate by Miles (2.24) is a good approximation to the exact formula (2.25), even in the presence of stratification. Furthermore, one can show that, as a function of u_*/c, $W_c''/|W_c'|$ is not very sensitive to stability. The wave-induced vertical velocity on the other hand is rather sensitive and is largest in stable stratification.

It is concluded that, in general, the growth rate of the waves depends on two dimensionless parameters, namely c_* and L_*. In terms of u_*/c, stable stratification gives a higher growth rate. This conflicts, to be sure, with one's intuition that unstable air flow should enhance the growth rate. A plot such as figure 2.4 may be misleading, however, because in nature, one observes that for the same wind speed steeper waves are found under unstable circumstances than under stable circumstances. Let us, therefore, study figure 2.5, where we have plotted $\beta(L)/\beta(\infty)$ as a function of fU_{10}/g (that is frequency made dimensionless with the wind speed U_{10}). At sufficiently high frequencies where the effects of stratification are rather small, we now indeed find larger growth rates in unstable situations. Thus, the increase in friction velocity is only partly compensated for by the effects of finite

Obukhov length. At low frequencies, the buoyancy effects remain dominant, however.

II.2.4 Effects of gustiness on wave growth

In the previous subsection we have seen that within a factor of two there is a fair agreement between Miles' theory and the field observations of Snyder *et al* (1981). This agreement is rather surprising, however, as obviously turbulent fluctuations might upset the rather delicate resonant interaction between mean flow and surface waves. Let us take a first step in understanding the effect of turbulence on the growth rate of surface gravity waves generated by a turbulent wind. To that end Miles' theory of wind-wave generation is extended by allowing the mean velocity profile in air to be a slowly varying function of time. We consider fluctuations in the wind field with a time scale longer than $1/\omega$ (where ω is a typical frequency of the gravity waves). This part of the turbulent spectrum we call gustiness.

By means of an elaborate analysis it can be shown that Miles' expression for growth of waves by wind also holds, in a good approximation, for a slowly varying wind profile. Details of this analysis, which relies on the use of the multiple time scale method, are given in Janssen (1986). In the following we shall restrict ourselves to neutrally stable flow. The rate of change of the wave spectrum F due to wind is then

$$\frac{\partial}{\partial t}F(\omega) = \gamma F(\omega), \tag{2.41}$$

where we recall that γ, the growth rate of the energy, is twice the growth rate of the amplitude:

$$\gamma = 2\gamma_a = -\pi\epsilon c|\chi_c|^2 \frac{W_c''}{|W_c'|}. \tag{2.42}$$

We have seen that for neutrally stable flow the relative growth rate γ/ω is a function of u_*/c only, which was illustrated by figure 2.3. To quantify the effect of gustiness a relatively simple fit of the growth rate is needed, which is valid for ocean waves. To that end, we take the empirical fit proposed by Snyder *et al* (1981)

$$\frac{\gamma}{\omega} = \max\left\{0.2\epsilon\left(28\frac{u_*}{c} - 1\right), 0\right\}, \tag{2.43}$$

adapted by Komen *et al* (1984) to accommodate friction velocity scaling.

To investigate the effect of gustiness one may regard the friction velocity u_* as a stochastic variable with a steady part $<u_*>$ and a fluctuating part δu_*.

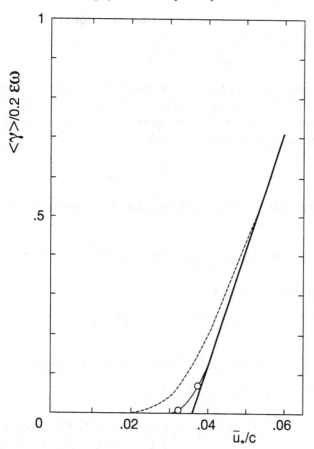

Fig. 2.6. Growth rates for different gustiness levels. The thin solid line corresponds to $\sigma_u/\bar{u}_* = 0.1$ and the dashed line was obtained for $\sigma_u/\bar{u}_* = 0.3$. The thick solid line describes wave growth in absence of wind gusts.

It is assumed that the statistics and the correlation time of the fluctuations are known. Thus,

$$u_* = \bar{u}_* + \delta u_*, \qquad (2.44)$$

so that

$$\gamma = \bar{\gamma} + b\gamma_1, \quad (b \ll 1), \qquad (2.45)$$

where the ensemble average $<b\gamma_1>$ is not zero because the growth rate γ is always positive. The growing waves act as a rectifier, as may be seen from (2.43) and for this reason gustiness will have a significant impact on wave growth.

Let us now consider the solution of the stochastic equation

$$\frac{\partial}{\partial t} F = (\bar{\gamma} + b\gamma_1)F. \tag{2.46}$$

When the correlation time τ_c of the random process $b\gamma_1$ is small, the evolution equation of the ensemble average of the wave spectrum, $<F>$, may be obtained immediately. Because growing waves act as a rectifier we already have a linear effect in b, thus, in lowest order we find

$$\frac{\partial <F>}{\partial t} = <\bar{\gamma} + b\gamma_1><F>. \tag{2.47}$$

Assuming Gaussian statistics for the fluctuation in the friction velocity, that is

$$p_{u_*}(x) = \frac{1}{\sigma_u\sqrt{2\pi}} \exp -\frac{(x - \bar{u}_*)^2}{2\sigma_u^2}, \tag{2.48}$$

the average growth rate becomes

$$\frac{<\bar{\gamma} + b\gamma_1>}{\epsilon\omega} = 5.6 \int_{X_0}^{\infty} dx p_{u_*}(x) \left[\frac{x}{c} - 0.036\right], \tag{2.49}$$

where $X_0 = 0.036c$. Evaluation of the integral gives the final result

$$\frac{<\bar{\gamma} + b\gamma_1>}{\epsilon\omega} = 5.6 \left[\frac{\sigma_u}{c\sqrt{2\pi}} \exp\left\{-\frac{1}{2\sigma_u^2}(0.036c - \bar{u}_*)^2\right\} + \right.$$
$$\left. \frac{1}{2}\left(\frac{\bar{u}_*}{c} - 0.036\right)\left(1 - \text{erf}\left(\frac{0.036c - \bar{u}_*}{\sigma_u\sqrt{2}}\right)\right)\right], \tag{2.50}$$

where

$$\text{erf}(z) = \frac{2}{\sqrt{\pi}} \int_0^z e^{-t^2} dt \tag{2.51}$$

is the error function. Equation (2.50) shows, that as expected, the effect of gustiness on wave growth is proportional to the variance σ_u of the friction velocity. It is, therefore, rather large, especially of course for the low frequency waves with $u_*/c = 0.036$. We have illustrated this in figure 2.6, where we have plotted the growth rate as a function of $\bar{u}_*/c$ for $\sigma_u/\bar{u}_* = 0.1$ and $\sigma_u/\bar{u}_* = 0.3$. It is concluded from this figure that especially in the later stages of wave growth, determined by the growth of low-frequency waves, the wave evolution will be affected by gustiness. This will be illustrated by numerical solution of the complete energy balance equation in § IV.5.2.

Finally we note that with this approach the mesoscale variability of the wind field can also be taken into account. Therefore, both temporal and spatial variability of the wind field may and should be treated, since the results may be affected significantly.

II.2.5 Effects of small-scale turbulence

Many authors have devoted their attention to the study of the effect of small-scale turbulence on wave growth. This line of research was started in the mid-1970s by Chalikov (1976) and Gent and Taylor (1976), when it was felt that the approach of Miles (1957) was inadequate because the effect of eddies on the wave-induced motion was disregarded. Chalikov and Makin produced a series of papers in which they address the growth of a single wave, the structure of air flow over a spectrum of waves and effects of atmospheric stability. Their work culminated in a determination of the drag coefficient over sea waves (Chalikov and Makin, 1991). Although more sophisticated turbulence closures were considered, they concluded that a simple mixing length model gave sufficiently reliable results.

More complicated turbulence models were advocated by Gent and Taylor (1976) and Al-Zanaidi and Hui (1984). The former used a one-equation model whereas the latter used a two-equation model of Saffman and Wilcox (1974), assuming, as always, a nonseparated boundary layer. Results, however, do not seem to differ much from the Russian group so their conclusion on the choice of turbulence model may be justified.

Thus far we have discussed numerical calculations of the growth rate of surface gravity waves. Using a simple eddy-viscosity model Jacobs (1987) obtained by means of matched asymptotic expansions using a drag coefficient as a small parameter a very elegant expression for the growth rate.

Van Duin and Janssen (1992) realized that Jacobs' method was not quite consistent. They noted that at least three layers (a viscous layer, an intermediate layer and an outer layer) were needed to obtain a valid asymptotic expansion. The end result was, however, identical to the one obtained by Jacobs.

The governing equations for air were taken as

$$\frac{\partial}{\partial t}\mathbf{u} + \mathbf{u} \cdot \nabla\mathbf{u} = -\frac{1}{\rho_a}\nabla p + \nabla \cdot v\left\{\nabla\mathbf{u} + (\nabla\mathbf{u})^T\right\}, \qquad (2.52)$$

where all symbols have been defined before, except for the superscript T which denotes transpose and v which is the kinematic viscosity.

The combined effects of molecular viscosity and turbulence are taken into account by assuming that the kinematic viscosity is given by

$$v = v_a + v_e, \qquad (2.53)$$

where v_a is the constant molecular viscosity and v_e is the eddy viscosity which may be space- and time-dependent. The turbulence was modelled using the eddy viscosity

$$v_e = \ell u_*, \tag{2.54}$$

where

$$\ell = \kappa(z - \eta), \tag{2.55}$$

with η the displacement of the air–water interface.

After a lengthy analysis, details of which can be found in van Duin and Janssen (1992), the growth rate of the waves is found to depend on the wind speed at height $1/k$ (see also Al-Zanaidi and Hui, 1984). In terms of the friction velocity one obtains

$$\frac{\gamma}{\omega} = 2\kappa\epsilon \left[C_D^{-\frac{1}{2}}(k) \left(\frac{u_*}{c}\right)^2 \cos\theta - \frac{u_*}{c} \right], \tag{2.56}$$

where θ is the angle between the wind and wave directions and $C_D(k)$ is the drag coefficient at height $1/k$. Using the Charnock relation (2.32) one has

$$C_D(k) = \left(\frac{\kappa}{\log(c^2/\alpha_{ch}u_*^2)} \right)^2. \tag{2.57}$$

Although (2.56) has been derived under a number of restrictive assumptions from an asymptotic analysis, the result itself, especially the functional form, is very useful. First of all, (2.56) is very similar to an expression for the growth rate suggested by Stewart (1974) on semi-intuitive grounds. Secondly, (2.56) may be used as the basis of a fit to the numerical results of Chalikov and Makin (1991); according to Burgers (private communication) a very good agreement is obtained. It is therefore concluded that (2.56) captures the essentials of the growth rate of waves by wind from models that include effects of small scale turbulence. For high-frequency waves, this expression shows, apart from a logarithmic dependence, the same scaling with friction velocity as found in Miles' theory. However, the present results differ significantly for the low-frequency waves ($c/u_* > C_D^{-1/2}(k)$, for $\theta = 0$) as these waves are damped while the Miles' theory always gives a positive growth rate. A similar conclusion holds true for waves travelling against the wind.

The present result for the growth rate, especially regarding the damping of the low-frequency waves, would give a completely different impact of gustiness on wave growth. Now, the wave system does not act as a rectifier hence the effect of gusts is only second order. It is therefore important to discuss the validity of mixing length modelling.

The turbulence model is valid if a typical eddy turnover time (such as the time for the turbulence to respond to a change in forcing) is much smaller than the period of the waves. In that event, changes induced by the water surface in the air flow are slow and modelling of the momentum transport by means of a model for a stationary turbulent flow seems justified. In other words, with ω the typical angular frequency of the wave and T_t the time scale of the turbulent flow, we have the condition

$$\omega T_t << 1. \tag{2.58}$$

At first sight one might think that this relation is best satisfied by long waves. However, it should be realized that the scale of the relevant eddies is determined by the height of the critical layer, which is strongly frequency-dependent. In fact, we take for the eddy turnover time $T_t = L_t/u_*$, where L_t is chosen to be equal to the critical height z_c (defined by $U(z_c) = c$) since in inviscid theory the critical layer plays a major role in the energy transfer from air to water. Using the usual logarithmic wind profile with Charnock's relation (2.32) for the roughness the above relation boils down to the following restriction on the ratio of phase speed to friction velocity

$$\frac{c}{u_*} << \frac{1}{\kappa} \log \left(\frac{25}{\alpha_{ch}} \right). \tag{2.59}$$

In other words, the turbulence model is only valid for high-frequency waves. Since according to (2.56) wave damping occurs for $c/u_* > C_D^{-1/2}(k)$ and since $C_D^{-1/2}(k)$ is usually larger than the right hand side of the condition (2.59) the result that, according to this theory, low-frequency waves are damped, is, to say the least, somewhat suspect.

Belcher and Hunt (1993) have made an attempt to infer the consequences of the limitations of mixing length modelling for wave growth. This followed earlier work on the flow over hills (Belcher *et al*, 1993). They argue that far away from the water surface, turbulence is slow with respect to the waves, so that large eddies do not have sufficient time to transport momentum. This then results in a truncation of the mixing length ℓ (2.55), because the large eddies are not effective in momentum transport on the time scale of the wave. The growth rate was determined for slowly moving waves and although from the asymptotic expansion it follows that their result is formally smaller than (2.56) it turns out that for practical values of the parameters the growth rate of Belcher and Hunt (1993) is somewhat larger. In the context of the truncated mixing length model of Belcher and Hunt it would be of interest to evaluate the damping rate of waves that propagate faster than the wind. Because of the truncation in mixing length one would expect a reduction of the damping rate compared to the result from the usual mixing length model (2.56). Unfortunately, the analysis of Belcher and Hunt is not valid for fast propagating waves.

Is there observational evidence of the damping of waves that propagate faster than the wind? The Bight of Abaco results, which are by now regarded as a classical work on the measurement of wind input, report insignificant damping rates in these circumstances. What about evidence for wave damping in an adverse wind? Observational evidence on this subject

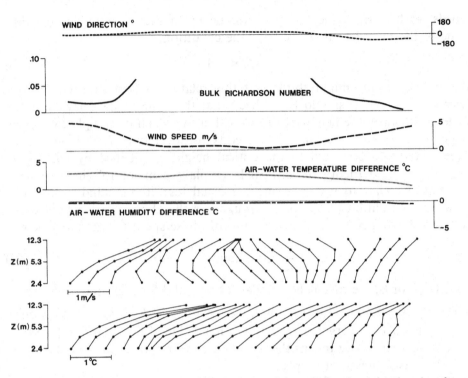

Fig. 2.7. Wind and temperature profiles measured over Lake Ontario showing formation and decay of a 'wave-driven wind'. The profiles are running averages over 30 min plotted at 10 min intervals. The humidity difference is expressed in buoyancy equivalent degrees Celcius. The time series of wind speed and air–water differences are obtained from measurements at the top level and the surface water temperature. (From Holland 1981.)

is scant. Dobson (1971) observed the pressure on the surface of a group of waves advancing against a light wind, and found damping coefficients about an order of magnitude smaller than growth rates under corresponding conditions. Stewart and Teague (1980) observed the decay of swell with decameter radar and found rates of about 15% of the corresponding growth rates. Their measurements do not yield the wind input directly, but rather the overall balance including nonlinear interactions and dissipation.

The laboratory measurements of Young and Sobey (1985) showed no significant wave surface pressure–slope correlation in an adverse wind and the weak attenuation they found was attributed to the wave-coherent tangential stress. On the other hand, Mizuno (1976) and Donelan (1983) from similar laboratory measurements found damping rates that were comparable with corresponding growth rates. A related question concerns the propagation of waves into an area of calm or light wind. The direct loss of momentum from the waves through interaction with the air above must, inevitably, lead to acceleration of the air flow in the direction of the waves. Such a 'wave-driven wind' has been reported by Harris (1966), Davidson and Frank (1973) and Holland (1981) *inter alia*. The observations from Holland (1981) in Lake Ontario are illustrated by figure 2.7. After the wind at 12 m height dies down, the increase of the wind near the surface is apparent, whereas the temperature profile suffers no similar change. It would appear that the waves generated by the previous 6 m/s wind impart momentum to the relatively quiet air surface layer. It is not clear, however, whether this momentum transfer from water to air is caused by a damping of the water waves. Currents caused by wave-induced motions and other causes may give rise to a tangential stress on the air resulting in wind as well.

Thus, in conclusion, one could summarize that theories including small-scale turbulence predict damping when waves propagate faster than the wind. However, these waves have a high critical layer so that relevant eddies are large and have large time scales. As a result (2.58) is violated and the theory is really outside its range of applicability. Observational evidence for wave damping under these conditions is inconclusive.

Evidence of damping of waves travelling against the wind is also conflicting, although Holland's results seem convincing and intriguing indeed. Perhaps one is dealing here with completely different circumstances as for waves propagating faster than the wind there may be a critical layer, while in the case of an adverse wind, this is obviously not so. The turbulence model discussed here may, therefore, be valid in cases of adverse wind.

It is not clear whether or not this effect is important for ocean wave prediction. More research into this matter is, however, clearly desirable.

II.2.6 Quasi-linear theory of wind-wave generation

In the previous subsections we have explored some of the linear theories of wind-wave generation. Normally, linearization is a good approximation in ocean wave dynamics. In case of Miles' theory we face, however, a complication because the dynamics inside the critical layer is not described by linear theory. This is evident from the delta-function behaviour of the vortex force (see the discussion on page 81) which implies that in linear theory the critical layer is regarded to be infinitely thin, although in practice the width of this layer will be finite. Inside the critical layer a pattern of closed streamlines will form (Kelvin's cat's eyes, cf. Lighthill, 1962 or Phillips, 1977). Neglecting turbulence for the moment, the closed streamline pattern will give rise to a smoothing of the vorticity distribution in the critical layer in such a way that for large times the curvature of the wind profile at the critical height vanishes. Hence, stabilisation of a single wave in an inviscid fluid would result and the transfer of momentum from air flow to the wave could be quenched. It is clear from this example of the evolution of a single wave in an inviscid fluid that it is important to investigate the possible consequences of wave–mean-flow interaction on the evolution of surface waves and in this section we present some of the results for a spectrum of surface gravity waves. We emphasize, however, that for a sufficiently broad wave spectrum wave–mean-flow interaction is of a different nature because we are dealing now with a continuum of critical layers with random phase. Thus, neighbouring critical layers will counteract each other so there will be a considerable reduction of the impact of critical layer dynamics on the evolution of the mean wind. The resulting effect will be determined in this section. With a spectrum of waves the curvature of the wind profile is also expected to vanish, but now in a layer above the surface. As a result the critical layers move up, which resolves the geometrical problem mentioned on page 73.

Clearly, gravity waves receive energy and momentum from the air flow and one should expect that this results in a slowing down of the air flow. In other words, surface gravity waves and their associated momentum flux may be important in controlling the shape of the wind profile over the oceans. In fact, the common belief in the field (Phillips, 1977) was that air turbulence was dominant in shaping the wind profile and the effect of surface gravity waves was considered to be small. However, Snyder *et al* (1981) found that the momentum transfer from wind to waves might be considerable so that the related wave-induced stress may be a substantial fraction of the total stress in the surface layer. This suggested that the velocity profile over sea waves may deviate from the usual profile of turbulent air flow over a flat plate. The consequence is that the drag coefficient at 10 m height should depend on the sea state. Experimental evidence for this was found by Donelan (1982) and later confirmed by Maat *et al* (1991) during the HEXOS (humidity exchange over sea) experiment.

This means that a theory has to be developed that takes the consequences of the growing waves on the mean flow into account. This theory was independently obtained by Fabrikant (1976) and Janssen (1982). These authors utilized an analogy that exists between resonant wave–mean-flow interaction in a fluid and the interaction of plasma waves and particles.

The linear interaction of plasma waves with particles was first investigated by Landau (1946). The rate of change of the energy of the particles was found to be proportional to the derivative of the particle-velocity distribution at the point of resonance, that is where the particle velocity equals the phase velocity of the plasma wave. For a plane parallel flow this change of energy is, as we have seen, proportional to the derivative of the vorticity at the critical height. Also, in a plane parallel flow, a pattern of closed streamlines is found near the point of resonance, similar to Kelvin's 'cat's-eye' pattern, and the same feature is found in the phase-space orbits of trapped particles in a given monochromatic plasma wave.

The linear theories of resonant interaction of gravity waves with a flow and plasma waves with particles are only valid on a short time scale, since in the course of time, nonlinear effects may become important, due to the growth of the waves. The study of nonlinear effects on the interaction of plasma waves and particles was started by Vedenov *et al* (1961) and Drummond and Pines (1962). One of the main results of these investigations was that plasma waves modify the particle distribution function, which may give rise to saturation of the plasma waves. Later these results were extended to include three-wave interactions and nonlinear wave–particle interactions (see also Davidson, 1972). The plasma waves were assumed to have a sufficiently broad spectrum such that the random phase approximation applies. (Incidentally, in the field of surface gravity waves, development of the nonlinear theory took place in reversed order. Phillips (1960) and Hasselmann (1960, 1962) initiated the theory of resonant four-wave interactions, while much later the possible relevance of wave–mean-flow interactions was recognized by Fabrikant and by Janssen.)

Here, we are concerned with the effects of water waves on the mean flow. A multiple time scale technique is used to obtain dynamical equations for the slowly varying energy density of the water waves and the wind velocity. The growth of the water waves due to atmospheric input occurs on a long time scale since this energy transfer is proportional to the ratio of air density to water density. Hence, we have at least two time scales, namely one related to the relatively rapid water wave oscillations and one of the order of the energy transfer time from air to the water waves. Another reason for the use of the multiple time scale method is that an iterative solution of a set of nonlinear equations (in this case the Euler equations plus boundary conditions) usually gives rise to secular terms in time. The introduction of different time scales then provides freedom to prevent secularity. The condition resulting from the elimination of secularity gives the equations for the slow time dependence of the wave energy and the air speed. In addition, we are concerned with a statistical description of the interaction of water waves and air, that is we consider the evolution in time of ensemble averages of quantities like the energy density. To this end, the nonlinear set of equation (2.1) is solved iteratively by means of a systematic expansion of the relevant quantities in powers of a small parameter. Finally, the appropriate averages are taken to obtain equations for the averaged quantities. Thus we consider a weakly nonlinear system for which the random-phase approximation is assumed to be valid (Hasselmann, 1967, Davidson, 1972).

We anticipate that there are at least two time scales determined by the ratio of air to water density ϵ. In order to proceed we also have to choose the order of magnitude of the amplitude of the oscillations. We will choose here $\epsilon^{\frac{1}{2}}$. The argument for this choice is as follows. Stewart (1967) observed that a substantial amount of energy is contained in the water waves. From this it may be inferred that

$$\rho_w <w^2> = \mathcal{O}(\rho_a U_{10}^2). \tag{2.60}$$

Here, the angle brackets denote as usual an ensemble average. Consequently, $<w^2> = \mathcal{O}(\epsilon U_0^2)$. It is, therefore, tempting to expand the elevation, the velocity and the pressure in powers of $\epsilon^{\frac{1}{2}}$. Thus,

$$\eta = \sum_{\ell=1} \epsilon^{\ell/2} \eta_\ell, \qquad \mathbf{u}_w = \sum_{\ell=1} \epsilon^{\ell/2} \mathbf{u}_{w,\ell}, \qquad p_w = \sum_{\ell=0} \epsilon^{\ell/2} p_{w,\ell},$$

$$\mathbf{u}_a = \mathbf{U} + \sum_{\ell=1} \epsilon^{\ell/2} \mathbf{u}_{a,\ell}, \qquad\qquad p_a = \epsilon \sum_{\ell=0} \epsilon^{\ell/2} p_{a,\ell}. \tag{2.61}$$

We remark that the series for the air pressure starts with a term $\mathcal{O}(\epsilon^2)$ since

$$p_a = \mathcal{O}(\rho_a U_a^2). \tag{2.62}$$

A straightforward iterative solution of a set of equations may, however, give rise to secular terms (in time, for example) in the series solution (Davidson, 1972). For this reason we introduce different time scales such that there is sufficient freedom to remove secularity. To that end it is sufficient to assume that average quantities such as $<w^2>$, depend on $\tau_0 = t, \tau_1 = \epsilon t, \ldots$. Hence,

$$\frac{\partial}{\partial t} <w^2> = \sum_{\ell=0} \epsilon^\ell \frac{\partial}{\partial \tau_\ell} <w^2> . \tag{2.63}$$

The τ_0 scale takes account of the relatively rapid wave oscillation, while growth of the waves due to atmospheric input occurs on the τ_1 scale, since this energy input is proportional to ϵ. It is the condition resulting from the elimination of secular behaviour on the short time scale τ_0 that gives us the slow time dependence of the wave energy density and the mean air speed U_0.

We limit the discussion on the effect of gravity waves on the mean flow to the case of constant density in air and water. Extension to the case of stratification in the atmosphere is, in principle, feasible, but has not yet been done.

Neglecting currents in the water, the water motion is assumed to be irrotational. For simplicity, we take infinite water depth so that the vertical velocity vanishes at $z \to -\infty$. Details of how to obtain the perturbation solution may be found in Janssen(1982). Here, we only briefly describe the main result. In the context of our assumptions we have energy and momentum transfer from the mean air flow to the surface gravity waves because of Miles' shear-flow instability. As the waves are growing an additional force is exerted on the mean flow which results in a deceleration of the mean flow. Therefore, considering only the effects of the waves for the moment, the equation for the mean flow U_0 becomes

$$\frac{\partial}{\partial t} U_0 = -\frac{\partial}{\partial z} <\delta u \, \delta w>, \qquad (2.64)$$

where $<\delta u \delta w>$ denotes the wave-induced stress and $\delta \mathbf{u}$ denotes the fluctuating part of the series for $\mathbf{u}_a$ given in equation (2.61). The point now is that in contrast with the usual fluid mechanics turbulence problem, an explicit expression for the wave-induced stress may be given, because we are dealing with dispersive waves (hence $c \neq c_g$). The result is an equation for U_0 of the diffusion type:

$$\frac{\partial}{\partial t} U_0 = D_W \frac{\partial^2}{\partial z^2} U_0, \qquad (2.65)$$

where the wave diffusion coefficient D_W is proportional to the surface elevation spectrum $F(k)$,

$$D_W = \frac{\pi c^2 k^2 |\chi|^2}{|c - c_g|} \frac{F(k)}{\rho_w g}, \qquad (2.66)$$

where F is the energy spectrum and the wavenumber k has to be expressed as a function of height through the resonance condition $W = 0$, c_g is the group velocity $\partial \omega / \partial k$ and χ is the normalized wave-induced vertical velocity, satisfying the Rayleigh equation. Equation (2.65) tells us that the air flow at a certain height z changes with time owing to resonant interactions of a water wave with frequency $\sigma = g/U(z)$. Hence, in this fashion there is possibly an energy transfer from the air flow U to the water waves, thus giving a rate of change of the spectrum given by equation (2.41).

To summarize our results, we obtain the following set of quasi-linear equations for the generation of water waves by wind:

$$\frac{\partial}{\partial t} F = -\pi \epsilon c |\chi_c|^2 \frac{W_c''}{|W_c'|} F,$$

$$\frac{\partial}{\partial t} U_0 = D_W \frac{\partial^2}{\partial z^2} U_0, \quad D_W = \frac{\pi c^2 k^2 |\chi|^2}{|c - c_g|} \frac{F}{\rho_w g}, \qquad (2.67)$$

$$W \triangle \chi = W'' \chi, \quad \chi(0) = 1, \quad \chi(\infty) = 0.$$

From the first equation of (2.67) we obtain the well-known result that only those waves are unstable for which the curvature U_0'' of the wind profile at the critical height is negative. The growth rate of the waves is, however, a function of time, as the wind profile depends on time according to the diffusion equation for U_0, possibly quenching the instability for large t.

It should be noted that Lighthill (1962), who discussed the physical interpretation of Miles' theory of wave generation by wind, obtained a similar result regarding the effect of a single wave on the wind profile. He did not realize, however, that the wind profile U_0 may be a slowly varying function of time. In addition, Fabrikant obtained a similar set of equations, although along different lines. Here we once again emphasize that, by means of the multiple time scale technique, equations for the slowly varying quantities F and U_0 are obtained from the requirement that there be no secularity of the first order quantities (such as u_2) on the fast time scale τ_0. This renders the series solution, given in (2.61), uniformly valid up to $\tau = \mathcal{O}(\epsilon^{-1})$.

The objection may be raised that the effect of Reynolds stresses on the wind profile and the growth of the waves has not been included. However, it is in the spirit of Miles' theory to consider only the effect of turbulence on the mean flow, disregarding possible interaction between the wave-induced oscillations and turbulence. Thus the mean flow equation becomes

$$\frac{\partial}{\partial t} U_0 = D_W \frac{\partial^2}{\partial z^2} U_0 + \frac{\partial}{\partial z} v_e(z) \frac{\partial}{\partial z} U_0, \qquad (2.68)$$

where D_W is the wave diffusion coefficient, while $v_e(z)$ is the eddy viscosity of the air. We emphasize that there is no real justification for (2.68). However, as we will see in a while, results compare well with quasi-linear models including the effects of turbulence on the mean flow and the wave-induced oscillations.

The effect of eddy viscosity is clear. If no water waves are present the well-known logarithmic wind profile is obtained in the steady state as $v_e \sim z$. In the presence of waves eddy viscosity is capable of maintaining the logarithmic wind profile if $D_W \ll v_e$. If, however, the waves are steep enough the effect of the waves on the wind profile may overcome eddy viscosity, especially in the layer just above the water waves.

In order to appreciate the effect of waves on the wind let us first disregard the effect of turbulence altogether, and investigate some properties of the quasi-linear theory of wind-wave generation. First of all we question whether or not the set of equations (2.67) admits a steady state. To that end, we derive an equation for the enstrophy of the mean flow. We differentiate the diffusion equation for U_0 with respect to height z to obtain

$$\frac{\partial}{\partial t} \left(\frac{\partial}{\partial z} U_0 \right) = \frac{\partial}{\partial z} \left[D_W \frac{\partial}{\partial z} \left(\frac{\partial}{\partial z} U_0 \right) \right], \qquad (2.69)$$

where we note that D_W is always positive. We next multiply (2.69) by $\partial U_0 / \partial z$ and integrate over z with the result

$$\frac{\mathrm{d}}{\mathrm{d}t} \int_0^\infty \mathrm{d}z \left(\frac{\partial U_0}{\partial z} \right)^2 = 2 \int_0^\infty \mathrm{d}z \frac{\partial}{\partial z} U_0 \frac{\partial}{\partial z} \left[D_W \frac{\partial}{\partial z} \left(\frac{\partial}{\partial z} U_0 \right) \right]$$

$$= 2 \int_0^\infty \mathrm{d}z \left[\frac{\partial}{\partial z} \left(D_W \frac{\partial}{\partial z} U_0 \frac{\partial^2}{\partial z^2} U_0 \right) - D_W \left(\frac{\partial^2 U_0}{\partial z^2} \right)^2 \right]. \qquad (2.70)$$

For appropriately chosen boundary conditions and by realizing that the wave diffusion coefficient should disappear at 0 and ∞, the perfect derivative on the right above integrates to zero, hence

$$\frac{d}{dt}\frac{1}{2}\int dz \left(\frac{\partial U_0}{\partial z}\right)^2 = -\int_0^\infty dz D_W \left(\frac{\partial^2 U_0}{\partial z^2}\right)^2. \qquad (2.71)$$

This equation states that the time derivative of the enstrophy of the mean flow, which is bounded from below by zero, is nonpositive. Hence, the mean air flow tends toward a condition where the right hand side disappears, which requires in the region where $D_W \neq 0$ that

$$\frac{\partial^2}{\partial z^2} U_0 = 0 \quad \text{as} \quad t \to \infty. \qquad (2.72)$$

Thus, for large times the wind profile becomes linear, implying that according to (2.67) the growth rate of the waves vanishes. Apparently, quasi-linear theory predicts a limitation of the amplitude of the initial unstable water waves for large times, that is the energy transfer from the air flow to the water waves is quenched. In practice, this effect is expected to occur for the high-frequency waves which have their critical layer just above the water surface. In passing it should be pointed out that (2.72) holds for a whole layer above the waves. As a consequence the critical layers move up. This is discussed in more detail in Janssen (1982).

We emphasize that the phrase 'large times' has a relative meaning. Large times means large compared to the period of the waves of concern. As the quenching of the Miles' instability is expected to occur primarily for the high-frequency waves, large times means, in practice, short compared to the typical evolution time of wind sea as this time scale is determined by the low-frequency waves.

Secondly, we note that the set of quasi-linear equations (2.67) admits an infinite set of balance equations, notably

$$\frac{d}{dt}\left[\rho_a \int_0^\infty f(U_0)dz + \int_0^\infty f'(U_0)\frac{F}{c}dk\right] = 0, \qquad (2.73)$$

where $f(U_0)$ is an arbitrary function of U_0, the prime denotes differentiation with respect to U_0 and in the second term on the left hand side $U_0 = c$. Equation (2.73) may be obtained by multiplication of the diffusion equation for U_0 by $f'(U_0)$; then integration with respect to z gives

$$\frac{d}{dt}\int_0^\infty f(U_0)dz = \int_0^\infty f'(U_0)D_W \frac{\partial^2}{\partial z^2}U_0 dz. \qquad (2.74)$$

In the integral on the right hand side we next convert to an integration over k via the resonance condition $U_0 = c(k)$, then using the expression for D_W

and the evolution equation for the wave spectrum, we finally arrive at the conservation law (2.73).

By an appropriate choice of $f'(U_0)$ we are able to express all moments of the spectrum F in terms of an integral of $f(U_0)$ in k-space. In particular, we obtain for $f'(U_0) = 1$, conservation of momentum,

$$\frac{\mathrm{d}}{\mathrm{d}t} \left[\rho_a \int_0^\infty U_0 \mathrm{d}z + \int_0^\infty \frac{F}{c} \mathrm{d}k \right] = 0 , \qquad (2.75)$$

and for $f'(U_0) = U_0$ conservation of mechanical energy,

$$\frac{\mathrm{d}}{\mathrm{d}t} \left[\frac{1}{2} \rho_a \int_0^\infty U_0^2 \mathrm{d}z + \int_0^\infty F \mathrm{d}k \right] = 0. \qquad (2.76)$$

From the conservation laws we see once more that for growing waves the wind profile changes in time.

The property of having an infinite set of conservation laws suggests that the quasi-linear set of equations admits exact solutions. This turns out to be not quite the case, but the asymptotic form of the wave spectrum may be obtained (Janssen, 1982). As a result the high-frequency part of the spectrum has an f^{-4} power law. We shall not give the details of this calculation as the quasi-linear model (2.67) only represents one aspect of the physics of wave evolution. Processes such as wave breaking and nonlinear interaction are relevant too, and it is the subtle balance between these three processes which will determine the spectral shape. However, the process of quenching of the Miles' instability is most certainly relevant, as will be seen in the next subsection. In general, the growth rate of the waves and the related momentum transfer from air to waves will be sea-state-dependent. As a consequence, the drag of air flow over sea waves will depend on the state of the waves.

II.2.7 Wave-induced stress and the drag of air flow over sea waves

In the previous subsection we have seen that the growth of waves depends on the sea state as the wind profile is affected by the presence of the surface gravity waves. Thus, the wind profile might be different compared with that over a flat plate. In this subsection we shall ask ourselves the question to what extent the wind profile will differ.

To be sure, observations from the field (Donelan, 1982, Maat *et al*, 1991) do indicate that the roughness length and the drag coefficient depend on the sea state. For wind sea a convenient measure of the stage of development of the waves is the so-called wave age c_p/u_*, where c_p is the phase speed of the peak of the spectrum and u_* the friction velocity. From Hasselmann *et al* (1973) it became evident that the wave spectrum for growing wind seas could be parametrized in terms of a single parameter, for example dimensionless fetch or dimensionless time. Here, we prefer to use the wave age, as with this parameter both fetch-limited and duration-limited situations can be dealt with. Typically, 'young' wind sea corresponds to a wave age of the order of 5 – 10, while 'old' wind sea has a wave age of the order of 25. Thus we shall study the dependence of, for example, the drag of air flow over sea waves on the wave age.

In order to have some idea about the order of magnitude of the effect of waves on the wind we shall determine the wave-induced stress using a simple model for the growth of waves by wind and compare the result with the total stress in the surface layer.

From the momentum balance (2.75) we infer that the wave momentum spectrum P is given by

$$P = \frac{F}{c} \tag{2.77}$$

and since the wave stress is given by the rate of change in time of wave momentum due to wind, we have

$$\tau_w = \int dk \frac{\partial}{\partial t} P \bigg|_{wind}. \tag{2.78}$$

We are mainly interested in the contribution to the stress of the low-frequency waves, as the effects of high-frequency gravity-capillary waves on the air flow will be modelled by a roughness length. For the rate of change of the energy spectrum due to wind, one can therefore use the empirical expressions (2.41) and (2.43). Since we are only interested in orders of magnitude at the moment, we shall use a very simple spectral shape

$$F(k) = \frac{1}{2} \rho_w g \alpha_p k^{-3}, \tag{2.79}$$

where α_p is the so-called Phillips constant. The k^{-3} power law was first proposed by Phillips (1958) and is based on wave breaking being the dominant limiting mechanism for the wave spectrum. We will see in a moment that the so-called Phillips constant is not really constant, but depends on wave age.

Substitution of (2.77) and (2.79) into (2.78) results in the following expression for the ratio of wave-induced stress τ_w to the total stress $\tau = \rho_a u_*^2$

$$\frac{\tau_w}{\tau} = \mu \alpha_p \left\{ 28 \left(\frac{c_p}{u_*} \right) - \frac{1}{2} \left(\frac{c_p}{u_*} \right)^2 \right\}. \tag{2.80}$$

This result holds for wind sea ($c_p/u_* < 28$).

The order of magnitude of the ratio of wave-induced stress to turbulent stress is then found to vary between 0.20 and 1 or even larger. However, this estimate of τ_w depends largely on the wave age dependence of the Phillips

constant. Following Snyder (1974) one would infer from the standard
JONSWAP fetch laws the relation

$$\alpha_p \sim (c_p/u_*)^{-2/3}. \tag{2.81}$$

It results in a wave-induced stress that is virtually independent of the waves,
as is shown in figure 2.8. As noted by Snyder (1974) and Hasselmann *et
al* (1973), the power law (2.81), which attempts to cover both laboratory data
and field data, does not fit the field data particularly well. In fact, there is
no reason to assume that both field and laboratory data may be resolved by
a single power law. Donelan *et al* (1985) show quite convincingly that field
and laboratory measurements belong to different families. A power law that
fits the JONSWAP data (and also the KNMI data, see Janssen *et al*, 1984) well
shows a much stronger dependence of α_p on wave age. It is given by

$$\alpha_p = 0.57(c_p/u_*)^{-3/2}. \tag{2.82}$$

The result is a much stronger wave age dependence of the wave-induced
stress, (see figure 2.8). We favour the power law (2.82) because, on the one
hand, (2.82) is in better agreement with the field data and, on the other
hand, the wave-induced stress is large for young wind sea $(c_p/u_* = 5)$ and
small for old wind sea $(c_p/u_* \simeq 30)$. The latter is in agreement with one's
intuition that air flow over young wind sea is rougher than over old wind
sea. Convincing experimental evidence for this may be found in the Lake
Ontario data of Donelan (1982), which were obtained under fetch-limited
conditions. These results are plotted in figure 2.9, which shows the drag
coefficient as a function of wind speed using the wave age as a label. Clearly,
the figure shows that air flow over old wind sea has a lower drag than over
young wind sea. For fixed wind speed, the drag coefficient varies due to the
wave age dependence by a factor of two.

It is conjectured that this variation of the drag coefficient by a factor of
two can only be explained if the wave stress has sufficiently sensitive wave
age dependence. Returning to figure 2.8, this again supports our choice of
the power law (2.82) for α_p.

On the other hand, figure 2.8 suggests several problems for young wind
sea as the wave stress could be larger than the total stress. In the steady state
one would expect that the total stress in the surface layer must be at least as
large as the wave stress since air viscosity will also diffuse momentum. With
the results of the previous subsection in mind we now realize the importance
of the quasi-linear effect. Namely, it enables us to obtain a consistent picture
of the air momentum balance over sea waves. For young wind sea, having
a large wave steepness, the wave diffusion coefficient D_W will be significant
resulting in a considerable reduction of the curvature in the wind profile.
Hence, the growth rate of the waves will be reduced in such a way that for

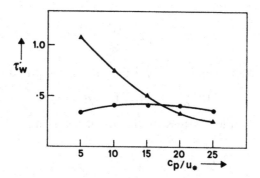

Fig. 2.8. Wave age dependence of wave-induced stress, normalized with stress for two different parametrizations of the Phillips' constant.

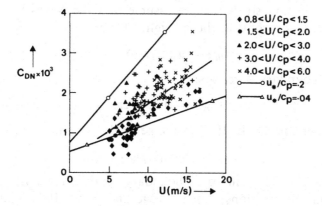

Fig. 2.9. The observed aerodynamic drag over surface waves as function of the wind speed at 10 meters, for different wave ages (after Donelan, 1982).

a steady state wind the wave stress is less than the total stress in the surface layer.

Let us therefore consider again the quasi-linear set of equations (2.67), including the effect of turbulence on the mean flow (2.68). For ease of reference, we reproduce these equations here:

$$\frac{\partial}{\partial t} U_0 = v \frac{\partial^2}{\partial z^2} U_0 + \frac{1}{\rho_a} \frac{\partial}{\partial z} \tau_{turb}, \quad v = v_a + D_W, \tag{2.83}$$

where

$$D_W = \frac{\pi c^2 k^2 |\chi|^2}{|c - c_g|} \frac{F}{\rho_w g}, \tag{2.84}$$

and

$$\left. \frac{\partial}{\partial t} F \right|_{wind} = \gamma F, \tag{2.85}$$

where

$$\gamma = -\pi \epsilon c |\chi_c|^2 \frac{W''}{|W_c'|}. \tag{2.86}$$

The wave-induced vertical velocity χ satisfies the Rayleigh equation

$$W \triangle \chi = W'' \chi, \quad \chi(0) = 1, \quad \chi(\infty) = 0. \tag{2.87}$$

Here, we model the turbulent stress by means of a mixing length model

$$\tau_{turb} = \rho_a \ell^2 |\frac{\partial}{\partial z} U_0| \frac{\partial}{\partial z} U_0, \tag{2.88}$$

with mixing length given by $\ell = \kappa z$ (κ is the von Kármán constant).

In order to be able to solve our problem we still have to specify two boundary conditions for the mean flow. For large heights we impose the condition of constant stress and we assume that the waves have no direct impact on the wind profile at those heights, hence

$$\tau_{turb} = \rho_a u_*^2 , z \to \infty. \tag{2.89}$$

At the lower boundary we choose

$$U_0(z) = 0 , z = z_0, \tag{2.90}$$

where following Charnock (1955), we take as the roughness length

$$z_0 = \alpha_{ch} u_*^2 / g, \tag{2.91}$$

with $\alpha_{ch} = 0.0144$. This choice of boundary condition requires some justification. From observations at sea (Large and Pond, 1982, Garratt, 1977, Smith, 1980 and Wu, 1982) it is known that even for old wind seas the aerodynamic drag is larger than the drag over a flat plate. This increase in drag must for the greater part be due to the momentum loss to the gravity-capillary waves as the longer waves have such small steepnesses (because the Phillips constant α_p is small for old wind sea) that their wave stress may be neglected. It is, therefore, important to take the momentum loss for the short waves into account. However, as no theory on the wave-induced stress of gravity-capillary waves is available, we have to rely on a parametrization. Since the phase speed of these short gravity-capillary waves is much smaller than, say, the wind speed at 10 m height, the air flow encounters a water surface with more or less stationary perturbations, that is the air flow 'feels' a water surface with a certain roughness. The choice of roughness length (2.91) is in agreement with the increase of drag coefficient with wind speed (Wu, 1982).

The set of equations (2.67), (2.88) – (2.91) describes the effect of the long gravity waves on the air flow. To close this set of equations we need to specify the wave spectrum F since the wave diffusion D_W depends on spectral shape. Clearly, the evolution of the wave spectrum and the wind speed are coupled

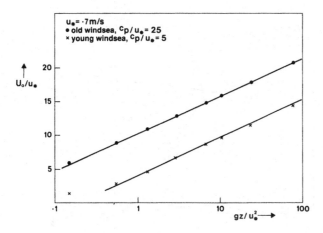

Fig. 2.10. Dimensionless wind speed U_0/u_* as a function of dimensionless height gz/u_*^2 for young and old wind sea.

and, in principle, one should solve the energy balance equation for the waves in tandem with the momentum equation for the wind. Here we shall not pursue this approach, but instead we assume that the wave spectrum is given by an empirical relation (the JONSWAP shape) and we shall concentrate on the effect of the waves on the wind. We shall, for simplicity, give the overshoot γ and the width parameter σ the constant values 3.3 and 0.10 respectively, whereas for the Phillips constant α_p we use the relation proposed by Snyder (1974), (equation (2.82)).

Given the spectral shape we shall search for steady-state solutions of the air flow above sea waves by means of an iteration method which initially takes $D_W = 0$ then calculates the wind profile to obtain the growth rate γ and the diffusion coefficient and so on. The rate of convergence of this procedure was judged by calculating the total stress

$$\tau_{tot} = \rho_a \left[\nu_a \frac{\partial}{\partial z} U_0 + \ell^2 | \frac{\partial}{\partial z} U_0 | \frac{\partial}{\partial z} U_0 \right] + \tau_w, \qquad (2.92)$$

where τ_w is given by (2.78), as in the steady state this is given by its asymptotic value for large z, $\tau_{tot} = \rho_a u_*^2$.

Details of the numerical procedure may be found in Janssen (1989b). Here we shall only give some of the results. The effect of the gravity waves on the wind profile is illustrated in figure 2.10, where we have plotted the dimensionless wind speed U_0/u_* as a function of dimensionless height gz/u_*^2 for young and old wind sea. It is clear from this figure that the long waves extract a significant amount of momentum from the air flow; however, young wind sea appears to be rougher than old wind sea. The wind profile is found to be approximately logarithmic except in a region close to the surface

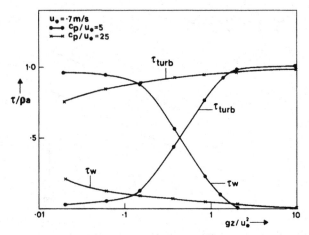

Fig. 2.11. The turbulent stress τ_{turb} and the wave stress τ_w as a function of height for young and old wind sea.

$(gz/u_*^2 < 1)$, where considerable deviations from the logarithmic profile are found. Even for the large friction velocity used ($u_* = 0.7$ m/s) this region corresponds to only a few centimetres above the sea surface. It is of interest to study the distribution of the stress in the surface layer over turbulence, the wave effect and viscosity. To this end we have plotted in figure 2.11 the turbulent stress τ_{turb} and the wave stress τ_w as a function of height for young and old wind sea. The viscous stress is not plotted because it is usually quite small, the reason is that by our choice of Charnock relation (2.91), the water surface is already rough for $u_* > 0.1$ m/s. For young wind sea we observe that at around $gz/u_*^2 \simeq 1$ the wave-induced stress becomes a considerable fraction of the total stress corresponding to the deviations from the logarithmic profile in figure 2.10. On the other hand, for old wind sea, the stress going into the long waves is only 35% of the total stress so that most of the stress is supplied to the very short gravity and capillary waves.

Referring again to figure 2.8, we have seen that the wave-induced stress obtained from parametric relations may be a considerable fraction of the total stress or even larger. In figure 2.12 we show what happens according to quasi-linear theory of wind-wave generation. Here, crosses denote the wave-induced stress calculated according to linear theory and diamonds show the results according to quasi-linear theory. We infer from figure 2.12 that for young wind sea the ratio τ_w/τ is reduced considerably, whereas for old wind sea this ratio hardly changes. Apparently, in equilibrium the curvature of the wind profile is considerably reduced for young wind sea in such a way that the ratio τ_w/τ remains less than unity. This must then be accompanied by a considerable reduction of the growth rate of the waves. This is illustrated

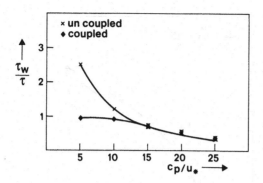

Fig. 2.12. Reduction of wave-induced stress due to quasi-linear theory.

in figure 2.13, where we plot the normalized growth rate γ/ω of the waves versus the inverse of the dimensionless phase speed c/u_*. Clearly, for fixed phase speed, the growth rate of the waves is larger for old wind sea than for young wind sea, at least if plotted as a function of dimensionless phase speed. This considerable reduction of the growth rate for young wind sea may be understood by realizing that for rough air flow and fixed dimensionless phase speed, the resonance between wave and air flow occurs at a larger height than for smoother air flow (see figure 2.10 using the resonance condition $U_0/u_* = c/u_*$). Consequently, using the logarithmic wind profile as an approximation, for fixed phase speed the quantity $W_c''/|W_c'| \simeq 1/z_c$ (where z_c is the critical height) is smaller for young wind sea than for old wind sea. This is, however, only a partial explanation because an excessive reduction of the growth rate γ/ω would result. By means of figure 2.10 one infers that at $c/u_* = 10$ this would result in a reduction of a factor of 6, whereas according to figure 2.12 there is only a reduction of a factor of 2. It turns out that the decrease in curvature of U_0 is accompanied by an increase of the wave-induced velocity in air, χ_c, such that the aforementioned reduction is partly compensated.

It is concluded that for young wind sea there is a strong two-way inter-action between wind and waves. This is, on the one hand, reflected by an air flow which is rougher than would be expected from Charnock's relation for the roughness length alone, and, on the other hand, a strong reduction of growth of the waves by wind. However, for old wind sea there is only a weak coupling between waves and wind. We emphasize that the reason for this is our choice of wave age dependence of the Phillips' constant. In other words, for young wind sea the high-frequency waves are much steeper than for old wind sea.

To conclude our discussion of results, we show the wave age dependence of the drag coefficient for two different friction velocities in figure 2.14. A sensitive dependence on wave age may be noted and it implies that we may

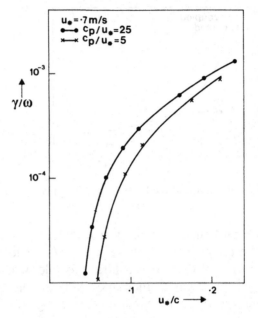

Fig. 2.13. Normalized growth rate γ/ω of the waves versus the inverse of the dimensionless phase speed c/u_*.

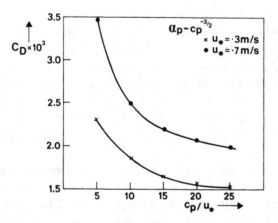

Fig. 2.14. The wave age dependence of the drag coefficient for two different friction velocities.

now give an explanation of the large scatter found in the field data for drag coefficient as a function of the wind speed at 10 m height. Returning to figure 2.9, we have shown lines of constant wave age c_p/u_* in the plot of drag coefficient versus wind speed. Clearly, the combination of Snyder's proposal for wave age dependence of the Charnock constant with quasi-linear theory explains the scatter in the field data.

We have seen that quasi-linear theory of wind-wave generation gives a better description of momentum transfer than the usual linear theory of wave growth since quasi-linear theory gives a drag coefficient that describes the sea state dependence of the drag. Linear theory cannot account for this. It is thus reassuring that we are able to give a description of the momentum transfer from air to water that is consistent with observations. Waves seem to play an important role in this process. It seems, therefore, to be a good idea to include this effect in applications such as storm-surge modelling, weather prediction and climate modelling. A more detailed discussion of this will follow in the following chapters. However, it should be stressed that the theory is still weak on a number of points. We had to parametrize the effects of gravity-capillary waves on the mean flow by means of a roughness length and we disregarded the effects of turbulence on the wave-induced flow. However, Jenkins (1992), including the effects of small scale turbulence, found very similar results for the drag coefficient as we found by means of quasi-linear theory.

Finally, it should be emphasized that a rather strong wave age dependence of the Phillips' parameter α_p is needed to explain within the framework of quasi-linear theory the sea state dependence of the surface stress over the ocean. Also, possible implications of the highly directional character of the wavenumber spectrum and the dependence of the growth rate of wind waves on the wind direction have not yet been discussed. These issues will be addressed in § III.3. First of all, it is shown that the parametrization of quasi-linear theory (using as input the measured wind speed and *in situ* frequency spectra) gives surface stresses which are in good agreement with observed surface stresses during HEXOS. Thus, in spite of the somewhat weaker observed sea state dependence of the Phillips' constant, a reliable calculation of the surface stress by means of quasi-linear theory turns out to be possible. Secondly, we will discuss in chapter III quasi-linear theory in the context of the third generation WAM model. Since the surface waves play an important role in the momentum transfer from air to water and since the growth rate of the waves depends on the friction velocity, a self-consistent determination of the surface stress is clearly desirable. The important issue is then whether in the context of the physics of the WAM model a sensitive sea state dependence of the Phillips' parameter and surface stress is still obtained. Indeed, this turns out to be the case.

As a last remark, it should be pointed out that according to our results for increasing wave age a smoother air flow is found. This inverse dependence of the drag coefficient on wave age has, however, a restricted range of validity. For extreme young wind sea ($c_p/u_* < 5$, a situation that seldom occurs in nature) the opposite trend is to be expected, as suggested by Nordeng (1991).

The reason for this is that under these extreme circumstances the scaling laws (2.81) and (2.82) are not valid. In fact, for zero wave age the Phillips' parameter is seen to become infinite. In its stead, a limitation of α_p is to be expected. This issue can, however, only be properly addressed in the context of a simulation with the WAM model. And indeed, when discussing in § III.6 the turning wind field case it is seen that after the wind has turned the drag coefficient drops considerably, simply because there are no high-frequency waves in the new wind direction. (This therefore corresponds to the case of extreme young wind sea.) As soon as waves have been generated in the new wind direction the drag coefficient increases until the waves reach a wave age of the order 5 – 10 after which the drag coefficient starts to decrease again.

II.2.8 Summary of conclusions

We have given an overview of our present (theoretical) knowledge of wind wave generation, with emphasis on the Miles instability. We have extended Miles' theory by including effects of stratification in the air and gustiness and we also discussed wave–mean-flow interaction which results in a sea state dependence of the momentum transfer from air to water.

We also discussed in some detail results from numerical models that include effects of small-scale turbulence. Results from these models regarding the growth rate of the gravity waves agree with Miles' theory for phase speeds smaller than some measure of the wind speed, but predict in contrast to Miles, damping of the waves when the phase speed is larger than the wind speed. Apart from the fact that these damping rates are quite small, it was pointed out that the usual mixing length model for turbulence is not valid for these long waves as the dominant eddies do not have sufficient time to transport a significant amount of momentum during a wave period. Presumably, mixing length modelling therefore overestimates the impact of turbulence on growth and damping of the long waves.

In our quasi-laminar approach we have been able to quantify three important physical effects: stratification, gustiness and sea state dependence of the wind profile. In § II.8, we will look for evidence of these processes in fetch-limited observations of wave growth. In chapter III we will discuss the inclusion of the wave age dependence of the wave growth in the wind input source term in the WAM model. In chapter IV (applications), we will discuss the sensitivity of wave model results to the computation of this source term (§ IV.5) by considering the consequences of stratification, gustiness and wave-age dependence in model simulations.

II.3 Wave–wave interaction

K. Hasselmann, P. A. E. M. Janssen and G. J. Komen

II.3.1 An example

So far we have neglected nonlinear terms in the water wave equations. This could be done because – on the average at least – ocean waves are not very steep, while it can be shown that nonlinear corrections are of the order $O(\epsilon) = O(ka)$ (for monochromatic waves at least). Before we make this statement more explicit we will consider a simple example of a nonlinear equation, namely the nonlinear Klein-Gordon equation:

$$\varphi_{tt} - \varphi_{xx} + \varphi = -4\epsilon\varphi^3. \tag{2.93}$$

This is one of the simplest nonlinear equations, with a so-called quartic interaction. The power of φ is three in equation (2.93), but one speaks nevertheless of a quartic interaction because the corresponding Lagrangian density

$$\mathscr{L} = \frac{1}{2}(\varphi_t^2 - \varphi_x^2 - \varphi^2) - \epsilon\varphi^4 \tag{2.94}$$

has a φ to the power four. A quartic rather than the simplest cubic coupling term is assumed as it turns out that this is the relevant coupling order for surface waves. In (2.93) ϵ (not to be confused with the steepness) is assumed to be small enough for a perturbation approximation to work.

In a naive perturbation approach one would write

$$\varphi = \varphi^{(0)} + \epsilon\varphi^{(1)} + \epsilon^2\varphi^{(2)} + \ldots \tag{2.95}$$

and on substitution in (2.93) one finds a hierarchy of equations:

$$\varphi_{tt}^{(0)} - \varphi_{xx}^{(0)} + \varphi^{(0)} = 0,$$

$$\varphi_{tt}^{(1)} - \varphi_{xx}^{(1)} + \varphi^{(1)} = -4\varphi^{(0)3}. \tag{2.96}$$

$$\ldots$$

In this approach the lowest order solutions are plane waves with $\omega^2 = 1 + k^2$. These force the higher order corrections, with forcing frequencies that are sums and differences of lower order ones. A problem then arises when resonance occurs, when the forcing frequency is identical to the free frequency. In such cases the forced modes grow linearly with time, as in the case of a simple harmonic oscillator:

$$\ddot{X} + \sigma^2 X = C e^{i\omega t}, \tag{2.97}$$

which has solutions

$$X \sim e^{i\omega t}, \tag{2.98}$$

when $\omega \neq \sigma$, but for $\omega = \sigma$,

$$X \sim te^{i\sigma t}. \tag{2.99}$$

This is called secular behaviour, since the amplitude grows with time. When an expansion term in (2.95) of order ϵ^n is secular the expansion necessarily breaks down after $t \sim 1/\epsilon^n$. Stokes encountered this problem in 1847. He showed that the problem could be resolved for monochromatic water waves by introducing a small correction to the frequency. This was possible because the secular change in the complex amplitude was 90° out of phase with the amplitude itself and could thus be interpreted as a rotation of the amplitude in the complex plane. The same problem arises when one deals with many wave components, but then the amplitudes change in magnitude as well as in phase and frequency renormalization is no longer sufficient to remove all secular behaviour.

This can be illustrated in a simple manner by considering more general solutions of equation (2.93) consisting of a superposition of many wave components. In this case the lowest order linear solution φ_0 takes the form (1.86)

$$\varphi^{(0)} \sim \sum A_n^{(0)} e^{i(k_n x - \omega_n t)}, \tag{2.100}$$

where $\omega_n^2 = 1 + k_n^2$ and the lowest order wave amplitudes $A_n^{(0)}$ are constant. We write A rather than a (as in chapter I) because in this section a will be used to denote normal-mode variables, including rapid time variations. It is also convenient to extend the summation over both positive and negative values of n to include the complex conjugate terms required for the representation of a real field φ, where we define $k_{-n} = -k_n$, $\omega_{-n} = -\omega_n$ and $A_{-n}^{(0)} = A_n^{(0)*}$. Substitution of this lowest order expression in (2.95) gives rise to higher pth order forced waves which are either off-resonance, and thus show a periodic time behaviour of the amplitude $A_n^{(p)}$, or are in resonance, and thus show secular behaviour, leading to a break-down of the expansion after a finite time. In either case the higher order amplitudes $A_n^{(p)}$ can no longer be regarded as independent of time. One obtains, for example, as the determining equation for the amplitude at the lowest interaction order $p = 3$

$$\ddot{A}_n^{(3)} - 2i\omega_n \dot{A}_n^{(3)} = -4\epsilon \sum_{k_a+k_b+k_c=k_n} A_a^{(1)} A_b^{(1)} A_c^{(1)} e^{i(\omega_n - \omega_a - \omega_b - \omega_c)t}. \tag{2.101}$$

If $\omega_n - \omega_a - \omega_b - \omega_c \neq 0$ the right hand side is rapidly varying with time, so that its dynamic effect on the evolution of $A_n^{(3)}$ is small. However, when resonance occurs,

$$k_a + k_b + k_c = k_n \text{ and } \omega_a + \omega_b + \omega_c = \omega_n, \text{ with } \omega_i^2 = 1 + k_i^2, \tag{2.102}$$

the forcing term on the right hand side of (2.101) becomes independent of time and produces a linear growth of the amplitude (note that the indices and thus the wavenumbers and frequencies in the resonance condition can have either sign).

For waves which interact only in resonance, the secular change in amplitude can be described by regarding the net amplitude $A_n = A_n^{(1)} + A_n^{(3)} + \dots$ as a slowly varying function of time. In this case one can write down the nonlinear interaction equations for the net amplitudes directly without introducing a perturbation expansion of the field (Whitham, 1974). However, the situation is a little more complicated in a continuum of waves, which is the appropriate description for ocean waves. Here we will be concerned with the continuous transition between resonant and nonresonant waves in the neighbourhood of resonance surfaces. We would therefore like to develop a general formalism for describing the nonlinear coupling between wave modes which applies for both resonant and nonresonant interactions. Moreover, the basic formalism should be free of other approximations, such as perturbation expansions. Once the basic wave–wave interaction equations have been derived in full generality, the solutions of the equations can then be determined in a second step by appropriate perturbation expansion techniques.

II.3.2 General formalism

The problem of the energy exchange between a continuum of interacting waves is not a problem which is specific only to ocean waves. It is a basic problem of physics which – following Peierls' (1929) classic paper on the heat conduction in solids – has been extensively studied in many applications in solid state physics, quantum field theory, plasma physics and in various areas of geophysics (see, for example, Hasselmann, 1966, 1967). There has evolved through this work a well-established formalism for treating wave–wave interactions (Peierls, 1955) which satisfies these requirements of full generality and can be directly applied to the present problem of ocean-wave coupling. We describe here briefly the general method, illustrating the approach first for the example of the nonlinear Klein-Gordon equation.

The basic idea is to express the field and field equations in terms of general time-dependent wave–amplitude coordinates. It is assumed that the field equations are homogeneous in a space $\mathbf{x}$ (which can be either the full three-dimensional space or, as in the case of surface waves, a two-dimensional subspace, or one-dimensional, as in our simple example of the nonlinear Klein-Gordon equation) and that the linearized field equations support a complete set of eigenmodes $\psi_{\mathbf{k}} \sim \exp(i\mathbf{k} \cdot \mathbf{x})$. The field φ is then expressed in terms of normal-mode coordinates $q_{\mathbf{k}}$ defined with respect to the set of eigenfunctions $\psi_{\mathbf{k}}$ (which are assumed to be complete and orthonormal with respect to a unit volume element of **x** space):

$$\varphi = \sum q_{\mathbf{k}} \psi_{\mathbf{k}}. \qquad (2.103)$$

For a real field φ, $q_{-\mathbf{k}} = q_{\mathbf{k}}^*$, $\psi_{-\mathbf{k}} = \psi_{\mathbf{k}}^*$.

In our example of the nonlinear Klein-Gordon equation, the eigenfunctions are simply the exponential functions $\exp(ik_n x)$, so that (2.103) reduces to the standard (spatial) Fourier representation of φ. For surface waves, $\psi_{\mathbf{k}}$ is a three-component vector field $\sim \exp(i\mathbf{k}.\mathbf{x})$ representing the three prognostic fields $\eta, u_1(z), u_2(z)$ (see chapter I).

Substituting (2.103) into the Lagrangian L of the system (the spatial integral of the Lagrangian density $\mathscr{L}$) yields

$$L = L_0 + L_{int}, \qquad (2.104)$$

where L_0 denotes the free-mode Lagrangian of the linearized system, given by

$$L_0 = \frac{1}{2} \sum (\dot{q}_{\mathbf{k}} \dot{q}_{-\mathbf{k}} - \omega_{\mathbf{k}}^2 q_{\mathbf{k}} q_{-\mathbf{k}}) \qquad (2.105)$$

and $L_{int} = L_{int}(q_{\mathbf{k}}, \dot{q}_{\mathbf{k}})$ is the interaction Lagrangian describing the coupling between the modes. The harmonic oscillator form (2.105) of the free-mode Lagrangian follows generally from the definition of a normal mode, if it is assumed that the system is symmetrical in time, so that for given **k** there exist two normal modes with positive and negative eigenfrequencies $\pm\omega_{\mathbf{k}}$.

For the nonlinear Klein-Gordon equation, the form (2.105) is immediately verified, while the interaction Lagrangian is given by

$$L_{int} = -\epsilon \sum_{\mathbf{k}_a + \mathbf{k}_b + \mathbf{k}_c + \mathbf{k}_d = 0} q_{\mathbf{k}_a} q_{\mathbf{k}_b} q_{\mathbf{k}_c} q_{\mathbf{k}_d}. \qquad (2.106)$$

The Lagrange equations

$$\frac{d}{dt} \frac{\partial L}{\partial \dot{q}_{\mathbf{k}}} = \frac{\partial L}{\partial q_{\mathbf{k}}} \qquad (2.107)$$

then yield as equations of motion a set of forced harmonic oscillator equations, which in the case of the nonlinear Klein-Gordon equation takes the form

$$\ddot{q}_{\mathbf{k}} + \omega_n^2 q_{\mathbf{k}} = -4\epsilon \sum_{\mathbf{k}_a + \mathbf{k}_b + \mathbf{k}_c = \mathbf{k}} q_{\mathbf{k}_a} q_{\mathbf{k}_b} q_{\mathbf{k}_c}. \qquad (2.108)$$

This is not yet the most suitable form for describing wave–wave interactions, as the coordinates $q_{\mathbf{k}}$ contain both positive and negative frequency components representing waves travelling in opposite positive and negative **k** directions. To separate the two wave components, one first reduces the single second order equation for $q_{\mathbf{k}}$ to two first order equations by transforming to a Hamiltonian system. Introducing the momentum variables $p_\mathbf{k} = \partial L/\partial \dot{q}_\mathbf{k}$, the Hamiltonian $H(p_\mathbf{k}, q_\mathbf{k}) = \sum p_\mathbf{k} \dot{q}_\mathbf{k} - L$ becomes

$$H = H_0 + H_{int}, \qquad (2.109)$$

where the free-mode Hamiltonian is given by

$$H_0 = \frac{1}{2} \sum (p_\mathbf{k} p_{-\mathbf{k}} + \omega_\mathbf{k}^2 q_\mathbf{k} q_{-\mathbf{k}}) \qquad (2.110)$$

and the nonlinear wave–wave coupling is described by the interaction Hamiltonian $H_{int}(q_\mathbf{k}, p_\mathbf{k})$. The Hamiltonian equations

$$\dot{q}_\mathbf{k} = \frac{\partial H}{\partial p_\mathbf{k}}, \qquad (2.111)$$

$$\dot{p}_{-\mathbf{k}} = -\frac{\partial H}{\partial q_{-\mathbf{k}}} \qquad (2.112)$$

yield the first order, two-component form of the forced harmonic oscillator equations

$$\dot{q}_\mathbf{k} = p_{-\mathbf{k}} + \frac{\partial H_{int}}{\partial p_\mathbf{k}}, \qquad (2.113)$$

$$\dot{p}_{-\mathbf{k}} = -\omega_\mathbf{k}^2 q_\mathbf{k} - \frac{\partial H_{int}}{\partial q_{-\mathbf{k}}}. \qquad (2.114)$$

Finally, the positive and negative frequency oscillations are separated by diagonalizing the linear part of equations (2.113) and (2.114) through a 45° rotation in the $(p_{-\mathbf{k}}, i\omega_\mathbf{k} q_\mathbf{k})$ plane. This yields the standard normal-mode coordinates:

$$a_\mathbf{k}^+ = \frac{1}{\sqrt{2\omega_\mathbf{k}}}(p_{-\mathbf{k}} - i\omega_\mathbf{k} q_\mathbf{k}), \qquad (2.115)$$

$$a_\mathbf{k}^- = \frac{1}{\sqrt{2\omega_\mathbf{k}}}(p_{-\mathbf{k}} + i\omega_\mathbf{k} q_\mathbf{k}). \qquad (2.116)$$

It is readily verified that the normal-mode coordinates satisfy the reality conditions

$$a_{-\mathbf{k}}^- = (a_\mathbf{k}^+)^*. \qquad (2.117)$$

As a function of $a_\mathbf{k}^+$, $a_{-\mathbf{k}}^-$ the free field Hamiltonian now takes the diagonal form

$$H_0 = \sum \omega_\mathbf{k} a_\mathbf{k}^+ a_{-\mathbf{k}}^- \qquad (2.118)$$

and the equations of motion become

$$\dot{a}_\mathbf{k}^s = -is\frac{\partial H}{\partial a_{-\mathbf{k}}^{-s}}, \qquad (2.119)$$

where $s = \pm$ is a sign index. Explicitly,

$$\dot{a}_{\mathbf{k}}^+ = -i\omega_{\mathbf{k}} a_{\mathbf{k}}^+ - i\frac{\partial H_{int}}{\partial a_{-\mathbf{k}}^-}, \tag{2.120}$$

$$\dot{a}_{\mathbf{k}}^- = i\omega_{\mathbf{k}} a_{\mathbf{k}}^- + i\frac{\partial H_{int}}{\partial a_{-\mathbf{k}}^+}. \tag{2.121}$$

Equations (2.120) and (2.121) are seen now to represent two nonlinearly coupled equations for two separate modes $a_{\mathbf{k}}^+, a_{\mathbf{k}}^-$ travelling in the positive and negative $\mathbf{k}$ directions, respectively (because of the reality condition (2.117), the negative-sign-index mode $a_{\mathbf{k}}^-$ travelling in the negative $\mathbf{k}$ direction is, of course, just the complex conjugate of the positive-sign-index mode $a_{-\mathbf{k}}^+$ travelling in the positive '$-\mathbf{k}$' direction). In quantum field theory, the variables $a_{-\mathbf{k}}^- = (a_{\mathbf{k}}^+)^*$ and $a_{\mathbf{k}}^+$ correspond to creation and annihilation operators respectively, for particles of momentum $\mathbf{k}$.

It is convenient in the following to revert to the more compact mode index notation used previously by writing simply for the mode travelling in the positive $\mathbf{k}$ direction $a_{\mathbf{k}}^+ = a_n$, where the index n runs over all discrete (directed) wavenumbers of the mode spectrum. The mode $a_{\mathbf{k}}^-$ travelling in the negative $\mathbf{k}$ direction is then the complex conjugate of the mode $a_{\bar{n}}$ associated with a different index $\bar{n}$ characterizing the negative wavenumber (we recall that negative indices $-n$ were introduced to represent the complex conjugate terms, not opposite propagation directions). Specifically, if $a_{-\mathbf{k}}^+ = a_{\bar{n}}$, we have $a_{\mathbf{k}}^- = (a_{-\mathbf{k}}^+)^* = a_{\bar{n}}^* = a_{-\bar{n}}$. In this notation, the normal-mode form of the equations of motion become finally, for positive n,

$$\dot{a}_n = -i\frac{\partial H}{\partial a_{-n}}, \tag{2.122}$$

or, with

$$H_0 = \sum_{n>0} \omega_n a_n a_{-n}, \tag{2.123}$$

$$\dot{a}_n = -i\omega_n a_n - i\frac{\partial H_{int}}{\partial a_{-n}}. \tag{2.124}$$

(The equations for the complex conjugate negative-index modes, which we will not need, are given by the complex conjugate set of equations, which are identical to (2.122) and (2.124) except for a change of sign of the right hand sides.)

It should be noted that the normal-mode variables a_n used here and in the rest of this section differ from the similarly denoted wave amplitudes a_n introduced in chapter I and used elsewhere in this monograph. The normal-mode variables here are normalized with respect to the action spectrum rather than the energy spectrum, and, in contrast to the wave amplitudes, they contain the eigenfrequency factor $\exp(-i\omega t)$.

The set of equations (2.122), or (2.124), represents the general normal-mode form of the field equations for an arbitrary nonlinear system which can be characterized by a Lagrangian or Hamiltonian which can be decomposed into a linear component and a nonlinear interaction term.

For the nonlinear Klein-Gordon equation, we find, for example,

$$H_{int} = \frac{\epsilon}{4} \sum_{\mathbf{k}_a+\mathbf{k}_b+\mathbf{k}_c+\mathbf{k}_d=0} (a_a - a_{\bar{a}}^*)(a_b - a_{\bar{b}}^*)(a_c - a_{\bar{c}}^*)(a_d - a_{\bar{d}}^*)/(\omega_a\omega_b\omega_c\omega_d)^{\frac{1}{2}},$$

(2.125)

where the sum runs over only positive indices $a, b, \ldots$ and the indices $\bar{a}, \bar{b}, \ldots$ refer to waves running in the negative $\mathbf{k}_a, \mathbf{k}_b, \ldots$ directions.

In the case of an arbitrary nonlinear system, it is assumed generally that the interaction Hamiltonian can be expanded in a power series with respect to the wave mode coordinates:

$$H_{int} = \sum_{a,b,c} D_{abc}\, a_a a_b a_c + \sum_{a,b,c,d} D_{abcd}\, a_a a_b a_c a_d + \ldots \qquad (2.126)$$

The coefficients $D_{...}$ are symmetrical in their indices and are nonzero only if the wavenumber sums of the coupled components vanish, $\mathbf{k}_a + \mathbf{k}_b + \mathbf{k}_c = 0$, $\mathbf{k}_a + \mathbf{k}_b + \mathbf{k}_c + \mathbf{k}_d = 0, \ldots$ (homogeneity condition). The symmetry will be found to be an important property in the discussion of the conservation properties of wave–wave interactions.

II.3.3 Application to ocean waves: the amplitude equations

In contrast to our example of the nonlinear Klein-Gordon equation with quartic interactions, ocean waves already interact at the lowest cubic order, $D_{abc} \neq 0$. However, resonance does not occur at this order, as it can be shown generally that the three-wave resonance conditions

$$\mathbf{k}_1 + \mathbf{k}_2 = \mathbf{k}_3, \quad \omega_1 + \omega_2 = \omega_3 \qquad (2.127)$$

cannot be satisfied for waves, such as surface gravity waves, whose dispersion curve has negative curvature. The lowest order resonance for gravity waves occurs for four-wave interactions, so that the theory of ocean-wave interactions is formally analogous to the quartic nonlinear Klein-Gordon equation. (However, resonant three-wave interactions occur in the short-wave gravity-capillary region of the spectrum. Although not discussed in this monograph, these are important for the energy balance of short waves which govern the microwave backscatter from the sea surface measured by remote sensing satellites, see §§ I.4 and V.3.)

The resonance conditions are discussed in some detail in Phillips (1977), who shows that the two-dimensional counterpart of the four-wave resonant interaction conditions (2.102) for gravity waves has solutions only for the sign combination

$$\mathbf{k}_1 + \mathbf{k}_2 - \mathbf{k}_3 = \mathbf{k}_4, \text{ and } \omega_1 + \omega_2 - \omega_3 = \omega_4, \qquad (2.128)$$

where the indices '1', '2',... are positive. (In the following, we shall use indices 1, 2, 3 and 4 generally to denote positive-index wave components, while wave indices which can take either sign will be denoted by indices $a, b, c, \ldots$ or $j, \ldots$.) The secular variations of the amplitudes due to these interactions are found to have an important influence on the energy balance of the wave spectrum.

The basic properties of resonant ocean-wave interactions were discovered during the fundamental investigations of Phillips (1960) and Hasselmann (1960, 1962). The latter author was the first to give a complete statistical theory of four-wave interactions for a homogeneous gravity wave field. The theory was later placed in the wider context of the general theory of wave–wave interactions which had been developed independently in other fields of physics (Hasselmann, 1966, 1967, 1968). Although we shall consider the application of the general theory here only to the special case of interactions within the gravity wave spectrum, it is useful to view this case in a wider context, as resonant wave–wave interactions involving surface gravity waves occur in many other geophysical applications, such as the generation of microseisms (by quadratic sum interactions of surface waves and by surface wave–bottom topography interactions), the generation of internal gravity waves (by quadratic difference interactions of surface waves), the scattering of ocean waves by bottom topography or oceanic turbulence and – as already mentioned – cubic gravity–capillary wave interactions (see also Hasselmann, 1966).

What is needed to apply the general theory now to ocean waves? First, one must compute the coupling coefficients $D_{abc}, D_{abcd}, \ldots$. Then one must solve the coupled-mode equations (2.124) to determine the rate of change of the spectrum.

The first step is simply a problem of algebra. If Lagrangian (particle) coordinates are used, the state of the fluid at any instant in time can be expressed in terms of the amplitudes $q_{\mathbf{k}}$ of the surface wave modes. The Lagrangian $L = T - U$ can then be determined by computing the kinetic and potential energies T and U, respectively, of the fluid motion. (The surface wave modes were defined in chapter I with respect to Eulerian coordinates, but to lowest linear order the two representations are identical, and the Lagrangian representation can be readily expanded to arbitrary nonlinear order.) The transformation to normal-mode coordinates a_n as described above then yields the coupling coefficients. Details are given in Hasselmann (1966).

Zakharov (1968) and Davidson (1972) have shown that for surface waves the step of computing the Lagrangian in the Lagrangian coordinate frame can be short-circuited by deriving directly a Hamiltonian in Eulerian coordinates (that this is possible is not obvious, as the transformation from Lagrangian to Eulerian coordinates is noncanonical; the approach would presumably not be applicable to other wave–wave interaction problems). The starting point is the Hamiltonian formulation of the boundary conditions at the surface (see chapter I):

$$\frac{\partial \eta}{\partial t} = \frac{\delta E}{\delta \psi}, \quad \frac{\partial \psi}{\partial t} = -\frac{\delta E}{\delta \eta}, \tag{2.129}$$

with

$$\psi(\mathbf{x}, t) = \phi(\mathbf{x}, z = \eta, t). \tag{2.130}$$

The idea is to solve the potential problem

$$\phi_{x_1 x_1} + \phi_{x_2 x_2} + \phi_{zz} = 0, \tag{2.131}$$

with boundary conditions (1.14), thereby expressing ϕ in terms of the canonical variables η and ψ. Once this is done one can express the energy E in terms of η and ψ, and the evolution of η and ψ follows from Hamilton's equations (2.129). Zakharov obtained the deterministic evolution equations for water waves by solving the potential problem (2.131) in an iterative fashion for small steepness ϵ. The standard form (2.124) can be obtained by transforming the Hamiltonian equations, with η, ψ decomposed into Fourier components $p_{\mathbf{k}}$, $q_{\mathbf{k}}$, to normal-mode coordinates a_n.

Zakharov simplified the standard form further by removing the nonresonant cubic interactions through a suitable nonlinear canonical transformation and retaining only the index sign combinations capable of satisfying the four-wave resonance conditions (2.128). One obtains then for the evolution of the Fourier components of the wave field the so-called Zakharov equation,

$$\frac{\partial}{\partial t} a_4 + i\omega_4 a_4 = -i \sum_{\mathbf{k}_1 + \mathbf{k}_2 - \mathbf{k}_3 - \mathbf{k}_4 = 0} T_{1234} \, a_1 a_2 a_3^*, \tag{2.132}$$

where

$$T_{1234} = 12 D_{1\,2\,-3\,-4} - 18i \sum_j \left\{ \frac{D_{1\,2\,-j} D_{j\,-3\,-4}}{\omega_j - \omega_1 - \omega_2} \right.$$

$$\left. -i \frac{D_{-3\,2\,-j} D_{j\,1\,-4}}{\omega_j + \omega_3 - \omega_2} - i \frac{D_{1\,-3\,-j} D_{j\,2\,-4}}{\omega_j - \omega_1 + \omega_3} \right\}. \tag{2.133}$$

Here the index '4' is positive, the sum in (2.132) is also restricted to positive

indices '1', '2' and '3' and the sum in (2.133) runs over both positive and
negative j (for each term in the sum only two possible values of j occur
because of the homogeneity condition that the wavenumber sums of the
interacting wave components must vanish).

The origin of the quadratic products of cubic coupling coefficients, which
arise through the elimination of the nonresonant cubic interactions, will
become clearer below. (Strictly speaking, the original Zakharov equation was
not Hamiltonian because the cubic interactions had been eliminated by a
transformation that was noncanonical. The correct canonical transformation
was found by Krasitskii (1990, 1992).)

The interaction coefficient T_{1234} was obtained by Zakharov (1968) and
Crawford *et al* (1981). (The original interaction coefficients D_{abc}, D_{abcd} were
given by Hasselmann (1962) for the general finite depth case; an error in one
of the finite depth terms was corrected by Herterich and Hasselmann (1980).)
The properties of the Zakharov equation (2.132) have been studied in great
detail by, for example, Crawford *et al* (1981) (for an overview, see Yuen and
Lake (1982)). Thus the nonlinear dispersion relation, first obtained by Stokes
(1847), follows from equation (2.132). Also, the Benjamin-Feir instability
(Brooke Benjamin and Feir, 1967) of a weakly nonlinear, uniform wave
train – or the more general sum-interaction instability of Hasselmann (1967),
in the four-wave version relevant for ocean waves – can be well described
by the Zakharov equation; the results on growth rates, for example, are
qualitatively in good agreement with the results of Longuet-Higgins (1978).
The Zakharov equation is therefore a good starting point to study the
properties of a random wave field.

II.3.4 The nonlinear energy transfer: general considerations

Once the interaction coefficients of the general mode interaction equations
(2.124) have been derived, for example in the specific Zakharov form (2.132),
the next step is to solve the equations to determine the evolution of the wave
spectrum. Again, a general formalism for deriving the nonlinear spectral
transfer rates from the basic mode coupling equations (2.124) has been
developed in other fields of physics; the analysis can be summarized in the
form of a simple set of interaction rules and diagrams (cf. Hasselmann,
1966). We give here both the standard derivation based on a direct mode–
mode interaction analysis and also a slightly modified derivation based on
the moment expansion method applied by Janssen (1991b).

Equation (2.132) describes the evolution of the amplitudes and phases of
the waves. For a statistical ocean-wave field we are interested, however, only
in the change of the statistical properties of the field. It was shown in § I.2.5
that in the linear approximation a dispersive wave field always develops
Gaussian statistics and can thus be completely described statistically by

its spectrum. As discussed below, the Gaussian property can be invoked to lowest order also for a wave field which is undergoing a slow change through weakly nonlinear resonant interactions, although this is less obvious. The relevant quantity which we wish to determine is thus (in the notation of the present section) the wave action spectrum

$$N_n = < a_n a_n^* >, \qquad (2.134)$$

where the angle brackets denote an ensemble average.

To determine the evolution of the spectrum we assume, for notational simplicity, that the wave field is strictly homogeneous in space. This assumption was, in fact, implicit in our analysis so far, in which the wave amplitudes were regarded as functions only of time, not of space. Once the nonlinear transfer rate has been computed under this restriction, however, the result can immediately be generalized to the general space-time-dependent case by invoking the two-scale concepts and propagating wave packet picture described in § I.2.6. The computed transfer rate can then be interpreted as a source term in the spectral transport equation.

In order to compute the rate of change of the wave spectrum we need to consider first the solution of the Zakharov equation. This can be determined by expanding the normal-mode amplitudes in a perturbation series with respect to ϵ,

$$a_n = \epsilon a_n^{(1)} + \epsilon^2 a_n^{(2)} + \epsilon^3 a_n^{(3)} + \dots. \qquad (2.135)$$

To lowest interaction order, the left hand side of equation (2.132) refers then to the third order component $a_4^{(3)}(\mathbf{k})$, while the right hand side contains only first order components $a_i^{(1)}(\mathbf{k})$. The equation can be integrated in the presence of the resonant secular terms only for finite times for which $a_4^{(3)}(\mathbf{k})$ can still be regarded as small compared with the first order terms. One obtains then as the integral of equation (2.132) to lowest (third) interaction order (assuming zero third order amplitudes as initial conditions)

$$a_4^{(3)} = -i \sum_{\mathbf{k}_1 + \mathbf{k}_2 - \mathbf{k}_3 - \mathbf{k}_4 = 0} T_{1234} \, \Delta(-\omega_1 - \omega_2 + \omega_3 + \omega_4, t) \, a_1^{(1)} a_2^{(1)} a_3^{(1)*}, \qquad (2.136)$$

where

$$\Delta(\omega, t) = \frac{1 - e^{-i\omega t}}{i\omega}. \qquad (2.137)$$

The function $\Delta(\omega, t)$ describes generally the response of a harmonic oscillator to periodic forcing, where the frequency ω denotes the difference between the forcing frequency and the eigenfrequency of the oscillator. In the present case it defines the response of the third order amplitudes to the forcing by cubic interaction products of first order wave components. For resonance, $\omega = \omega_1 + \omega_2 - \omega_3 - \omega_4 = 0$, equation (2.137) yields $\Delta(\omega, t) = t$. Off resonance,

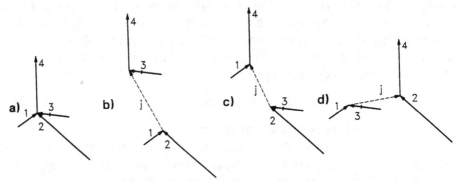

Fig. 2.15. Interaction diagrams summarizing the generation of third order resonant wave perturbation 4 through nonlinear coupling between three wave components 1, 2 and 3. Arrows represent wavenumber vectors. Complex conjugate wave amplitudes are indicated by cross-bars. Full arrows represent free-wave components, dashed arrows nonresonant forced waves. *Panel a*: direct quartic coupling. *Panels b,c,d*: Successive cubic coupling leading to generation of intermediate second order nonresonant forced waves of (positive or negative) index *j*.

for $\omega \neq 0$, we have a bounded response,

$$\Delta(\omega, t) = \Delta_{nr}(\omega) = \frac{1}{i\omega} \quad \text{for nonresonance} \qquad (2.138)$$

(for a strictly nonresonant response, in which one is not concerned with the transition to resonance, the homogeneous-solution component $-e^{-i\omega t}/i\omega$ of Δ, which was included to remove the singularity at $\omega = 0$, can be dropped).

It is useful to represent the solution (2.136) in graphical form by the interaction diagram of figure 2.15a. Interaction diagrams are a useful tool for summarizing and identifying the different terms in the perturbation expansion, while at the same time depicting the wavenumbers (momentum) of the modes involved in an interaction; drawing interaction diagrams saves a lot of writing of algebra. The first order wave components which generate the perturbation are represented as arrows, defining the wavenumber vectors of the components which converge at a vertex, while the resultant perturbation is shown as an arrow leaving the vertex. The outgoing component is drawn as a full arrow if the wave is forced in resonance and by a dashed line for off-resonant forcing. (Figure 2.15a thus represents a resonant interaction; a nonresonant interaction, represented by an outgoing dashed arrow, would, of course, also have been possible.) Complex conjugate (negative index) components are indicated by a cross-bar on the arrow. In computing the resultant wavenumber vector of the perturbation as the vector sum of the ingoing components, these vectors receive a negative sign. The resultant perturbation is given by the product of the ingoing wave components, the coupling coefficient associated with the vertex (the relevant indices can be read off from the structure of the diagram) and the Δ-function response function (whose frequency argument is also apparent from the diagram). The net third order wave amplitude (2.136) is given by the sum over all possible third order diagrams, i.e. by the sum over all possible ingoing wave components in figure 2.15a (for the complete set of all resonant and nonresonant third order components, one would need also to include the other nonresonant sign combinations which were excluded from the Zakharov equation, but these are not relevant for the present discussion).

If the general form (2.124) of the mode interaction equations is used without removal of the cubic interaction terms, the additional interaction diagrams shown in figures 2.15b – d also arise at third order. These involve two successive nonresonant cubic interactions which were absorbed into the quartic coefficient T_{1234} in the Zakharov equation (2.133). The coupled mode equations must then first be solved at lowest cubic interaction order to determine the second order nonresonant forced wave components $a_j^{(2)}$ generated by quadratic interactions between two first order components. The cubic interaction equations are then solved again at the next perturbation order to obtain a resonant third order component $a_4^{(3)}$ by quadratic interactions between $a_j^{(2)}$ and a third first order wave component. Explicitly, the interaction diagrams figures 2.15b – d correspond to the second, third and fourth terms in (2.133).

In general, multiple-vertex diagrams are constructed using the same rules as single-vertex diagrams, with the additional rule that the outgoing component of a vertex is now allowed to occur as an ingoing component of a vertex at the next interaction order. The order of a multiple-vertex interaction diagram is given simply by the number of ingoing first order components. Thus in figure 2.15b, for example, the second order nonresonant component j generated in a cubic interaction by the quadratic product of the two ingoing first order components 1 and 2 appears at the next perturbation order as an ingoing component in a further cubic interaction in which j and the first order complex conjugate component 3 couple quadratically to generate the outgoing third order resonant component 4. The relevant response functions at the two vertices are $\Delta(\omega_j - \omega_1 - \omega_2, t) = \Delta_{nr}(\omega_j - \omega_1 - \omega_2)$ (for the first nonresonant vertex) and $\Delta(\omega_3 + \omega_4 - \omega_1 - \omega_2, t)$ (for the second resonant vertex). The numerical coefficients in (2.133) arise from the various possible permutations in the positions of the indices in the interaction coefficients $D_{12-3-4}, D_{12-j}, \ldots$.

In the special case of a discrete set of interacting wave components, as referred to, for example, in our discussion of the nonlinear Klein-Gordon equation, the components can be regarded as either in resonance or off resonance, so that only the secular or stationary form of $\Delta(\omega, t)$ will apply. However, in the case of a random wave field, consisting of a continuum of wave components, we must consider the continuous transition between resonance and nonresonance which occurs when we carry out integrations across the resonant hypersurfaces within the multi-dimensional wavenumber product space of all possible resonant and nonresonant interactions. To compute these integrations we will make use of the following asymptotic distribution properties of $\Delta(\omega, t)$ with respect to integrations over ω for large t (for the later discussion of the Gaussian hypothesis and its relation to irreversibility we also give the corresponding relations for large negative t):

$$|\Delta(\omega, t)|^2 \to 2\pi|t|\delta(\omega) \text{ for } t \to \pm\infty, \qquad (2.139)$$

$$\Delta(\omega, t) \to \pm\pi\delta(\omega) \text{ for } t \to \pm\infty, \qquad (2.140)$$

(irrelevant asymptotic principal value terms have been ignored).

We note that while the maximal value of $|\Delta(\omega, t)|^2$ on the resonance surface $\omega = 0$ itself is proportional to t^2, the integral across a resonance surface is proportional only to $|t|$. This is because the magnitude of the resonance factor $\Delta(\omega, t)$ increases as $|t|$ but its half-width decreases inversely proportional to $|t|$. For the same reason the integral of $\Delta(\omega, t)$ across a resonance surface yields a constant δ-function contribution for large $|t|$.

In applying the solution of the Zakharov equation to determine now the evolution of the spectrum it is clear that we will encounter the basic closure problem of nonlinear statistical systems, i.e. the turbulence problem. The rate of change of the second moments in which we are interested is determined formally by the third moments, whose rate of change, in turn, is determined by the fourth moments, and so on, in an infinite hierarchy of equations. However, weakly interacting, dispersive wave systems represent an exceptional case in the sense that for such systems it has been demonstrated that closure can be achieved rigorously, to the lowest resonant interaction order, through the Gaussian (or, as will be shown, apparently Gaussian) property of linear dispersive wave fields. For a Gaussian field the moments of all orders can be expressed in terms of the second moments. The odd moments vanish and the even moments decompose into the sums of all combinations of products of the second moments, i.e. of the spectra.

In the statistical theory of the evolution of a weakly nonlinear wave field, the Gaussian property of the wave field therefore plays a central role in enabling the closure of the hierarchy of moment equations. As discussed in § I.2.5, for a linear wave field, the Gaussian property can be explained physically by the Central Limit Theorem, using the picture of a random wave field consisting of a superposition of a very large number of statistically independent wave packets which have propagated into a given area of the ocean from different distant regions. Formally, it can be shown that for a linear dispersive wave field with an arbitrary initial state for which the moments tend to zero for large separations of the position vectors, the non-Gaussian components of the moments, the cumulants, degenerate asymptotically into a rapidly oscillating fine structure with respect to wavenumber. Since this cannot be resolved by finite fixed-bandwidth spectral analysis techniques, the field can be regarded effectively as Gaussian (Hasselmann, 1967). Do these results continue to apply also in the weakly nonlinear case? This has been the subject of some debate in the field of geophysical fluid dynamics (largely, however, as the result of inadequate knowledge of earlier fundamental work in statistical physics).

It is tempting to assume (and has sometimes been erroneously argued) that the Gaussian hypothesis remains valid in a weakly nonlinear system because the tendency to a Gaussian field is a first order linear process, so that it must necessarily dominate over the weak nonlinear coupling, which generates only small higher order cumulants. However, this argument fails to recognize that the linear dispersive mixing does not simply cause the cumulants to tend to zero, but merely transforms them into a finite-amplitude but nonresolvable fine structure.

From statistical mechanics it is known that fine structures are associated with the origin of irreversibility. Under certain conditions they can be neglected when going forward in time, but they cannot, in general, be neglected when reconstructing the past history. Without proof we shall make the assumption that the cumulants can be neglected, i.e. that the field is Gaussian when going forward in time, and will derive from this hypothesis the Boltzmann integral expression (2.145) for the nonlinear transfer. However, since the Boltzmann equation is irreversible (the left hand side changes sign, but the right hand side remains unchanged on a change of sign in time and the horizontal coordinates), while the original equations from which it was derived were reversible, it can be concluded that the Gaussian hypothesis cannot then also be applicable when going backwards in time. We will come back to this point after we have derived the nonlinear transfer equation (2.145).

II.3.5 The nonlinear energy transfer: the transfer integral

After these general comments, we turn now to the derivation of the evolution equation for the second moment $< a_n a_n^* >$ from the solution of the deterministic evolution equation (2.132) for the amplitudes. To this end we expand the action spectrum in a perturbation series

$$N_n = N_n^{(2)} + N_n^{(4)} + N_n^{(6)} + \dots, \tag{2.141}$$

where

$$N_n^{(2)} = < a_n^{(1)} a_n^{(1)*} >, \tag{2.142}$$

$$N_n^{(4)} = < a_n^{(2)} a_n^{(2)*} > + < a_n^{(1)} a_n^{(3)*} > + < a_n^{(3)} a_n^{(1)*} >, \tag{2.143}$$

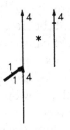

Fig. 2.16. Interaction diagrams describing the (nonsecular) contribution $< a_4^{(3)}.a_4^{(1)*} >$ of the lowest order modification N_4 of the action spectrum. The diagram on the left is in resonance, but is in quadrature with the first order amplitude and yields only a frequency shift.

$$N_n^{(6)} = < a_n^{(3)} a_n^{(3)*} > + < a_n^{(2)} a_n^{(4)*} > + < a_n^{(4)} a_n^{(2)*} >$$

$$+ < a_n^{(1)} a_n^{(5)*} > + < a_n^{(5)} a_n^{(1)*} > . \qquad (2.144)$$

There are now two ways of proceeding to derive the rate of change of the action spectrum. We can substitute the solutions of the Zakharov equation directly into the action expressions (2.143) – (2.144) and then form the time derivative of the resultant higher order secular terms. Or, alternatively, we can form the time derivative of the action expressions (2.143) – (2.144) first. This yields a sum of quadratic products, each of which consists of a time-differentiated and a non-time-differentiated amplitude factor. One can then substitute into these products the relevant expressions for each of the two factors as given by the Zakharov equation (2.132) and the solution (2.136) of the Zakharov equation, respectively. We discuss both approaches, each of which provides a different view of the energy transfer mechanism. The first method illustrates more clearly the nature of the secular changes occurring in the action spectrum and the two-timing limit involved in the derivation. The second approach, which is in the spirit of the moment expansion method of Janssen (1991b), saves one time integration. The relation between the two methods (which are, of course, basically equivalent) becomes clear if the structure of the coupling is presented in the form of interaction diagrams.

Following the first approach, it can be readily seen that the lowest order interaction expression $N_n^{(4)}$ in (2.143) yields no secular contribution to the action. The first term on the right hand side vanishes, since in the transformed Zakharov frame there exist no second order amplitude perturbations. The remaining two complex conjugate terms contain a single third order amplitude factor which is, in fact, in resonance. The relevant pair of interaction diagrams characterizing the product $< a_4^{(3)} \times a_4^{(1)*} >$ (where the index n in (2.143) is taken as 4) is shown in figure 2.16. The Gaussian hypothesis has been invoked in collapsing the mean product of the four wave components

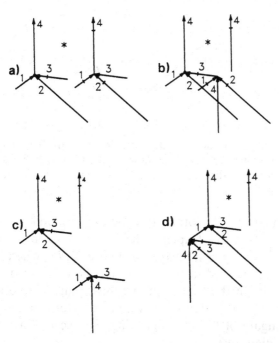

Fig. 2.17. Pairs of interaction diagrams representing the quadratic products of perturbation amplitudes associated with the lowest order secular change in the action spectrum N_4 (equation (2.144)). *Panel a*: energy (action) gain of outgoing component 4 in the basic interaction diagram a of figure 2.18. *Panels b,c,d*: Energy loss (or, for a complex conjugate ingoing component 4, figure 2.18b: energy gain) of ingoing component 4 in basic interaction diagrams b, c and d of figure 2.18.

appearing in the two diagrams into the product of the means of two complex conjugate pairs (i.e '1' = '3' and '2' = '4', or equivalent perturbations). The generated third order component $a_4^{(3)}$ is seen to be always in resonance, independent of the input component $a_1^{(1)}$. However, the secular change $a_4^{(3)}$ is 90° out of phase with the first order component $a_4^{(1)}$ with which it must be correlated (cf. (2.132), (2.137)). Thus the interaction yields only a frequency shift. This represents the generalization of Stokes' single-wave analysis to a continuous spectrum. The term has no impact on the energy transfer. Similar internal pairings at a quartic vertex also occur at higher order. They can be shown to have no influence on the energy transfer and will be discarded in the following.

The lowest order secular change in the spectrum is described by the term $N_n^{(6)}$, equation (2.144). The first term $a_4^{(3)} a_4^{(3)*}$ (again setting $n = 4$) contains a quadratic Δ factor which, according to equation (2.139), yields a resonant frequency δ-function factor proportional to t for large t. Physically, this term represents the energy gained by component 4 in the Zakharov equation

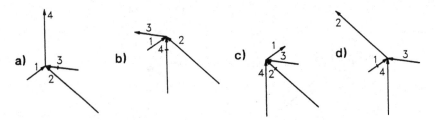

Fig. 2.18. The four basic interaction diagrams determining the energy transfer between the components of a resonantly interacting quadruplet. The diagrams can be interpreted also as collision diagrams in the Hasselmann particle picture.

due to the forcing from the other three wave components 1, 2 and 3 of a resonantly interacting quadruplet. The interactions involved in the pair of components $a_4^{(3)}, a_4^{(3)*}$ are indicated in figure 2.17a, where the Gaussian hypothesis has been invoked in requiring that the total set of six first order amplitudes occurring in the two diagrams must pair off into three complex conjugate pairs 1, 2 and 3. Thus the second diagram of the pair must be the complex conjugate of the first (ignoring, for the reasons given, internal pairings within a diagram).

The next two terms on the right hand side of equation (2.144) vanish again, since the second order amplitudes are zero. The last two terms describe the energy lost (or, as will be found for the case of an ingoing complex conjugate amplitude, gained) by each of the three ingoing components of a resonant interaction quadruplet (cf. figure 2.15a). This term is a little more cumbersome to compute, as it requires extending the perturbation expansion of the solution of the Zakharov equation one iteration order further. It also yields a linear term in time with a corresponding resonant frequency δ-function factor. The relevant pairs of interaction diagrams (in which the second diagram in a pair is simply the trivial first order component $a_4^{(1)*}$) of the last term in (2.144) are indicated in figure 2.17b – d. The second-to-last term in (2.144) is given by the complex conjugate set of diagrams.

We note that all higher-order interaction diagrams of figure 2.17 are constructed to yield an outgoing component 4, since we wish to determine in (2.144) the secular change of the action spectrum for the prescribed spectral component 4. Physically, however, as pointed out, the diagrams b, c and d of figure 2.17 correspond to the balancing energy loss (or gain) of each of the three ingoing components 1, 2 and 3 of the basic interaction diagram figure 2.15a. The cubic product of these components yielded the driving term which generated the energy gain of the component 4 in the pair of interaction diagrams shown in figure 2.17a. In order to denote the balancing loss terms of the ingoing components as an energy change of the prescribed wave component 4, the indices of the basic interaction diagram of figure 2.15a must be permuted, as shown in the four basic interaction diagrams of figure 2.18 (figure 2.18a is a repeat of figure 2.15a). Although all four basic interaction diagrams of figure 2.18 refer to the same resonant interaction quadruplet, it is important to recognize that the energy transfer is actually the result of four different basic interactions, characterized by the four different combinations of three ingoing wave components and one outgoing component. Our method of analysis has focussed on only one particular component of the quadruplet, component 4. This is, in fact, not a very efficient procedure for computing the energy transfer numerically. We shall discuss below a more symmetrical view of the nonlinear interactions in which the energy (action) changes of all four components involved in the four basic interactions of a resonant quadruplet are considered simultaneously.

If the interactions are sufficiently weak, the time t can now be chosen such that 1) it is sufficiently large that the asymptotic δ-function relation (2.137) is applicable, and 2) it is sufficiently small that the changes in the spectrum due to the secular terms can still be regarded as perturbations. The detailed implications for the weakness of the interaction needed to satisfy these two so-called two-timing conditions will not be discussed here, but it is clear that they depend not only on the smallness of the perturbation parameter ϵ, but also on the shape of the spectrum: the representation of the finite-width response function $|\Delta(\omega, t)|^2$ for finite times t as a δ-function will break down earlier for a very narrow spectrum than for a broad spectrum. In the limit of a discrete line spectrum, the theory is no longer applicable.

Assuming, however, that both requirements can be satisfied (which is, in fact, always the case for typical ocean-wave spectra), we can now form the time derivative of the perturbation $N_4^{(6)}$ of the action and interpret this as the rate of change of the linear action spectrum. This interpretation is justified, since the secular changes occur only in the free waves which define the linear action spectrum. The resultant equation is the Boltzmann equation for four-wave resonant interactions (reverting now to continuous notation):

$$\frac{d}{dt} N_4 = 4\pi \int |T_{1234}|^2 \delta(\mathbf{k}_1 + \mathbf{k}_2 - \mathbf{k}_3 - \mathbf{k}_4) \delta(\omega_1 + \omega_2 - \omega_3 - \omega_4)$$
$$\times [N_1 N_2 (N_3 + N_4) - N_3 N_4 (N_1 + N_2)] \, d\mathbf{k}_1 \, d\mathbf{k}_2 \, d\mathbf{k}_3. \quad (2.145)$$

This equation is valid for typical ocean-wave spectra in the energy-containing range of the spectrum relevant for wave models. However, it breaks down for interactions between very short waves and long waves, when the wavelengths of the short waves become comparable with the heights of the long waves. In this range a two-scale formalism is appropriate in which the short waves are described by a spectral action balance equation, the interaction with the long waves arising through the refraction of the short waves by the long-wave orbital velocity field (cf. Longuet-Higgins and Stewart, 1961, Hasselmann, 1971b). The hydrodynamic two-scale model has been extensively used together with the analagous microwave two-scale model in studies of microwave backscatter from the sea surface (cf. Keller and Wright, 1975, Hasselmann *et al*, 1985). However, the interactions are thought to be less important for the long-wave energy balance and are not included in the WAM model.

The nonlinear transfer expression (2.145) also breaks down in the limit of very shallow nondispersive water waves. The nonlinear transfer generally increases as the water depth decreases (cf. § III.3.2 and equation (3.57)). Ultimately, the basic weak-interaction conditions are no longer satisfied when the characteristic time scale of the transfer is no longer significantly larger than the characteristic wave period. The increase in the transfer rate is mainly due to the transition towards a nondispersive system. In the nondispersive shallow water limit, the argument of the frequency δ-function in the transfer integral becomes stationary on the resonance surface and the integral diverges.

In fact, for nondispersive waves, three-wave resonant interactions between waves propagating in the same direction become possible. However, the weak-interaction formalism is not applicable to these interactions either, since the interacting wave packets do not run apart and the mixing argument essential for the justification of the Gaussian spectral closure hypothesis, as discussed below, is no longer valid (if the Gaussian hypothesis is nevertheless formally invoked, the secular terms are found to grow as $t^{3/2}$ rather than t). In the nondispersive shallow water limit the interacting waves become phase locked and generate strongly nonlinear spilling breakers, hydraulic jumps etc.

In the second approach for the derivation of the Boltzmann equation, the rate of change of the action spectrum is determined by differentiating the quadratic product (2.142) and using the Zakharov equation (2.132) to describe the rate of change of the individual amplitudes:

$$\frac{\mathrm{d}}{\mathrm{d}t} N_4 = \frac{1}{2}(<\dot{a}_4 a_4^*> + <a_4 \dot{a}_4^*>)$$

$$= -\frac{i}{2} \sum_{1,2,3} T_{1234} <a_1 a_2 a_3^* a_4^*> + \text{c.c.} \qquad (2.146)$$

We arrive again at essentially the same interaction diagrams as before. However, the analysis is now a little simpler, as we have already introduced the time derivative and therefore need compute the solution of the Zakharov equation only to the first iteration order, not also to the second.

The lowest order fourth-moment expression on the right hand side of (2.146) is of fourth order (all amplitudes first order). This vanishes, however, because the first term and the indicated second complex conjugate term are imaginary and therefore cancel in the sum. Although there is resonant

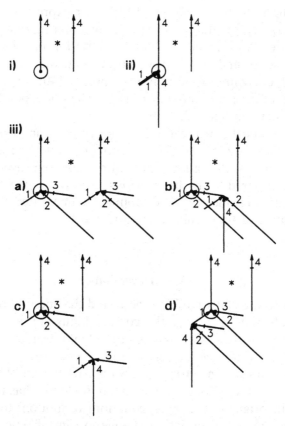

Fig. 2.19. Interaction diagrams for the rate of change of action spectrum N_4. *Panel (i)*: second order (trivial, zero rate of change); *Panel (ii)*: fourth order (lowest nontrivial order: Stokes' frequency shift, no rate of change of spectrum); *Panel (iii)*: sixth order, finite rates of change arising from four different interaction combinations (cf. figure 2.17). Notation as in figure 2.15, with the additional symbol $\odot$ at a vertex to denote the time derivative (coupling without the time integration factor Δ).

forcing at this order, producing a nonzero secular change in $a_4^{(3)}$ and thus a finite time-mean term $\dot{a}_4^{(3)}$, the resonant response is in quadrature with $a_4^{(1)}$. We recognize again the generalized Stokes' frequency shift term. The corresponding pair of interaction diagrams representing the product $T_{1234} < (a_1^{(1)} a_2^{(1)} a_3^{(1)*}) \times a_4^{(1)*} >$, with the Gaussian pairing conditions '1' = '3', '2' = '4', is shown in figure 2.19(ii). Here and in the following interaction diagrams, a circled dot in the diagram vertex indicates a time derivative term (i.e. the factor Δ is missing: the diagram represents the right hand side of the Zakharov equation itself rather than its time integral).

However, at the next interaction order $dN_4^{(6)}/dt$ we obtain for the fourth moments on the right hand side of (2.146) four expressions of the general form $< a_1^{(3)} a_2^{(1)} a_3^{(1)*} a_4^{(1)*} >$, the four expressions differing only in the position of the third order term in the product. Substitution of the solution (2.136) into these moments and application of the Gaussian closure hypothesis together with the asymptotic response relation (2.140) for large (positive) t then yields the Boltzmann equation (2.132). The corresponding interaction diagrams are shown in figure 2.19(iii).

The equivalence of the two methods of derivation is immediately apparent on comparing figures 2.16 and 2.17 with figure 2.19. The occurrence of a 'dot' at a vertex means that a time integration has been saved in the present computation compared with the previous derivation (the additional time integration in the first method is, of course, compensated later by forming the time derivative of the secular growth terms).

II.3.6 Irreversibility

We return now to the Gaussian hypothesis and the question of irreversibility. We have already remarked that the correct 'Gaussian' assumption is that one can take the statistics as Gaussian, but only if one goes forwards in time. To illustrate this further, we look now in more detail at what happens if we assume that the statistics are truly Gaussian, irrespective of the direction of time. As pointed out, the Boltzmann equation no longer has the time reversal symmetry of the original frictionless equations of motion: the left hand side of the equation changes sign with a change in sign of time, while the right hand side remains unchanged, provided the spectrum is not modified. We must therefore have introduced an additional hypothesis which destroys this symmetry. But the only additional hypothesis that we introduced was the Gaussian hypothesis, and if this is now assumed to apply rigorously, it is invariant with respect to time reversals. What is the explanation of this paradox?

Let us inspect the nature of the invariance or lack of invariance more carefully. Consider a statistical ensemble of individual sea state realizations $\Psi(\mathbf{x}, z, t)$ which at some initial time $t = 0$ is rigorously Gaussian. Since the Euler equations are invariant with respect to reversals of the time and horizontal coordinate axes, the ensemble of reflected sea state realizations $\Psi'(\mathbf{x}, z, t) = \Psi(-\mathbf{x}, z, -t)$ will then also represent a legitimate set of dynamically consistent sea state realizations. Both sea state ensembles are characterized initially by the same spectra, $N'(\mathbf{k}, t = 0) = N(\mathbf{k}, t = 0)$, while at later or earlier times, $N'(\mathbf{k}, t) = N(\mathbf{k}, -t)$ (a wave travelling backwards in time in some negative $\mathbf{x}$ direction is the same as a wave travelling forwards in time in the positive $\mathbf{x}$ direction). But since both ensembles are characterized by the same spectrum N_0 initially, it follows from the Boltzmann equation, which applies for both ensembles, that $N'(\mathbf{k}, t) = N(\mathbf{k}, t)$, or

$$N(\mathbf{k}, t) = N(\mathbf{k}, -t). \qquad (2.147)$$

However, the solutions of the Boltzmann equation clearly do not have this time reversal symmetry property.

The formal resolution of this contradiction is readily found if we return to our two-timing derivation. We used the asymptotic δ-function forms (2.139), (2.140) of the response function Δ for large positive t. We could equally well have applied the forms for large negative t. This would have yielded the Boltzmann equation with the opposite sign. Thus if the sea state is truly Gaussian at time $t = 0$, it follows from our two-time scale analysis that the time derivative of the spectrum must be discontinuous at $t = 0$; the time curve exhibits a cusp at $t = 0$, so that the evolution of the spectrum is indeed symmetrical with respect to the past and the future, in accordance with (2.147).

But this is clearly unrealistic and in contradiction with the philosophy of our analysis. We have assumed that the Gaussian hypothesis can be applied not only at one particular time $t = 0$ but for all times and that the rate of change of the spectrum computed by carrying out the two-timing limit in the forward direction yields an integro-differential equation (or, in the general context of the spectral transport equation, a source function) which is valid for all times, without a discontinuous derivative (cusp) at some particular time. Thus the only logical resolution of the irreversibility paradox is that the sea state can not be regarded as strictly Gaussian to lowest interaction order. The Boltzmann equation is, in fact, based on the time-asymmetrical hypothesis that the sea state can be regarded as Gaussian when considering the evolution of the spectrum forwards in time, but it cannot then be regarded as (even approximately) Gaussian when reconstructing the past.

At the level of the two-timing analysis considered here, in which slow time is considered only locally, the validity of the Gaussian assumption when going forwards in time cannot be proven and must be regarded simply as a hypothesis. However, in view of the fundamental importance of weak interaction theory for statistical mechanics in general, considerable efforts have been expended to prove the validity of this assumption. Prigogine (1962), for example, has shown through a complex nonlocal analysis, in which the evolution of the wave field is expanded in an infinite series with respect to slow time, that the local rate of change equation (2.145) is indeed valid to lowest order in the nonlinearity parameter for all times $t > t_0$, provided the wave field was truly Gaussian at some much earlier initial time t_0.

Although the rigorous justification of the Gaussian hypothesis is a difficult mathematical problem, it can be readily understood physically by invoking Boltzmann's original arguments, applied now to wave packets rather than true particles. A resonant wave packet quadruplet interacts only during the limited time period in which all packets overlap in a common finite area of the ocean. Since they have propagated into the interaction area from different regions of the ocean, it can be assumed that they are statistically independent at the beginning of their interaction. After they have interacted, they are weakly correlated and propagate then away again into different regions of the ocean. However, any attempt to measure residual correlations between sets of wave components after they have resonantly interacted must be unsuccessful. This would require a search for very small effects at very large separation distances: the cumulants degenerate into a nonmeasurable fine structure. The nonresolvable fine-structure is none the less important if one wishes to reconstruct the past: if the paths of wave packets are traced backwards to the region in which they interacted, it cannot be assumed that within the interaction region they are statistically independent. The asymmetry between going forwards and going backwards in time and thus the irreversibility of the Boltzmann equation resides ultimately in the time asymmetrical assumption that at some very much early initial time all wave packets were genuinely uncorrelated for large separation distances. The Gaussian hypothesis can then be applied for all later times when going forwards in time, but not when going backwards towards the initial state – in accordance with the formal mathematical derivation.

One could also consider other time asymmetrical hypotheses. For example, it is probable that the results of Prigogine could be derived using the weaker assumption that initially the cumulants of the wave field were not zero but smooth in wavenumber space, i.e that the wave field was statistically independent at large spatial separations (cf. Benney and Saffman (1966) and discussion in Hasselmann (1967)). Alternatively, it could be assumed that the waves are continually generated and dissipated by mechanisms which are uncorrelated at large separations. But in all cases an 'arrow of time' has to be introduced through an additional assumption which cannot be deduced from the basic reversible equations of motion alone.

In the remaining part of this section we will discuss some general properties of the Boltzmann equation (2.145). Its numerical integration will be discussed later in chapter III. The resulting physical balance between wind input, nonlinear interaction and dissipation will be discussed in § III.3.3.

II.3.7 General conservation properties of the nonlinear transfer

Preservation of the nonnegative nature of N

We note that the Boltzmann integral has the property – as it must – that it can never give rise to negative spectral densities. Before a spectral density becomes negative, it must first become zero. But for zero N, the spectral product in the integral (2.145) and also the remaining quadratic factor are positive. Thus dN/dt is positive, and N will become positive rather than negative.

Conservation laws

The evolution equation (2.145) admits three conservation laws, which can be verified by explicit substitution:

a conservation of action

$$\frac{d}{dt}\int dk N(\mathbf{k}) = 0.$$

(2.148)

b conservation of momentum

$$\frac{d}{dt}\int d\mathbf{k}\ \mathbf{k}N(\mathbf{k}) = 0.$$

(2.149)

c conservation of energy

$$\frac{d}{dt}\int d\mathbf{k}\ \omega N(\mathbf{k}) = 0.$$

(2.150)

This means that a wave field cannot gain or loose energy through the four-wave resonant interaction. Growth or dissipation of wave action, momentum or energy must therefore take place through other processes such as wind input, whitecapping or bottom interaction.

Implications for energy transfer

The conservation of two scalar quantities, energy and action, implies an important general property of the energy transfer. A similar relation holds for the energy transfer in a two-dimensional turbulence spectrum in the atmosphere, which also conserves two scalar quantities, energy and enstrophy. The one-dimensional energy transfer

$$\frac{d}{dt}N(\omega) = \int \frac{d}{dt}N(\mathbf{k})k\frac{dk}{d\omega}d\theta,$$

(2.151)

where θ is the wave propagation direction, must have at least three lobes of different sign. It cannot have a two lobe structure representing, for example, an energy cascade from low to high frequencies, as in three-dimensional turbulence (negative lobe at low frequencies, positive lobe at high frequencies). The ratio energy/action = ω increases monotonically with

frequency. Thus if the net action lost in the negative low-frequency lobe
balances the action gained in the high-frequency lobe, the energy lost in
the low-frequency lobe must necessarily be smaller than the energy gained
in the high-frequency lobe, so that energy is not conserved. The nonlinear
energy transfer in a surface wave spectrum does indeed have a three lobe
structure, which has important implications for the growth of the spectrum,
to be discussed in § III.3.3.

II.3.8 Interaction symmetries, principle of detailed balance and wave–particle analogy

Since the conservation of integral action, momentum and energy applies
for an arbitrary spectrum, it must clearly apply not only to the integral
quantities, but to each resonant interaction quadruplet separately. These
local balance properties were not immediately apparent, however, in our
derivation of the nonlinear transfer integral, which focussed on the energy
change of only one of the four components of a quadruplet.

The local symmetry properties of the nonlinear transfer integral can be
recognized more easily by writing the integral in the form

$$\frac{\mathrm{d}}{\mathrm{d}t} N(\mathbf{k}) = \int \delta(\mathbf{k} - \mathbf{k}_4) Q \mathrm{d}\Omega, \tag{2.152}$$

where

$$\begin{aligned} Q &= 4\pi |T_{1234}|^2 \delta(\mathbf{k}_1 + \mathbf{k}_2 - \mathbf{k}_3 - \mathbf{k}_4)\delta(\omega_1 + \omega_2 - \omega_3 - \omega_4) \\ &\quad \times [N_1 N_2 (N_3 + N_4) - N_3 N_4 (N_1 + N_2)], \end{aligned} \tag{2.153}$$

and

$$\mathrm{d}\Omega = \mathrm{d}\mathbf{k}_1 \ \mathrm{d}\mathbf{k}_2 \ \mathrm{d}\mathbf{k}_3 \ \mathrm{d}\mathbf{k}_4 \tag{2.154}$$

denotes the eight-dimensional phase space element of all possible interacting
quadruplets (without regard to the resonance side conditions, which are
expressed separately by the δ-functions in (2.153)). Thus $Q\mathrm{d}\Omega$ represents the
increment of action exchanged by quadruplet interactions per increment $\mathrm{d}\Omega$
of interaction phase space.

The factor $\delta(\mathbf{k} - \mathbf{k}_4)$ in (2.152) filters out the contribution from the $\mathbf{k}_4$
component of the quadruplets. From the symmetrical form of (2.153) it
follows that we could equally well have filtered out the contribution from
the components $\mathbf{k}_3$ or, with an opposite sign, the components $\mathbf{k}_1$ or $\mathbf{k}_2$:
in all three cases we would have recovered the same transfer integral, but
with a permutation of the index notation. Expressed differently: for any
given quadruplet configuration of resonantly interacting wavenumbers, the
integration over the eight-dimensional interaction space Ω covers the same
quadruplet configuration eight times, through the eight possible permutations

of wavenumber indices (interchange of 1 and 2, interchange of 3 and 4 and interchange of the pairs 1, 2 and 3, 4) which define the same quadruplet configuration. All eight permutations yield the same incremental action change for any given wavenumber component, except for a change in sign when pairs 1, 2 and 3, 4 are interchanged. Thus the components $\mathbf{k}_4$ and $\mathbf{k}_3$ both gain and the components $\mathbf{k}_1$ and $\mathbf{k}_2$ both lose the same common action increment $Q d\Omega$ from the quadruplet interactions of any of the eight phase space elements $d\Omega$ defining a given quadruplet. This is referred to as the principle of detailed balance.

It follows further from the resonance conditions that each component i of a quadruplet interaction must then also gain or lose a corresponding increment $Q d\Omega\, \omega_i$ and $Q d\Omega\, \mathbf{k}_i$ of energy and momentum, respectively. The principle of detailed balance and the conservation of energy, momentum and action for each elementary interaction quadruplet is seen to be a consequence of the symmetry of the expression (2.153) for Q, which, in turn, follows from the symmetry of the coupling coefficients T_{1234} (the symmetry of the coupling coefficients and the principle of detailed balance were, in fact, first derived for ocean waves by Hasselmann (1960), by an inverse argument, starting from the conservation of energy and momentum, before the Lagrange-Hamiltonian formalism was applied to ocean waves).

The principle of detailed balance holds generally for interactions in elementary particle physics, and also at the atomic level for chemical interactions. It is useful to pursue the particle analogy further. The interactions of a resonant quadruplet can be regarded as a collision process between 'particles' (wave packets) in which two particles 1 and 2 are annihilated (created) and two particles 3 and 4 are created (annihilated). In fact, the nonlinear transfer integral (2.145) can be derived from the standard Feynman diagram rules for interacting bosons in the classical limit of high boson particle densities.

In Hasselmann (1966, 1968), the nonlinear transfer integrals describing the energy transfer for wave–wave interaction processes in classical random wave fields were summarized by a set of rules applied to 'collisions' between particles in a slightly modified particle picture. The collision diagrams in this particle picture are analogous to (but differ in detail from) Feynman diagrams. They are used in the present context only as a convenient tool for summarizing the structure of the final Boltzmann integral representing the net result of resonant wave–wave interactions. In contrast to the interaction diagrams used above, collision diagrams do not describe the detailed structure of the interactions between wave-mode amplitudes but rather the rate of change of the action spectra resulting from the amplitude interactions.

In the Hasselmann particle picture, the action spectrum is interpreted (as in chapter I) as the number density of particles (wave packets). The nonlinear energy transfer is interpreted as the result of particle collisions. The side condition is introduced that all collision processes can lead to only one outgoing component. The ingoing components of a collision can have either positive or negative action (i.e. the ingoing components can be either particles or 'anti-particles', the number densities of particles and 'anti-particles' being the same). The probability of a collision is proportional to the product of the number densities of the ingoing components. The same collision cross-section applies for all n combinations of $(n-1)$ ingoing components and one outgoing (positive particle) component for a given set of n resonantly interacting components. With the aid of these simple rules and a general expression for the collision cross-section as a function of the interaction coefficients, the nonlinear transfer integral for an arbitrary wave–wave interaction process can immediately be written down.

The relevant collision diagrams for the case of surface wave interactions are shown in figure 2.18. In the Hasselmann particle picture the two basic boson interactions $1,2 \rightarrow 3,4$ and $3,4 \rightarrow 1,2$ of the normal quantum theoretical particle picture are replaced by the four interactions $1,2,-3 \rightarrow 4$; $1,2,-4 \rightarrow 3$; $3,4,-1 \rightarrow 2$ and $3,4,-2 \rightarrow 1$. The diagrams of figure 2.18 were presented before as interaction diagrams of the Zakharov equation in the context of the perturbation expansion of the action. The close relation in the present particle picture between collision diagrams and interaction diagrams is apparent.

The particle picture illustrates a basic difference between the conservation of energy and momentum and the conservation of action in four-wave interactions. The conservation of energy and momentum applies for all wave–wave interaction processes and follows generally from the invariance of the Lagrangian or Hamiltonian with respect to time and space translations. The conservation of action, however, holds only for wave–wave interactions in which equal numbers of particles are created and annihilated. It follows from the invariance of the Lagrangian with respect to a common phase shift of the wave components. This is valid for four-wave interactions with the sign combinations of the Zakharov equation, but not, for example, for three-wave interactions.

The principle of detailed balance and the wave–particle analogy are useful not only for summarizing the structures of Boltzmann integrals for a variety of different geophysical applications (cf. Hasselmann, 1966), but also for numerical computations. Thus the principle of detailed balance was applied by Hasselmann and Hasselmann (1985b) and Snyder *et al* (1993) in developing symmetrized techniques for integrating the Boltzmann equation, leading to improved conservation properties and a considerable saving in computer time.

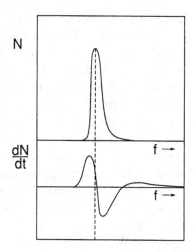

Fig. 2.20. The shape of dN/dt for a peaked skew action spectrum.

II.3.9 Nonequilibrium entropy

The appropriate nonequilibrium entropy associated with (2.145) is

$$S = \int d\mathbf{k} \ln N(\mathbf{k}). \qquad (2.155)$$

From the Boltzmann equation it can be shown that

$$\frac{d}{dt} S \leq 0, \qquad (2.156)$$

with the equality holding when

$$\frac{1}{N_1} + \frac{1}{N_4} = \frac{1}{N_2} + \frac{1}{N_3}, \qquad (2.157)$$

for all possible quadruplets satisfying the resonance conditions. In other words in the absence of input and dissipation, equation (2.145) tells us that a homogeneous random wave field will evolve in an irreversible manner to an equilibrium spectrum satisfying (2.157).

The general solution of (2.157) is

$$N(\mathbf{k}) = (a + \mathbf{b} \cdot \mathbf{k} + c\omega)^{-1}, \qquad (2.158)$$

where a, $\mathbf{b}$ and c are constants. The condition that N must be positive requires $\mathbf{b} = 0$. For $a = 0$, one obtains a uniform (white) energy spectrum ωN in wavenumber space, while for $c = 0$ one obtains a constant number density N, the well-known equilibrium solution of the Liouville equation for a system governed by Hamiltonian equations of motion.

In figure 2.20 the shape of dN/dt is schematically given for a peaked, skew action spectrum, such as is typically observed in practice (see § II.7).

The most pronounced feature of the nonlinear interaction is the transfer of action (and energy) to lower frequencies. In addition, there is a transfer to high frequencies, without which the three conservation laws could not be satisfied. The search for 'thermal equilibrium solutions' and the study of spectral evolution on the basis of equation (2.145) – of considerable theoretical interest – are only relevant for 'closed' systems in which the energy is conserved. They are not applicable to numerical wave prediction because ocean waves are not a closed system, since they exchange energy with the atmosphere and the underlying ocean. In one special situation, namely in the absence of wind, ocean waves approximate a closed system, but in that case the (swell) waves become so nearly linear that the time scale of the four-wave interaction is large with respect to the travel time of the waves across the basin. For growing wind waves the resulting action spectra are typical nonequilibrium spectra. We will discuss this in more detail in the next chapter after we have completed our discussion of the source terms and the numerical integration of (2.145).

II.3.10 The equilibrium energy cascade

While an equilibrium solution of the nonlinear transfer equation in the thermodynamic sense of a closed system with no asymptotic exchange of energy within the system is irrelevant in practice, a statistical equilibrium in which a constant flux is maintained through the system comes closer to the real situation. This was first proposed by Zakharov (1968), using dimensional arguments similar to the familiar Kolmogorov inertial cascade concept of isotropic turbulence. He showed that for a constant energy input to the wave spectrum at low wavenumbers and an isotropic dissipative sink at very high wavenumbers, the nonlinear interactions would adjust the energy spectrum $F(\mathbf{k})$ in the region between the spectral source and sink to a constant-flux isotropic equilibrium form $\sim k^{-4}$. In practice, the energy input by the wind and the dissipation by whitecapping do not appear to be sufficiently separated in the wavenumber domain to apply Zakharov's argument rigorously, and an isotropic spectrum is also not observed. Nevertheless, numerical experiments with a wave model based on the exact nonlinear transfer expression (Komen *et al*, 1984) indicate that although the directional distribution remains anisotropic, the spectra tend to adjust to a form in which the directionally averaged one-dimensional spectra are close to k^{-3} (corresponding to a k^{-4} two-dimensional spectrum), with a rather constant flux down the spectrum. Moreover, it is found that the nonlinear flux is generally rather strong, so that relatively small deviations from the k^{-3} form are sufficient to generate divergent fluxes which can balance nonzero input and dissipation source functions in the cascade region. Thus although the requirements for Zakharov's constant flux cascade are not satisfied formally, it appears that the concept is nevertheless quite useful as a first approximation and explains why – although observed two-dimensional spectra are far from isotropic – the one-dimensional spectra tend to be fairly close to a k^{-3} power law.

II.4 Wave dissipation by surface processes

M. Donelan and Y. Yuan

The least understood aspect of the physics of wave evolution, as it pertains to spectral modelling, is the dissipation source function. Waves may lose energy continuously by viscous dissipation and by the highly intermittent process of wave breaking. The continuous slow drain of wave energy by viscosity is well understood and easily calculated. While it is an important sink for very short gravity and gravity-capillary waves, it is insignificant for the longer waves that are handled explicitly in large scale numerical models. Understanding and modelling the wave breaking process are thus of critical importance in achieving an accurate representation of the principal sink function in the energy balance equation.

There have been several attempts to explore the dynamics of deep water wave breaking by both experimental studies and the mathematical development of theoretical concepts. Prominent among these are the laboratory experiments of Duncan (1981) and Rapp and Melville (1990) and the theoretical and numerical work of Longuet-Higgins and Cokelet (1978) and Longuet-Higgins (1988). In each case the focus has been on the dynamics of a single breaking event. Here we seek a spectral representation of the energy transfer rate associated with wave breaking and it is not at all clear how the one approach can be used to advance our understanding of the other. Lacking a complete dynamical theory of the rate of energy transfer from and within a spectrum of waves, progress in numerical wave modelling has, nonetheless, been made via several different approximate theories of wave dissipation via breaking. There are essentially three types of these theoretical models; we will designate them as follows:

(i) Direct equivalent pressure working or 'whitecap' model.

(ii) Instability or 'quasi-saturated' model.

(iii) Steepness limited or 'probability' model.

In the following we give a brief description of these different models and end with a comparison of their spectral distribution in typical seas.

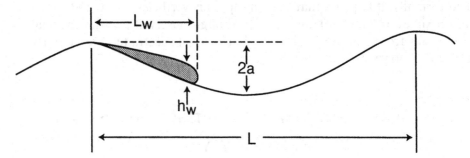

Fig. 2.21. Geometric similarity of whitecap and wave.

II.4.1 Whitecap model

The mathematical development of this model is due to Hasselmann (1974).
The assumptions are that the whitecaps can be treated as a random distribution of perturbation forces, which are formally equivalent to pressure pulses,
and that the scales of the whitecaps in space and time are small compared
to the wavelength and period of the associated wave. The development of
the theory involves two steps. First, it is shown generally that all processes
which are weak-in-the-mean, even if they are strongly nonlinear locally,
yield source functions which are quasi-linear to lowest interaction order:
the source function consists of the spectrum at the wavenumber considered
multiplied by a factor which is a functional of the entire wave spectrum
(and which can depend also on additional external factors, such as the wind
speed). This derivation was presented in the general context in § I.2.8. Secondly, an attempt is made to determine the factor for the special case of
whitecapping. This involves a number of rather complex approximations
based on the assumed space- and time-scale inequalities of the whitecaps
and the energy containing waves of the spectrum. We offer here a simple
physical interpretation of the 'whitecap' model of dissipation.

The existence of an attenuation factor in this model implies that the whitecaps are preferentially situated on the forward faces of the waves, thereby
exerting a downward pressure on the upward moving water and hence doing
negative work on the wave. This, indeed, is a common observation, as is
the fact that the extent of the whitecap, along the direction of travel of
the wave, varies in proportion to the length of the wave. Duncan (1981)
produced continuous breakers in a laboratory and observed that these steady
'whitecaps' and the underlying waves were in geometric similarity. Actual
whitecaps are not steady, but they do persist for a reasonable fraction (10%
to 30%) of the underlying wave period and in that time their shape and
position on the wave are maintained. We assume, therefore, that whitecap

and underlying wave are in geometric similarity (figure 2.21):

$$L_w/L = \text{const},$$
$$h_w/a = \text{const}. \qquad (2.159)$$

The pressure exerted by the whitecap on the surface of the wave is p_w

$$p_w \sim \rho_w g h_w \sim \rho_w g a. \qquad (2.160)$$

Thus the drag or negative momentum transfer to the wave is

$$-\frac{\rho_w g}{2c}\frac{dE}{dt} = \overline{p\frac{\partial \eta}{\partial x}} \sim \rho_w g a \times ka = 2\rho_w g k E. \qquad (2.161)$$

Hence

$$\frac{dE}{dt} \sim -\omega E. \qquad (2.162)$$

This yields a direct dissipation source function that is linear in both the spectral density and the frequency. There is, however, a further attenuation mechanism associated with the passage of a whitecapping wave through a field of smaller waves. It is well known (e.g. Banner *et al*, 1989) that the breaking of large scale waves causes rapid attenuation of short waves in its wake. In a continuous spectrum this may be represented by requiring the dissipation source function to depend on frequency relative to the peak. Both of these mechanisms (the pressure-induced decay and the attenuation of short waves by the passage of large whitecaps) are sensitive to the extent of breaking, i.e. whitecap coverage, which itself depends on the overall steepness of the wave field. Thus, combining these three processes, we have a dissipation function of the form used in the WAM model (Komen *et al*, 1984):

$$S_{ds1} = -C_{ds1}\left(\frac{\hat{\alpha}}{\hat{\alpha}_{PM}}\right)^m \left(\frac{\omega}{\bar{\omega}}\right)^n \omega F(\mathbf{k}), \qquad (2.163)$$

where

$$\hat{\alpha} = m_0 \bar{\omega}^4/g^2 \qquad (2.164)$$

and C_{ds1}, m and n are fitting parametrs, $\bar{\omega}$ is the mean radian frequency and $(\hat{\alpha}/\hat{\alpha}_{PM})$ is a measure of the overall steepness of the wave field. The expression (2.163) corresponds to the result of Hasselmann (1974) when $n = 1$.

II.4.2 Quasi-saturated model

Phillips (1985) and Donelan and Pierson (1987) have taken a different approach. While the expression they derive for the whitecapping dissipation is still basically consistent with the general quasi-linear form (2.163) for a

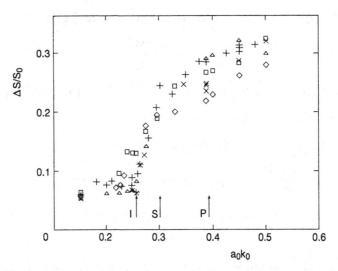

Fig. 2.22. Excess momentum lost from the wave field due to breaking as a function of the amplitude parameter, a_0k_0. The symbols correspond to different spectral bandwidths. The arrows show the amplitudes for incipient breaking (I), single spilling breaker (S) and single plunging breaker (P). Reproduced, with permission, from Melville and Rapp (1985). Copyright Macmillan Magazines Limited.

process which is weak-in-the-mean (§ I.2.8), the proportionality factor, γ, is estimated differently from Hasselmann (1974). Instead of assuming that the space and time scales of the whitecapping process are small compared with the characteristic wavelengths and periods of the waves for which the dissipation is being determined – i.e. that the whitecapping process is highly local in physical space and time – they make the complementary assumption that whitecapping is essentially local in wavenumber space.

They picture the wave breaking process as highly nonlinear in the wave steepness, having no effect until some limiting steepness is achieved when the wave form becomes unstable and spills or plunges forward, producing a whitecap at large and intermediate scales or a micro-breaker at very small scales. At the end of the breaking process, which typically takes a largish fraction of a wave period, a substantial energy loss is endured. Rapp and Melville (1990), in a carefully controlled laboratory experiment, caused packets of waves to interact and break with subsequent loss of up to 30% of the momentum per breaker (figure 2.22). Pierson *et al* (1992) examine the spectral changes in a similar experiment and demonstrate that, although the energy transfer associated with the breaking of a group is not localized at the peak of the group, most of the energy loss is concentrated there (figure 2.23). Consequently, a wave breaking event may not be viewed as a strictly local delta-function disturbance, which would have a white spectral

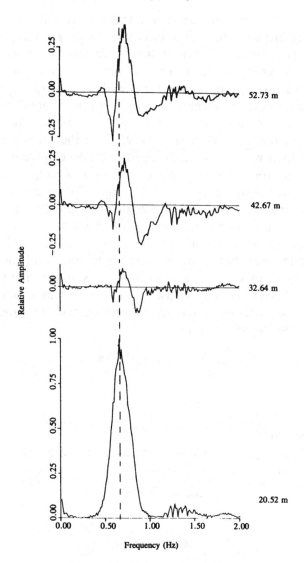

Fig. 2.23. Fourier coefficient (amplitude) spectrum at 20.52 m fetch (bottom) and spectral differences between this fetch and greater fetches (in 10 m increments) for the case of maximum steepness of 0.3. Wave breaking occurs between the top and second panels.

distribution in the wavenumber domain. As pointed out by Donelan *et al* (1972) wave breaking occurs as a wave passes through the peak of the group envelope. Once the wave breaks the whitecap so formed travels on the forward face of the wave (i.e. at the phase speed) for a good fraction of a wave period. Two wave periods later (for gravity waves) another wave breaks at the peak of the same group and this may repeat for several waves before the energy lost from the group has diminished its envelope sufficiently. Both the phase coherence of whitecap and wave and the repeated breaking engendered by the group structure tend to produce a localized adjustment to the spectrum in the wavenumber domain. For a narrow spectrum such as may be characteristic of recent swell or a strongly forced wind sea, the adjustment will be narrowly focussed in wavenumber or frequency space as in the experiments of Pierson *et al* (1992). On the other hand, a broad spectrum, associated with waves approaching full development, corresponds to less defined groupiness in the wave field and correspondingly greater spread in the 'spectral signature' of a whitecap. The spectrum, however, is smooth and continuous so that one may still associate the breaking energy loss at a particular wavenumber with that wavenumber, or at least a range of wavenumbers surrounding it. Phillips (1985) has argued that in the equilibrium range wind input, wave–wave interaction and dissipation are all of roughly equal importance, and since there is no internal wavenumber scale, the ratios of the three terms must be constant. From a balance with the known form of the wave–wave interaction term, he finds that the dissipation rate is cubic in the 'degree of saturation', $B(\mathbf{k})$.

$$S_{ds2} = -C_{ds2}\omega k^{-4}B^3(\mathbf{k}), \tag{2.165}$$

where

$$B(\mathbf{k}) = k^4 F(\mathbf{k}) \tag{2.166}$$

and C_{ds2} is a constant.

The form (2.165), (2.166) is consistent formally with the general quasi-linear form (2.163)

$$S_{ds} = -\gamma F(\mathbf{k}). \tag{2.167}$$

if we set

$$\gamma = C_{ds2}\omega k^8 F^2(\mathbf{k}). \tag{2.168}$$

In contrast to (2.163), the factor γ is determined now not by the integral properties of the wave field but by the local spectrum itself – in accordance with the hypothesis that whitecapping can be regarded as a local process in wavenumber space.

It appears unlikely, however, that the form (2.165) holds generally for arbitrary spectra. Consider, for example, two wave spectra A and B, each of which represents a narrow banded swell. Assume that the spectra are identical with respect to their total energy, mean frequency, rms slopes and angular distribution, differing only in their frequency band widths δf. The rate of dissipation of the total swell energy computed according to (2.165) differs in the two cases by the ratio $(\delta f_A/\delta f_B)^{-2}$. Yet in physical space the only difference between the two swell systems is that the scales of the envelopes of the slowly modulated swell amplitudes differ. The local Rayleigh statistics of all wave properties: the wave amplitudes, slopes, accelerations etc., which presumably govern the dynamics and statistics of local breaking events, are identical. It is therefore difficult to see any physical reason why the whitecapping dissipation should be different for the two cases. Implicit in this argument is, of course, the assumption that the wave breaking at the crest of some particular wave is not influenced by a possible breaking event on the adjacent wave crest a wavelength away - i.e. that whitecapping is, in fact, local in physical space rather than wavenumber space.

Dissipation expressions which are proportional to some higher power of the local wave spectrum have been proposed by other authors, for example by Snyder *et al* (1992) in a simple model to study the effects of nonlinearities of the dissipation source function on the energy balance of the spectrum. Although questionable in full generality, such expressions can nevertheless be justified for particular situations. If it can be argued that the factor γ is determined by an interaction integral which is dominated by the interactions in the neighbourhood of the given wavenumber $\mathbf{k}$, and if the spectrum is furthermore slowly varying within this region of integration, or follows a power law, the factor γ can be indeed be expressed as a function of the local spectral density and local wavenumber (the narrow-band spectrum in the counterexample just discussed clearly violates these conditions).

The available experimental evidence has not yet been analysed in sufficient detail to decide whether whitecapping should be regarded to first order as local in physical space and time or as local in wavenumber space – or whether neither assumption is an acceptable first approximation. The general quasi-linear form (2.163) can accommodate either hypothesis. Although the dissipation source function (3.50) used in the WAM model, cycle 4, is in accordance with an integral form for γ, the expression for γ is determined empirically and differs in detail from the integral form originally proposed by Hasselmann (1974). It would be useful to apply systematic inverse modelling techniques (to be discussed in chapter VI) to the available data, augmented by additional data on swell dissipation now becoming available through satellites to try to resolve this question.

II.4.3 Probability model

This approach is based on the Stokes' limiting criterion that waves will break when the downward acceleration at the crest exceeds $g/2$. Longuet-Higgins (1969) was the first to examine wave breaking from the point of view of probability theory. The breaking criterion based on Stokes' limiting wave form is:

$$a_b\omega^2 = g/2 \qquad (2.169)$$

or

$$a_b = a\left\{\frac{1}{2}g/(a\omega^2)\right\}, \qquad (2.170)$$

where a_b is the amplitude above which waves break. The critical amplitude a_b may also be viewed as the amplitude to which a wave is reduced in the breaking process to bring the acceleration at the crest down to $g/2$.

Using the assumption that wave breaking reduces waves of height greater than a_b to height a_b exactly, Longuet-Higgins (1969) deduced the expected value of the loss of energy per wave cycle as

$$\triangle E = \frac{1}{2}\rho_w g \int_{a_b}^{\infty}(a^2 - a_b^2)p(a)\mathrm{d}a, \qquad (2.171)$$

where $p(a)$ is the probability density function of wave heights.

The evaluation of a_b according to (2.170) requires information on the joint probability density function of wave height and frequencies. In 1969 there were no theoretical or observational descriptions of this joint probability density function. Consequently, based on the narrow banded assumption Longuet-Higgins replaced the wave frequency in (2.170) with the mean frequency $\bar{\omega}$:

$$\bar{\omega} = \left[\int_0^{\infty}\omega^2 F(\omega)\mathrm{d}\omega\Big/\int_0^{\infty}F(\omega)\mathrm{d}\omega\right]^{\frac{1}{2}}. \qquad (2.172)$$

Under the assumption of a Gaussian, stationary and narrow banded sea the wave heights (crest to following trough) are Rayleigh distributed and (2.171) may be evaluated to yield the energy loss per wave cycle:

$$\tilde{\omega} = \frac{\triangle E}{E} = \exp(-g^2/8\bar{\omega}^4 m_0), \qquad (2.173)$$

where m_i is the ith moment of the spectrum.

Yuan *et al* (1986) extended the approach of Longuet-Higgins (1969) to remove the restriction to narrow banded spectra. Again the surface elevation $\eta(t)$ is assumed to be Gaussian and stationary and breaking occurs wherever the vertical acceleration of the surface exceeds $g/2$. The action of breaking is to reduce the surface elevation in the ratio of $g/2$ to the actual vertical accelerations. The surface thus limited by breaking $\eta_b(t)$ is given by:

$$\eta_b(t) = \eta(t)\left\{\frac{g/2}{|\eta''(t)|}H[\,|\eta''(t)| - g/2\,] + H[\,g/2 - |\eta''(t)|\,]\right\} \qquad (2.174)$$

in which $H[\,]$ is the Heaviside unit step function. This amounts to clipping the crests of the steepest waves so that vertical accelerations are limited to $g/2$.

The spectrum of the 'broken waves' corresponding to (2.174) is then evaluated via the covariance of $\eta_b(t_1 = t)$ with $\eta_b(t_2 = t + \tau)$. After considerable manipulation (Yuan et al, 1986) the spectrum of the broken waves is then expressed in terms of the observed spectrum of surface elevation:

$$F_b(\omega) = A_1^2 \left[1 - \left(\frac{\omega}{\omega_1}\right)^2\right]^2 F(\omega), \tag{2.175}$$

where

$$\omega_1^2 = \left(\frac{A_1}{A_2}\right)\left(\frac{m_4}{m_2}\right) \gg \omega_p^2 \tag{2.176}$$

and

$$A_1 = (2\pi)^{\frac{1}{2}} L E_1(L^2/2) + 2\int_0^L Z(x)dx,$$

$$A_2 = (2\pi)^{\frac{1}{2}} L E_1(L^2/2), \tag{2.177}$$

$$L^{-1} = \frac{m_4^{\frac{1}{2}}}{g/2} \ll 1$$

is the normalized rms acceleration,

$$E_1(x) = \int_x^\infty \eta^{-1} e^{-\eta} d\eta \tag{2.178}$$

and

$$Z(x) = \frac{1}{(2\pi)^{\frac{1}{2}}} \exp\left(-\frac{x^2}{2}\right). \tag{2.179}$$

The dissipation source function may be deduced from the observed spectrum and the reduced spectrum of the broken wave field with an additional assumption regarding the time scale of the process of breaking. Longuet-Higgins implicitly assumes that the appropriate time scale is one wave period and here we take it to be the mean period of wave maxima:

$$T_p = 2\pi(m_2/m_4)^{\frac{1}{2}}. \tag{2.180}$$

Using the asymptotic expansion of the exponential integral function $E_1(L^2/2)$ we have the dissipation source function:

$$S_{ds3} = -C_{ds3} \left(\frac{m_0\omega_z^4}{g^2}\right)^{\frac{1}{2}} \exp\left\{-\frac{1}{8}(1 - \epsilon^2)\frac{g^2}{m_0\omega_z^4}\right\} \omega_z \left(\frac{\omega}{\omega_z}\right)^2 F(k), \tag{2.181}$$

where $\omega_z = (m_2/m_0)^{\frac{1}{2}}$ is the average zero crossing frequency and

$$\left(\frac{m_0\omega_z^4}{g^2}\right)^{\frac{1}{2}} = m_0^{\frac{1}{2}}k_z \qquad (2.182)$$

is proportional to the significant slope (Huang et al, 1981);

$$\epsilon^2 = 1 - \frac{m_2^2}{m_0 m_4} < 1 \qquad (2.183)$$

is a measure of the spectral bandwidth, and C_{ds3} is a constant.

The idea of using a limiting configuration to establish the rate of energy loss through breaking is an attractive one. The calculation depends on knowing the probability density of the limited (by breaking) waves as well as that which would be obtained if no breaking occurred. The latter is inaccessible from data, but may be approximated using a theoretical distribution, where one exists. This was the approach taken by Longuet-Higgins (1969) in his treatment of energy loss in a narrow banded spectrum. The Rayleigh distribution provided the 'ground state' from which the distribution of waves, limited by breaking could be subtracted. He used a mean frequency to establish the breaking criterion, yielding an overall energy loss with accompanying spectral distribution.

The approach taken by Yuan et al (1986), outlined above, is somewhat different. Here the idea was to obtain a spectral representation for wave breaking and so it was necessary to relax the narrow band approximation. Yuan et al use the observed spectrum and consider the changes that will be brought about by limiting the wave height in accordance with the Stokes criterion. If indeed the wave heights are 'hard-limited' in this way, then the observed spectrum would, in fact, correspond to the 'broken wave' spectrum in the development of Yuan et al. This approach, then, would not be fruitful since, by definition, no wave could be observed in excess of the limiting criterion and no information could be obtained on the energy available for loss through breaking.

Fortunately, waves are not 'hard-limited' in this way, because in a real sea, wave amplitudes are being modulated continuously by the underlying group structure. Imagine for the moment a train of perfect Stokes waves whose amplitudes are being very slowly (of order ten wave periods) increased. Ignoring energy loss through the propagation of parasitic capillaries, one might expect that at the point of exceedence of $g/2$ all the crests will begin spilling forward gently and the amplitudes will be hard-limited. This corresponds to the incipient breakers of the Banner (1990a) experiments in which a steady wave on a moving stream could be made arbitrarily steep by altering the depth or angle of attack of an underwater obstacle. In a real sea, the wave amplitudes change rapidly (of order one wave period) and the onset of breaking is rapid for waves which are momentarily well above the critical amplitude and correspondingly slower for waves marginally above it.

Two extremes of this are commonly observed in tanks and at sea. On the one hand, if a wave steepens very abruptly (short, high group structure) it will sometimes produce a distinct plunging breaker with the characteristic cylinder of enclosed air as the tip touches the surface. In this case the energy loss is very rapid, the group envelope is greatly attenuated and a sequence of two such plungers would be a rare event indeed. On the other hand, slowly modulated waves (long, moderate group structure) may sometimes momentarily exceed the limiting height without breaking at all. Once breaking begins it continues for some time (of the order of one wave period) and the amplitude of the underlying waves continuously diminishes as energy is fed to the turbulence.

Thus, the observed distribution of wave heights and periods includes three classes of waves: those too gentle to break, those that have just broken, and those whose amplitudes are larger than some limiting value and which are breaking or about to break. The difference between the probability distributions of all waves and of only the first two classes is a measure of the available energy to be dissipated. This is the basis for the approach of Yuan *et al* (1986).

The remaining question is the time scale associated with the loss of energy by wave breaking. Observations in tanks and in nature suggest that a typical active whitecap exists for a time between 1/4 and 1/2 of a wave period. During its active stage, the whitecap travels along with the phase speed of the underlying wave and as it goes, it leaves behind a patch of foam. The rate of advance of the forward edge of the whitecap is a good measure of the phase speed of the dissipating wave, while the longitudinal extent of the final patch of foam yields the distance travelled during its active stage. Observations of this sort would be very valuable in assigning a time scale to the process of wave dissipation leading to a loss of available wave energy as deduced from the probability approach of Yuan *et al* (1986). Longuet-Higgins (1969) implicitly assigned a time scale of one average wave period, while Yuan uses the peak period of the spectrum. It would seem that the local period would be more appropriate.

II.4.4 Summary

In the foregoing, we have described three distinct approaches to estimating the dissipation source function. It is worth remarking that they all depend on an incomplete description of the physics of the process of wave breaking and are, consequently, only approximate. Let us examine the sequence of events leading up to and during breaking to see how the various approaches fit into nature's scheme of things. Wave energy is continually being drained by viscosity. This process is on sound theoretical footing and is appropriately included in the model of Donelan and Pierson (1987) among others. When a wave becomes very steep, but not yet in the process of breaking, it radiates energy via 'parasitic capillaries' which themselves lose energy quickly to viscosity. This process is ignored in all the models. Once a wave starts breaking it loses energy in at least two ways: fluid is injected into the whitecap to become turbulent and the turbulence itself interacts with the orbital flow on the forward face of the wave. The presence of the whitecap on the forward face of the wave provides a downward pressure in the right phase to extract energy from the wave. This last effect is the basis of the Hasselmann (1974) model. Both the approach by Phillips (1985) and that by Yuan *et al* (1986) are not specific about the physics of energy loss but, instead, impose a limit on the allowed steepness of the waves. Phillips argues that this limit is maintained by a highly nonlinear process, which is associated with a localized region of wavenumbers around a particular breaking wave. The physics in the probabilistic approach of Yuan *et al* lies in setting the limiting criteria and determining the time scale of energy loss. In its present form the Stokes limit is applied and the time scale in chosen arbitrarily. An improvement on the choice of both limit and time scale requires considerable research into breaking as an unsteady phenomenon. Finally, we compare the form of the dissipation functions given by these three approaches. In particular, we are interested in the spectral distribution of dissipation and its sensitivity to the overall wave steepness. In figure 2.24 the three dissipation functions are compared for two spectra corresponding to full development and to strongly forced conditions ($U/c_p = 3$). For S_{ds1} the fitting parameters C_{ds1}, m and n are chosen to agree with Komen *et al* (1984); i.e. $3.33 \times 10^{-5}, 2$ and 1 respectively. The proportionality constants for the other two dissipation functions C_{ds2} and C_{ds3} were selected to yield the same peak value of the dissipation for the fully developed case: $C_{ds2} = 6.3$; $C_{ds3} = 700$.

S_{ds1} and S_{ds3} are both linear in spectral density and quadratic in frequency and, consequently, are identical in shape. However, their sensitivities to the overall steepness parameter are very different. S_{ds1} is quadratic in the steepness parameter and shows about two orders of magnitude enhancement at the peak of the strongly forced case over the fully developed value at the same wavenumber. Roughly a factor of six of the increase is due to the enhanced peak of the spectrum itself; the rest being due about equally to the increased peak frequency and to the overall steepness $\hat{\alpha}$. By contrast, S_{ds3} shows an increase of six orders of magnitude, almost all due to the exponential dependence on the significant slope (or, almost equivalently, the overall steepness).

The quasi-saturated model, S_{ds2} is different from the other two in both shape and sensitivity to the steepness. Since it is cubic in the degree of

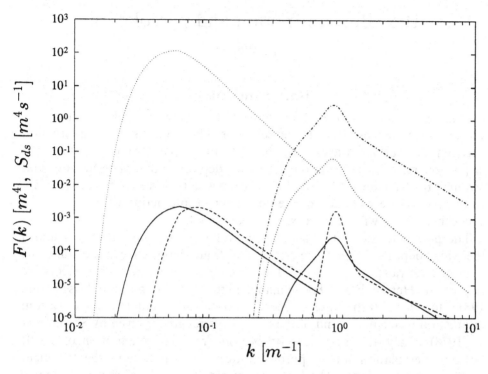

Fig. 2.24. A comparison of three dissipation source functions for a fully developed spectrum (peaks to the left of centre of the figure) and a strongly forced spectrum ($U/c_p = 3$). In both cases the wind speed is 10 m/s and the form of the elevation spectrum is as given by (2.233). Wavenumber spectra (dotted lines); whitecap model (solid lines); quasi-saturated model (dashed lines); probability model (dash-dotted lines).

saturation, its peak is shifted towards higher wavenumbers for full development, and much less so for the strongly forced case. In addition, the dissipation rate is cubic in the spectral density and acts to limit the spectral peak enhancement.

The evolution of the spectrum depends sensitively on the choice of source functions (Young and Banner, 1992). It remains to be seen which, if any, of the above dissipation source functions yields realistic results when tested in a model free of imposed constraints and driven by an appropriate input function.

II.5 Bottom friction and percolation

S. L. Weber

II.5.1 Introduction

In shallow water the orbital motions of the water particles, induced by surface waves, extend down to the sea floor. This gives rise to an interaction between the surface waves and the bottom. Apart from this effect the dispersion relation depends on the water depth, so that (i) refraction and shoaling become important and (ii) the deep water source terms are modified. In this section we will concentrate on the direct interaction with the bottom. The other effects will be discussed in chapter III.

The parameter kh, where k is the modulus of the wavenumber $\mathbf{k}$ and h the water depth, measures the strength of finite depth effects. We study the intermediate range $0.7 < kh < 3.0$. For $kh < 0.7$ strong nonlinear effects are important (Herterich and Hasselmann, 1980, Hasselmann and Hasselmann, 1981); for $kh > 3.0$ the water is deep. An overview of different wave–bottom interaction mechanisms and of their relative strengths is given by Shemdin *et al* (1978). They are: scattering on bottom irregularities, motion of a soft bottom, percolation into a porous bottom and friction in the turbulent bottom boundary layer. The first process results in a local redistribution of wave energy by scattering of wave components on, for example, mesoscale sand ripples or coral reefs. The last three are dissipative. Their strength depends on the bottom conditions. Bottom motion is only significant for a mud bottom. Percolation increases with the permeability of the bed material, whereas friction increases with the height of the roughness elements on the sea bed. A process not considered by Shemdin *et al* (1978) is sediment suspension under intense wave activity.

In this section we will discuss bottom friction and percolation. In the next section we will consider bottom motion and bottom scattering.

II.5.2 Dissipation over sand bottoms

The question of how to model bottom dissipation is answered by the bottom conditions. This seems to complicate things considerably in the case of a sand bottom. Percolation is important for coarse sand, but friction dominates for fine sand or when sand ripples are present. Bouws *et al* (1985) investigated an extensive shallow water data set (TMA) and did not find a dependence of the wave spectrum on differences in the bottom conditions. This gave rise to the (in fact unjustified) hypothesis that bottom dissipation is not of primary importance (Resio, 1987).

Waves in shallow water cause an uneven distribution of the pressure on

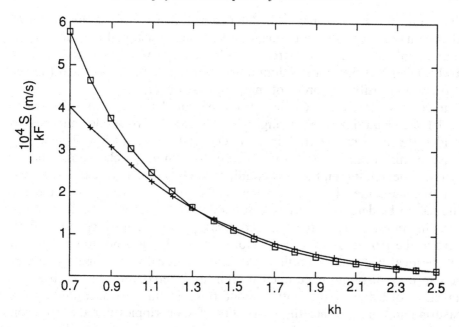

Fig. 2.25. The dissipation rate $-S/F$ divided by wavenumber k as a function of nondimensional water depth kh in the case of: percolation ($C_p/\cosh^2 kh$, with $C_p = 0.0006$ m/s) (crosses); friction ($C_f/\sinh 2kh$, with $C_f = 0.001$ m/s) (squares).

the bottom. If the bottom is porous, the water contained in the sediments moves in the direction of decreasing pressure. This motion of the water inside the bottom, which is obviously alternating with the wave motion above the bottom, implies a loss of energy. The phenomenon has been studied by Reid and Kajiura (1957). For a deep layer of isotropic sand the energy dissipation due to percolation is (Shemdin *et al*, 1978):

$$S_p(\mathbf{k}) = -C_p \frac{k}{\cosh^2 kh} F(\mathbf{k}), \tag{2.184}$$

where F is the variance spectrum of the surface elevation. The permeability coefficient C_p is typically 0.0006 m/s for fine sand with a mean sediment grain diameter $D_m = 0.25$ mm and $C_p = 0.01$ m/s for coarse sand with $D_m = 1$ mm.

The dissipation due to bottom friction is given by:

$$S_f(\mathbf{k}) = -C_f \frac{k}{\sinh 2kh} F(\mathbf{k}), \tag{2.185}$$

where to a first order approximation the frictional dissipation coefficient C_f can be taken as constant (see §§ II.5.6 – II.5.7). The coefficient C_f is typically 0.001 – 0.01 m/s, depending on the bed and flow conditions. The dissipation rates (2.184) and (2.185) are given in figure 2.25 for a situation where D_m

= 0.25 mm and C_f has a typical flat bed value. In this example the two processes have about the same magnitude. Their functional dependence on kh differs only in the case of extreme low-frequency swell. For coarse sand or gravel (higher C_p) percolation dominates, whereas for a rippled bed (higher C_f) friction is easily one order of magnitude stronger.

The coefficients C_p and C_f have been measured in the laboratory under well-defined conditions. Applying S_p or S_f in a wave model one has to consider the bottom material in a model grid-area, which will generally consist of mixed sand, gravel and possibly some mud patches. Sand ripples may occur depending on the flow conditions. Therefore, in practice, even the order of magnitude of the area-mean coefficient is often not clear and the value has to be determined by inverse modelling from the wave dissipation. From this point of view the two processes cannot be distinguished. With regard to the TMA study we can therefore remark that it is not to be expected that the local bed material will show up in the locally measured spectrum. One can only distinguish between friction and percolation on the basis of detailed knowledge of the bottom conditions in an area surrounding the measuring station. Neither the possibility of sand ripples nor the variation of the bed material over the fetch were considered in the TMA analysis.

Observations indicate that sand ripples are generally present on the continental shelves (Komar *et al*, 1972, Dingler and Inman, 1976, Amos *et al*, 1988, Drake and Cacchione, 1992). These can be generated by wave motion, by stationary currents or by both. Therefore, we hypothesize that friction is the dominant bottom dissipation process over the continental shelves. Under specific conditions (river estuaries with a large outflow of mud, coral reefs) other mechanisms should be considered.

II.5.3 Basic parameters of the bottom boundary layer

The viscosity of the water, however small it may be, imposes the no-slip condition at the bottom. This gives rise to a thin boundary layer, where the horizontal velocity increases rapidly from zero at the bottom to a finite value, determined by the outer flow, at the top. The flow is hydrodynamically rough when the roughness length k_N, which can be associated with the roughness elements on the sea bed, is much larger than the viscous length scale v/u_*, where v is the kinematic viscosity of the water and u_* the turbulent friction velocity (see, for example, Sleath, 1984). Flow over sand is almost always hydrodynamically rough in nature.

The roughness length is determined by the bed material, which may consist of different sands (ranging from fine to coarse), gravel, shells, rock etc. These materials are moved by bioturbation and the water motion. Sand ripples may be created within a few wave cycles by storm waves (Dingler and Inman, 1976) or they may be tide-generated. They erode slowly and are only

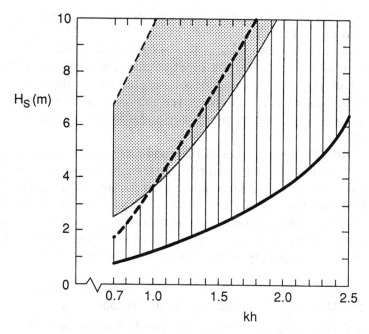

Fig. 2.26. The sand-ripple regime as a function of the significant wave height H_S and the nondimensional water depth kh for two example peak frequencies $f_p = 0.12$ Hz (wind sea) and $f_p = 0.06$ Hz (swell). A flat sea bed corresponds to a rms bottom orbital velocity $(U_w^b)_{rms}$ smaller than 15 cm/s for wind sea (bold solid line) and 25 cm/s for swell (simple solid line). Sheet flow occurs for $(U_w^b)_{rms}$ larger than 70 cm/s for both wind sea (bold dashed line) and swell (simple dashed line). In the intermediate ripple regime (hatched for wind sea and grey for swell) sand ripples are formed. The surface wave parameters have been used to estimate $(U_w^b)_{rms}$ (see Weber, 1991a).

destroyed by extreme waves, which cause the sand to move in layers (sheets) above the sea bed (Amos *et al*, 1988).

Ripple growth and decay are processes on extremely short time and length scales which cannot be resolved explicitly by a wave model. However, it is important to distinguish between different ripple regimes. The motion of the bed material is determined by the ratio of the bottom stress to the restoring force of gravity (Sleath, 1984). This ratio can be estimated from the root-mean-square free-stream bottom orbital velocity $(U_w^b)_{rms}$ using typical values of the sediment grain diameter $D_m = 0.25$ mm and sediment density $\rho_s = 2.65 \times 10^3$ kg m^{-3}. Dingler and Inman (1976) give limits for the different ripple regimes in terms of $(U_w^b)_{rms}$, which were applied by Weber (1991a) to two cases of extremely shallow water waves. Approximating the bottom velocity in terms of the significant wave height, the wave frequency and kh, the ripple regime can be expressed in terms of surface wave parameters. Results are shown in figure 2.26.

In the case of a flat bed (low $(U_w^b)_{rms}$) the bed roughness scales with D_m. Friction is then small compared to the other source terms (Weber, 1991a). For sheet flow (high $(U_w^b)_{rms}$) the roughness is determined by the flow itself (Wilson, 1989). Sheet flow has up till now never been considered as a bottom friction process, although it can be significant under high waves in not too deep water (Staub *et al*, 1985).

Critical Reynolds numbers in terms of the nondimensional bottom roughness are given by Jonsson (1980). In the examples given in figure 2.26 the flow is always fully turbulent if it exceeds the limit for ripple formation. For flat beds the Reynolds number is well below its critical value and the flow is laminar or in the transition regime. In the following we will restrict our attention to rough turbulent flows in the intermediate ripple regime. The roughness length is then typically of the order of centimetres.

II.5.4 Dissipation as a function of the bottom stress

In this section we will derive an expression for the dissipation due to bottom friction, considering the general case of a random wave field superimposed on a mean current. Our approach is based on Kajiura (1968), who treated a single wave, and on Christoffersen and Jonsson's (1985) extension of his argument to the combined monochromatic wave–current flow.

The linear momentum equation for boundary layer flow is:

$$\frac{\partial \mathbf{u}}{\partial t} + \frac{1}{\rho_w} \nabla p = \frac{1}{\rho_w} \frac{\partial \tau}{\partial z}, \qquad (2.186)$$

with ρ_w the density of water, $\mathbf{u}$ the Reynolds-averaged horizontal velocity, p the Reynolds-averaged pressure and τ the turbulent stress in the boundary layer. For notational convenience the density will be absorbed in the stress in the following, so that τ always has the dimension of a velocity squared.

One can formally divide $\mathbf{u}, p$ and τ into a steady, ensemble-averaged part and a remainder, for example:

$$\tau = \tau_c + \tau_w, \qquad \text{with} \begin{cases} \tau_c &= \; <\tau>, \\ \tau_w &= \; \tau - <\tau>, \end{cases} \qquad (2.187)$$

where $<\;>$ denotes averaging over the random wave phase. The velocity and the pressure consist of a linear superposition of a pure current motion and a pure random wave motion. However, the stress is nonlinear and τ_c and τ_w depend on the combined motion. It is assumed that τ_w is zero outside the wave boundary layer, which is a thin sublayer of the current boundary layer.

The evolution of the wave momentum can be found by averaging (2.186) over the random wave phase and subtracting the averaged equation from the full equation (2.186):

$$\frac{\partial}{\partial t}\mathbf{u}_w + \frac{1}{\rho_w}\nabla p_w = \frac{\partial}{\partial z}\tau_w. \qquad (2.188)$$

As the pressure gradient is independent of height inside the wave boundary layer, it can be replaced by its value at the top:

$$\frac{\partial}{\partial t}(\mathbf{u}_w - \mathbf{U}_w^b) = \frac{\partial}{\partial z}\tau_w, \qquad (2.189)$$

with $\mathbf{U}_w$ the free-stream orbital velocity. The superscript b denotes the value at the bottom $z = -h$, which is equal to the value at the top of the boundary layer for the inviscid wave flow.

The energy dissipation of the wave motion due to bottom friction can be obtained by multiplying (2.188) by $\mathbf{u}_{w\mathbf{k}}$, the orbital velocity of the wave component with wavenumber $\mathbf{k}$, integrating over the depth and taking the average over the random wave phase:

$$
\begin{aligned}
g\frac{\partial}{\partial t}F(\mathbf{k}) &= <\int_{-h}^{-h+\delta} \mathbf{u}_{w\mathbf{k}}\cdot\frac{\partial}{\partial z}\tau_w\,\mathrm{d}z> \\
&= <\int_{-h}^{-h+\delta} \left\{\mathbf{U}_{w\mathbf{k}}^b + (\mathbf{u}_{w\mathbf{k}} - \mathbf{U}_{w\mathbf{k}}^b)\right\}\cdot\frac{\partial}{\partial z}\tau_w\,\mathrm{d}z> \\
&= <\mathbf{U}_{w\mathbf{k}}^b\cdot\tau_w>|_{-h}^{-h+\delta} + <\int_{-h}^{-h+\delta}\tau_w\cdot\frac{\partial}{\partial z}\mathbf{U}_{w\mathbf{k}}^b\,\mathrm{d}z> \\
&\quad + <\int_{-h}^{-h+\delta}(\mathbf{u}_{w\mathbf{k}} - \mathbf{U}_{w\mathbf{k}}^b)\frac{\partial}{\partial t}(\mathbf{u}_w - \mathbf{U}_w^b)\,\mathrm{d}z>,
\end{aligned}
$$

where δ is the thickness of the wave boundary layer. The second term vanishes because $\mathbf{U}_{w\mathbf{k}}^b$ is constant in the wave boundary layer and the third term vanishes because of periodicity of the motion. We finally find:

$$S_f(\mathbf{k}) = -\frac{1}{g}<\mathbf{U}_{w\mathbf{k}}^b\cdot\tau_w^b>. \qquad (2.190)$$

The spectral energy dissipation thus depends on the (known) free-stream orbital bottom velocity and the (basically unknown) turbulent bottom stress. One generally assumes that the stress can be parametrized in terms of the bottom roughness and parameters of the inviscid wave–current flow. A slightly different derivation of (2.190) is given by Hasselmann and Collins (1968).

Direct measurements of the bottom stress induced by random waves alone or superimposed on a mean current do not exist. Field measurements are always taken well above the wave boundary layer (see, for example, Huntley and Hazen, 1988). Laboratory experiments are conducted for a single wave or for a small number of low-steepness wave components (swell spectra). In some experiments the roughness elements are glued to the bottom in such a way that the bed material and the flow are not in equilibrium (Sleath, 1987, Jensen *et al*, 1989), which makes it difficult to apply the results to field conditions. In many experiments not the turbulent stress itself but the amplitude decrease of the surface wave is measured. Interpretation of the data then relies on the parametrization assumed for the bottom stress and, in the case of more wave components, on the validity of neglecting nonlinear interactions. A useful overview of present knowledge on monochromatic wave cases is given by Jonsson (1980) and on combined wave–current cases by Christoffersen and Jonsson (1985).

Parametrization of the bottom stress in the case of random waves has to rely on what is known about monochromatic waves. This seems to be an acceptable approximation as the bottom velocity spectrum is generally very narrow and single-peaked. Results from purely oscillatory flow were already applied in § II.5.3.

II.5.5 Parametrization of the bottom stress

There are numerous approaches to solving the closure problem of the turbulent boundary layer equations, ranging from a simple empirical formula to higher order closure schemes. One approach is the eddy-viscosity model, which has been widely applied to stationary flows in the atmosphere or the ocean. In this model the turbulent stress is parametrized in analogy to the viscous case:

$$\tau = v_e \frac{\partial \mathbf{u}}{\partial z}, \tag{2.191}$$

with $v_e \sim \tau^{1/2}$ the so-called eddy-viscosity coefficient. For stationary flows the magnitude of the stress τ can equivalently be expressed by a drag law of the form:

$$\tau = C_D(z)U^2, \tag{2.192}$$

where U denotes the magnitude of the velocity at a reference height z and the drag coefficient C_D is a function of the ratio of the surface roughness to the reference height. In the following the top of the wave boundary layer will be taken as the reference height.

The eddy-viscosity concept was first applied to the oscillatory boundary layer by Kajiura (1968). Models for the combined monochromatic wave–current flow have been developed by Grant and Madsen (1979) and Christoffersen and Jonsson (1985). Their approaches are similar. We will discuss the main features of the eddy-viscosity model of Christoffersen and Jonsson (1985).

In the case of a combined wave–current motion it is possible to split

(2.191) into a wave part and a current part, as defined in (2.187):

$$\tau = \tau_c + \tau_w = v_e \frac{\partial \mathbf{u}_c}{\partial z} + v_e \frac{\partial \mathbf{u}_w}{\partial z}. \tag{2.193}$$

The stress is thought to be determined by the current alone in the so-called current boundary layer and by the current and the wave motion in an interior boundary layer close to the sea bed, the wave boundary layer. We write:

$$v_e = v_{ec} \sim (\tau_c^b)^{1/2} \qquad \text{outside the wave boundary layer,} \tag{2.194}$$

$$v_e = v_{ew} \sim <(\tau^b)^2>^{1/4} \quad \text{inside the wave boundary layer.} \tag{2.195}$$

Here we have slightly modified Christoffersen and Jonsson's definition in order to account for the random wave field. An essential assumption is that v_{ew} depends only on the vertical coordinate, not on the wave phase. Trowbridge and Madsen (1984) show (for monochromatic flow) that the wave energy dissipation is adequately described using a time-independent eddy-viscosity coefficient.

Solving the boundary layer equations with appropriate boundary conditions yields for τ_c a drag law of the form (2.192), with $U = U_c^b$. For τ_w there are two possibilities. The first is to retain a spectral description. This yields drag laws for each spectral component. The second is to represent at one step or another in the computations the frequency range by, for example, the peak frequency. This results in an integral drag law of the form:

$$\tau_w^b = C_D (U_w^b)_{rms} \mathbf{U}_w^b. \tag{2.196}$$

Note that the wave part and the current part of the bottom stress are related to the wave part and the current part of the free-stream bottom velocity, respectively, by (2.192) and (2.196). The influence of the wave flow and the steady flow on each other is represented by C_D, which can be enhanced due to the combined motion compared to the pure wave or pure current case. Applying this parametrization to a number of cases of combined flow it is found that the steady current feels the wave motion much more strongly than the wave motion feels the current. Christoffersen and Jonsson (1985) show that their model compares favourably with laboratory measurements.

Hasselmann and Collins (1968) offered a completely different approach to model the bottom stress induced by a random wave field superimposed on a steady current. They proposed as a generalization of (2.192) a drag law which relates the total bottom stress to the total velocity at a reference height, for example the top of the wave boundary layer:

$$\tau^b = C_D \mid \mathbf{U}^b \mid \mathbf{U}^b. \tag{2.197}$$

Note that (2.197) is nonlinear in the wave phase, whereas (2.196) is linear.

The value of the drag coefficient has to be estimated from measurements of wave dissipation.

Let us illustrate the differences between (2.196) and (2.197) by an example. For computational simplicity we will consider a case where $\mathbf{U}_w^b$ and $\mathbf{U}_c^b$ are in the same direction. Furthermore we will assume that $U_c^b > U_w^b$, so that $\mathbf{U}_c^b + \mathbf{U}_w^b > 0$ always and τ_w^b can easily be computed from (2.197). According to the eddy-viscosity model the wave motion does not feel the current for $U_c^b < U_w^b$. Therefore this is not a serious restriction. Substitution of (2.196) in the dissipation expression (2.190) yields:

$$S_f(\mathbf{k}) = -\frac{1}{g}C_D(U_w^b)_{rms} < (U_{w\mathbf{k}}^b)^2 > . \qquad (2.198)$$

On the other hand, combining (2.197) with (2.190) results in

$$S_f(\mathbf{k}) = -\frac{2}{g}C_D U_c^b < (U_{w\mathbf{k}}^b)^2 > . \qquad (2.199)$$

Both expressions contain the mean square of the bottom velocity which can be associated with the wave component with wavenumber $\mathbf{k}$. Rewriting this factor in terms of the surface elevation spectrum one finds from (2.198)

$$S_f(\mathbf{k}) = -2C_D(U_w^b)_{rms}\frac{k}{\sinh 2kh}F(\mathbf{k}). \qquad (2.200)$$

This expression is equal to (2.185) with $C_f = 2C_D(U_w^b)_{rms}$. From (2.199) one also rederives (2.185), but with $C_f = 4C_D U_c^b$. In the two-layer eddy-viscosity model C_f is thus a linear function of the rms bottom velocity, whereas in the one-layer drag law it is proportional to the current velocity. Moreover, in the former C_D depends on the combined flow as stated earlier, whereas in the latter the value of C_D is not specified.

II.5.6 Models of bottom friction

After the pioneering work of Putnam and Johnson (1949) a small number of expressions have been proposed to model the dissipation of random ocean waves due to friction. In the following we outline these; S_f has always the general form (2.185). Depending on the parametrization used for τ^b the dissipation coefficient C_f is a function of the bottom roughness, integral spectral parameters, the frequency or direction of a wave component.

Hasselmann and Collins (1968) proposed a *drag law* of the form (2.197). This results in a complicated expression for C_f which depends on the drag coefficient, on the bottom velocity spectrum and on the direction of a wave component relative to the primary axis of the bottom velocity spectrum. The latter feature appears because τ^b is taken to be nonlinear in the random wave phase (Weber, 1991b). Due to the formulation of the drag law there is a strong dependence on the tidal current. Collins (1972) approximates the full coefficient by leaving out the dependence on the direction of a wave component.

In Hasselmann *et al* (1973) the Hasselmann and Collins theory was applied to measurements of swell decay. The drag coefficient was basically assumed to be constant. The computed values of C_f ranged over two orders of magnitude and no correlation was found with the tidal phase or velocity or swell parameters like the wave height, direction or frequency. Although these results were quite discouraging, the mean value (over the data set) of C_f performs quite well in operational wave forecasts (Janssen *et al*, 1984). For extreme storm waves the mean value is too low (Bouws and Komen, 1983).

With hindsight two comments can be made on the JONSWAP results. The first is that the wave orbital velocities were mostly in the range 0.1 – 5 cm/s. This is too low to generate turbulence in the wave boundary layer (Jonsson, 1980), so that the wave–bottom interaction would be very weak. An alternative explanation of the observed decay is discussed in § II.7. For a few of the JONSWAP cases the wave motion yields Reynolds numbers above the critical value and bottom velocities of the same order of magnitude as the tidal current. The tidal signal was not visible in these cases. From the eddy-viscosity model it follows that this is to be expected as the tidal current does not modify the wave energy dissipation, when the current velocity and the orbital bottom velocity are of the same order of magnitude. This seems to indicate that a two-layer eddy-viscosity model performs better for combined wave–current flow than a one-layer drag law.

An *eddy-viscosity model* for random wave dissipation, with a *monochromatic approximation* of the spectral frequency range by the peak frequency, has been proposed independently by Madsen *et al* (1989) and Weber (1989, 1991a). This approximation is valid for self-similar spectra, which can be represented by one spectral parameter. Their models differ in details. Applying the approximate eddy-viscosity model to an extreme storm and to a swell case yields values of C_f which agree with measurements of wave decay (Weber, 1991a).

A *spectral eddy-viscosity model* has been derived by Weber (1991b). This model can be applied in complicated situations where the spectrum loses its self-similar form and double peaks appear, like swell interacting with wind sea or quickly turning winds. The coefficient C_f depends on the bottom velocity spectrum, the bottom roughness and the frequency of a wave component.

II.5.7 Evaluation

Tests with regional versions of the WAM model (see chapter IV) have shown that the mean JONSWAP value of $C_f = 0.008$ m/s is adequate for moderate storms, but that C_f should depend on the wave field for extreme storm events (Cavaleri *et al*, 1989, Weber, 1991a). Hindcasts of one storm with different dissipation expressions show that the dependence of C_f on the direction (drag law (2.197)) or frequency (spectral eddy-viscosity model) of a wave component has no impact on the wave height or mean frequency (Weber, 1991a), although these features may change the details of the spectral shape. A developing wave field reaches equilibrium almost immediately after the wave motion starts to feel the bottom and bottom friction becomes significant in the wave energy balance (Weber, 1988). The value of C_f therefore does not change much during a storm, but it varies from event to event depending on the wind conditions. The eddy-viscosity model (2.200), equivalent to Collins' (1972) drag law approximation, seems to be a reasonable alternative to using a constant value for C_f. The drag coefficient C_D can be approximated from Grant and Madsen's (1979) model for monochromatic flow or the numerical fit for C_f computed from the approximate spectral model by Weber (1989, 1991a) can be used. The latter is given by:

$$C_f = \exp(-8.34 + 6.34 z_b^{0.08})(U_w^b)_{rms}. \qquad (2.201)$$

Here C_f depends explicitly on $z_b = k_N \omega_p / (U_w^b)_{rms}$, the nondimensional bottom roughness. A constant roughness value $k_N = 0.04$ m is adequate and compatible with the flow conditions for a range of swell and wind sea spectra (Weber, 1991a). The influence of the tidal current on the wave energy dissipation is generally small enough to be neglected.

II.5.8 Conclusion

Evidently, there have been advances since the 1970s in modelling bottom friction: application of the eddy-viscosity concept, theories on wave–current interaction and consideration of sand motion have let to a better under-standing. In particular, it is now understood that the effect of currents on wave dissipation is relatively small. Also, the limits of bottom friction models have been defined: the motion should be intense enough to generate turbulence, but not so intense that sheet flow occurs. New insights and val-idation of these models must now come from measurements of the bottom stress under a random wave field and of the spectral evolution in shallow water.

II.6 Bottom scattering and bottom elasticity

L. Cavaleri

In this section we will discuss bottom scattering and bottom elasticity. The first process is conservative, the second is not.

Bottom scattering

A surface wave field produces on the bottom a pressure and a horizontal velocity field. Because of the differential attenuation with depth of the different frequencies, the signal on the seabed is shifted towards lower frequencies, that is towards the longer wavelengths.

The bottom is rarely flat. If rocky, we can expect irregular, more or less pronounced, corrugations; if sandy, various wavelengths will be present, from the order of centimetres (ripples) up to that of kilometres (sandwaves, see Sleath, 1984). If their statistical characteristics do not vary much in space, they can be represented by a two-dimensional spectrum. The possibility exists that the two spectra, waves and bottom, interact with each other. The phenomenon has been studied theoretically by Hasselmann (1966) and was used by Long (1973) to justify the swell attenuation measured during the JONSWAP experiment. Given proper resonance conditions, the interaction between the two spectra causes a scattering of wave energy, and consequently, in the case of swell, a decrease of the overall energy propagating towards the shore. The results of Long (1973) suggest that a 0.60 m rms oscillation of the bottom depth would have been enough to justify the JONSWAP experimental results.

Practical application of scattering theory in wave modelling is hindered by two serious difficulties. The first is the heavy computational requirement for the calculation, which is acceptable for an experiment, but inhibits routine application. The other is the need to know the two-dimensional bottom spectrum. This information is rarely available. The problem is rendered more serious for sandy bottoms, when the profile may change in time without necessarily keeping its statistical characteristics.

It is therefore not surprising that the approach by Long (1973) has not received much following. Its potential importance has been clearly shown by Davies and Heathershaw (1983) with some wave tank tests. Sinusoidal waves of different frequencies were generated at one end of the tank. A section of the bottom was shaped as a sequence of sinusoidal oscillations, their number varying from four to ten in the different experiments. Directional spectra were measured along the tank by several gauges. The results, shown in figure 2.27, are very remarkable. The process is very selective in frequency,

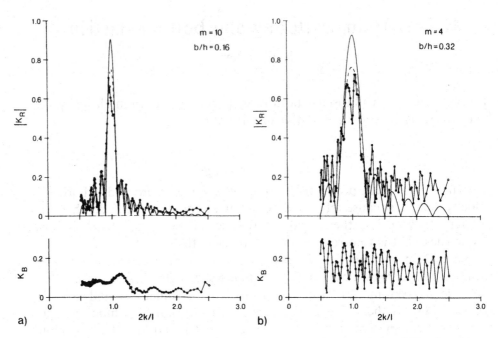

Fig. 2.27. Wave reflection coefficient in a wave tank (after Davies and Heathershaw, 1983).

and four oscillations are sufficient to reflect 80% of the incoming wave, if its wavelength doubles that of the bottom oscillation. In this one-dimensional case we are effectively dealing with the Bragg effect (see, for example, Elmore and Heald, 1969). The larger the number of oscillations, the more selective is the process. On top of this there is a very good agreement between theory and experiment.

Comparable results are very difficult to achieve in the field, because it is difficult to obtain a sufficiently accurate evaluation of the actual directional distribution of energy in the bottom and wave spectrum. This information is essential because the consequences of scattering are very sensitive to the details of the distribution. Taking advantage of a simpler situation, a field verification has been attempted by Ewing and Pitt (1982) with the analysis of swell data recorded by a pitch-roll buoy on the west coast of Scotland. In this area the bottom is characterized by bottom ridges parallel to the coast, with an intermediate distance of the same order of magnitude as the local swell wavelength. The measurements were available at only one point, so the only evidence of backscatter could be looked for in the possible presence of energy moving away from the coast. Using the sensitive directional analysis scheme of Long and Hasselmann (1979), Ewing and Pitt (1982) have indeed found indications of a back flow of energy from the coast. However, the uncertainties associated with the variations in the bottom depth impeded Ewing and Pitt from drawing any final conclusion.

We end this discussion with a historical note. In the ancient Republic of Venice it was common practice to discharge into the sea the debris of construction or demolition work. Divers in front of the Venetian littoral have often reported the presence of remnants of these discharges, distributed along lines parallel to each other and to the coast, with an interval of 50 – 60 m. This distance is about half the wavelength of the sirocco storm waves that propagate perpendicular to the coast in this area. We like to think that the ancient seamen and engineers had empirically discovered the efficiency of the Bragg effect and devised a simple and efficient defence for their precious coastline.

Bottom elasticity

In standard wave modelling the seabed is considered to be rigid, except for the possible horizontal transport of sand. For a rocky bottom the hypothesis is obviously correct. In case of sand a certain degree of elasticity is present, and the seabed reacts with small vertical movements to the variation of pressure associated with the overlying wave field. These movements have been measured by Rosenthal (1978) and Forristall and Reece (1985). They found the motion to be quite limited in amplitude (order of one centimetre at most), and substantially in phase with the pressure wave at the bottom. The related energy absorption turned out to be negligible, the seabed acting as an undamped spring.

The picture changes completely if the bottom is characterized by a viscous mud layer. In this case a bottom wave, slowed down by viscosity, is *pulled* by the surface wave. The viscous resistance leads to a large phase difference between the forcing function (the pressure wave) and the resulting oscillation (the mud wave), with a consequent strong absorption of energy. The results can be quite spectacular and the process dominant in the control of the evolution of the wave conditions. With large water waves at the surface in relatively shallow water, the amplitude of the bottom wave can be quite large, up to the level of metres.

There have been several approaches to the problem, examples being Hsiao and Shemdin (1980), Kirwan (1985) and a number of papers by Yamamoto, the last one being Yamamoto and Tori (1986). In general terms the problem is solved similarly to elementary wave theory (§ I.2.2). The mud wave is represented by a stream function. For our present purposes the relevant result is the modification of the dispersion relation (Hsiao and Shemdin, 1980) into

$$k = \frac{\omega^2}{g} \frac{1 + \tanh kh\,\Omega}{\tanh kh + \Omega}, \qquad (2.202)$$

where k is the wavenumber, σ the angular frequency, g gravitational acceleration and h the water depth. Ω is expressed as

$$\Omega = r\frac{(m^2 - k^2)A}{(m^2 + k^2)B - 2k^2},$$ (2.203)

where r is the ratio between water and mud density and A and B are functions of k, of the thickness H of the mud layer and of a function m defined by

$$m = k\left[1 - \frac{\omega^2}{k^2(J - i\omega v)}\right]^{\frac{1}{2}}.$$ (2.204)

Here J is the ratio between the shear modulus G and the density of the mud and v is the kinematic viscosity.

From (2.202) it is evident that the bottom motion affects the wavenumber, i.e. the wavelength and the phase speed of the surface wave. Expression (2.202) reduces to the usual dispersion relationship (1.55) when at least one of the following conditions is satisfied:

$H = 0$ no mud layer,
$G \to \infty$ or $v \to \infty$ a solid bottom,
$r = 0$ very large mud density.

In general, the wavenumber k is a complex number. When writing $k = k_r + ik_i$, the real part k_r is connected to the actual wavelength $L = 2\pi/k_r$ and k_i is the attenuation coefficient appearing in the energy dissipation law:

$$\dot{E} = -2k_i c_g E.$$ (2.205)

For wave modelling purposes the linearity of the process allows independent application of (2.205) to each single wave component in the spectrum.

In practice, (2.202) is solved numerically. Hsiao and Shemdin (1980) have explored different combinations of the controlling factors (depth of the layers, viscosity and density of mud, shear modulus). Their main findings can be summarized as follows:

– The effect of the mud layer decreases with increasing water depth or, which is the same, with decreasing wavelength.
– The effect decreases with increasing stiffness of the mud.
– The wavelength decreases. The difference is easily 10 or 20%, and may reach 50% in spectacular cases.
– Dissipation is higher with thicker mud layers.
– Dissipation increases with increasing mud viscosity, but only up to a certain point, after which the excessive viscosity slows down the mud motion and decreases the dissipation.

A practical difficulty for the application of (2.205) is the evaluation of the dissipation coefficient k_i, which may change as a function of the wave conditions. We will come back to this in § IV.6, where we will discuss the practical application of dissipation by a viscous muddy layer.

II.7 Interactions with ice

D. Masson

The most northerly and southerly boundaries of global wave models correspond to the polar ice packs, whose extensions undergo large seasonal variations. From the point of view of wave modelling, the solid ice pack can be conveniently treated as land, with a fully absorbing property and no energy input to the sea. However, for a significant fraction of the seasonal ice cover, the sea surface is only partially covered by dispersed ice floes; the so-called marginal ice zone (MIZ) can extend over wide areas between the solid ice and the open ice-free sea.

When the wind blows from the solid ice pack towards the open sea, waves growing inside the MIZ are scattered by the floes, and their spectral characteristics modified. In such a situation, a practical consideration is to determine whether or not wave generation within the ice cover must be included in wave climate studies and forecasts. The development of ocean waves in the MIZ has been studied by Masson and LeBlond (1989). For wave generation by the wind to be of significance, the fraction of the area covered by ice, f_i, has to be relatively small. Thus the region of interest is restricted to the outer portion of the ice cover where the ice field is composed of randomly distributed small floes with no preferred shape. Accordingly, a sparse, homogeneous, random distribution of rigid, noncolliding ice floes with the shape of truncated cylinders is assumed over the sea surface. Each mass of floating ice endeavours to follow the displacement of the supporting water surface within limits imposed by its rigidity and inertia. The associated scattering effect tends to decrease the energy content of the wave field and to spread the wave energy out over a broader range of directions.

The evolution of the wave spectrum is obtained by integration of the energy balance equation, with the usual wind input, S_{in}, dissipation, S_{ds} and nonlinear interactions term, S_{nl}, modified to include the effect of the partial ice cover. First, because waves can be neither generated by the wind nor dissipated by the usual breaking mechanisms in the fraction of the sea surface covered by ice, both S_{in} and S_{ds} are reduced by a factor $(1 - f_i)$. In addition, a new term, S_{ice}, is added which parametrizes the directional scattering and extra dissipation of energy by the ice. The determination of this additional source term consists of analysing the scattering of a random wave field incident on a random array of ice floes. Assuming that each component of the wave spectrum interacts independently with the floes, the directional scattered spectrum is then obtained, for each frequency, by summing the contributions from the incident and scattered waves of all directions.

In order to determine S_{ice}, the scattering of one spectral component by a single floe is first examined. A floating object both diffracts and scatters incoming wave energy in all directions. Assuming the fluid to be incompressible and the flow irrotational, the problem reduces to the determination of a velocity potential, ϕ, which satisfies the Laplace equation, $\nabla^2\phi(x, y, z; t) = 0$, where (x, y, z) are Cartesian coordinates with the z axis positive upwards and $z = 0$ the plane of the undisturbed free surface, and the x axis in the direction of the incident wave. The fluid motion is assumed harmonic in time, and to be the sum of the incident wave, ϕ_i, and a disturbance due to the presence of the body, ϕ_s:

$$\phi = (\phi_i + \phi_s)e^{-i\omega t}, \qquad (2.206)$$

where ω is the angular frequency. The linearized potential for a plane wave of amplitude a is

$$\phi_i = -\frac{iga}{\omega}\frac{\cosh[k(z+h)]}{\cosh(kh)}e^{ikx}, \qquad (2.207)$$

where k is the wavenumber with $k = \omega^2/g\tanh(kh)$, and h the water depth. The disturbance potential can be treated linearly as the sum of the contributions from the diffraction of the incident wave on a fixed body, and the scattering by a body forced to oscillate in still water. At a large distance, r, from the body, this potential takes the form

$$\phi_s = -\frac{iga}{\omega}\frac{\cosh[k(z+h)]}{\cosh(kh)}D(\omega, \theta)r^{-1/2}e^{ikr}. \qquad (2.208)$$

This potential describes cylindrical waves radiating away from the floe, and for which the scattering coefficient, $D(\omega, \theta)$, gives the distribution of scattered wave energy around the floe. For a given floe geometry, the scattering coefficient is a function of the horizontal scale of the floe relative to the incident wavelength only. In general, the amplitude of the scattered waves is maximum when the incident wavelength is comparable to the floe diameter.

To obtain the wave field resulting from the scattering of each spectral component on the whole array of floes, a statistical average of the scattered waves is taken over the domain using the theory of wave propagation and scattering in random media. First, the total wave field is divided into an average (coherent) field and a fluctuating (incoherent) field. Because of the randomness of the ice floe distribution, the phases of the waves scattered by different floes are not correlated. At any point in the ice field, the total wave energy is then the sum of the energy associated with the coherent field and with the fluctuating field. In terms of the surface displacement, $\eta(\mathbf{r}_a, t)$, at a point $\mathbf{r}_a$ between ice floes, this can be expressed, neglecting multiple scattering, as

$$< |\eta(\mathbf{r}_a, t)|^2 > = |< \eta(\mathbf{r}_a, t) >|^2 + \int\int |v_s^a|^2 < |\eta(\mathbf{r}_s, t)|^2 > \rho_e(\mathbf{r}_s)d\mathbf{r}_s, \quad (2.209)$$

where $|v_s^a|^2 < |\eta(\mathbf{r}_s, t)|^2 >$ is a symbolic notation to indicate the scattered wave energy at $\mathbf{r}_a$ due to a scatterer located at $\mathbf{r}_s$. The 'effective' number density, $\rho_e(\mathbf{r}_s)$, gives an estimate of the number of floes per unit area radiating waves at $\mathbf{r}_a$ without being shaded by other floes. Using the far-field expression in the case of single scattering of equation (2.208), the distribution of wave energy over direction after scattering can be expressed in terms of the scattering coefficient $D(\omega, \theta)$.

When a wave spectrum, $F(\omega, \theta)$, enters an ice field, the scattered spectrum, $F_{ice}(\omega, \theta)$, can be obtained by summing, for each spectral component, the contributions coming from waves of the same frequency and incident from all directions according to

$$F_{ice}(\omega, \theta_i) = F(\omega, \theta_j)(T_{ij})_\omega \quad (2.210)$$

with each element $(T_{ij})_\omega$ of the transfer function matrix taking the form

$$(T_{ij})_\omega = A^2[\beta|D(\omega, \theta_{ij})|^2\Delta\theta$$

$$+ \delta(\theta_{ij})(1 + |\alpha D(\omega, 0)|^2) + \delta(\pi - \theta_{ij})|\alpha D(\omega, \pi)|^2], \quad (2.211)$$

where $\theta_{ij} = |\theta_i - \theta_j|$, $\Delta\theta$ is the angular interval of the spectrum, δ is the Dirac function, and the scattering parameters β and α are functions of ω. In equation (2.211), the wave energy after scattering is composed of the incident wave energy reduced by a factor A^2 to account for energy lost within the ice field by various dissipative processes (e.g. ice deformation), the energy associated with the two waves of the coherent field travelling in the $\pm$ incident direction, and the directional scattering contributions with intensity apportioned as $|D(\omega, \theta)|^2$.

The evolution of a spectrum incident on a partial ice cover is obtained by integration of the energy balance equation using the exact-NL model (see chapter III for its definition) in which the effect of the ice is included as described above. When the diameter of the floes is comparable to the incident wavelength, the ice cover is very efficient in spreading out the wave energy over all directions. As the spectrum tends rapidly to isotropy, the energy balance is drastically modified, with a much reduced energy transfer rate from the atmosphere, along with a significant decrease in nonlinear energy transfer. Therefore, the directional scattering of the waves by the ice floes prevents the normal growth of wave energy and the decrease in peak frequency. Applying these results to the problem of wave generation by an offshore wind starting to blow over the MIZ, one can see that as the young waves approach the ice edge, they encounter gradually smaller floes which would cause eventually, through scattering, a rapid inhibition of the wave growth. The ability of an offshore wind to generate a significant wave field in the MIZ is then severely limited by the partial ice cover.

Note that, while these results indicate that the local development of a significant wave field within the MIZ is practically impeded by the partial ice cover, they do not exclude the propagation of swell in the area. Long waves for which the wavelength is much larger than the ice floe diameter could travel through a field of broken floes without being seriously affected. Perrie and Toulany (1991) examined the effect of the ice cover on the predictions of the WAM model during the LEWEX experiment simply by treating the ice edge as the land boundary. On one particular occasion where relatively long waves were travelling through an ice tongue, the measurements were in better agreement with the model results when the ice cover was ignored, suggesting that the long waves were not significantly influenced by the presence of the ice.

II.8 Growth curve observations

K. K. Kahma and C. J. Calkoen

Of the three source terms in the radiative transfer equation the best-known one is the nonlinear transfer term S_{nl}. Less well known is the input term S_{in}, and the formulation for the dissipation term S_{ds} proposed by Komen *et al* (1984) actually contains tuning parameters which have been adjusted so that the growth matches the experimentally measured wave growth. Experimental growth curves are therefore important in verifying and adjusting the source functions.

II.8.1 Theory

The Kitaigorodskii (1962, 1970) similarity law has provided a powerful tool in analysing wave data. In principle a large number of variables may control the wave growth. Kitaigorodskii suggested that in the idealized situation of duration-limited waves (when a uniform and steady wind has blown over unlimited ocean for time t after a sudden onset) the following set of variables should be considered

$$F(\omega, U_\infty, g, t, \Omega, \nu_a, \nu_w, \rho_a, \rho_w, \gamma), \qquad (2.212)$$

where γ is the surface tension divided by density, ρ_a and ρ_w are air and water densities respectively, v_a and v_w the kinematic viscosities of air and water, g is the acceleration of gravity and Ω the Coriolis parameter. Using assumptions of the nature of the wave motion and the mechanism of the wave growth he found that after the initial stage of wave growth the energy containing part of the spectrum is controlled mainly by the variables ω, U_∞, g, t. This gives the similarity laws for the wave spectrum $F(\omega)$, the total variance m_0 and the peak (angular) frequency of the spectrum ω_p:

$$Fg^3/U_\infty^5 = \text{function}(\omega U_\infty/g, gt/U_\infty), \tag{2.213}$$

$$m_0 g^2/U_\infty^4 = \text{function}(gt/U_\infty), \tag{2.214}$$

$$\omega_p U_\infty/g = \text{function}(gt/U_\infty). \tag{2.215}$$

Conditions for duration-limited growth are difficult to fulfill and, from the point of view of the analysis of experimental data, two other idealized cases are more important. One is the limiting case of fully developed waves when a uniform and steady wind has blown over an unlimited ocean long enough for the wave field to become independent of time. In this case the relations are simplified to

$$Fg^3/U_\infty^5 = \text{function}(\omega U_\infty/g), \tag{2.216}$$

$$m_0 g^2/U_\infty^4 = \text{constant}, \tag{2.217}$$

$$\omega_p U_\infty/g = \text{constant}. \tag{2.218}$$

The situation that can best be approximated experimentally is the fetch-limited case, when a uniform steady wind has blown from a straight shoreline long enough for the wave field at distance X from the upwind shore to become independent of time. This gives the well-known relations

$$Fg^3/U_\infty^5 = \text{function}(\omega U_\infty/g, gX/U_\infty^2), \tag{2.219}$$

$$m_0 g^2/U_\infty^4 = \text{function}(gX/U_\infty^2), \tag{2.220}$$

$$\omega_p U_\infty/g = \text{function}(gX/U_\infty^2). \tag{2.221}$$

The scaling laws are based on assumptions some of which are known to be only approximately valid. In addition, variables not discussed by Kitaigorodskii (1962, 1970), such as those describing the stability of the atmosphere or the roughness over the shore, can cause deviations to the scaling law.

In the derivation of the scaling law, Kitaigorodskii characterized the turbulent wind in the boundary layer by U_∞, the wind speed at an altitude no longer affected by the energy and momentum transfer from the atmosphere to the waves. In practice the hypothetical U_∞ has to be replaced by some measured value. There are still conflicting opinions as to whether the best approximation is the friction velocity u_*, or the wind speed U_{10}, or some other parameter, such as the wind speed at half the dominant wave length $U(\lambda/2)$ as proposed by Donelan and Pierson (1987).

Because there is a significant change in the surface roughness at the shore, the surface wind and u_* are functions of fetch, at least near the shore. Using the estimates of Taylor and Lee (1984) and utilizing an empirical approximation of the total source term, Donelan *et al* (1992) have demonstrated that this change in the wind speed is sufficiently large to affect the universality of the fetch-limited scaling laws significantly. It seems, however, that under certain assumptions the local u_* could give scaling laws that are less affected by this problem. These assumptions require that u_* is measured (or calculated) at the same fetch as the waves. If u_* is not measured, the calculation algorithm should accurately model the change of u_* with fetch and wind speed. Furthermore, if data sets are to be compared, a common algorithm for u_* must be used.

II.8.2 Observations

In the wave model comparison (SWAMP, 1985) large differences were found among fetch-limited growth curves produced by the various models. At the time this was unexpected, as it was asserted that all models participating in SWAMP had been calibrated against both fetch-limited and duration-limited data. Similar differences, however, can be seen in the experimental data. A striking example is shown in figure 2.28, the difference between the growth from JONSWAP (Hasselmann *et al*, 1973) and data collected in the Bothnian Sea (Kahma, 1981a), the latter showing double energy compared with the former.

One of the recommendations of the SWAMP group was that existing experimental data sets should be collected into a database, which could then be reanalysed. When this was done, the study (Kahma and Calkoen, 1992) confirmed the earlier conclusion on the subject (Kahma, 1986) that no single factor seems to be responsible for the large difference between the original JONSWAP relation and the Bothnian Sea relation. However, by reanalysing the JONSWAP wind data and dividing the data according to stability the authors were able to give a consistent explanation for the differences between the Bothnian Sea data, the JONSWAP data, and the data measured in Lake Ontario (Donelan *et al*, 1985).

Kahma and Calkoen (1992) found that the data from Lake Ontario during unstable stratification show the same rapid growth as the Bothnian Sea data, the latter being measured entirely in unstable stratification. The growth of energy was approximately linear with fetch in both data sets (figure 2.29).

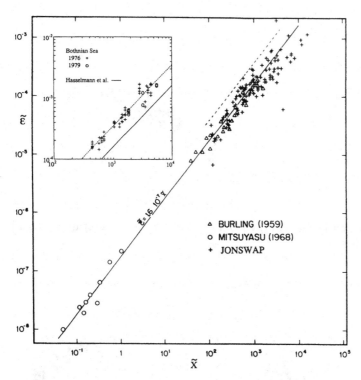

Fig. 2.28. An illustration of the differences in empirical growth relations: The relations $\tilde{\epsilon} = 1.6 \times 10^{-7}\tilde{X}$ (solid line), reported in Hasselmann *et al* (1973), and $\tilde{\epsilon} = 3.5 \times 10^{-7}\tilde{X}$ (dashed), found in the Bothnian Sea experiments of 1976 and 1979 (Kahma, 1981a,b), are plotted on the data of figure 2.10 in Hasselmann *et al* (1973). In the inset the same relations are plotted on the Bothnian Sea data.

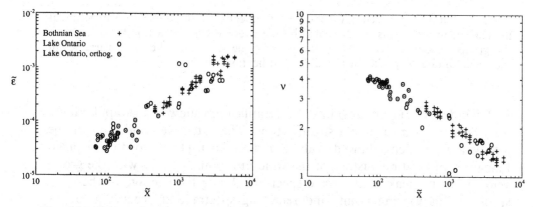

Fig. 2.29. Dimensionless energy $\tilde{\epsilon}$ and peak frequency ν in the Bothnian Sea data and the Lake Ontario data in unstable stratification.

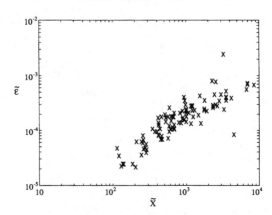

Fig. 2.30. Dimensionless energy from the JONSWAP data set using the wind speed given in Muller (1976).

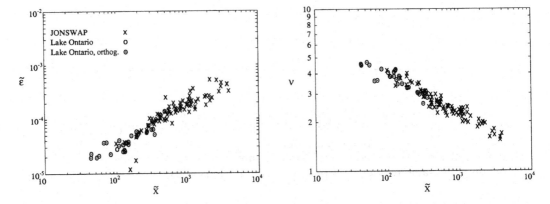

Fig. 2.31. Dimensionless energy $\tilde{\varepsilon}$ and peak frequency ν in the JONSWAP experiment in steady and increasing wind, scaled by the average wind at a point moving with the group velocity of the wave at the peak of the spectrum; the same variables in the Lake Ontario experiment in stable stratification.

The data from Lake Ontario in stable stratification showed less rapid growth with fetch than in unstable stratification. The extensive wind data of the JONSWAP experiment allowed Kahma and Calkoen (1992) to do a detailed reanalysis of the time history of the wind at a point moving with the group velocity of the peak of the wave spectrum. Using the average of the wind speed at this varying point, and removing spectra that according to this criterion were not measured in steady or increasing (with fetch) wind, they found that the scatter was substantially reduced (cf. figures 2.30 and 2.31).

The atmospheric stratification during the JONSWAP experiment was mainly stable. When the part of data from Lake Ontario that had been measured in stable stratification was compared with the reanalysed JONSWAP data, it was found that they agreed well (figure 2.31). The growth of energy with fetch in stable stratification was slower than linear. In the case of U_{10} scaling the energy of the unstable stratification groups is as much as 1.7 times the energy of the stable stratification groups. When only the Lake Ontario data are considered, the difference is somewhat smaller, about 1.5 times. Both these differences between stable and unstable groups are statistically highly significant. On the other hand, among groups of the same stability no statistically significant differences can be found by this approach. The ratio between the stable groups is 1.04 and between the unstable groups 1.07.

It has been argued that the fundamental parameter describing the wind input to the waves is the total surface stress, and that the differences observed in the growth of fetch-limited waves result from the inability of U_{10} to describe that. If u_* is a good stability-independent replacement for U_∞ these differences should disappear when it is used in the Kitaigorodskii scaling law. When there are significant differences in atmospheric stability, differences in the wave growth relations are indeed reduced when U_{10} is replaced by u_* calculated using the stability correction (e.g. Liu and Ross, 1980, Kahma and Calkoen, 1992). The benefit is less clear when the data sets do not have large stability differences. Discrepancies between different data sets have been reduced when friction velocity scaling has been used (Janssen *et al*, 1987a), but within individual data sets having only small stability differences the improved correlation between dimensionless energy and dimensionless peak frequency could equally well be attributed to spurious correlation, because u_* usually has a wider scatter than U_{10}.

For this study the data have also been scaled by two alternative friction velocities. Both are different from the one used by Kahma and Calkoen (1992) to show what are the possibilities to reduce the scatter by different algorithms to calculate u_*. They are both corrected for stability using the Monin-Obukhov similarity theory (Monin and Yaglom 1971; § I.3.7). The first, u_*, was calculated using the wind-speed dependent drag coefficient (WAMDI, 1988):

$$C_D = 0.8 \times 10^{-3} + 0.065 \times 10^{-3} \times \max\{U_{10}, 7.5 \text{ m/s}\}. \qquad (2.222)$$

This drag coefficient is essentially the one used by Bouws (1986) as well as Janssen *et al* (1987a) when they reported reduced scatter in growth relations when U_{10} was replaced by u_*. In addition the data have been scaled by u_* calculated using a wave-dependent z_0 proposed by Donelan (1990):

$$z_0/\sqrt{m_0} = 5.53 \times 10^{-4}(U_{10}/c_p)^{2.66}. \qquad (2.223)$$

The analogous relation with the right hand side in terms of u_* has also been considered (Voorrips, private communication), but this did not lead to significant differences in the scatter. When the friction velocities are

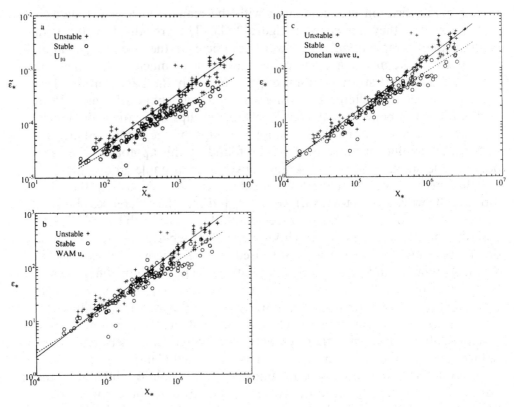

Fig. 2.32. Dimensionless energy in unstable and stable stratification: a) scaling wind U_{10}; b) scaling wind u_* calculated using a wind-dependent drag coefficient; c) scaling wind u_* calculated using a wave-dependent z_0.

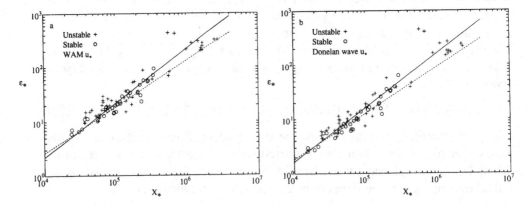

Fig. 2.33. Dimensionless energy in the Lake Ontario experiment in unstable and stable stratification. The regression lines are those of the previous figure: a) scaling wind u_* calculated using a wind-dependent drag coefficient; b) scaling wind u_* calculated using a wave-dependent z_0.

calculated with the help of (2.223) the difference between the unstable and stable stratification groups is reduced to 1.43 or 1.47, respectively (figure 2.32). Between the stable groups the ratio is 1.05 or 1.07, and between the unstable groups it is 1.02 or 1.03. The stability difference is smaller in the Lake Ontario data (figure 2.33) than between the Bothnian Sea data and the JONSWAP data. However, even in the Lake Ontario data the remaining ratio with u_* scaling is 1.21 or 1.22, respectively, and both are statistically significant at the 98% confidence level. As far as the scatter and the difference between the stability groups are concerned, the two friction velocities give essentially the same results as the friction velocity used in Kahma and Calkoen (1992). Indeed, no significant improvement in scatter could be expected by introducing the wave age, since this parameter is highly correlated with the dimensionless fetch. On the other hand, the best fit power laws are very different. Consequently, different growth curves will result from different algorithms for friction velocity u_*.

The fetch-limited growth relations using U_{10} and u_* are:

Stable stratification: (number of observations 108)

$$\tilde{\epsilon} = 9.3 \times 10^{-7} \times \tilde{X}^{0.77}, \tag{2.224a}$$

$$v = 12 \times \tilde{X}^{-0.24}, \tag{2.224b}$$

$$\epsilon_* = 2.1 \times 10^{-3} \times X_*^{0.79}, \tag{2.224c}$$

$$v_* = 2.3 \times X_*^{-0.25}. \tag{2.224d}$$

$$\epsilon'_* = 7.3 \times 10^{-4} \times (X'_*)^{0.85}, \tag{2.224e}$$

$$v'_* = 3.0 \times (X'_*)^{-0.26}. \tag{2.224f}$$

The quantities labelled with an asterisk have been obtained with (2.222); for the quantities labelled both with an asterisk and a prime the wave-dependent drag (2.223) has been used.

Unstable stratification: (number of observations 94)

$$\tilde{\epsilon} = 5.4 \times 10^{-7} \times \tilde{X}^{0.94}, \tag{2.225a}$$

$$v = 14 \times \tilde{X}^{-0.28}, \tag{2.225b}$$

$$\epsilon_* = 4.7 \times 10^{-4} \times X_*^{0.95}, \tag{2.225c}$$

$$v_* = 3.6 \times X_*^{-0.28}. \tag{2.225d}$$

$$\epsilon'_* = 1.6 \times 10^{-4} \times X'_*, \tag{2.225e}$$

$$v'_* = 4.7 \times (X'_*)^{-0.3}. \tag{2.225f}$$

Composite dataset:

$$\tilde{\epsilon} = 5.2 \times 10^{-7} \times \tilde{X}^{0.9}, \tag{2.226a}$$

$$v = 13.7 \times \tilde{X}^{-0.27}, \tag{2.226b}$$

$$\epsilon_* = 6.5 \times 10^{-4} \times X_*^{0.9}, \tag{2.226c}$$

$$v_* = 3.08 \times X_*^{-0.27}. \tag{2.226d}$$

$$\epsilon'_* = 2.2 \times 10^{-4} \times (X'_*)^{0.96}, \tag{2.226e}$$

$$v'_* = 4.2 \times (X'_*)^{-0.29}. \tag{2.226f}$$

Here $\tilde{X} = gX/U_{10}^2$ or $X_* = gX/u_*^2$ is the dimensionless fetch, $\tilde{\epsilon} = g^2 m_0/U_{10}^4$ or $\epsilon_* = g^2 m_0/u_*^4$ is the dimensionless energy per unit area and $v = \omega_p U_{10}/g$ or $v_* = \omega_p u_*/g$ is the dimensionless peak frequency. Davidan (1990) also calculated wave-age-dependent u_* scaling from a large number of datasets and obtained the relation $\epsilon_* = 10^{-4} \times X_*$.

II.8.3 Is there evidence for u_* scaling?

We have seen that the difference between stable and unstable observation is partially removed by using u_* as a scaling parameter. However, Kahma and Calkoen (1992) present evidence against the idea that using u_* scaling could remove all of the differences in the growth rate related to the atmospheric stability. They show that the experimental results concerning the effect of stability on the momentum retained by waves (Donelan, 1979) contradict the view that the use of u_* instead of U_{10} could alone bring the data sets into line. One could claim that the difference between stability groups in figures 2.32b and c results from the imperfections of the models that were used to calculate u_*, but this argument cannot be applied to the estimates of the fraction of momentum retained by waves in Donelan (1979), because he calculated them directly by measuring the waves at two places and measured the atmospheric stress by the correlation method. We should therefore keep open the possibility that the reduced differences between unstable and stable groups when u_* is used instead of U_{10} are merely a consequence of the fact that, other conditions being equal, the wave energy and u_* at a given wind speed are both larger in unstable stratification than in stable stratification.

II.8.4 Conclusion

The theoretical basis for similarity in fetch-limited growth is not without assumptions. In practice, one expects deviations from exact scaling because winds (stresses) are not constant and homogeneous, coasts are not straight and other parameters like gustiness, surface roughness and air/sea temperature difference may play a role. The case for u_* scaling has not been proved. Under certain assumptions it would appear to be a better replacement for U_∞ than U_{10} and is therefore the one adopted in this book. It is well to remember that if u_* is not measured (the usual case) the algorithm chosen to estimate it may itself be source of considerable error.

II.9 Observations of spectral shape

M. Donelan

Oceanic wave spectra generally contain a combination of local wind-generated seas mixed with swells from one or more distant storms. Even the local wind sea is seldom the product of a steady and homogeneous wind system. Consequently, the idea of a universal description of oceanic wave spectra based on local wind conditions is a rather limited concept. A global numerical wave model that considers conditions everywhere on the oceans and that contains a complete description of the balance of sources and sinks is ultimately required to describe conditions at any particular space-time point. Nonetheless, it is instructive to examine the development of wave spectra under restrictive conditions of stationarity and homogeneity and in which the fetch is clearly defined. Under these conditions, one may apply the concept of similarity in the spectral shape in which the spectrum may be concisely described by a small set of parameters. This is best illustrated by superimposing suitably normalized spectra (requiring two parameters, i.e. the coordinates of the spectral peak) as in figure 2.34 (U_c is the 10 m wind projected on the mean propagation direction of the waves at the peak of the spectrum; c_p is as usual the phase velocity of the wave at the peak). These spectra are drawn from both laboratory and field data and cover a very wide range of (nondimensional) fetches corresponding to very actively growing waves to full development, or more precisely, conditions in which the phase speed of the waves at the spectral peak approach the wind speed.

The concept of an equilibrium range in the high wavenumber waves is central to the idea of spectral similarity. However, the theoretical development (Phillips, 1958) based on a fully saturated short wave spectrum appears not to be supported by observations (Forristall, 1981, Kahma, 1981b, Donelan *et al*, 1985). In the 'energy containing' region of the equilibrium range of the spectrum ($1.5\omega_p < \omega < 3\omega_p$) the slope is well described by an ω^{-4} power law (figure 2.35). At lower frequencies more parameters are required to describe the spectral peak, and at higher frequencies the observed 'frequency-of-encounter' spectrum is not a simple function of the wavenumber spectrum because the orbital velocities of the long waves cause Doppler shifting of the short-wave spectrum (Kitaigorodskii *et al*, 1975, Ataktürk and Katsaros, 1987). Toba (1973), on dimensional grounds, has argued for an equilibrium range that is quasi-saturated and dependent on u_*. In a pioneering study of the directional properties of ocean waves using a floating buoy, Longuet-Higgins *et al* (1963) documented the increased spreading of the higher-frequency waves. Subsequent studies (Hasselmann *et al*, 1973, Mitsuyasu *et al*, 1975, D.E. Hasselmann *et al*, 1980, Donelan *et al*, 1985) amply

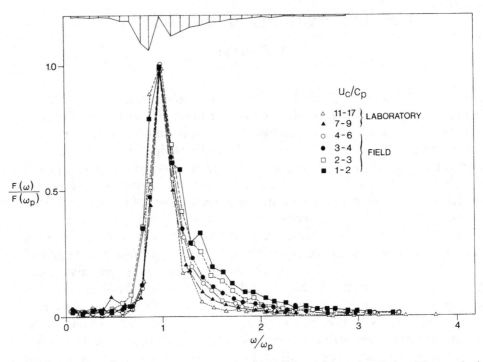

Fig. 2.34. Normalized frequency spectra grouped into classes by U_c/c_p. The vertical bars at the top of the figure are an estimate of the 90% confidence limits based on the standard error of the mean. (From Donelan *et al*, 1985.)

confirm this universal characteristic of developing wind seas. Donelan *et al* (1985) suggest that the progressive spreading of the directional spectrum with relative frequency ω/ω_p would tend to whiten the equilibrium range of the frequency spectrum, leading to a description of the quasi-saturated range with $\omega^{-5}(\omega/\omega_p)$ frequency dependence. The equivalent wavenumber dependence is $k^{-4}(k/k_p)^{\frac{1}{2}}$. Kitaigorodskii (1983) and Phillips (1985) offer theoretical arguments for an ω^{-4} equilibrium range based on an energy balance (see also the discussion in § II.3).

For frequencies above $3\omega_p$ there is little direct evidence of the wavenumber spectrum but it cannot continue with a slope of -3.5 for this would imply short wave spectral levels far higher than those observed even in very strong forcing in laboratory tanks. Furthermore, the mean square slope would greatly exceed optical and radar estimates of mean square slope (Jackson *et al*, 1992). Kitaigorodskii (1983) has postulated such a break in the spectral slope from quasi-saturated ($k^{-3.5}$) to fully saturated (k^{-4}) and evidence is accumulating although it cannot yet be said to be definitive. At even higher wavenumbers near k_m (capillary–gravity transition) microwave radar

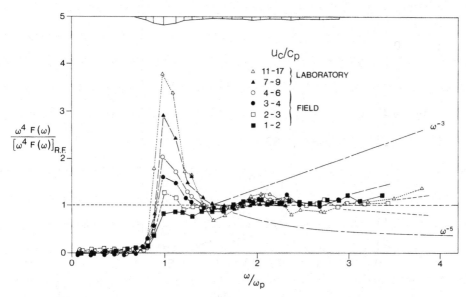

Fig. 2.35. Frequency spectra times ω^4 normalized by the rear face $\left[\omega^4 F(\omega)\right]_{rf}$ which is the average of $\omega^4 F(\omega)$ in the region of $(1.5\omega_p < \omega < 3\omega_p)$. The lines corresponding to ω^{-5} and ω^{-3} are also shown $(-\,-\cdot\,-\,-)$. The effect of a 10 cm/s ambient current with or against the waves is also shown $(-\,-\ \ -\,-)$, as is the effect of wind drift in a 10 m/s wind $(\cdot\cdot\cdot)$. The spectra are grouped in classes of U_c/c_p. (From Donelan *et al*, 1985.)

reflectivity indicates strong wind sensitivity. If the radar response is largely via first order Bragg scattering then the spectrum in the capillary–gravity range must also be unsaturated and wind sensitive. Finally at very high wavenumbers the spectrum must be entirely suppressed by viscous damping. All of this adds up to a rather complex spectral dependence on wavenumber, wind speed, fetch, gravity, surface tension and viscosity as the minimum set of governing variables. However, in this section we shall concern ourselves with the spectral evolution of gravity waves in the energy containing region $0 < \omega < 3\omega_p$ and therefore we may ignore surface tension and viscosity.

Fetch-limited spectra

The similarity of wind wave spectra supports the idea that the development of the spectrum in an offshore wind may be described in terms of a small set of external variables. The appropriate variables — fetch X, friction velocity u_* and the gravitational acceleration g — lead to Kitaigorodskii's (1962) 'similarity hypothesis' in which appropriately nondimensionalized spectral properties are functions of the nondimensional fetch, Xg/u_*^2. These ideas were explored experimentally by several authors, both in laboratory tanks

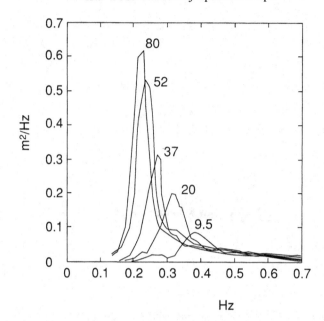

Fig. 2.36. Evolution of wave spectra with fetch for offshore winds (11–12 h, Sept. 15, 1968). The spectra are labelled with the fetch in kilometres. (From Hasselmann *et al*, 1973.)

and in the field: Mitsuyasu (1968, 1969), Mitsuyasu *et al* (1971), Liu (1981). During the Joint North Sea Wave Project (JONSWAP, Hasselmann *et al*, 1973), Kitaigorodskii's similarity hypothesis was revisited in a well-designed field experiment, and led to a convenient spectral description of fetch-limited wind seas. The JONSWAP findings have significantly advanced our ideas on the dynamical balance in a wind sea and opened the way to more sophisticated wave prediction modelling.

The development of the spectrum in an offshore wind is illustrated in figure 2.36 (from Hasselmann *et al*, 1973). The spectral peak moves towards lower frequencies, as the fetch increases from 9.5 to 80 km, leaving behind a quasi-saturated high-frequency tail. The spectral peak rises well above the extrapolated tail so that the energy at a particular frequency rises to a maximum as fetch increases and falls again as the peak frequency reduces further. This 'enhanced' peak (see also figure 2.35) contributes a substantial amount to the total variance of surface elevation and is a sensitive indicator of the changes in the balance of the source functions as the spectrum evolves. A description of the fully developed spectrum by Pierson and Moskowitz (1964) contained a single parameter, α, describing the spectral level on the high-frequency tail. From observations, the condition of full development is said to be obtained when the phase velocity corresponding to the peak frequency exceeds the wind speed (measured at 10 m height) by 20%. Hasselmann *et al* (1973) introduced three new parameters to describe the height, γ and width, σ_a and σ_b of the enhanced peak. The parameters σ_a and σ_b refer to the width of the asymmetrical peak below and above the peak frequency respectively. The spectra were fitted to the frequency power law ω^{-5} at high frequencies and no consistent dependence of γ and σ_a and σ_b on nondimensional fetch was found. In spite of the poor fit in the equilibrium range, the JONSWAP prescription is still widely used.

For easy reference the JONSWAP spectral formulae are reproduced here:

$$F(f) = \alpha_j g^2 (2\pi)^{-4} f^{-5} \exp\left\{-\frac{5}{4}\left(\frac{f_p}{f}\right)^4\right\} \gamma_j^\Gamma, \tag{2.227}$$

where

$$\Gamma = \exp\left\{-(f - f_p)^2 / 2\sigma_j^2 f_p^2\right\}, \tag{2.228}$$

with

$$\alpha_j = 0.076 \left(\frac{Xg}{U_{10}^2}\right)^{-0.22}, \tag{2.229}$$

$$\gamma_j = 3.3, \tag{2.230}$$

$$\sigma_{ja} = 0.07, \tag{2.231}$$

$$\sigma_{jb} = 0.09. \tag{2.232}$$

Acknowledging the now established ω^{-4} form for the equilibrium range of wind sea spectra, Donelan *et al* (1985) fitted a set of spectra obtained from Lake Ontario to ω^{-4} at high frequencies. They found that the spectral parameters could be related to the wind forcing parameter U_c/c_p, where U_c is the component of the 10 m height wind vector in the mean direction of propagation of the waves at the spectral peak and c_p is the corresponding phase speed. The frequency spectrum $F(\omega)$ is given by:

$$F(\omega) = \alpha_d g^2 \omega^{-5} (\omega/\omega_p) \exp\left\{-\left(\frac{\omega_p}{\omega}\right)^4\right\} \gamma_d^\Gamma, \tag{2.233}$$

where

$$\Gamma = \exp\left\{-(\omega - \omega_p)^2 / 2\sigma_d^3 \omega_p^2\right\}, \tag{2.234}$$

with

$$\alpha_d = 0.006(U_c/c_p)^{0.55} \qquad \text{for } 0.83 < U_c/c_p < 5, \tag{2.235}$$

$$\sigma_d = 0.08 \left[1 + 4/(U_c/c_p)^3\right] \qquad \text{for } 1 < U_c/c_p < 5, \tag{2.236}$$

$$\gamma_d = \begin{cases} 1.7 & \text{for } 0.83 < U_c/c_p < 1, \\ 1.7 + 6.0 \, \log(U_c/c_p) & \text{for } 1 \le U_c/c_p < 5. \end{cases} \tag{2.237}$$

The spreading of the directional spectrum $F(\omega, \theta)$ was found to depend principally on ω/ω_p:

$$F(\omega, \theta) = \frac{1}{2} F(\omega) \beta \, \text{sech}^2 \beta \left\{ \theta - \bar{\theta}(\omega) \right\}, \qquad (2.238)$$

where $\bar{\theta}$ is the mean wave direction and

$$\beta = \begin{cases} 2.61(\omega/\omega_p)^{+1.3} & \text{for } 0.56 < \omega/\omega_p < 0.95, \\ 2.28(\omega/\omega_p)^{-1.3} & \text{for } 0.95 < \omega/\omega_p < 1.6, \\ 1.24 & \text{otherwise.} \end{cases} \qquad (2.239)$$

In the energy containing range of the spectrum both (2.227) and (2.233) fit observed spectra equally well. The latter gives a better fit in the equilibrium range $1.5 \, \omega_p - 3 \, \omega_p$, but this is not surprising because the original JONSWAP fit only applied to the energy containing range. Sometimes a modified JONSWAP spectrum with an f^{-4} has been used. Note that the WAM model, to be discussed in chapter III, imposes an f^{-5} tail for $f > 2.5 \, \bar{f} \simeq 3 \, f_p$, which is consistent with the above discussion.

II.10 Observations of the directional response to turning winds

L. Holthuijsen

The response of waves to a varying wind speed in terms of wave energy and frequency is fairly well understood, partly because standard geophysical conditions have been used to study such response. An example is the fetch-limited development of total wave energy and peak frequency (Hasselmann *et al*, 1976). To achieve a similar degree of understanding for the directional response, a corresponding parametrization is often used of the mean wave direction relaxing towards the local wind direction. This was done, for example, by SWAMP (1985).

II.10.1 Observation and presentation techniques

The response of waves in turning-wind situations in open sea has mostly been studied with heave, pitch and roll buoys (dedicated buoys or meteorological buoys with a 2.5 – 10 m diameter discus-shaped hull, see § I.4 and, for example, van der Vlugt, 1984). These can be used to estimate integral directional parameters (Longuet-Higgins *et al*, 1963 and Kuik *et al*, 1988) or the buoy data can be used to reconstruct a low-resolution image of the directional energy distribution (see, for example, Longuet-Higgins *et al*, 1963

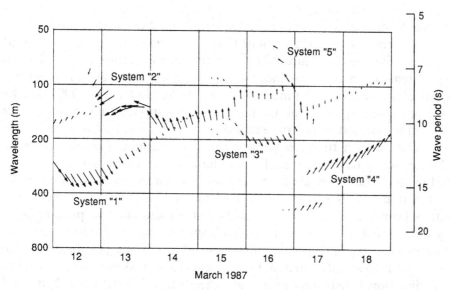

Fig. 2.37. The five component wave systems of LEWEX at location 50° N, 45° W as hindcasted by the WAM model.

and Krogstad *et al*, 1988). Imaging techniques and array techniques provide high-resolution estimates of the two-dimensional spectrum (e.g. synthetic aperture radar, Brüning *et al*, 1988 or wave gauge arrays, Donelan *et al*, 1985).

The most complete presentation of the data is a contour line plot of the two-dimensional spectrum in frequency-direction (f, θ) space or wavenumber-direction (k, θ) space or wavenumber vector $(\mathbf{k})$ space. Gerling (1991, 1992) developed a method to reduce such spectra to component wave systems. The method, which is explained in more detail in § V.4.3, characterizes each of these wave systems with a modal wavenumber and propagation direction estimate, and with an estimate of the significant wave height that would be attributed to just that wave system alone. An advantage of this description over integral measures such as total significant wave height, average frequency, or average direction, is that it avoids mixing of spectral information due to distinct meteorological events. Typically, the spectral signature of each such event will be a distinct spectral peak. Another algorithm tracks the evolution of peaks through either a temporal or spatial variation of spectra. Parameters describing component wave systems can then be determined as temporal or spatial functions. This analysis procedure has been usefully applied to time series of WAM and other model and buoy spectra from the Labrador Extreme Waves Experiment (LEWEX) (Beal, 1991). In that eight day dataset, the wave field variability was well described as the evolution of six wave systems, and the principal model–buoy differences were time translations of the partitioned significant wave height curves corresponding to each system. Corrections to the hindcast wind field are suggested from these differences. Figure 2.37 shows the time evolution of five of these wave systems from the WAM model. Plotting the mean wave direction per frequency as a function of time (see, for example, Guillaume, 1990) may be seen as a reduction of the above technique. A continued reduction of this presentation method is the presentation of the frequency-integrated mean wave direction as a function of time (see, for example, Guillaume, 1990). Such time series have been used to evaluate wave hindcasts using conventional methods based on circular statistics (see, for example, Mardia, 1972). The directional error of the WAM model in the study of Guillaume (1990) has thus been found to be of the order of 12° (compared to pitch-and-roll buoy data). Such time series have also been used to estimate the response of the mean wave direction in turning-wind situations (see below). The response of higher order directional moments (width, skewness and kurtosis of the directional distribution) has not been studied extensively. Kuik and Holthuijsen (1981) and Masson (1990) observed that the directional width increases when the wind turns. The degree of widening in such situations appears to depend on the rate of turning (the higher frequencies retaining their directional spreading more than the lower frequencies).

II.10.2 Observation conditions

To obtain time scales of the wave directional response under turning winds, observations have been acquired in the North Sea (D.E. Hasselmann *et al*, 1980, Günther *et al*, 1981, Holthuijsen *et al*, 1987 and van Vledder and Holthuijsen, 1993) and the North Atlantic Ocean (Allender *et al*, 1983, Masson, 1990). The significant wave height varied typically from 0.5 to 7.5 m and the wind speed at 10 m elevation (U_{10}) from 4 to 19 m/s. The anemometers were usually close to the sea surface (buoy mounted) but occasionally at large elevations (around 100 m at offshore industry platforms) requiring considerable extrapolation to 10 m elevation. The observations were usually taken in fetch-unlimited conditions or at least in fairly homogeneous wave conditions in relatively deep water (relative to the observed wave lengths) in turning-wind situations (step-like or slowly rotating).

II.10.3 Results of the observations

The time scale of the locally induced turning of the mean wave direction θ_0 towards the local wind direction θ_w is defined in terms of a relaxation

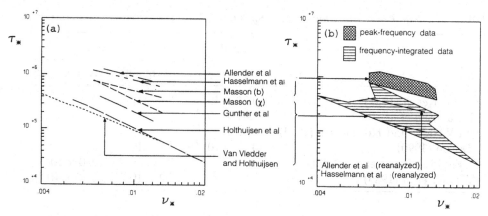

Fig. 2.38. Dimensionless time scale as a function of wave age from various sources, partly reinterpreted (panel a) and reanalysed (panel b).

model:

$$\frac{\partial \theta_0}{\partial t} = \frac{1}{\tau} \sin(\theta_w - \theta_0), \tag{2.240}$$

in which the mean wave direction may be defined in different ways and which may be considered either frequency-dependent or frequency-independent (that is frequency-integrated). Theoretically the waves should turn towards the mean direction of the total source function (see, for example, Holthuijsen *et al*, 1987):

$$\frac{\partial \theta_0}{\partial t} = \frac{\cos(\theta_0) \int_0^{2\pi} \int_0^\infty \cos(\theta) S(f,\theta) df d\theta}{\cos(\theta_S) \int_0^{2\pi} \int_0^\infty \cos(\theta) F(f,\theta) df d\theta} \sin(\theta_S - \theta_0), \tag{2.241}$$

with the mean wave direction θ_0 defined as in Holthuijsen *et al* (1987):

$$\theta_0 = \tan^{-1} \frac{\int_0^{2\pi} \int_0^\infty \sin(\theta) F(f,\theta) df d\theta}{\int_0^{2\pi} \int_0^\infty \cos(\theta) F(f,\theta) df d\theta} \tag{2.242}$$

and θ_S identically defined as the mean direction of $S(f,\theta)$ (with $F(f,\theta)$ replaced with $S(f,\theta)$). However, θ_S is an unobservable quantity and it has been replaced with the wind direction (eventually the two will be equal).

The observations are usually selected to retain waves which respond to the local wind. This is achieved by selecting certain meteorological events or wave fields with certain wave parameter values (the wave age, for example) or by filtering in frequency space. However, the observations which remain after such a selection are still affected to some extent by radiative effects of inhomogeneities in the wave field. These effects can be hindcasted and some observations have been corrected and selected accordingly (van Vledder and Holthuijsen, 1993).

Table 2.1. *Values of the time scale coefficients b and χ from various sources.*

	χ value	b value	
		original	re-analysed
Günther *et al* (1981)	0.21×10^{-2}	–	–
Holthuijsen *et al* (1987)	0.41×10^{-2}	–	–
van Vledder and Holthuijsen (1993)	0.57×10^{-2}	–	–
Masson (1990)	0.12×10^{-2}	3.1×10^{-5}	–
D. E. Hasselmann *et al* (1980)	–	2.0×10^{-5}	6.0×10^{-5}
Allender *et al* (1983)	–	1.7×10^{-5}	14.0×10^{-5}

(i) Frequency-independent observations

The technique used to estimate the response time scale of the frequency-integrated mean wave direction is usually based on a finite difference version of the above relaxation model. This time scale can be normalized with the wind speed at 10 m elevation (U_{10}) or the friction velocity $u_* = \sqrt{C_D} U_{10}$ (where C_D is the drag coefficient) and the gravitational acceleration g to obtain the dimensionless time scale $\tilde{\tau} = g\tau/U_{10}$ or $\tau_* = g\tau/u_*$. Under the assumption that the locally generated spectrum can be described with only a few parameters, Günther *et al* (1981) found from a parametrization of the energy of the waves that this dimensionless time scale should depend on the wave age, v as $\tilde{\tau} = \chi^{-1}v^{-2}$ (dimensionless peak frequency, $v = f_p U_{10}/g$). With this dependency, observed values of τ, f_p and U_{10} can be transformed to estimates of the coefficient χ, although Masson (1990) uses a regression technique. The mean values of χ of Günther *et al* (1981), Holthuijsen *et al* (1987), Masson (1990) and van Vledder and Holthuijsen (1993) are given in table 2.1.

The corresponding relationships in terms of τ_* and $v_* = f_p u_*/g$ are given in figure 2.38a (obtained with an average value of $C_D = 0.001736$ for wind speeds of 10 and 20 m/s (Wu, 1982)). The best-fit through the data of van Vledder and Holthuijsen (1993, not shown here) gives a power of v_* somewhat higher (-1.7) than -2 as in the expression of Günther *et al* (1981).

(ii) Frequency-dependent observations

In addition to observing the response time scale of the frequency-integrated mean wave direction Masson (1990) observed the response time scale of the mean wave direction at each frequency. Such observations have also been carried out by D.E. Hasselmann *et al* (1980) and Allender *et al* (1983).

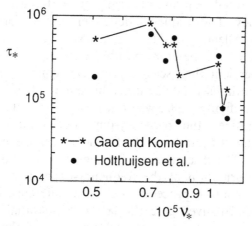

Fig. 2.39. Dimensionless time scale as a function of wave age from one set of observations: linear estimate (Holthuijsen et al, 1987) and nonlinear estimate (Gao and Komen, 1993).

In their analysis the correspondingly frequency-dependent time scale τ_f is assumed to be dependent on frequency as $\tau_f = (2\pi b f)^{-1}$. They estimated b values with a regression technique correlating the observed rate of turning $\partial\theta_0/\partial t$ with $2\pi f \sin(\theta_w - \theta_0)$. They also stratified the results in classes of the local (in frequency space) wave age U_{10}/c (c is the phase speed of the wave component). The b values are given in table 2.1. The stratification (not shown) shows that no correlation with the local wave age U_{10}/c exists.

II.10.4 Discussion

The scatter in the individual data sets is large, as can be seen from table 2.1 and figure 2.39. This is at least partly due to the simple parametrization in terms of wave age. Masson (1990) provides empirical evidence that the time scale is also correlated with the rate of wave turning. This is supported theoretically with parametrizations of the spectral energy balance by Gao and Komen (1993) and Janssen (private communication, 1991). A nonlinear relaxation model may therefore be more appropriate than the above linear model. A reanalysis of the data of Holthuijsen *et al* (1987) by Gao and Komen with their (nonlinear) time scale estimate indeed improves the correlation with wave age (figure 2.39).

From the definition (2.242) of θ_0 it can be seen that in case of narrow spectra, the value of the mean wave direction is dominated by the spectral peak. Moreover, for such spectra the nonlinear wave–wave interactions are more active near the spectral peak than elsewhere. It may therefore be assumed that the directional response of locally generated waves is dominated by the peak frequency. The frequency-dependent expressions of D.E. Hasselmann *et al* (1980), Allender *et al* (1983) and Masson (1990), when applied to the peak frequency, can therefore be used to compare their results with the frequency-independent time scale τ. For the required transformations, van Vledder and Holthuijsen (1993) used the same value for C_D as above. The result is shown in figure 2.38a, where it is obvious that the observations separate into two families: the frequency-dependent data as applied to the peak frequency on the one hand and the frequency-independent data on the other (figure 2.38b). The discrepancies between the two families may well be due to the transformation of the frequency-dependent observations or to a difference in estimation procedures or to a difference in the selection of the observations. To investigate the last two potential sources of discrepancy van Vledder and Holthuijsen (1993) reanalysed the published time series of D.E. Hasselmann *et al* (1980) who published about 1/3 of their original time series and Allender *et al* (1983) who published all of their original time series, with the above finite difference technique instead of the regression technique. The results of this reanalysis is a value of b which is a factor of 8 larger than the original value for Allender *et al* (1983) and a factor of 3 larger for D.E. Hasselmann *et al* (1980), see table 2.1. These differences are appreciable and they bring the data of D.E. Hasselmann *et al* (1980) and Allender *et al* (1983) within the range of Günther *et al* (1981) and Holthuijsen *et al* (1987) as shown in figure 2.38b. The effect of different selection criteria seems to be marginal (van Vledder and Holthuijsen, 1993).

II.10.5 Conclusion

The present state-of-the-art of observations of waves turning to the local wind direction is unsatisfactory. This may be due to several factors. One is the basic parametric approach that has been used but which has been abandoned with the advent of third generation wave modelling. Another factor is the choice of parameters within the parametric approach. For instance, a different definition of time scale may be called for, since empirical and theoretical considerations indicate that a nonlinear relaxation model may be better than the presently used linear relaxation model. Still another factor is the choice of estimation procedure, which seems to have an appreciable effect on the estimates of the observed time scales.

II.11 Observations of wave growth and dissipation in shallow water

G. J. Komen

II.11.1 Growth curves

In previous sections finite depth effects have been encountered, such as refraction, shoaling and modification of the deep water source terms. In particular, in § II.5 dissipation at the bottom has been discussed from a theoretical point of view. It is interesting now to look in some more detail at observations of waves in shallow water.

Let us begin with a few facts. The climatology of the North Sea indicates that extreme wave heights in the shallow southern part are significantly lower than in the deep northern part (see figure 2.40). Going from deep to shallow the extreme wave height with a return period of 10 years varies between 16 and 7 metres, a large and significant variation. But this is not exclusively due to finite depth effects. First of all the extreme mean wind speed also varies, and this contributes to the effect. A rough, first estimate of the effect would follow from the incorrect assumption that extreme waves occur in infinite fetch situations. For these the extreme wave heights scale with the friction velocity squared:

$$H_S^\infty = H_*^\infty u_*^2/g, \qquad (2.243)$$

with H_*^∞ a constant. The observed extreme wind speed varies from 36 m/s in the north to 28 m/s in the south (Korevaar, 1990). By converting this to friction velocity and using (2.243) one would seem to explain most of the effect. However, in practice, for extreme wind speeds infinite fetch is never realized, due to the limited size of midlatitude depressions. This is fortunate for users of the sea – a wind speed of 36 m/s corresponds with an equilibrium wave height of approximately 50 m – and it means that extreme waves are always very young. As a result the extreme dimensionless wave height increases more slowly than u_*^2. Therefore, although the wind climatology is partly responsible for the wave height reduction in the southern North Sea, it cannot account for all of the variation and wave dynamical effects must also play a role. One possible cause is the close proximity of coasts in shallow water. In general, this leads to fetch limitations, which naturally lead to reduced wave heights, but again this is not likely to play a significant role, as extreme wave events occur in the case of north-westerly storms, for which the North Sea basin has no coastal fetch limitations. Another mechanism, refraction, can lead to important spatial wave height variations, but the spatial *averages* which occur in climatological studies are much less

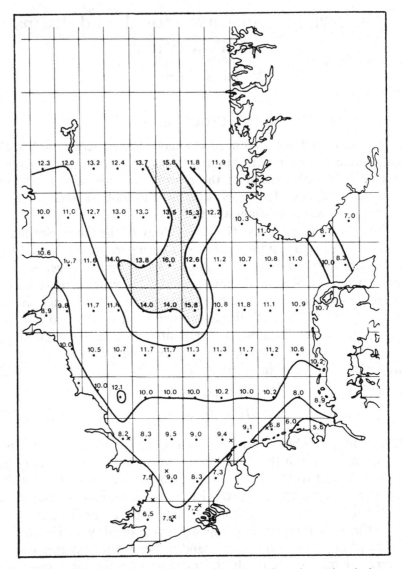

Fig. 2.40. Extreme significant wave heights in metres (based on visual observations) with a return period of 10 years (after Korevaar, 1990). Reprinted by permission of Kluwer Academic Publishers.

sensitive to its effects. Experimentally not much is known about the wind input and whitecapping-dissipation source terms in shallow water. However, simulations with parametrizations suggest that they are only mildly sensitive to depth variations. Therefore, it is tentatively concluded that it is the bottom source term that is mainly responsible for the reduced extreme wave heights in shallow water.

More evidence for the effect of a bottom source term comes from hind-casts with numerical wave prediction models. Deep water models simply overpredict in shallow water. A good example is the so-called Texel storm (Bouws and Komen, 1983, Golding, 1978). This storm occurred on 3 January 1976. It was an extreme storm with its centre located at the Dogger Bank. Wind speed (~ 25 m/s) and direction (300°) were fairly constant. Waverider measurements showed an increase of wave height until a maximum average level of 6.8 m was reached. This value can be well reproduced by numerical models, provided a bottom source term is included. Bouws and Komen (1983) showed that its effect on the total energy balance is of the same magnitude as that of other source terms. Another storm which illustrates the point is the Shetland storm (to be discussed later, in chapter IV). The so-called SWIM study (SWIM, 1985), in which three second generation models (GONO, BMO and HYPA) were compared also illustrated the important effect of the bottom source term. In a similar numerical study Graber and Madsen (1988) showed that in addition to depth the configuration of the sea bed, i.e. the presence of ripples or bed forms, fine sand or coarse sand, also influenced the growth curves and the attainable maximum wave heights.

The types of consideration given above led to the introduction of the concept of growth curves in shallow water. Holthuijsen (1980) gave a fair account of the state of the art at that time. Many authors (Bretschneider, 1954, 1973, Groen and Dorrestein, 1976, Krylov *et al*, 1976, Ou and Tang, 1974, Roest, 1960, Thijsse, 1949) introduced the concept of maximum di-mensionless wave height in depth-limited wave growth. A typical example is

$$\tilde{H}_\infty = 0.283 \tanh(0.530 \tilde{h}^{0.750}), \qquad (2.244)$$

with $(\tilde{\cdot}) = (\cdot)g/U_{10}^2$ (Bretschneider, 1973). Most authors did not give a very precise definition of $\tilde{H}_\infty$, and this may have led to some confusion. In general the maximum wave height will depend not only on friction velocity or wind speed, but also on the upwind depth distribution and the sea bottom morphology. This has not always been acknowledged. Also, there is a lack of field measurements that could illuminate these effects. We choose to define $\tilde{H}_\infty$ in shallow water as the maximum wave height that would be obtained in idealized conditions: infinite ocean, constant homogeneous winds over the whole basin and a flat bottom with some uniform roughness. It is extremely difficult if not impossible to find a basin that satisfies all of these conditions for large (and infinite) fetch, so observations are never made in idealized conditions. Deep water wave growth observations show scatter because of non-ideal coastal geometry, and spatial and temporal variation of the generating winds. Shallow water observations have additional scatter due to up-wind depth variations, due to variations in the composition of the sea bed and to interaction with the tides. The observations made by the above

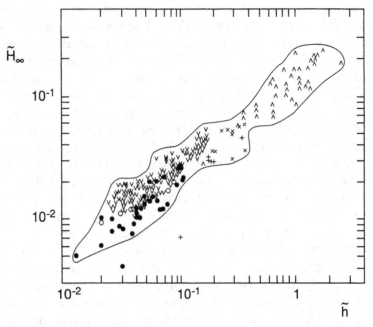

Fig. 2.41. Dimensionless wave height as a function of dimensionless depth for different sets of observations (after Holthuijsen, 1980).

mentioned authors are shown in figure 2.41. The scatter does not come as a surprise.

After the appropriate definition of the concept of depth-limited maximum wave height has been given, one can start to consider the effect of fetch limitation. As in deep water, the wave height grows with fetch until the maximum value is reached. A fit to the observed behaviour is given in the Shore Protection Manual (SPM, 1977, 1984). In this fit H_S is scaled with U_{10}. When analysing observations in Lake Marken, Bouws (1986) rescaled the SPM growth curve in terms of u_*, simply substituting $u_*/\sqrt{C_D}$ for U_{10}, taking for C_D a value of 1.775×10^{-3}, corresponding to a wind speed of 15 m/s. This gives the following result:

$$H_* = H_*^\infty \tanh(0.14 X_*^{0.42}/H_*^\infty),$$
$$H_*^\infty = 159 \tanh(4.6 \times 10^{-3} h_*^{0.75}), \qquad (2.245)$$

with the usual scaling relations $(\ )_* = (\ )g/u_*^2$. For $h \to \infty$ and small fetch this curve reduces to

$$H_* = 0.14 X_*^{0.42}, \qquad (2.246)$$

in reasonable agreement with the (neutrally stable) deep water growth curves of § II.8, which can be rewritten as $H_* = 0.18 X_*^{0.4}$

II.11.2 Spectral shape

As in deep water, it is difficult to make a general statement about the spectral shape, because spatial and temporal variation of the wind and geometric factors may lead to very complex wave spectra. However, Kitaigorodskii (1983) argued that the high-frequency tail should be expressed in terms of the wavenumber k rather than the frequency, and that there should be a universal form

$$F(k) = \text{constant}/k^3 \tag{2.247}$$

with the constant not dependent on depth. One can relate the frequency spectrum and the wavenumber spectrum in the usual way

$$F(k)\,\mathrm{d}k = F(f)\,\mathrm{d}f \tag{2.248}$$

to obtain

$$F(f,h) = \text{const}\frac{1}{k(f,h)^3}\frac{1}{(\partial f/\partial k)(f,h)}. \tag{2.249}$$

One may then eliminate the constant to obtain the formal relationship

$$F(f,h) = r(f,h)F(f,\infty) \quad k \gg k_{peak}, \tag{2.250}$$

with

$$r(f,h) = \frac{k(f,\infty)^3\;(\partial f/\partial k)(f,\infty)}{k(f,h)^3\;(\partial f/\partial k)(f,h)}. \tag{2.251}$$

In practice this factor is not far from unity in the tail, because high-frequency waves, in general, do not 'feel' the bottom.

In deep water wave spectra have often been fitted to the JONSWAP shape. This was first of all a data-reduction technique, but it was also assumed at times that fetch-limited wave growth (the case of 'pure wind-sea') could be fully described in terms of JONSWAP spectra. Third generation modelling exercises have confirmed that this is the case but only in an approximate sense. Bouws *et al* (1985) have extended this concept to shallow water. Their results derive from the analysis of spectra from different data sets (T = Texel refers to the 'Texel storm' discussed above; M = Marsen = Marine Remote Sensing Experiment and A = Arsloe = Atlantic Ocean Remote Sensing Land-Ocean Experiment). They found that in shallow water spectra can be best fit by the TMA spectral shape which is

$$F_{TMA}(f;h) = r(f,h)F_J(f), \quad 0 < f < \infty. \tag{2.252}$$

The extension of the proportionality to lower-frequencies reduces the spectral level of the spectrum near the spectral peak, but it cannot be justified by the theory of Kitaigorodskii. For the Texel storm, Bouws (private communication) has shown that a JONSWAP fit can be made equally well; in that case the peak enhancement is smaller than in deep water.

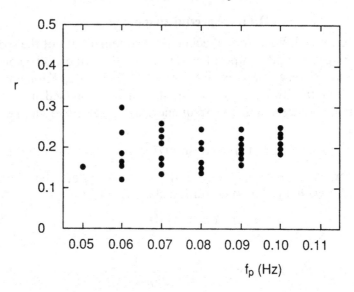

Fig. 2.42. Mean reduction factor (between stations K13 and EURO) of ten swell cases versus peak frequency (after Yan and Bouws, 1987).

In third generation wave modelling there is no need to compare model predictions with parametrized spectra. Therefore, the issue of spectral shape has lost much of its meaning in this context. A well-defined problem, namely what is the spectral evolution in idealized depth- and fetch-limited wave growth, cannot be investigated experimentally, because of the afore-mentioned difficulty in finding a suitable basin. However, it can be easily simulated in a numerical model.

II.11.3 Swell dissipation

One might hope to learn more about the effect of the bottom source term by studying the dissipation of swell when it runs into shallow water. Observations show that long swell waves begin to dissipate energy long before they break near the shore. The first quantitative observations were made during JONSWAP (Hasselmann *et al*, 1973). The observed damping showed very large scatter, but the mean effect could be described in terms of an empirical bottom source term

$$S_J = -\frac{c}{g^2}\frac{\omega^2}{\sinh^2 kh}F(f,\theta). \qquad (2.253)$$

This agrees with the expression given in § II.5 if you identify C_{bot} with $2c/g$. The mean observed value during JONSWAP was $c = 0.038$ m^2 s^{-3}.

Yan and Bouws (1987) considered swell with wave periods of 10 – 20 seconds, generated by gales in the Norwegian Sea and the northern North Sea, and they analysed their height reduction r during propagation through the southern part, where they feel the bottom. Their results, given in figure 2.42 can be well described by the following source term

$$S_{bot} = S_J - \frac{\alpha k}{\cosh^2 kh} F(\omega), \qquad (2.254)$$

where the second term represents percolation. They obtained $\alpha = 0.0018$ m/s. However, the results show a lot of scatter, which may have been caused by the neglect of refraction. Also, the two terms in (2.254) have a rather similar k dependence, so their independent estimate can not be very accurate.

II.11.4 Concluding remark

Observational knowledge of wave dissipation by bottom processes in shallow water is remarkably sparse. This is caused in part because, in general, observations have not been made or analysed with a clear idea about the underlying mechanism, such as friction by the viscous boundary layer, turbulence or sheet flow, and the physical nature of the bottom (flat sand, bottom ripples, ...). Another complication comes from the fact that many processes operate simultaneously (refraction, interaction with tides, temporal and spatial wind variability), and this hampers a quantitative analysis. We believe that in order to make significant progress new measurements will be necessary. To analyse these we will need a comprehensive numerical model that can use the data in an inverse modelling approach, in which all relevant processes are considered simultaneously, which will be discussed in §§ VI.6 and VI.7.

Chapter III

Numerical modelling of wave evolution

P. A. E. M. Janssen *et al*

III.1 Introduction

The principles of wave prediction were already well known at the beginning of the sixties (§ I.1). Yet, none of the wave models developed in the 1960s and 1970s computed the wave spectrum from the full energy balance equation. Additional *ad hoc* assumptions have always been introduced to ensure that the wave spectrum complies with some preconceived notions of wave development that were in some cases not consistent with the source functions. Reasons for introducing simplifications in the energy balance equation were twofold. On the one hand, the important role of the wave–wave interactions in wave evolution was not recognized. On the other hand, the limited computer power in those days precluded the use of the nonlinear transfer in the energy balance equation.

The first wave models, which were developed in the 1960s and 1970s, assumed that the wave components suddenly stopped growing as soon as they reached a universal saturation level (Phillips, 1958). The saturation spectrum, represented by Phillips' one-dimensional f^{-5} frequency spectrum and an empirical equilibrium directional distribution, was prescribed. Nowadays it is generally recognized that a universal high-frequency spectrum (in the region between 1.5 and 3 times the peak frequency) does not exist because the high-frequency region of the spectrum not only depends on whitecapping but also on wind input and on the low-frequency regions of the spectrum through nonlinear transfer. Furthermore, from the physics point of view it has now become clear that these so-called first generation wave models exhibit basic shortcomings by overestimating the wind input and disregarding nonlinear transfer.

The relative importance of nonlinear transfer and wind input became more evident after extensive wave growth experiments (Mitsuyasu, 1968, 1969, Hasselmann *et al*, 1973) and direct measurements of the wind input to the waves (Snyder *et al*, 1981, D.E. Hasselmann *et al*, 1986). This led to the development of second generation wave models which attempted to simulate properly the so-called overshoot phenomenon and the dependence of the high-frequency region of the spectrum on the low frequencies. However, restrictions resulting from the nonlinear transfer parametrization effectively required the spectral shape of the wind sea spectrum to be prescribed. The specification of the wind sea spectrum was imposed either at the outset in the formulation of the transport equation itself (parametrical or hybrid models) or as a side condition in the computation of the spectrum (discrete models). These models therefore suffered basic problems in the treatment of wind sea and swell. Although, for typical synoptic-scale wind fields the evolution towards a quasi-universal spectral shape could be justified by two scaling arguments (Hasselmann *et al*, 1976), nevertheless complex wind seas generated by rapidly varying wind fields (in, for example, hurricanes or fronts) were not simulated properly by the second generation models.

The shortcomings of first and second generation models have been documented and discussed in the SWAMP (1985) wave-model intercomparison study. The development of third generation models was suggested in which the wave spectrum was computed by integration of the energy balance equation, without any prior restriction on the spectral shape. As a result the WAM group was established, whose main task was the development of such a third generation wave model. In this chapter we shall describe the latest version of the so-called WAM model.

In the previous chapter we have given an extensive overview of what is presently known about the physics of wave evolution, in so far as it is relevant to a spectral description of ocean waves. Thus, we described in detail our knowledge of the generation of ocean waves by wind and the impact of the waves on the air flow, we discussed the importance of the resonant nonlinear interactions for wave evolution and we presented our knowledge on dissipation of wave energy by whitecapping and bottom friction.

In this chapter we will be engaged in the task of making optimal use of our knowledge of wave evolution in the context of numerical modelling of ocean waves. However, in order to be able to develop a numerical wave model that produces forecasts in a reasonable time, compromises regarding the functional form of the source terms in the energy balance equation have to be made. For example, a traditional difficulty of numerical wave models has been the adequate representation of the nonlinear source term S_{nl}. Since the time needed to compute the exact source function expression greatly exceeds practical limits set by an operational wave model, some form of parametrization is clearly necessary. Likewise, the numerical solution of the momentum balance of air flow over growing ocean waves, as presented in the previous chapter, is by far too time consuming to be practical for numerical modelling. It is therefore clear that a parametrization of the functional form of the source terms in the energy balance equation is a necessary step to develop an operational wave model.

The programme of this chapter is as follows. In § III.2 we discuss the kinematic part of the energy balance equation, that is, advection in both deep and shallow water, refraction due to currents and bottom topography. The next section, § III.3, is devoted to a parametrization of the input source term and the nonlinear interactions. The adequacy of these approximations is discussed in detail, as is the energy balance in growing waves. Next, in § III.4 we discuss the numerical implementation of the model. We distinguish between a prognostic part of the spectrum (that part that is explicitly calculated by the numerical model) (WAMDI, 1988), and a diagnostic part.

The diagnostic part of the spectrum has a prescribed spectral shape, the level of which is determined by the energy of the highest resolved frequency bin of the prognostic part. Knowledge of the unresolved part of the spectrum allows us to determine the nonlinear energy transfer from the resolved part to the unresolved part of the spectrum. The prognostic part of the spectrum is obtained by numerically solving the energy balance equation. The choice of numerical schemes for advection, refraction and time integration is discussed. The integration in time is performed using a semi-implicit integration scheme in order to be able to use large time steps without incurring numerical instabilities in the high-frequency part of the spectrum. For advection and refraction we have chosen a first order, upwinding flux scheme. Advantages of this scheme are discussed in detail, especially in connection with the so-called garden sprinkler effect (see SWAMP, 1985, p144). Alternatives to first order upwinding, such as the semi-Lagrangian scheme which is gaining popularity in meteorology, will be discussed as well.

The following section, § III.5, is devoted to software aspects of the WAM model code with emphasis on flexibility, universality and design choices. A brief summary of the manual accompanying the code is given (Günther *et al*, 1991). Finally, § III.6 gives a summary of a number of simple tests we have performed highlighting the importance of fetch, turning wind fields, shallow water effects, swell dissipation and directional spreading.

III.2 The kinematic part of the energy balance equation

P. A. E. M. Janssen, K. Hasselmann and S. Hasselmann

In this section we shall briefly discuss some properties of the energy balance equation in the absence of sources and sinks. Thus, shoaling and refraction – by bottom topography and ocean currents – are investigated in the context of a statistical description of gravity waves.

Let x_1 and x_2 be the spatial coordinates and k_1, k_2 the wave coordinates, and let

$$\mathbf{z} = (x_1, x_2, k_1, k_2) \tag{3.1}$$

be their combined four-dimensional vector. Then the most fundamental form of the transport equation for the action density spectrum $N(\mathbf{k}, \mathbf{x}, t)$ without the source term can be written in the flux form (see also (1.187))

$$\frac{\partial}{\partial t} N + \frac{\partial}{\partial z_i}(\dot{z}_i N) = 0, \tag{3.2}$$

where $\dot{\mathbf{z}}$ denotes the propagation velocity of a wave group in the four-dimensional phase space of $\mathbf{x}$ and $\mathbf{k}$. This equation holds for any field $\dot{\mathbf{z}}$, and also for velocity fields which are not divergence-free in four-dimensional phase space. In the special case when $\mathbf{x}$ and $\mathbf{k}$ represent a canonical vector pair – this is the case, for example, when they are the usual Cartesian coordinates – the propagation equations for a wave group (also known as Hamilton's equations of motion) read:

$$\dot{x}_i = \frac{\partial}{\partial k_i} \Omega, \qquad (3.3a)$$

$$\dot{k}_i = -\frac{\partial}{\partial x_i} \Omega, \qquad (3.3b)$$

where Ω denotes the dispersion relation

$$\omega = \Omega(\mathbf{k}, \mathbf{x}, t) = \sigma + \mathbf{k} \cdot \mathbf{U}, \qquad (3.4)$$

with σ the so-called intrinsic frequency

$$\sigma = \sqrt{gk \tanh(kh)}, \qquad (3.5)$$

where the depth $h(\mathbf{x}, t)$ and the current $U(\mathbf{x}, t)$ may be slowly-varying functions of $\mathbf{x}$ and t.

The Hamilton equations have some intriguing consequences. First of all, equation (3.3a) just introduces the group speed $\partial\Omega/\partial k_i$ while (3.3b) expresses conservation of the number of wave crests. Secondly, the transport equation for the action density may be expressed in the advection form

$$\frac{d}{dt} N = \frac{\partial N}{\partial t} + \dot{z}_i \frac{\partial}{\partial z_i} N = 0, \qquad (3.6)$$

as, because of (3.3a,b), the field $\dot{\mathbf{z}}$ for a continuous ensemble of wave groups is divergence-free in four-dimensional phase space,

$$\frac{\partial}{\partial z_i} \dot{z}_i = 0. \qquad (3.7)$$

Thus, along a path in four-dimensional phase space defined by the Hamilton equations (3.3a,b), the action density N is conserved. This property only holds for canonical coordinates for which the flow divergence vanishes (Liouville's theorem – first applied by Dorrestein (1960) to wave spectra). Thirdly, the analogy between Hamilton's formalism of particles with Hamiltonian H and wave groups obeying the Hamilton equations of motion should be pointed out. Indeed, wave groups may be regarded as particles and the Hamiltonian H and angular frequency Ω play similar roles. Because of this similarity Ω is expected to be conserved as well (under the restriction that Ω does not depend on time). This can be verified by direct calculation of the rate of change of Ω following the path of a wave group in phase space,

$$\frac{\mathrm{d}}{\mathrm{d}t}\Omega = \dot{z}_i\frac{\partial}{\partial z_i}\Omega = \dot{x}_i\frac{\partial}{\partial x_i}\Omega + \dot{k}_i\frac{\partial}{\partial k_i}\Omega = 0. \tag{3.8}$$

The vanishing of $\mathrm{d}\Omega/\mathrm{d}t$ follows at once upon using the Hamilton equations (3.3a,b). Note that the restriction of no time dependence of Ω is essential for the validity of (3.8), just as the Hamiltonian H is only conserved when it does not depend on time t. The property (3.8) will play an important role in our discussion of refraction.

We now turn to the important case of spherical coordinates. When one transforms from one set of coordinates to another there is no guarantee that the flow remains divergence-free. However, noting that equation (3.2) holds for any rectangular coordinate system, the generalization of the standard Cartesian geometry transport equation to spherical geometry (see also Groves and Melcer, 1961 and WAMDI, 1988) is straightforward. To that end let us consider the spectral action density $\hat{N}(\omega,\theta,\phi,\lambda,t)$ with respect to angular frequency ω and direction θ (measured clockwise relative to true north) as a function of latitude ϕ and longitude λ. The reason for the choice of frequency as the independent variable (instead of, for example, the wavenumber k) is that for a fixed topography and current the frequency Ω is conserved when following a wave group, therefore the transport equation simplifies. In general, the conservation equation for $\hat{N}$ thus reads

$$\frac{\partial}{\partial t}\hat{N} + \frac{\partial}{\partial \phi}(\dot{\phi}\hat{N}) + \frac{\partial}{\partial \lambda}(\dot{\lambda}\hat{N}) + \frac{\partial}{\partial \omega}(\dot{\omega}\hat{N}) + \frac{\partial}{\partial \theta}(\dot{\theta}\hat{N}) = 0, \tag{3.9}$$

and since $\dot{\omega} = \partial\Omega/\partial t$ the term involving the derivative with respect to ω drops out in case of time-independent current and bottom. The action density $\hat{N}$ is related to the normal spectral density N with respect to a local Cartesian frame (x,y) through $\hat{N}\mathrm{d}\omega\mathrm{d}\theta\mathrm{d}\phi\mathrm{d}\lambda = N\mathrm{d}\omega\mathrm{d}\theta\mathrm{d}x\mathrm{d}y$, or

$$\hat{N} = NR^2\cos\phi, \tag{3.10}$$

where R is the radius of the earth. Substitution of (3.10) into (3.9) yields

$$\frac{\partial}{\partial t}N + (\cos\phi)^{-1}\frac{\partial}{\partial\phi}(\dot{\phi}\cos\phi N) + \frac{\partial}{\partial\lambda}(\dot{\lambda}N) + \frac{\partial}{\partial\omega}(\dot{\omega}N) + \frac{\partial}{\partial\theta}(\dot{\theta}N) = 0, \tag{3.11}$$

where, with c_g the magnitude of the group velocity,

$$\dot{\phi} = (c_g\cos\theta + \mathbf{U}\,|_{north})R^{-1}, \tag{3.12a}$$

$$\dot{\lambda} = (c_g\sin\theta + \mathbf{U}\,|_{east})(R\cos\phi)^{-1}, \tag{3.12b}$$

$$\dot{\theta} = c_g\sin\theta\tan\phi R^{-1} + (\dot{\mathbf{k}}\times\mathbf{k})k^{-2}, \tag{3.12c}$$

$$\dot{\omega} = \partial\Omega/\partial t \tag{3.12d}$$

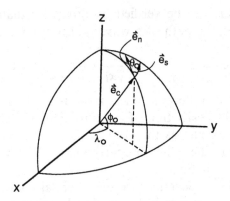

Fig. 3.1. Wave propagation on a globe.

represent the rates of change of the position and propagation direction of a wave packet. Equation (3.11) is the basic transport equation which we will use in the numerical wave prediction model. The remainder of this section is devoted to a discussion of some of the properties of (3.11). We first discuss some peculiarities of (3.11) for the infinite depth case in the absence of currents and next we discuss the special cases of shoaling and refraction due to bottom topography and currents.

Great circle propagation on the globe

From (3.12a–d) we infer that in spherical coordinates the flow is not divergence-free. Considering the case of no depth refraction and no explicit time dependence, the divergence of the flow becomes

$$\frac{\partial}{\partial\phi}\dot{\phi} + \frac{\partial}{\partial\lambda}\dot{\lambda} + \frac{\partial}{\partial\theta}\dot{\theta} + \frac{\partial}{\partial\omega}\dot{\omega} = c_g\cos\theta\tan\phi R^{-1} \neq 0, \qquad (3.13)$$

which is nonzero because the wave direction, measured with respect to true north, changes while the wave group propagates over the globe along a great circle. That wave groups propagate along a great circle may be seen as follows.

The great circle path of a wave group on the spherical earth may be written in the form (WAMDI, 1988)

$$\mathbf{x} = R(\mathbf{e}_c\cos(c_g t/R) + \mathbf{e}_s\sin(c_g t/R)), \qquad (3.14)$$

where

$$\mathbf{e}_c = \begin{pmatrix} \cos\phi_o\cos\lambda_o \\ \cos\phi_o\sin\lambda_o \\ \sin\phi_o \end{pmatrix} \qquad (3.15)$$

and

$$\mathbf{e}_s = \begin{pmatrix} - & \cos\theta_o \sin\phi_o \cos\lambda_o - \sin\theta_o \sin\lambda_o \\ - & \cos\theta_o \sin\phi_o \sin\lambda_o + \sin\theta_o \cos\lambda_o \\ & \cos\theta_o \cos\phi_o \end{pmatrix} \qquad (3.16)$$

are orthogonal unit vectors pointing in the direction of the location and velocity respectively of the wave group at time $t = 0$; $\phi_o, \lambda_o, \theta_o$ denote the initial spherical coordinates and propagation direction of the wave group at time $t = 0$ (see figure 3.1 for a detailed picture). The vector $\mathbf{e}_s$ can be readily recognized as the wave group propagation direction by considering first the special case $\theta_o = 0, \lambda_o = 0$ and then rotating about the x-axis ($\theta_o \neq 0$) and subsequently about the z-axis ($\lambda_o \neq 0$) to recover the general case. Differentiating (3.14) with respect to t and setting $t = 0$ yields the propagation equation for latitude ϕ and longitude λ (of course, with current $\mathbf{U} = 0$). To derive the refraction equation (3.12) we note from figure 3.1 that

$$\cos\theta = \frac{1}{c_g}(\frac{d\mathbf{x}}{dt} \cdot \mathbf{e}_n), \qquad (3.17)$$

where

$$\mathbf{e}_n = \begin{pmatrix} - & \sin\phi \cos\lambda \\ - & \sin\theta \sin\lambda \\ & \cos\phi \end{pmatrix} \qquad (3.18)$$

is the local northward-pointing horizontal unit vector. Differentiating (3.16) we obtain

$$\frac{d\theta}{dt}\sin\theta = -\frac{1}{c_g}\left\{\frac{d^2\mathbf{x}}{dt^2} \cdot \mathbf{e}_n + \frac{d\mathbf{x}}{dt} \cdot \frac{d}{dt}\mathbf{e}_n\right\}. \qquad (3.19)$$

Using (3.14) the acceleration of the wave group is given as

$$\frac{d^2}{dt^2}\mathbf{x} = -\left(\frac{c_g}{R}\right)^2 \mathbf{x}, \qquad (3.20)$$

and, therefore, the first term in curly brackets vanishes because $\mathbf{x}$ is always orthogonal to $\mathbf{e}_n$ (see also figure 3.1). Hence,

$$\frac{d\theta}{dt} = -\frac{1}{c_g \sin\theta}\frac{d\mathbf{x}}{dt} \cdot \frac{d}{dt}\mathbf{e}_n, \qquad (3.21)$$

and using (3.12) and (3.12) we find

$$\frac{d\theta}{dt} = c_g \sin\theta \tan\phi R^{-1}, \qquad (3.22)$$

which is identical to equation (3.12) in the absence of depth refraction. We emphasize that the refraction as given by (3.22) is entirely due to the change in time of the local northward-pointing vector $\mathbf{e}_n$. This type of refraction is therefore entirely apparent and only related to the choice of coordinate system.

Shoaling

Let us now discuss finite depth effects in the absence of currents by considering some simple topographies. We first of all discuss shoaling of waves for the case of wave propagation parallel to the direction of the depth gradient. In this case, depth refraction does not contribute to the rate of change of wave direction $\dot{\theta}$ because, with equation (3.3b), $\mathbf{k} \times \dot{\mathbf{k}} = 0$. In addition, we take the wave direction θ to be zero so that the longitude is constant ($\dot{\lambda} = 0$) and $\dot{\theta} = 0$. For time-independent topography (hence $\partial\Omega/\partial t = 0$) the transport equation becomes

$$\frac{\partial}{\partial t}N + (\cos\phi)^{-1}\frac{\partial}{\partial\phi}(\dot{\phi}\cos\phi N) = 0, \tag{3.23}$$

where

$$\dot{\phi} = c_g \cos\theta R^{-1} = c_g/R, \tag{3.24}$$

and the group speed only depends on latitude ϕ. Restricting our attention to steady waves we immediately find conservation of the action density flux in the latitude direction, or,

$$\frac{c_g \cos\phi}{R}N = \text{const.} \tag{3.25}$$

If, in addition, it is assumed that the variation of depth with latitude occurs on a much shorter scale than the variation of $\cos\phi$, the latter term may be taken constant for present purposes. It is then found that the action density is inversely proportional to the group speed c_g,

$$N \sim 1/c_g \tag{3.26}$$

and if the depth is decreasing for increasing latitude, conservation of flux requires an increase of the action density as the group speed decreases for decreasing depth. This phenomenon, which occurs in coastal areas, is called shoaling. Its most dramatic consequences may be seen when tidal waves, generated by earthquakes, approach the coast resulting in tsunamis. It should be emphasized though, that in the final stages of a tsunami the kinetic description of waves, as presented here, breaks down because of strong nonlinearity.

Refraction

The second example of finite depth effects that we discuss is refraction. We again assume no current and a time-independent topography. In the steady state the action balance equation becomes

$$(\cos\phi)^{-1}\frac{\partial}{\partial\phi}\left(\frac{c_g}{R}\cos\theta\cos\phi N\right) + \frac{\partial}{\partial\lambda}\left(\frac{c_g\sin\theta}{R\cos\phi}N\right) + \frac{\partial}{\partial\theta}(\dot\theta_o N) = 0, \quad (3.27)$$

where

$$\dot\theta_o = \left(\sin\theta\frac{\partial}{\partial\phi}\Omega - \frac{\cos\theta}{\cos\phi}\frac{\partial}{\partial\lambda}\Omega\right)(kR)^{-1}. \quad (3.28)$$

In principle, equation (3.27) can be solved by means of the method of characteristics. We will not give the details of this, but we would like to point out the role of the $\dot\theta_o$ term for the simple case of waves propagating along the shore. Consider, therefore, waves propagating in a northerly direction (hence $\theta = 0$) parallel to the coast. Suppose that the depth only depends on longitude such that it decreases towards the shore. The rate of change of wave direction is then positive as

$$\dot\theta_o = -\frac{1}{kR\cos\phi}\frac{\partial}{\partial\lambda}\Omega > 0, \quad (3.29)$$

since $\partial\Omega/\partial\lambda < 0$. Therefore, waves which are propagating initially parallel to the coast will turn towards the coast. This illustrates that, in general, wave rays will bend towards shallower water resulting in, for example, focussing phenomena and caustics. In this way a sea mountain plays a similar role for gravity waves as a lens for light waves.

Current effects

Finally, we discuss some current effects on wave evolution. First of all, a horizontal shear may result in wave refraction; the rate of change of wave direction follows from (3.27) by taking the current into account,

$$\dot\theta_c = \frac{1}{R}\left(\sin\theta\left[\cos\theta\frac{\partial}{\partial\phi}U_\phi + \sin\theta\frac{\partial}{\partial\phi}U_\lambda\right] - \frac{\cos\theta}{\cos\phi}\left[\cos\theta\frac{\partial}{\partial\lambda}U_\phi + \sin\theta\frac{\partial}{\partial\lambda}U_\lambda\right]\right),$$

$$(3.30)$$

where U_ϕ and U_λ are the components of the water current in latitudinal and longitudinal directions. Considering the same example as in the case of depth refraction, we note that the rate of change of the direction of waves propagating initially along the shore is given by

$$\dot\theta_c = -\frac{1}{R\cos\phi}\frac{\partial}{\partial\lambda}U_\phi, \quad (3.31)$$

which is positive for an along-shore current which decreases towards the coast. In that event the waves will turn towards the shore.

The most dramatic effects may be found, when the waves propagate against the current. For sufficiently large current, wave propagation is prohibited and wave reflection occurs. This may be seen as follows. Consider waves propagating to the right against a slowly varying current U_o. At $x \to -\infty$ the current vanishes, decreasing monotonically to some negative value for $x \to +\infty$. Let us generate at $x \to -\infty$ a wave with a certain frequency value Ω_0. Following the waves, we know from (3.8), that for time-independent circumstances the angular frequency of the waves is constant, hence for increasing strength of the current the wavenumber increases as well. Now, whether the surface wave will arrive at $x \to +\infty$ or not depends on the magnitude of the dimensionless frequency $\Omega_0 U_m/g$ (where U_m is the maximum strength of the current); for $\Omega_0 U_m/g < 1/4$ propagation up to $x \to +\infty$ is possible, whereas in the opposite case propagation is prohibited. Considering deep water waves only, the dispersion relation reads

$$\Omega = \sqrt{gk} - kU_o \qquad (3.32)$$

and the group velocity $\partial\Omega/\partial k$ vanishes for $k = g/4U_o^2$ so that the value of Ω at the extremum is $\Omega_c = g/4U_o$. At the location where the current has maximum strength the critical angular frequency Ω_c is the smallest. Let us denote this minimum value of Ω_c by $\Omega_{c,min}(= g/4U_m)$. If now the oscillation frequency $\Omega_0 < \Omega_{c,min}$ is in the entire domain of consideration, then the group speed is always finite and propagation is possible (this of course corresponds to the condition $\Omega_0 U_m/g < 1/4$), but in the opposite case propagation is prohibited beyond a certain point in the domain. What actually happens at that critical point is still under debate. Because of the vanishing group velocity, a large increase of energy at that location may be expected suggesting that wave breaking plays a role. On the other hand, it may be argued that near such a critical point the usual geometrical optics approximation breaks down and that tunnelling and wave reflection occurs (Shyu and Phillips, 1990). A kinetic description of waves which is based on geometrical optics then breaks down as well. This problem is not solved in the WAM model. In order to avoid problems with singularities and nonuniqueness (note that for finite Ω_c one frequency Ω corresponds to two wavenumbers) we merely transform to the intrinsic frequency σ (instead of frequency Ω) because a unique relation between σ and wavenumber k exists.

Concluding remark

To conclude this section we note that a global third generation wave model solves the action balance equation in spherical coordinates. By combining previous results of this section, the action balance equation reads

$$\frac{\partial}{\partial t}N + (\cos\phi)^{-1}\frac{\partial}{\partial\phi}(\dot{\phi}\cos\phi N) + \frac{\partial}{\partial\lambda}(\dot{\lambda}N) + \frac{\partial}{\partial\omega}(\dot{\omega}N) + \frac{\partial}{\partial\theta}(\dot{\theta}N) = S, \quad (3.33)$$

where

$$\dot{\phi} = (c_g \cos \theta + \mathbf{U}\,|_{north})R^{-1}, \qquad (3.34a)$$

$$\dot{\lambda} = (c_g \sin \theta + \mathbf{U}\,|_{east})(R \cos \phi)^{-1}, \qquad (3.34b)$$

$$\dot{\theta} = c_g \sin \theta \tan \phi R^{-1} + \dot{\theta}_D, \qquad (3.34c)$$

$$\dot{\omega} = \partial \Omega / \partial t \qquad (3.34d)$$

and

$$\dot{\theta}_D = \left(\sin \theta \frac{\partial}{\partial \phi} \Omega - \frac{\cos \theta}{\cos \phi} \frac{\partial}{\partial \lambda} \Omega \right)(kR)^{-1}, \qquad (3.35)$$

and Ω is the dispersion relation given in (3.4). Before discussing possible numerical schemes to approximate the left hand side of equation (3.33) we shall first of all discuss the parametrization of the source term S, where S is given by

$$S = S_{in} + S_{nl} + S_{ds} + S_{bot}, \qquad (3.36)$$

representing the physics of wind input, wave–wave interactions and dissipation due to whitecapping and bottom friction.

III.3 Parametrization of source terms and the energy balance in a growing wind sea

P. A. E. M. Janssen, K. Hasselmann, S. Hasselmann and G. J. Komen

In this section we will be faced with the task of providing an efficient parametrization of the source terms as they were introduced in chapter II. The need for a parametrization is evident when it is realized that both the exact versions of the nonlinear source term and the wind input require, per grid point, one minute of CPU time on the fastest computer that is presently available. In practice, a typical one day forecast should be completed in a time span of the order of two minutes, so it is clear that compromises have to be made regarding the functional form of the source terms in the action balance equation. Even optimization of the code representing the source terms by taking as most inner do-loop, a loop over the number of grid points (thus taking optimal advantage of vectorisation) is not of much help here as the gain in efficiency is at most a factor of 10 and as practical applications usually require in the order of 2000–4000 grid points. In § III.3.1 we therefore discuss a parametrization of wind input and dissipation while § III.3.2 is devoted to a discrete-interaction operator parametrization of the nonlinear interactions. The adequacy of the approximation for wind input and nonlinear transfer is discussed as well. Dissipation owing to bottom friction is not discussed here because the details of its parametrization were presented in chapter II. We merely quote the main result (2.185),

$$S_{bot} = -C_{bot}\frac{k}{\sinh(2kh)}N, \qquad (3.37)$$

where the constant $C_{bot} = 0.038/g$. The relative merits of this approach were discussed as well.

Finally, in § III.3.3, in which we study the energy balance equation in growing wind sea, the relative importance of the physics source terms will be addressed.

III.3.1 Wind input and dissipation

In chapter II we have presented results of the numerical solution of the momentum balance of air flow over growing surface gravity waves, summarizing a series of studies by Janssen (1989b), Janssen *et al* (1989a) and Janssen (1991a). The main conclusion was that the growth rate of the waves generated by wind depends on a number of additional factors, such as the atmospheric density stratification, wind gustiness and wave age. So far systematic investigations of the impact of the first two effects have not been made. In § II.8 we saw that stratification effects observed in fetch-limited wave growth can be partly accounted for by scaling with u_* (which is consistent with the theoretical results of § II.2). The remaining effect is still poorly understood, and is therefore ignored in the standard WAM model, but it will be discussed as a possible extension of the model in § IV.5.3. In this section we focus on the dependence of wave growth on wave age, and the related dependence of the aerodynamic drag on the sea state, which effect is fully included in the WAM model.

A first attempt to parametrize the interaction between wind and waves was presented by Janssen *et al* (1989b). There are, however, two problems with this approach. First of all, in the linear approximation, the growth rate of the waves due to wind was given by the Snyder *et al* (1981) expression which gives too low an input for high-frequency waves. This may underestimate the wave-induced stress. Secondly, the stress related to the gravity–capillary waves (parametrized by the Charnock relation) was just added to the long-wave stress, which is incorrect as there is an interaction between the two. A more realistic parametrization of the interaction between wind and wave was given by Janssen (1991a), a summary of which is given below.

The basic assumption Janssen (1991a) made, which was corroborated by his numerical results of 1989, was that even for young wind sea the wind profile has a logarithmic shape, though with a roughness length that depends on the wave-induced stress. As shown by Miles (1957), the growth rate of gravity waves due to wind then only depends on two parameters, namely

$$x = (u_*/c)\cos(\theta - \phi) \quad \text{and} \quad \Omega_m = \kappa^2 g z_0/u_*^2, \qquad (3.38)$$

where κ^2 is the von Kármán constant. As usual, u_* denotes the friction velocity, c the phase speed of the waves, ϕ the wind direction and θ the direction in which the waves propagate. The so-called profile parameter Ω_m characterizes the state of the mean air flow through its dependence on the roughness length z_0. Thus, through Ω_m the growth rate depends on the roughness of the air flow, which, in its turn, depends on the sea state. A simple parametrization of the growth rate of the waves follows from a fit of numerical results presented in chapter II. One finds

$$\frac{\gamma}{\omega} = \epsilon\beta x^2, \qquad (3.39)$$

where γ is the growth rate, ω the angular frequency, ϵ the air–water density ratio and β the so-called Miles' parameter. In terms of the dimensionless critical height $\mu = kz_c$ (with k the wavenumber and z_c the critical height defined by $U_0(z = z_c) = c$) Miles' parameter becomes

$$\beta = \frac{\beta_m}{\kappa^2}\mu \ln^4(\mu), \quad \mu \le 1, \qquad (3.40)$$

where β_m is a constant. In terms of wave and wind quantities μ is given as

$$\mu = \left(\frac{u_*}{\kappa c}\right)^2 \Omega_m \exp(\kappa/x), \qquad (3.41)$$

and the input source term S_{in} of the WAM model is given by

$$S_{in} = \gamma N, \qquad (3.42)$$

where γ follows from (3.39) and with N the action density spectrum.

The stress of air flow over sea waves depends on the sea state and from a consideration of the momentum balance of air it is found that the kinematic stress is given as (Janssen, 1991a)

$$\tau = (\kappa U(z_{obs})/\ln(z_{obs}/z_0))^2, \qquad (3.43)$$

where

$$z_0 = \hat{\alpha}\tau/g\sqrt{1-y}, \quad y = \tau_w/\tau. \qquad (3.44)$$

Here, z_{obs} is the mean height above the waves and τ_w is the stress induced by gravity waves (the 'wave stress')

$$\tau_w = \epsilon^{-1}g \int d\omega d\theta\, \gamma N\mathbf{k}. \qquad (3.45)$$

The frequency integral extends to infinity, but in its evaluation only an f^{-5} tail of gravity waves is included and the higher level of capillary waves is treated as a background small-scale roughness. In practice, we note that the wave stress points in the wind direction as it is mainly determined by the high-frequency waves which respond quickly to changes in the wind direction.

The relevance of relation (3.44) cannot be overemphasized. It shows that the roughness length is given by a Charnock relation (Charnock, 1955)

$$z_o = \alpha \tau / g. \tag{3.46}$$

However, the dimensionless Charnock parameter α is not constant but depends on the sea state through the wave-induced stress since

$$\alpha = \hat{\alpha} / \sqrt{1 - \frac{\tau_w}{\tau}}. \tag{3.47}$$

Evidently, whenever τ_w becomes of the order of the total stress in the surface layer (this happens, for example, for young wind sea) a considerable enhancement of the Charnock parameter is found, resulting in an efficient momentum transfer from air to water. The consequences of this sea-state-dependent momentum transfer will be discussed in § IV.5.1.

This finally leaves us with the choice of two unknowns namely $\hat{\alpha}$ from (3.47) and β_m from (3.40). The constant $\hat{\alpha}$ was chosen in such a way that for old wind sea the Charnock parameter α has the value 0.0185 in agreement with observations collected by Wu (1982) on the drag over sea waves. It should be realised though, that the determination of $\hat{\alpha}$ is not a trivial task, as beforehand the ratio of wave-induced stress to total stress is simply not known. It requires the running of a wave model. By trial and error the constant $\hat{\alpha}$ was found to be $\hat{\alpha} = 0.01$.

The constant β_m is chosen in such a way that the growth rate γ in (3.39) is in agreement with the numerical results obtained from Miles' growth rate. For $\beta_m = 1.2$ and a Charnock parameter $\alpha = 0.0144$ we have shown in figure 3.2 the comparison between Miles' theory and (3.39). In addition observations as compiled by Plant (1982) are shown. Realizing that the relative growth rate γ/f varies by four orders of magnitude it is concluded that there is a fair agreement between our fit (3.39), Miles' theory and observations. We remark that the Snyder *et al* (1981) fit to their field observations (see also equation (2.43)), which is also shown in figure 3.2, is in perfect accordance with the growth rate of the low-frequency waves although growth rates of the high-frequency waves are underestimated. Since the wave-induced stress is mainly carried by the high-frequency waves an underestimation of the stress in the surface layer would result.

We conclude that our parametrization of the growth rate of the waves is in good agreement with the observations. The next issue to be considered is how well our approximation of the surface stress compares with observed surface stress at sea. Fortunately, during HEXOS (Katsaros *et al*, 1987) wind

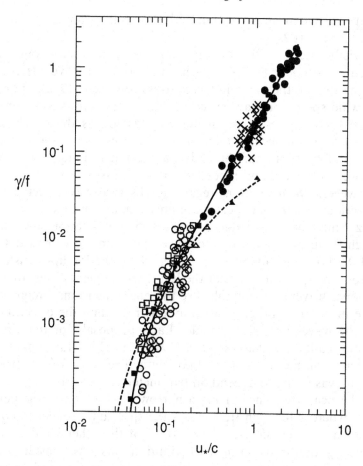

Fig. 3.2. Comparison of theoretical growth rates with observations compiled by Plant (1982). Full line: Miles' theory; full dots: parametrization of Miles' theory (3.39); dashed line: the fit by Snyder *et al* (1981).

speed at 10 m height, U_{10}, surface stress τ and the one-dimensional frequency spectrum were measured simultaneously so that our parametrization of the surface stress may be verified experimentally.

A first attempt towards verification of some of the consequences of the quasi-linear theory of wind-wave generation was made by Maat *et al* (1991) using HEXOS data. They used measured spectra and the observed friction velocity u_* to determine the wave age parameter c_p/u_* (where c_p is the phase speed of the peak of the spectrum). For wind sea this wave age is a good measure of the sea state. (Recall that young wind sea corresponds to a wave age of the order of 10 while old wind sea has a wave age of the order of 25.) In this fashion they were able to relate the drag coefficient $C_D = \tau/U_{10}^2$ to the wave age. They confirmed that, indeed, the drag of air flow over sea waves is sea-state-dependent in agreement with the results from quasi-linear theory as presented in § II.2.

Let us now briefly discuss a direct comparison between observations and the parametrized quasi-linear theory by using the observed wave spectrum for the determination of the wave-induced stress (Janssen, 1992). For a given observed wind speed and wave spectrum, the surface stress is obtained by solving (3.43) for the stress τ in an iterative fashion as the roughness length z_o depends, in a complicated manner, on the stress. Since the surface stress was measured by means of the eddy correlation technique, a direct comparison between observed and modelled stress is possible.

There are, however, two restrictions regarding observed wave spectra obtained by a waverider. Waverider spectra are only reliable up to a frequency of about 0.5 Hz. Since the wave-induced stress is mainly determined by the medium- to high-frequency range of the wave spectrum we extended the spectra beyond 0.5 Hz by means of an f^{-5} tail, where the Phillips constant was obtained from the spectral values of the last three frequency bins. In addition, conventional waveriders only observe the one-dimensional frequency spectrum, hence in order to progress assumptions regarding the directional distribution of the waves have to be made. Early proposals for the directional distribution made by Mitsuyasu *et al* (1975) and D.E. Hasselmann *et al* (1980) were based on heave-pitch-roll buoy data. The width of the directional distribution was found to depend on parameters such as the frequency normalized by the peak frequency giving a narrow distribution at the peak of the spectrum, whereas for high frequencies the spectrum broadened considerably. Donelan *et al* (1985) (see also § II.9) on the other hand found that the broadening of the directional distribution was less considerable. Based on data obtained from an array of wave staffs, they found, for wind generated deep water waves, the spectral shape given in equation (2.233). We have adopted this spectral shape because it is based on the most thorough analysis of directional data and because it resembles the wave model spectra closely at high frequencies. It was assumed that for the HEXOS data, the mean wave direction coincided with the local wind direction. In order to be sure of this, great care was taken to select wind sea cases by only considering spectra that had a peak frequency larger than the Pierson-Moskowitz frequency. In addition, multi-peaked spectra were rejected.

In figure 3.3 we have plotted modelled against observed friction velocities. Fitting the squares with a linear regression line we obtain a slope of 0.96 and an intercept of 0.05 indicating that there is a good agreement between observed stress and the stress resulting from the parametrized quasi-linear theory. In this regard, it may be wondered whether a sea-state-dependent roughness length can produce better results than the sea-state-independent roughness length, as suggested by Charnock (1955). We therefore determined the stress in the surface layer with the roughness length $z_o = \alpha_{ch}\tau/g$, where,

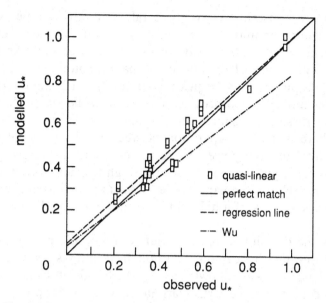

Fig. 3.3. Comparison of modelled and observed friction velocities.

as suggested by Wu (1982), we took $\alpha_{ch} = 0.0185$. The resulting regression line is given in figure 3.3 as well; it has a slope of 0.78 and an intercept of 0.06, indicating that a sea-state-dependent roughness length formulation is to be preferred.

It is therefore concluded that the parametrized version of quasi-linear theory gives realistic growth rates of the waves and a realistic surface stress. However, the success of this scheme for wind input critically depends on a proper description of the high-frequency waves. The reason for this is that the wave-induced stress depends in a sensitive manner on the high-frequency part of the spectrum. Noting that for high frequencies the growth rate of the waves (3.39) scales with wavenumber as

$$\gamma \sim k^{3/2}, \tag{3.48}$$

and the whitecapping dissipation of § II.4 scales as

$$\gamma_d \sim k, \tag{3.49}$$

an imbalance in the high-frequency wave spectrum may be anticipated. Eventually, wind input will dominate dissipation due to wave breaking, resulting in energy levels which are too high when compared with observations. Janssen *et al* (1989b) realized that the wave dissipation source function has to be adjusted in order to obtain a proper balance at the high frequencies. The dissipation source term of chapter II is thus extended as follows

$$S_{ds} = -C_{ds} <\omega> (<k>^2 m_0)^2 [(1-\delta)k/<k> +\delta(k/<k>)^2]N, \tag{3.50}$$

where C_{ds} and δ are constants, m_0 is the total wave variance per square metre, k the wavenumber and $< \omega >$ and $< k >$ are the mean angular frequency and mean wavenumber, respectively. In practice, we take $C_{ds} = 4.5$ and $\delta = 0.5$. The choice of the above dissipation source term over the one presented in chapter II may be justified as follows. In chapter II, it is argued that whitecapping is a process that is weak-in-the-mean, therefore, the corresponding dissipation source term is linear in the wave spectrum. Assuming that there is a large separation between the length scale of the waves and the whitecaps, the power of the wavenumber in the dissipation term is found to be equal to one. For the high-frequency part of the spectrum, however, such a large gap between waves and whitecaps may not exist, allowing the possibility of a different dependence of the dissipation on wavenumber.

This concludes the description of the input source term and the dissipation source term due to whitecapping. Although an impressive amount of experimental evidence for wave growth exists, which indicates that we have modelled the wind input source term reasonably well, in sharp contrast to this hardly any experimental evidence exists regarding the dissipation of a random sea due to whitecapping. The experimental determination of the growth rate due to wind is far from straightforward because it involves the determination of a small phase difference between the pressure fluctuation and the derivative of the surface elevation. But, within a factor of two, we know at least how the growth rate depends on the ratio of friction velocity to phase speed. Also, the dependence of the surface stress on the sea state is reasonably well known. Unfortunately, a direct determination of the dissipation of a random sea is simply not known. The only hope is that nature presents us with case studies where the effects of dissipation may be studied in isolation. Such a case might be the dissipation of swell with a small wave steepness, because in that event, effects of wind input and nonlinear wave–wave interaction may be disregarded. From the simple tests in § III.6 we will see, however, that the dissipation time scale of swell is a few days, so that, in practice, it will be very difficult to obtain reliable evidence on the dissipation source term.

Presently, the only way out of this is to take the functional form for the dissipation in (3.50) for granted and to tune the constants C_{ds} and δ in such a way that the action balance equation (3.33) produces results which are in good agreement with data on fetch-limited growth and with data on the dependence of the surface stress on wave age. In addition, a reasonable dissipation of swell should be obtained. We decided to follow this method and, after an extensive tuning exercise, the constants C_{ds} and δ were given the values 4.5 and 0.5 while the constant $\hat{\alpha}$ in the Charnock parameter was given the value 0.01. A more rational approach towards tuning is discussed in chpater VI, where it is applied to an earlier version of the WAM model.

III.3.2 Nonlinear transfer

In chapter II we have derived the source function S_{nl} describing the nonlinear energy transfer from first principles. Inspecting equation (2.145) in some detail, it is clear that the evaluation of S_{nl} requires an enormous amount of computation. Even with the present-day computing power a wave model based on the exact representation of the nonlinear interactions is not feasible. Therefore, some form of parametrization of S_{nl} is needed.

In the past several attempts have been made to address this problem. The approach used by Barnett (1968) and Ewing (1971) replaced the nonlinear transfer for any given spectrum by the transfer of a scaled reference spectrum of prescribed (Pierson-Moskowitz) shape. The drawback of these approaches is, however, that the influence of the spectral shape on the strength and distribution of the nonlinear transfer is disregarded. From wave growth experiments (Mitsuyasu, 1968, 1969, Hasselmann *et al*, 1973) it was found that, in the initial stages of wave growth, the spectral shape differed significantly from the Pierson-Moskowitz shape which holds for old wind sea. Typically, the spectral peak was more narrow and enhanced compared with the one for old wind sea. The consequence was that, for young wind sea, the nonlinear transfer was an order of magnitude larger than for old wind sea. Thus, the role of the nonlinear transfer in controlling the shape and rate of evolution of the wind sea spectrum was more clearly recognised. As the nonlinear transfer for a Pierson-Moskowitz spectrum is not representative enough for the wide variety of wind sea and mixed wind sea/swell spectra, alternative parametrizations of the nonlinear transfer had to be developed, which have the same number of degrees of freedom as the spectrum itself.

From the work of Hasselmann and Hasselmann (1985b) it became evident that it is not sensible to give a parametrization of the nonlinear transfer which, for a given wave spectrum, mimics the exact nonlinear transfer as well as possible. The reason for this is that a small change in the wave spectrum results in a significant change in the nonlinear transfer. This sensitive dependence of nonlinear transfer on the spectral shape suggests that parametrizations can only be tested reliably by actually incorporating them in a wave model and verifying that the wave growth simulated by the model agrees in standard test cases with the growth curves obtained with the same model but using computations of the exact nonlinear transfer.

Various alternative parametrizations were investigated by S. Hasselmann *et al* (1985). First of all, two parametrizations were considered which are a straightforward extension of the technique used by Barnett (1968) and Ewing (1971). In principle, the exact nonlinear transfer is calculated for a 'representative' set of wave spectra characterized by a limited number of spectral-shape parameters. In the wave model, the nonlinear transfer for a given spectrum is replaced by the stored exact transfer of a member of the spectral set whose shape closely resembles the given model spectrum. However, it was found that instabilities were unavoidable in this parametrical approach. This was because the nonlinear transfer did not have enough degrees of freedom to match the number of degrees of freedom of the spectrum, so that the spectrum could always 'break out' into 'uncontrolled' regions of phase space. The method therefore cannot be applied for wave spectra that fall outside the chosen 'representative' set of wave spectra for which the exact nonlinear transfer has been precalculated.

Accordingly, two further approximations were considered in which the nonlinear transfer is approximated by general nonlinear operator expressions. In the first operator parametrization, which we call the diffusion approximation or local interaction approximation, the nonlinear energy transfer is given as a fourth order, cubic diffusion operator. Although it is not considered sufficiently accurate for quantitative calculations in wave models, the diffusion operator is useful for understanding the role of the nonlinear transfer in the energy balance of waves.

The local interaction approximation assumes that the scattering coefficient σ_B is strongly peaked near the central interaction point $\mathbf{k}_1 = \mathbf{k}_2 = \mathbf{k}_3 = \mathbf{k}_4 = \mathbf{k}$ and that by comparison the action density spectrum is slowly varying. Consistently, using these assumptions, the local interaction expansion yields a cubic, fourth order diffusion operator of the form (Hasselmann and Hasselmann, 1981)

$$\frac{\partial}{\partial t} N \bigg|_{NL} = \frac{\partial^2}{\partial k_i \partial k_j} \left(D_{ijmn} \left[N^2 \frac{\partial^2 N}{\partial k_m \partial k_n} - 2N \frac{\partial N}{\partial k_m} \frac{\partial N}{\partial k_n} \right] \right), \qquad (3.51)$$

where the diffusion tensor D is given by

$$D_{ijmn} = 2^{-10} \int d\mathbf{k}' d\mathbf{k}'' \sigma_B \delta(\omega_1 + \omega_2 - \omega_3 - \omega_4)$$
$$\times (k'_m k'_n - k''_m k''_n)(k'_i k'_j - k''_i k''_j), \qquad (3.52)$$

with

$$\mathbf{k}_1 = \frac{1}{2}(\mathbf{k} + \mathbf{k}'), \quad \mathbf{k}_2 = \frac{1}{2}(\mathbf{k} - \mathbf{k}'), \qquad (3.53)$$

$$\mathbf{k}_3 = \frac{1}{2}(\mathbf{k} + \mathbf{k}''), \quad \mathbf{k}_4 = \frac{1}{2}(\mathbf{k} - \mathbf{k}''). \qquad (3.54)$$

A comparison of (3.51) with exact computations is given in figure 3.4. From this figure we see that expressions of the form (3.51) do not reproduce the source function S_{nl} for both growing and fully developed wind sea, quantitatively. Typically, the transfer rate for the Pierson-Moskowitz spectrum is seen to be too low by a factor of two, while a sharply peaked JONSWAP

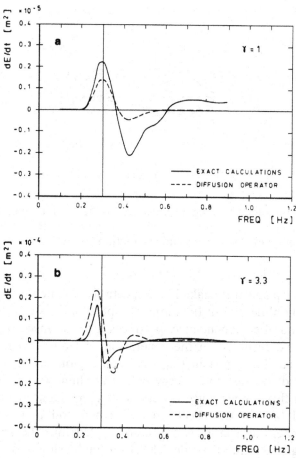

Fig. 3.4. Comparison of the exact one-dimensional nonlinear transfer S_{nl} with the local-interaction approximation for a Pierson-Moskowitz spectrum (panel a) and JONSWAP spectrum (panel b).

spectrum has too high a transfer rate. The diffusion operator gives a good approximation to the contributions in the vicinity of the central interaction point (Hasselmann and Hasselmann, 1985b), nevertheless, contributions further away from the central interaction point cannot be neglected. Significant contributions to the nonlinear transfer come from the intermediate wavenumber quadruplets whose separations are probably slightly too large to be adequately represented by a local approximation.

The nonlinear diffusion operator is, however, helpful in understanding the mechanism by which the nonlinear wave–wave interactions maintain the shape of the spectrum. One would expect that the diffusive character of the nonlinear transfer would give rise to a smoothing of the wave spectrum

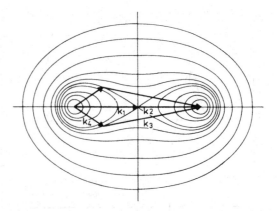

Fig. 3.5. The two interaction configurations used in the discrete-interaction approximation. Contour lines represent the possible and points of the vectors $\mathbf{k}_1$ and $\mathbf{k}_4$ for any interaction quadruplet in the full interaction space.

rather than a sharp spectral peak. This expectation is true for linear diffusion operators. The nonlinearity of the diffusion operator, however, together with its strong wavenumber dependence, produces a pronounced inhomogeneity of the action flux, resulting in the generation of frontal-type structures in wavenumber flux. Thus, just in the transition region between the peak of the spectrum and the very low-energy region at slightly lower wavenumber a strong energy flux convergence occurs. This convergence causes the front on the forward face of the spectrum to steepen and to migrate towards lower wavenumbers as energy is continuously fed into the front from the high-energy, high-wavenumber side. The front finally slows down and decays when it has moved sufficiently far from the wavenumber region in which the spectrum receives energy from the wind.

Let us discuss now a second operator parametrization, namely the discrete interaction approximation which is similar to the diffusion approximation but overcomes the shortcomings of this approach. To this end, S. Hasselmann *et al* (1985) constructed a nonlinear interaction operator by considering only a small number of interaction configurations consisting of neighbouring and finite distance interactions. It was found that, in fact, the exact nonlinear transfer could be well simulated by just one mirror-image pair of intermediate range interactions configurations. In each configuration, two wavenumbers were taken as identical $\mathbf{k}_1 = \mathbf{k}_2 = \mathbf{k}$. The wavenumbers $\mathbf{k}_3$ and $\mathbf{k}_4$ are of different magnitude and lie at an angle to the wavenumber $\mathbf{k}$, as required by the resonance conditions. The second configuration is obtained from the first by reflecting the wavenumbers $\mathbf{k}_3$ and $\mathbf{k}_4$ with respect to the $\mathbf{k}$-axis (see also figure 3.5). The scale and direction of the reference wavenumber are allowed to vary continuously in wavenumber space.

The simplified nonlinear operator is computed by applying the same symmetrical integration method as is used to integrate the exact transfer integral (see also Hasselmann and Hasselmann, 1985b), except that the integration is taken over a two-dimensional continuum and two discrete interactions instead of five-dimensional interaction phase space. Just as in the exact case the interactions conserve energy, momentum and action.

For the configurations

$$
\begin{aligned}
\omega_1 &= \omega_2 = \omega, \\
\omega_3 &= \omega(1 + \lambda) = \omega_+, \\
\omega_4 &= \omega(1 - \lambda) = \omega_-,
\end{aligned}
\tag{3.55}
$$

where $\lambda = 0.25$, satisfactory agreement with the exact computations was achieved. From the resonance conditions the angles θ_3, θ_4 of the wavenumbers $\mathbf{k}_3(\mathbf{k}_+)$ and $\mathbf{k}_4(\mathbf{k}_-)$ relative to $\mathbf{k}$ are found to be $\theta_3 = 11.5°, \theta_4 = -33.6°$.

The discrete interaction approximation has its most simple form for the rate of change in time of the action density in wavenumber space. In agreement with the principle of detailed balance (see also § II.3), we have

$$
\frac{\partial}{\partial t}
\begin{pmatrix} N \\ N_+ \\ N_- \end{pmatrix}
=
\begin{pmatrix} -2 \\ +1 \\ +1 \end{pmatrix}
Cg^{-8}f^{19}[N^2(N_+ + N_-) - 2NN_+N_-]\Delta\mathbf{k},
\tag{3.56}
$$

where $\partial N/\partial t$, $\partial N_+/\partial t$, $\partial N_-/\partial t$ are the rates of change in action at wavenumbers $\mathbf{k}, \mathbf{k}_+, \mathbf{k}_-$ due to the discrete interactions within the infinitesimal interaction phase-space element $\Delta\mathbf{k}$ and C is a numerical constant. The net source function S_{nl} is obtained by summing equation (3.56) over all wavenumbers, directions and interaction configurations.

For a JONSWAP spectrum the approximate and exact transfer source functions are compared in figure 3.6. The nonlinear transfer rates agree reasonably well, except for the strong negative lobe of the discrete-interaction approximation. This feature is, however, less important for a satisfactory reproduction of wave growth than the correct determination of the positive lobe which controls the down shift of the spectral peak.

The usefulness of the discrete-interaction approximation follows from its correct reproduction of the growth curves for growing wind sea. This is shown in figure 3.7 where a comparison is given of fetch-limited growth curves for some important spectral parameters computed with the exact nonlinear transfer, or, alternatively, with the discrete-interaction approximation. The one-dimensional spectra are shown in figure 3.8. Evidence of the stronger negative lobe of the discrete interaction approximation is seen through the somewhat smaller values of the Phillips constant α_p. The broader spectral shape seen in figure 3.8 corresponds with the smaller values of peak enhancement γ for the parametrized case. On the other hand, the agreement

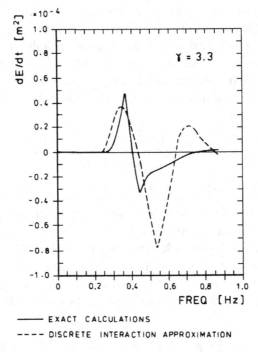

Fig. 3.6. Comparison of the exact one-dimensional nonlinear transfer S_{nl} with the discrete-interaction approximation for a JONSWAP spectrum.

of the more important scale parameters, the energy ϵ_* and the peak frequency ν_*, is excellent (note that as always an asterix denotes nondimensionalisation of a variable through g and the friction velocity u_*).

The above analysis is made for deep water. Numerical computations by Hasselmann and Hasselmann (1981) of the full Boltzmann integral for water of arbitrary depth have shown that there is an approximate relation between transfer rates in deep water and water of finite depth: for a given frequency-direction spectrum, the transfer for finite depth is identical to the transfer for infinite depth, except for a scaling factor R:

$$S_{nl}(\text{finite depth}) = R(\bar{k}h)S_{nl}(\text{infinite depth}), \tag{3.57}$$

where $\bar{k}$ is the mean wavenumber. This scaling relation holds in the range $\bar{k}h > 1$, where the exact computations could be closely reproduced with the scaling factor

$$R(x) = 1 + \frac{5.5}{x}\left(1 - \frac{5x}{6}\right)\exp\left(-\frac{5x}{4}\right), \tag{3.58}$$

with $x = (3/4)\bar{k}h$. This approximation is used therefore in the WAM model.

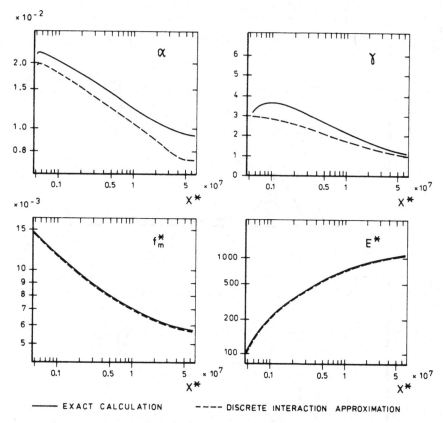

Fig. 3.7. Comparison of fetch-growth curves for spectral parameters computed using the exact form and the discrete interaction approximation of S_{nl}. All variables are made dimensionless using u_* and g.

III.3.3 The energy balance in a growing wind sea

Having discussed the parametrization of the physics source terms we now proceed with studying the impact of wind input, nonlinear interaction and whitecap dissipation on the evolution of the wave spectrum for the simple case of a duration-limited wind sea. To this end we numerically solved equation (3.33) for infinite depth and a constant wind of approximately 18 m/s, neglecting currents and advection. Typical results are shown in figure 3.9 for a young wind sea ($T = 3$ h) and in figure 3.10 for an old wind sea ($T = 96$ h). In either case the directional averages of S_{nl}, S_{in} and S_{ds} are shown as functions of frequency. First of all we observe that, as expected from our previous discussions, the wind input is always positive, and the dissipation is always negative, while the nonlinear interactions show a three lobe structure of different signs. Thus, the intermediate frequencies receive

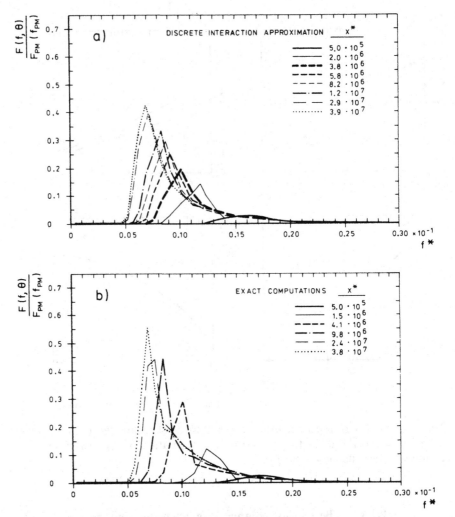

Fig. 3.8. The growth of fetch-limited one-dimensional spectra computed using the discrete interaction approximation (panel a) and the exact form of S_{nl} (panel b).

energy from the airflow which is transported by the nonlinear interactions towards the low and high frequencies.

Concentrating for the moment on the case of young wind sea, we immediately conclude that the one-dimensional frequency spectrum in the 'high'-frequency range must be close to f^{-4}, because the nonlinear source term is quite small (see the discussion in § II.3.10 on the energy cascade caused by the four-wave interactions and the associated equilibrium shape of the spectrum). We emphasize, however, that because of the smallness of S_{nl} it cannot be concluded that the nonlinear interactions do not control

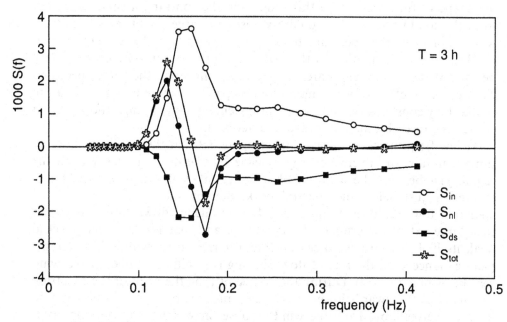

Fig. 3.9. The energy balance for young duration-limited wind sea.

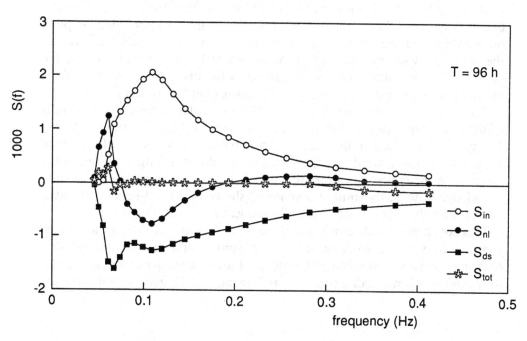

Fig. 3.10. The energy balance for old wind sea.

the shape of the spectrum in this range. On the contrary, a small deviation from the equilibrium shape would give rise to a large nonlinear source term which will drive the spectrum back to its equilibrium shape. The role of wind input and dissipation in this relaxation process can only be secondary because these source terms are approximately linear in the wave spectrum. The combined effect of wind input and dissipation is more of a global nature in that they constrain the magnitude of the energy flow through the spectrum (which is caused by the four-wave interactions).

At low frequencies we observe from figure 3.9 that the nonlinear interactions maintain an 'inverse' energy cascade by transferring energy from the region just beyond the location of the spectral peak (at $f \simeq 0.12$ Hz) to the region just below the spectral peak, thereby shifting the peak of the spectrum towards lower frequencies. This frequency downshift is, however, to a large extent, determined by the shape and magnitude of the spectral peak itself. For young wind sea, having a narrow peak with a considerable peak enhancement, the rate of downshifting is significant while for old wind sea this is much less so. During the course of time the peak of the spectrum gradually shifts towards lower frequencies until the peak of the spectrum no longer receives input from the wind because these waves are running faster than the wind. Under these circumstances the waves around the spectral peak are subject to a considerable dissipation so that their wave steepness becomes reduced. Consequently, because the nonlinear interactions depend on the wave steepness, the nonlinear transfer is reduced as well. The peak of the positive low-frequency lobe of the nonlinear transfer remains below the peak of the spectrum, where it compensates the dissipation. As a result, a quasi-equilibrium spectrum emerges. The corresponding balance of old wind sea is shown in figure 3.10. The nature of this balance depends on details of the directional distribution (see Komen *et al*, 1984 for additional details). The question of whether an exact equilibrium exists appears of little practical relevance. For old wind sea the timescale of downshifting becomes much larger than the typical duration of a storm. Thus, although from the present knowledge of wave dynamics it cannot be shown that wind-generated waves evolve towards a steady state, for all practical purposes they do!

This concludes our discussion of the parametrization of the physics source terms. Before presenting detailed results of some synthetic cases we shall first describe the numerical scheme we have used to solve the action balance equation, after which we shall discuss the architecture of the WAM model software.

III.4 Numerical scheme

P. A. E. M. Janssen, S. Hasselmann, K. Hasselmann and G. J. Komen

In this section we discuss the numerical aspects of the solution of the action balance equation as implemented in the WAM model.

Although, thus far, we have discussed the transport equation for gravity waves for the action density, because this is the most natural thing to do from a theoretical point of view, the actual WAM model is formulated in terms of the frequency-direction spectrum $F(f, \theta)$ of the variance of the surface elevation. The reason for this is that in practical applications one usually deals with surface elevation spectra, because these are measured by buoys. The relation between the action density and the frequency spectrum is straightforward. It is given by

$$F(\omega, \theta) = \sigma N(\omega, \theta), \tag{3.59}$$

where σ is the intrinsic frequency (see also equation (3.4)). This relation is in accordance with the analogy between wave packets and particles, since particles with action N have energy σN and momentum kN.

The continuous wave spectrum is approximated in the numerical model by means of step functions which are constant in a frequency-direction bin. The size of the frequency-direction bin depends on frequency. A distinction is being made between a prognostic part and a diagnostic part of the spectrum. The prognostic part of the spectrum has KL directional bands and ML frequency bands. These frequency bands are on a logarithmic scale, with $\Delta f / f = 0.1$, spanning a frequency range $f_{max} / f_{min} = (1.1)^{\text{ML}-1}$. The logarithmic scale has been chosen in order to have uniform relative resolution, and also because the nonlinear transfer scales with frequency. The starting frequency may be selected arbitrarily. In most global studies the starting frequency f_o was 0.042 Hz, the number of frequencies ML was 25 and the number of directions KL was 12 (30° resolution). For closed basins, such as the Mediterranean Sea where low-frequency swell is absent, a choice of starting frequency f_o of 0.05 Hz is sufficient. In chapter IV results of a 15° resolution model will be discussed as well.

Beyond the high-frequency limit f_{hf} of the prognostic region of the spectrum, an f^{-5} tail is added, with the same directional distribution as the last band of the prognostic region. The diagnostic part of the spectrum is therefore given as

$$F(f, \theta) = F(f_{hf}, \theta) \left(\frac{f}{f_{hf}} \right)^{-5} \quad \text{for} \quad f > f_{hf}. \tag{3.60}$$

The high-frequency limit is set as

$$f_{hf} = \min\{f_{\max}, \max(2.5 <f>, 4f_{PM})\}. \tag{3.61}$$

Thus, the high-frequency extent of the prognostic region is scaled for young waves by the mean frequency $<f>$ and for more developed wind seas by the Pierson-Moskowitz frequency f_{PM}. A dynamic high-frequency cut-off, f_{hf}, rather than a fixed cut-off at $f_{\max}$ is necessary to avoid excessive disparities in the response time scales within the spectrum.

A diagnostic tail needs to be added for $f > f_{hf}$ to compute the nonlinear transfer in the prognostic region and also to compute the integral quantities which occur in the dissipation source function. Tests with an f^{-4} tail show that (apart from the calculation of the wave-induced stress) the results are not sensitive to the precise form of the diagnostic tail. The contribution to the total energy from the diagnostic tail is normally negligible. Because observations seem to favour an f^{-5} power law (Birch and Ewing, 1986, Forristall, 1981, Banner, 1990b; but see also the discussion in § II.9) this power law is used for the high-frequency part of the spectrum.

The prognostic part of the spectrum is obtained by numerically solving the energy balance equation. We will now discuss the different numerical schemes and time steps that are used to integrate the source functions and the advective terms of the transport equation.

III.4.1 Implicit integration of the source functions

An implicit scheme was introduced for the source function integration to enable the use of an integration time step that was greater than the dynamic adjustment time of the highest frequencies still treated prognostically in the model. In contrast to first and second generation wave models, the energy balance of the spectrum is evaluated in detail up to a high cut-off frequency. The high-frequency adjustment time scales are considerably shorter than the evolution time scales of the energy-containing frequency bands near the peak of the spectrum, in which one is mainly interested in modelling applications. Thus, in the high-frequency region it is sufficient to determine the quasi-equilibrium level to which the spectrum adjusts in response to the more slowly changing low-frequency waves, rather than the time history of the short time scale adjustment process itself. A time-centred implicit integration scheme whose time step is matched to the evolution of the lower frequency waves meets this requirement automatically: for low-frequency waves, the integration method yields, essentially, the same results as a simple forward integration scheme (but is of second rather than first order), while for high frequencies the method yields the (slowly changing) quasi-equilibrium spectrum (WAMDI, 1988).

The implicit second order, centred difference equations (leaving out the advection terms) are given by

$$F_{n+1} = F_n + \frac{\Delta t}{2}(S_{n+1} + S_n), \tag{3.62}$$

where Δt is the time step and the index n refers to the time level.

If S_{n+1} depends linearly on F_{n+1}, equation (3.62) could be solved directly for the spectrum F_{n+1} at the new time step. Unfortunately, none of the source terms are linear. We therefore introduce a Taylor expansion

$$S_{n+1} = S_n + \frac{\partial S_n}{\partial F}\Delta F + \cdots. \tag{3.63}$$

The functional derivative in (3.63) (numerically a discrete matrix M_n) can be divided into a diagonal matrix Λ_n and a nondiagonal residual N_n,

$$\frac{\partial S_n}{\partial F} = M_n = \Lambda_n + N_n. \tag{3.64}$$

Substituting (3.63) and (3.64) into (3.62), realizing, in addition, that the source term S may depend on the friction velocity u_* at time level $n+1$, we obtain

$$\left[1 - \frac{1}{2}\Delta t\{\Lambda_n(u_*^{n+1}) + N_n(u_*^{n+1})\}\right]\Delta F = \frac{1}{2}\Delta t(S_n(u_*^n) + S_n(u_*^{n+1})) \tag{3.65}$$

with $\Delta F = F_{n+1} - F_n$. A number of trial computations indicated that the off diagonal contributions were generally small if the time step was not too large. Disregarding these contributions, the matrix on the left hand side can be inverted, yielding for the increment ΔF,

$$\Delta F = \frac{1}{2}\Delta t(S_n(u_*^n) + S_n(u_*^{n+1}))\left[1 - \frac{1}{2}\Delta t\Lambda_n(u_*^{n+1})\right]^{-1}. \tag{3.66}$$

For a typical test case, good agreement was obtained between an explicit integration with a time step of 3 minutes and the implicit scheme with only diagonal terms for time steps up to about 20 minutes.

III.4.2 Advective terms and refraction

The advective and refraction terms in the energy balance equation have been written in flux form. We shall only consider, as an example, the one-dimensional advection equation

$$\frac{\partial}{\partial t}F = -\frac{\partial}{\partial x}\Phi, \tag{3.67}$$

with flux $\Phi = c_g F$, since the generalization to four dimensions λ, ϕ, θ and ω is obvious. Two alternative propagation schemes were tested, namely a first order upwinding scheme and a second order leap frog scheme (for an account of the numerical schemes of the advection form of the energy balance equation see WAMDI, 1988). The first order scheme is characterized by a higher numerical diffusion, with an effective diffusion coefficient $D \sim \Delta x^2/\Delta t$, where Δx denotes grid spacing and Δt is the time step. For numerical stability the time step must satisfy the inequality $\Delta t < \Delta x/c_g$, so that $D > c_g \Delta x$. The advection term of the second order scheme has a smaller, inherent, numerical diffusion, but suffers from the drawback that it generates unphysical negative energies in regions of sharp gradients. This can be alleviated by including explicit diffusion terms. In practice, the explicit diffusion required to remove the negative side lobes in the second order scheme, is of the same order as the implicit numerical diffusion of the first order scheme, so that the effective diffusion is generally comparable for both schemes.

As shown in WAMDI (1988) both schemes have similar propagation and diffusion properties. An advantage of the second order scheme is that the lateral diffusion is less dependent on the propagation direction than in the first order scheme, which shows significant differences in the diffusion characteristics for waves travelling due south-north or west-east compared with directions in between. In general, the differences between the model results using first or second order propagations methods were, however, small.

Historically, the main motivation for considering the second order scheme in addition to the first order scheme was not to reduce diffusion, but to be able to control it. In contrast to most other numerical advection problems, an optimal propagation scheme for a spectral wave model is not designed to minimize the numerical diffusion, but rather to match it to the finite dispersion associated with the finite frequency-direction spectral resolution of the model (SWAMP, 1985, appendix B). In this context, it should be pointed out that an ideal propagation scheme would give poor results for sufficiently large propagation times, since it would not account for the dispersion associated with the finite resolution in frequency and direction (the so-called garden sprinkler effect). Now, the dispersion due to the different propagation velocities of the different wave components within a finite frequency-direction bin increases linearly with respect to propagation time or distance, whereas most propagation schemes yield a spreading of the wave groups which increases with the square root of the propagation time or distance. However, Booij and Holthuijsen (1987) have shown that linear spreading rates may be achieved by introducing a variable diffusion coefficient proportional to the age of the wave packets. Thus far, this idea has not been tested in the context of a third generation wave model (but see Chi Wai Li, 1992).

To summarize our discussion, we have chosen the first order upwinding scheme because it is the simplest scheme to implement (requiring less computer time and memory) and because in practice it gives reasonable results. Applied to the simple advection scheme in flux form (3.67) we obtained the

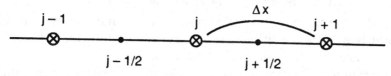

Fig. 3.11. Definition of grid points for first order upwinding scheme.

following discretization, where for the definition of grid points we refer to figure 3.11.

The rate of change of the spectrum ΔF_j in the jth grid point is given by

$$\Delta F_j = -\frac{\Delta t}{\Delta x}(\Phi_{j+1/2} - \Phi_{j-1/2}), \qquad (3.68)$$

where Δx is the grid spacing and Δt the propagation time step, and

$$\Phi_{j+1/2} = \frac{1}{2}[v_j + |v_j|]F_j + \frac{1}{2}[v_j - |v_j|]F_{j+1}, \qquad (3.69)$$

where $v_j = 0.5(c_{g,j} + c_{g,j+1})$ is the mean group velocity and the flux at $j-1/2$ is obtained from (3.69) by replacing $j+1/2$ with $j-1/2$. The absolute values of the mean speeds arise because of the upwinding scheme. For example, for flow going from the left to the right the speeds are positive and, as a consequence, the evaluation of the gradient of the flux involves the spectra at grid points $j-1$ and j.

We finally remark that one could consider using a semi-Lagrangian scheme for advection. This scheme is gaining popularity in meteorology because it does not suffer from the numerical instabilities which arise in conventional discretization schemes when the time step is so large that the Courant-Friedrich-Lewy (CFL) criterion is violated. The wave model community has, so far, not worried too much about this problem because advection is a relatively inexpensive part of the computations. In addition, in most applications, the propagation time is larger or equal to the source time step, which is usually 20 min. According to the CFL criterion, short propagation time steps (less than, say, 10 min) are only required for very high resolution ($\Delta x < 20$ km). But in these circumstances the advection will induce changes in the physics on a short time scale, so that it is advisable to decrease the source time step accordingly. Therefore, in the WAM model, the source time step is always less than or equal to the propagation time step.

III.4.3 Boundary conditions and grid nesting

Normally the wave model grid is surrounded by land points. Therefore, the natural boundary conditions are no energy flux into the grid and free advection of energy out of the grid at the coast line.

The generation and propagation of ocean waves covers a wide range of space and time scales. In the open ocean, the scale of a wind sea system is determined by the size of a depression, which typically has scales of the order of 1000 km. On the other hand, near the coast, the scale of a wave system is determined by the coastal geometry and bottom topography, which have usually much smaller scales. A wave model which covers all scales uniformly is not practicable because of computer limitations. In addition, running a high-resolution wave model for the open ocean seems a waste of computer time.

There are several ways out of this problem. One approach would be to run a wave model with a variable grid, having a high resolution whenever needed (for example near the coast) and having a coarse resolution in the open ocean. So far this approach has not been followed. The WAM model was developed with the practical application in mind of running a global ocean wave model at ECMWF and running limited area models at the European National Weather Centres. Therefore, preference was given to another solution, in which one has the option to run the model on nested grids. This gives the opportunity to use results of a coarse mesh model from a large region in a fine mesh regional model. Several successive levels of nesting may be necessary. The two-dimensional spectra computed by the coarse mesh model are saved at grid points which are on the boundary of the limited area, high-resolution grid. These spectra are then interpolated in space and time to match the high resolution at the grid boundaries. It should be pointed out, however, that a straightforward linear interpolation of spectra gives problems because the interpolated spectra are usually not well balanced, resulting in their rejection when used as boundary conditions for a fine mesh run. To circumvent this problem, the following interpolation procedure is used. Instead of linearly interpolating the spectra from the adjacent points of the coarse grid directly, we rescale these spectra in such a way that the rescaled spectrum has the same mean frequency, mean wave direction and wave energy as found from a linear interpolation of these mean quantities to the fine mesh grid point. The wave spectrum at the fine mesh grid point is then found by linearly interpolating the rescaled spectra. This procedure seems to give satisfactory results and an example of a three level nested grid application is discussed in chapter IV.

III.5 The WAM model software package

P. A. E. M. Janssen, H. Günther, S. Hasselmann and L. Zambresky

The WAM model software that has been developed over a period of seven years is fairly general. Spectral resolution and spatial resolution are flexible and the model can be run globally or regionally with open and closed boundaries. Open boundaries are important in case one wishes to use results from a coarse resolution run as boundary conditions for a fine mesh, limited area run. Options such as shallow water, depth refraction or current refraction may be chosen. In this subsection we shall briefly describe the wave model software with emphasis on flexibility and universality. Before doing this, we shall first discuss some design choices.

The model was developed with an important application in mind, namely for predicting operationally waves over the whole globe. With a modest spatial resolution of 3° (resulting in approximately 4000 grid points) and 25 frequencies and 12 directions, it follows that about 1.2 million equations have to be solved. Since the most expensive part of the numerical code, the nonlinear source term, cannot be vectorized, vectorization is achieved over the grid points, which are placed in the innermost loop. In order to make this loop as long as possible, a mapping from the two-dimensional spherical grid to a one-dimensional array is performed. If there are no limitations to the amount of internal memory of the computer, the most efficient procedure is to convert the entire global grid to a single one-dimensional array. In practice, however, there may be restrictions on the amount of memory to be used. For example, in the early days of the WAM model development, the model was tested on a Cray 1S with an internal memory of only 750,000 words. Clearly, the full model grid would not fit into this small memory. It was, therefore, decided to split up the globe in blocks of NIBLO grid points. Typically, NIBLO = 512. The blocks are set up in such a way that the north and south boundaries are either land or open ocean, whereas the east and west boundaries are land, periodic (this occurs, for instance, in the southern ocean), or open ocean (for nesting). In order to allow waves to propagate across the north or south boundaries of a block, the blocks overlap by a number of latitudes, depending on the propagation scheme. Since we have chosen a first order upwinding scheme, which involves only two neighbouring grid points, the number of overlapping latitudes is two. The computations are done from the last but one southerly latitude to the last but one most northerly latitude (see also figure 3.12 and Günther *et al*, 1991). Although loading only one block at a time circumvents the problem of the limited memory, the drawback of this approach is that extensive input-output (IO) operations are needed. After performing the computations on block IG, the

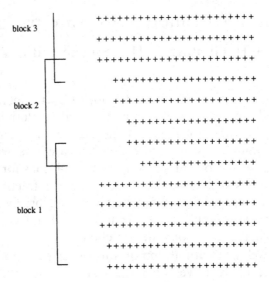

Fig. 3.12. Structure of the model grid.

results have to be written to disc and the results of the previous time step of block IG + 1 have to be read before the computations on block IG + 1 may be started. To avoid waiting for IO, an IO scheme is used that allows for simultaneously writing results of block IG − 1, reading results of the previous time step of block IG + 1, while performing calculations on block IG. The block structure combined with this IO scheme yields a very efficient and flexible wave model code. With the preprocessing program PREPROC one may lay out a block structure according to one's own choice. The latest generation of computers, such as the CRAY-YMP, allows the whole globe to be loaded into the core of the computer, hence one may choose a one block structure. This is, however, not always the optimal choice when high resolution applications are considered and/or when the wave model is coupled to another model, for example an atmospheric model or a storm-surge model.

This completes our discussion regarding the design of the WAM model. The remainder of this subsection is devoted to the model system. A more detailed description of this may be found in the manual WAModel cycle 4 by Günther *et al* (1991).

The model system consists of three parts:

(i) pre-processing programs,
(ii) processing programs,
(iii) post-processing programs.

The WAM model is designed to run as a module of a more general system or as a stand-alone program. It is set up for a Cray computer with a Unicos operating system, but CDC Cyber 205 modifications are available as well. The pre-processing programs generate the model grid, bathymetry-dependent dispersion relation, etc. Post-processing programs are provided for archiving and for further analysis of the model output.

III.5.1 Pre-processing programs

Two pre-processing programs are provided:

(i) PREPROC,
(ii) PRESET.

The program PREPROC generates all time-independent information for the wave model. Starting from a regional or global topographic data set the model grid is created in the form required for the model. The standard model grid is a rectangular latitude-longitude grid, but Cartesian grids can also be chosen. Frequency, angular and group velocity arrays are generated. If the current refraction option is activated, PREPROC expects a current data set and interpolates the data onto the model grid. A number of model constants and matrices, such as the surface stress as a function of wind speed and wave-induced stress, are pre-computed and stored together with the model grid, frequency and angular information and the currents in two output files. If nested grids are generated, the information for the output, input and interpolation of boundary spectra are presented and stored in separate files for the coarse and fine mesh models.

PRESET generates an initial wave field for a wave model cold start. Controlled by the user input of PRESET, either the same initial JONSWAP spectrum is used at all ocean grid points, or the initial spectra are computed from the local initial winds, according to fetch laws with a $\cos^2$ directional distribution.

III.5.2 Processing programs

There are two processing programs, namely

(i) CHIEF,
(ii) BOUINT.

The program BOUINT is used for nested grids. It interpolates, in time and space, the output spectra from a coarse grid model run onto the fine grid along the boundaries of the coarse and fine grids.

CHIEF is the shell program of the stand-alone version of the wave model which calls the subroutine version WAVEMDL of the wave model. All time-dependent variables and user defined parameters are set, the wind fields are transformed to the wave model grid and the transport equation is integrated over a chosen period. The program uses the output files of PREPROC as set-up files and the files generated by PRESET or a previous model run as initial values. A wind input file has to be provided by the user. All additional information must be defined in the user input file. The model can be integrated with independently chosen propagation, source term and wind time steps, under the restriction that all time step ratios must be an integer, or the inverse of an integer.

A number of model options and parameters may be selected by the user in the program input. The following model options are implemented:

(i) Cartesian or spherical propagation,
(ii) deep or shallow water,
(iii) with or without depth refraction or with depth and current refraction,
(iv) nested grids,
(v) time interpolation of winds, or no time interpolation,
(vi) model output at regular intervals, or through a list,
(vii) printer and/or file output of selected parameters.

All run-time-dependent files are fetched dynamically and follow a fixed file name convention. The user has control over directory names and paths through the model input. If selected, model results are saved in four files. These files contain:

(i) gridded output fields of significant wave height, mean wave direction, mean frequency, friction velocity, wave direction, peak frequency, drag coefficient and normalized wave-induced stress,
(ii) gridded output field of swell parameters such as wave height, swell direction, mean wind-wave direction and mean swell frequency,
(iii) spectra at selected grid points,
(iv) swell spectra at selected grid points.

A comprehensive view of the program CHIEF, which is clearly the most important part of the model system, is given in the flow chart of figure 3.13. We need not discuss details here and only highlight the main points. The subroutine INITMDL is only called once. It reads the necessary input generated by PREPROC and PRESET (or by a previous model run) and sets up the necessary information for the model run. PREWIND deals with reading of the winds provided by the user and the transformation to the wave model grid. If required, time interpolation is performed. Finally, the subroutine WAMODEL integrates the energy balance equation. The physics of the wave model is contained in the subroutines PROPAGS and IMPLSCH which are called in a loop over the blocks of the wave model grid. PROPAGS deals with propagation and refraction, whereas IMPLSCH performs the implicit integration in time of the source terms S_{in} (SINPUT), S_{nl} (SNONLIN), S_{ds} (SDISSIP) and S_{bot} (SBOTTOM). The remaining subroutines in WAMODEL are related to the generation of output files or restart files.

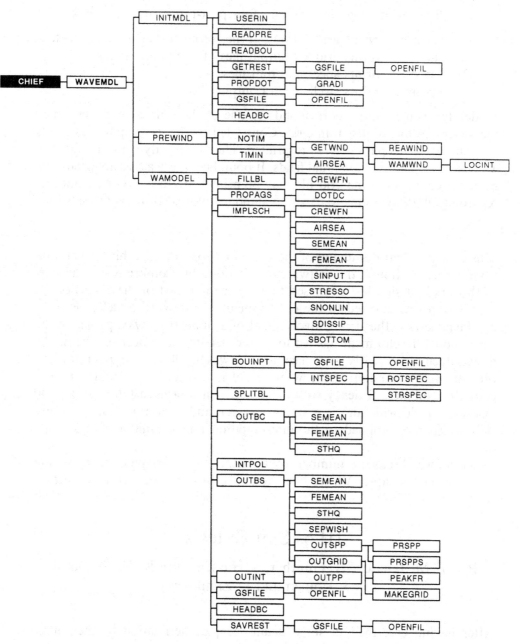

Fig. 3.13. Flow chart of program CHIEF.

III.5.3 Post-processing programs

The standard set of programs contains four post-processing programs:

1. PGRID   prints gridded output files of mean sea state parameters,
2. PSWGRID   prints gridded output files of swell parameters,
3. PSPEC   prints spectra output files,
4. PSWSPEC prints swell spectra output files.

Evidently, each program corresponds to one of the four output files which are generated by the program CHIEF. Controlled by the user input, the results of a chosen set of parameters are printed. The files are dynamically fetched. The user may choose individual fields. If boundary spectra files are produced, both the course and fine grid file may be printed by PSPEC. As no standard, regarding plotting, seems to exist, no standard plot software is available.

This concludes our discussion of the software aspects of the third generation WAM model. Although the software description only comprises a small part of this book, it should be realised that the greater part of the efforts of the WAM group was devoted to the development of the WAM model code. One can imagine how the strong involvement of a number of WAM people in the wave model development has led to heated debates on aspects of the model design during the yearly WAM meetings. However, all this has paid off. The present cycle 4 version of the WAM model is a beautiful looking FORTRAN code. It combines efficiency with flexibility. It has been installed at over 40 institutes world wide and is used for research and operational applications. It has also been applied to the interpretation and assimilation of satellite data.

Before we discuss a number of applications in chapter IV and data assimilation in chapter VI, we first present results of some synthetic cases.

III.6 Simple tests

P. A. E. M. Janssen, H. Günther, S. Hasselmann, K. Hasselmann,
G. J. Komen and L. Zambresky

After having described in some detail the parametrization of the source terms, the numerical treatment of the advection and refraction terms, the numerical integration scheme and the wave model source code, we will now discuss results obtained with the WAM model.

We will postpone the presentation of realistic hindcasts to chapter IV. Here we will be concerned with the results of some simple, perhaps somewhat

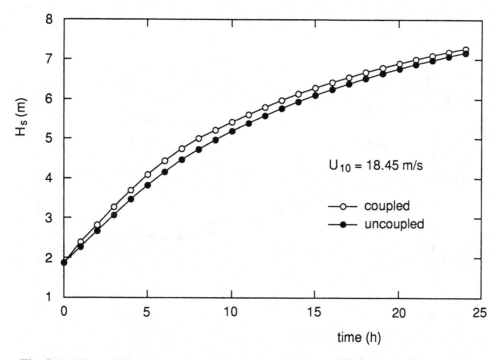

Fig. 3.14. Time dependence of wave height for a reference run and a coupled run.

artificial, cases. We first present results obtained with a single-grid-point version of the model, thereby concentrating on the source terms of the energy balance equation. The generation of wind waves and the dissipation of swell will be investigated. We will compare the results of cycle 4, which has a quasi-linear wind input, with results obtained from a linear wind input term. In addition, we will study the evolution in time of the wave spectrum when the wind turns by 90°.

Secondly, including advective terms, the dependence of the wave spectrum on orthogonal fetch will be studied for the case of a constant wind blowing offshore. Model results will be compared with observations obtained during the JONSWAP (Hasselmann *et al*, 1973) field campaign and with the observations discussed in § II.8. Finally, the dependence of the growth curve on water depth will be studied in some detail.

Duration-limited growth

Let us first discuss the numerical experiments we have carried out with the one-grid-point version of the WAM model and which we implemented on a personal computer (AT with coprocessor). We report only results on the time

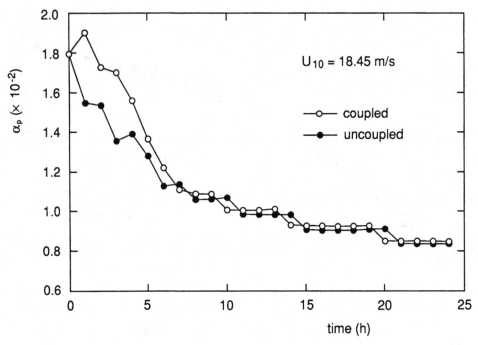

Fig. 3.15. Time dependence of Phillips' parameter α_p for a reference run and a coupled run.

dependence of wave height, Phillips' parameter α_p, wave stress and drag coefficient. They are displayed in figures 3.14 – 3.17.

The initial condition was a JONSWAP spectrum with peak frequency f_p = 0.20 Hz, Phillips' parameter α_p = 0.018, overshoot parameter γ = 3.0 and spectral width σ = 0.08. The directional spreading was given by the usual $\cos^2$ distribution. The energy balance equation was integrated for 12 directions and 25 frequencies (on a logarithmic scale with a starting frequency of 0.0418 Hz with a fractional increase of 10%), using the implicit integration scheme. The integration step was 3 min, whereas wind and waves were coupled every 15 min, so that there was sufficient time for the air flow to relax to a new equilibrium. According to the quasi-linear theory of the wind input source term, the surface stress depends on the sea state, giving a much rougher air flow for young wind sea than for old wind sea. For young wind waves one would, therefore, expect a slowing down of the air flow. A proper description of this deceleration of the flow may, however, only be accomplished in a coupled atmosphere–ocean–wave model. Such tests cannot be regarded as straightforward and since we would like to concentrate here on aspects of the wave physics, we shall make the simplifying assumption that the wind speed is constant. The wind speed U_{10} was chosen to be 18.45

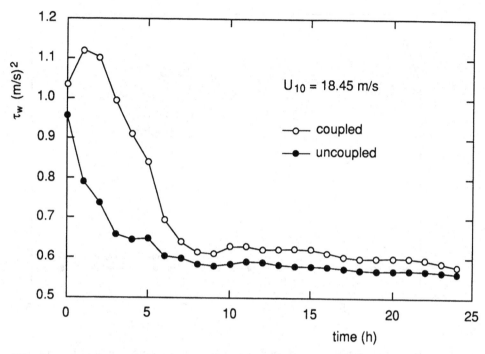

Fig. 3.16. Time dependence of wave-induced stress for a reference run and a coupled run.

m/s, corresponding, in the absence of waves, to a friction velocity u_* of 0.85 m/s (with the Charnock parameter $\alpha_{ch} = 0.0185$). Results with a coupled atmosphere–ocean–wave model will be reported in chapter IV, where we shall concentrate on the impact of sea-state-dependent roughness length on the evolution of a depression.

The first experiment which we performed was a reference run with the wave model without any coupling between wind and waves. This was achieved by simply disregarding the effect of waves on the roughness length (see also equation (3.44)). The constant $\hat{\alpha}$ was given the value 0.0185, thus we have a drag coefficient in agreement with the observations of Wu (1982). The corresponding results in the figures 3.14 – 3.17 are denoted by a •.

Next, we coupled the waves and the air flow as in cycle 4 of the WAM model. The effect of waves on the roughness length was taken into account through equation (3.44). The results for the coupled run in the figures 3.14 – 3.17 are denoted by a ∘. We infer from these figures that two-way interaction has some impact on the results for wave height and wave stress. The Phillips parameter α_p behaves in a satisfactory manner and attains after 24 h the limiting value of 0.008. Furthermore, the impact of the sea state on the aerodynamic drag is shown in figure 3.17. Evidently, the reference run

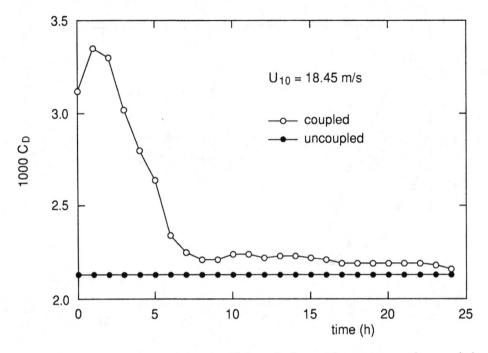

Fig. 3.17. Time dependence of drag coefficient C_D for a reference run and a coupled run.

shows no dependence of the drag coefficient on duration, whereas in the coupled run, a considerable variation with time is found, because the drag coefficient depends on the sea state through the wave-induced stress (the time dependence of which is given in figure 3.16).

To conclude the presentation of results of this coupled experiment, the time evolution of the wave spectrum is shown in figure 3.18. The combination of a somewhat stronger dissipation at high frequencies with a stronger interaction between wind and waves for young wind sea, gives a more pronounced overshoot than in the initial WAM model results (see also WAMDI, 1988).

Turning wind

Thirdly, results of the WAM model for a turning wind field are shown. After 6 h the wind is turned by 90°. The typical response of the two-dimensional wave spectrum is shown in figure 3.19, while the evolution of the drag coefficient is shown in figure 3.20. It is interesting to note the significant reduction in drag at $T = 7$ h and the subsequent increase. Also, the second maximum in the drag coefficient is lower than the first maximum. These features may be explained as follows: as soon as the wind turns, new wind

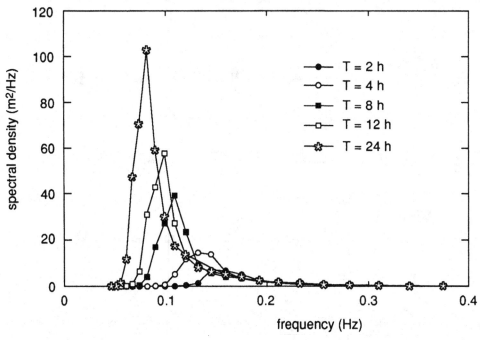

Fig. 3.18. Evolution in time of the one-dimensional frequency spectrum for the coupled run.

sea is generated in the new wind direction. In comparison with the case of no swell, the build up of new wind sea is slowed down by the presence of swell because the quasi-linear dissipation (3.50) is much larger. This explains the sudden drop at $T = 7$ h and the lower secondary maximum in the drag coefficient. Since this turning wind field case simulates, in a simple way, what would happen near an atmospheric front, this delicate behaviour of the aerodynamic drag is of special interest. In addition, we conclude from this numerical simulation, that in complicated mixed wind sea and swell cases, no clear stratification of the drag coefficient in terms of the wave age c_p/u_* is to be expected. The reason for this is that the wave stress is determined by the high-frequency part of the spectrum and that in mixed wind sea and swell cases, no definite relation between wave age and the high-frequency part of the spectrum exists. A similar conclusion may be drawn from Donelan's observations (1982) presented in § II.2. For a more complete discussion see Janssen (1991a; the present results differ in some details, because after publication of the original results van Vledder found a programming error in the treatment of nonlinear energy transfer from the diagnostic part of the spectrum to the prognostic part. Correction of this

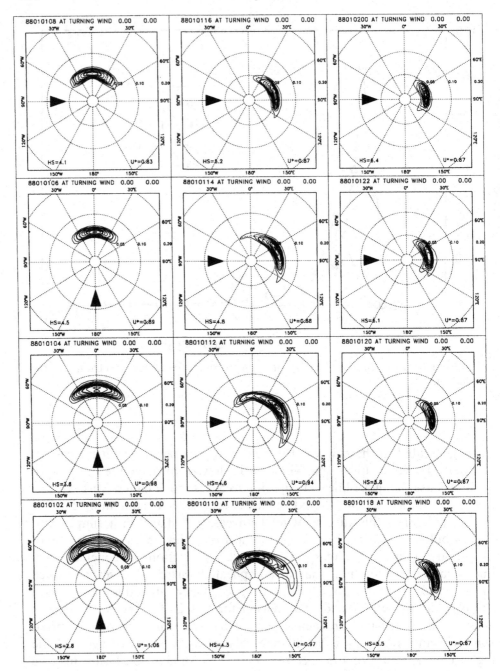

Fig. 3.19. Evolution of two-dimensional spectrum for a turning wind field. Wind is turned after 6 h by 90°.

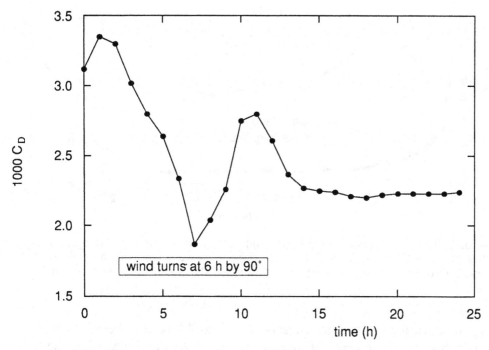

Fig. 3.20. Time dependence of drag coefficient for turning wind field case.

error gave a somewhat higher Phillips' constant and a slightly higher drag coefficient for young wind sea.)

Swell decay

We conclude the presentation of the numerical experiments of the one-grid-point version of the WAM model with a discussion of the behaviour of swell. To this end we ran the model for six days, the first two days with a wind speed of 18.45 m/s, after which the wind dropped to a value of 5 m/s. In figure 3.21 the time evolution of wave height and peak frequency is shown. The first day after the wind drop shows a significant decrease of wave height due to whitecap dissipation. Later the wave height hardly changes. This indicates that the dissipation, which depends on the mean steepness of the waves, has decreased considerably. Even after the wind has dropped, a shift in peak frequency, caused by the nonlinear interactions, may be noted. This peak frequency shift occurs, however, on a much longer time scale than during the first two days. This is to be expected, because the mean steepness of the waves decreases fairly quickly after the wind has dropped, thereby reducing the effect of the nonlinear interactions.

We should emphasize that both the rate of decrease of swell wave height

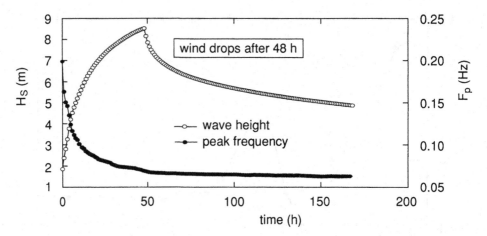

Fig. 3.21. Time evolution of wave height and peak frequency over a six day period. After two days the wind drops. Notice the decay in wave height and the slight shift in peak frequency during the last four days when the waves are considered to be swell.

and the downshift of the peak frequency of swell have not been observed, simply because clear cases of swell decay have, to our knowledge, never been documented. An extensive set of swell decay measurements for swell propagating across the Pacific (Snodgrass *et al*, 1966) yielded negligible dissipation for very long, low amplitude swell, but the data were inadequate to quantify the decay which presumably occurred near the source regions. The main interest of the surface waves community has always been in the generation of waves by wind. Now it is time to try to observe the (possible) decay of swell, because this will provide an additional test for our knowledge of wave dissipation, the process which, as we have emphasized over and over again, is the least well known. Figure 3.21 shows that the time scale of swell decay is rather long. For this reason, the observation of swell decay has not been an easy task in the past. Fortunately, with the provision of global two-dimensional wave spectral data from the SAR wave mode data of ERS-1, excellent data on the propagation and decay of swell over trans-oceanic distances are now available. First investigations (Brüning *et al*, 1993a,b) indicate a rather good agreement between observed and WAM-predicted swell decay rates.

Fetch-limited wave growth

After completing the discussion of the single-grid-point cases, we now continue by considering advection as well. We shall concentrate on what has become known as the SWAMP2 case (SWAMP, 1985). This corresponds to the

JONSWAP (Hasselmann *et al*, 1973) case of a constant wind blowing orthogonally offshore. The purpose of this numerical experiment is to compare model results with observations. Unfortunately, most observed growth curves are presented in terms of U_{10} scaling, since the surface stress is usually not measured. The WAM model physics is, on the other hand, formulated in terms of the friction velocity, because this is the only relevant air flow quantity which is independent of height. Thus, scaling laws for the variance m_0, peak frequency f_p and the Phillips parameter α_p, are based on u_*-scaling. For example, for wind sea, the fetch dependence of the dimensionless energy $\epsilon_* = g^2 m_0 / u_*^4$, is given by the universal law

$$\epsilon_* = \epsilon_*(X_*), \tag{3.70}$$

where $X_* = gX/u_*^2$ is the dimensionless fetch. Writing this law in terms of the wind speed U_{10} we have

$$\tilde{\epsilon} = C_D^2 \epsilon_* (C_D^{-1} \tilde{X}) \tag{3.71}$$

and this law is not universal because the drag coefficient $C_D = \tau / U_{10}^2$ is not a universal function of fetch. As explained already in chapter II, C_D is for wind sea, not only a function of wave age c_p/u_* (which depends solely on X_*), but also a function of $g z_{obs}/u_*^2$ (with $z_{obs} = 10$ m). See, for this, figure 2.14. The important consequence is therefore that, when scaling in terms of U_{10} is performed, a family of growth curves is found. It may be pointed out that the occurrence of a family of growth curves is easily understood when it is realized that the height $z_{obs} = 10$ m is really arbitrary and bears no relation to any length scale in the physical system.

Nevertheless, observed growth curves are usually expressed in terms of a wind speed measured at a certain height. In order to make a comparison between modelled and observed growth rates possible, use of the drag law has to be made. This can be done in principle. It was found to be more convenient, however, to run the WAM model with the proper wind speed and to obtain, in this way, the modelled growth curve with U_{10} scaling. This was done for a wind speed of 8 m/s, which is about the average wind speed of the JONSWAP experiment. For this relatively small wind speed, the wind generated waves have fairly high frequency, thus we chose a starting frequency $f = 0.15$ Hz. As usual, the number of frequency bins was 25 and the number of directions was 12.

For simplicity, we took an infinitely long coast line so that wave propagation along the shore is not relevant. Thus, one only needs to calculate the evolution of the wave spectrum along a line perpendicular to the coast. The grid step was taken to be 5 km, while the wind time step was 900 s and the source time step was 180 s. In order to prevent numerical instabilities of the advection scheme, a small propagation step (= 180 s) was used.

Results of this high-resolution version of the WAM model and observed

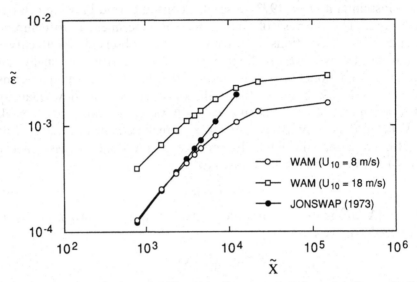

Fig. 3.22. Fetch dependence of dimensionless energy $\tilde{\epsilon}$ for a wind speed of 8 m/s. The JONSWAP fetch law is also shown. Finally, the growth curve for a wind speed of approximately 18 m/s is shown.

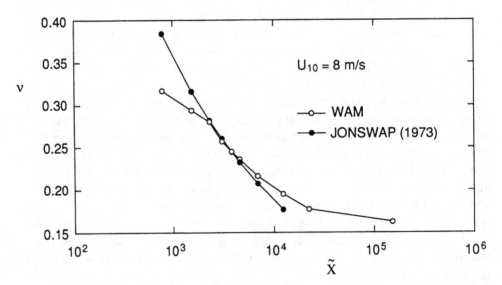

Fig. 3.23. Fetch dependence of dimensionless peak frequency $v = U_{10}f_p/g$. The JONSWAP fetch law is also shown.

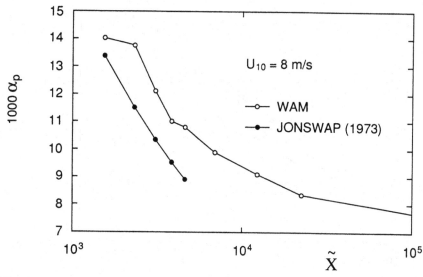

Fig. 3.24. Fetch dependence of Phillips' parameter α_p. The JONSWAP fetch law is plotted as well.

fetch laws from JONSWAP (as obtained by Günther, 1981), are compared in the figures 3.22 – 3.24, where we show the fetch dependence of dimensionless energy $\tilde{\epsilon}$, dimensionless peak frequency $\nu = U_{10}f_p/g$ and Phillips' parameter α_p. The JONSWAP fetch laws are only plotted between dimensionless fetch $\tilde{X} = 10^3$ and 10^4 because that is their range of validity. It is concluded that there is a fair agreement between model and observations. We emphasize that, the agreement between observed and modelled peak frequency and Phillips' parameter is especially impressive, as these quantities are plotted on a linear scale; this is in contrast to the observations presented in JONSWAP which are plotted on a logarithmic scale. Regarding the comparison of observed and modelled wave height, we note that model results do not scale with the wind speed at 10 m height. To make this point clear, we have added in figure 3.22 a graph of dimensionless wave height versus dimensionless fetch, for a wind speed of 18.45 m/s. Significant differences with the growth curve obtained with a 8 m/s wind speed should be noted. As already explained, the reason for this difference is that the WAM model physics is based on u_*-scaling and that the mean drag coefficient of the respective cases differs by a factor of 1.5. Obviously, since the WAM model physics is formulated in terms of the friction velocity, scaling with the help of u_* is to be preferred.

We would like to stress again that the scaling of wave growth with friction velocity is based on sound theoretical arguments. We therefore strongly suggest that observations of fetch-limited wave growth are analysed using

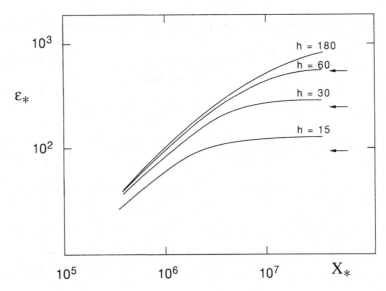

Fig. 3.25. Fetch dependence of dimensionless energy ϵ_* for various depths. The arrows indicate the asymptotic values of Bouws' fit, equation (2.245).

u_* scaling. This, of course, requires the reliable determination of the surface stress, but nowadays this is not an impossible task, as was shown during HEXOS (Smith *et al*, 1992). Alternatively, for wind sea, the observed HEXOS relation between drag coefficient and wave age and stress may be used to convert observed winds to surface stresses.

Shallow water

After having discussed in some detail the deep-water fetch laws, we conclude the presentation of the special cases by considering briefly the effect of shallow water (no refraction). We chose a wind speed of 18.45 m/s and a grid step of 10 km. All other parameters were the same, except that, because of the high wind speed, we took the first frequency bin to be at 0.0418 Hz. Results for dimensionless energy as function of dimensionless fetch for different depth are shown in figure 3.25. For comparison with empirical data Bouws' fit (2.245) for infinite fetch is indicated by arrows (for 15, 30 and 60 m). The agreement is quite remarkable. Figure 3.26 shows the dependence of dimensionless peak frequency on fetch and depth. The effects of finite depth are evident from these figures. For shallow water the wave height is reduced and the peak frequency is increased. The reason for this is that the long waves are more affected by bottom friction than the short waves, which reduces the energy levels of the long waves. These effects are most pronounced when one has both shallow water and a large dimensionless

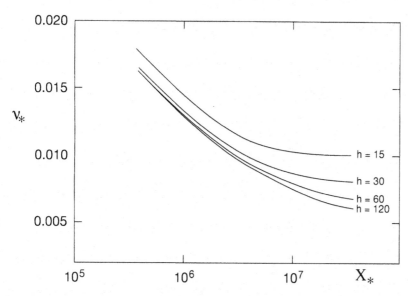

Fig. 3.26. Fetch dependence of dimensionless peak frequency ν_* for various depths.

fetch. It should be emphasized, however, that this rarely occurs in nature. Nevertheless, as discussed in § II.11, one may find a considerable reduction of wave height, in extreme conditions by a factor of two.

This concludes the discussion of the special cases. To summarize this chapter, we have given a rather thorough discussion of the numerical wave prediction model WAM. We have highlighted some properties of the conservation of action, we have considered the parametrization of the source terms and we have discussed the integration scheme and the advection scheme. After presenting important aspects of the design of the numerical code, we investigated a number of special cases and, in passing, had a rather thorough discussion on the appropriate scaling of the fetch laws. In the next chapter we will investigate the performance of the WAM model in realistic cases.

Chapter IV

Applications to wave hindcasting and forecasting

L. Cavaleri *et al*

IV.1 Introduction

L. Cavaleri

In the previous chapters we have been dealing with theoretical, physical and numerical aspects of wave modelling, and the WAM model in particular. It is now time to face practical problems and to turn to applications. The WAM model is one of the most widely used wave models in the world, both for forecasting and hindcasting on global and regional scales. There is therefore a wealth of applications and experience which can be used to discuss and to judge the behaviour of the model.

The purpose of applications

Model applications serve two purposes: a practical one and a scientific one. In practical applications the model is accepted as a reliable tool. As such it can be used either in real-time, to forecast sea conditions for ship routing, offshore operations and for coastal protection, or in a hindcasting mode for computation of the sea state during a particular event and to determine wave climatology and extremal statistics. For scientific purposes, once properly verified, the model is, within the limits derived from its formulation, a representation of the real world. It can therefore be used for numerical experiments to simulate field experiments under conditions that would be difficult to find or expensive to organize. The model is a tool for the better understanding of the relevant phenomena and how they interact. It can also be used to study the sensitivity of the results to a change in the input conditions.

Clearly, before using model results for practical purposes, we need to validate the model in extended comparisons with measured data under careful control of the input data, such as the wind, bathymetry and geometry of the basin. We refer to this phase as *validation-refutation*. During this phase we test the model in various geophysical conditions. At the same time we try to single out parts of the model that should be improved by studying the impact that different features have on the wave response. Some of the more important features, which receive attention in subsequent sections, are the effect of inaccuracies in the wind, the influence of different source functions, refraction by bottom topography, interaction with currents and numerical aspects.

Refutation of the model, when results are incorrect, is potentially one of the most effective ways of learning about the physics of the model. As long as the results are good there is little to search for. Incorrect results, on the other hand, force us to look for a deeper understanding and may lead to improvement. Since a third generation wave model is based on physical principles, it is particularly useful for this purpose. As a consequence, there is a direct link between practical applications and scientific progress.

The results of the WAM model in this respect have been both encouraging and disappointing. They are encouraging because in all applications the WAM model has performed very well, even with regard to details of the two-dimensional spectrum (where this has been measured, for example by the ERS-1 SAR). In none of the many intercomparisons with measured data have we been able to identify basic shortcomings of the model.

For the same reason the results are also disappointing: the many applications have not led to further improvement. We have some idea where the model might be improved. For example, in the case of bottom friction, there are indications that the present formulation may be inadequate under severe conditions (see § IV.6). But in such cases further data and research are needed before a change in the model appears warranted. The better we approximate the behaviour of nature, the more difficult each further improvement becomes. From this it has become clear that we will be able to diagnose (rather than just test) model performance in complex real-life wind situations only if:

— we have adequate statistics (from hindcasts and from real-time applications over several years),
— we have a large number of well-documented cases at hand, particularly at the regional scale,
— the wind fields are well documented,
— we have measurements of two-dimensional wave spectra,
— we apply sophisticated inverse modelling techniques.

We are only now just beginning to satisfy these conditions, for example through satellite wind-wave measurements and through the adjoint techniques described in chapter VI. But since these matters were not yet mature at the time of writing of the book, we will instead illustrate the impact of the wind field and the various source functions, refraction terms and numerics in a variety of hindcasting and forecasting applications. The purpose is to give the reader a feeling for the way a wave model operates and the various problems one encounters in applying the model. The chapter may also serve as a reference manual to be consulted when one intends to make one's own application, because it illustrates the pitfalls that may be encountered in different situations. Both global and regional applications will be shown but, since the focus is on the physics of the model, there will be some emphasis on regional applications. The global applications, for which satellite data are very important, will be discussed in more detail in chapter V (global satellite wave measurements).

Scale of the model

A numerical wave model is usually formulated in general terms, without any specific reference to the dimensions of the application space. However, a model and the grid itself must be consistent with the physics reflected in its equations and with the scale of the phenomena we want to represent. The results lose reliability at small scales, when other phenomena, not considered in the model, can be significant for the evolution of the field. A common example is the lack of any consideration of diffraction, present in the lee of an island or a peninsula or, at an even smaller scale, in a harbour. Clearly a model built with an oceanic perspective is not suitable for dealing with this problem.

On larger scales economical and technical reasons have naturally led to an extension of the grid to the full global scale. For practical reasons large scale models still have a fairly coarse resolution, but they are nevertheless well suited to deal with large-scale phenomena. Typical examples are the large extra-tropical winter storms of the North Atlantic Ocean, originating at high latitudes on the American coast, which gradually move to the east towards Europe, and Pacific swell, which can travel many thousands of kilometres for several days before running onto one of the surrounding coasts.

A coarse-resolution wave model can only be applied meaningfully when sharp gradients in the field are absent. If such gradients arise, as in the case of a hurricane or often close to the coast, the model results provide only an approximate, smoothed description of the actual situation. If in such cases a detailed analysis is required, it is necessary to focus on the area of interest by using a smaller and finer grid. A limited area model (LAM) allows a more accurate study of a given event in a given area. However, for the full exploitation of its capability the input information from the boundary and, in particular, the meteorological model must be sufficiently accurate. If not, we increase the resolution, but we do not improve the accuracy of the results.

The ideal situation for a LAM is a closed basin, without external influences, or, if open to the ocean, a case with offshore flowing information (a growing sea under an offshore wind is the obvious example). Alternatively, boundary information can be obtained by embedding (or 'nesting') the LAM into a large-scale model. If all input information is available with adequate accuracy, a LAM is particularly suitable for studying the physics of a model, because the limited area and the improved input conditions allow a detailed analysis of the results and the investigation of the effects of a particular phenomenon.

Present day computer power is not sufficient for running high-resolution global wave models, and even if it were, this would not always be worthwhile, because one does not require high-resolution results everywhere. For this reason it is convenient to split the applications into large-scale (coarse) and small-scale (fine). Their respective characteristics and possibilities can then be summarized as follows.

Large-scale wave models

– Suitable for the analysis and investigation of large-scale phenomena. In general, the computation of the sea state, even if only for a limited region, requires a proper assessment of what is happening elsewhere in the basin.
– Needed for large-scale applications, such as ship routing, interpretation and assimilation of satellite sea surface observations or coupling with a global atmospheric model.
– Useful for a general estimate of the overall model performance. However, the validation statistics must be interpreted with care and stratified with respect to properly chosen criteria. For instance the bias or the rms error of the H_S estimate in the Pacific Ocean does not tell whether this is due to wrong generation or to incorrect advection and dispersion of swell, nor does it say whether the error is associated more with high or with low values of H_S. A proper analysis of these possibilities requires conditional statistics and is often not very conclusive without further information. The provision of global wave height data from satellite altimeter and two-dimensional wave spectra from satellite SARs should increase the power of this approach significantly.
– Not suitable for a detailed study of a limited area or for the phenomeno-logical analysis of a given event with high space and time variability.

Limited area wave models

– Focus attention on a limited area and take the local geometry and the local variability into account, so giving, with respect to large-scale models, a better description of what is going on.
– Suitable for detailed case studies and for the intercomparison of different approaches.
– Sensitive to the specification of the model and, as such, suitable for studying its physics and the related sensitivity of the results.
– Require a higher accuracy in the input and in the definition of the boundary conditions in view of their higher accuracy.
– If nested within a larger scale model, must use grids which extend well beyond the area of interest.

Characteristics of a wave model

When considering the application of a wave model to a specific case, we still have a considerable freedom in choosing the characteristics of the model. The choice will depend on the use we have in mind for the results and on the resulting constraints on the various parameters of the model.

The basic elements to consider are the following:

Grid: the extension will depend on the area of interest and on the meteorological and oceanographic phenomena we expect to encounter. The grid step size will be a trade-off between the desire for high resolution and accuracy on one side, and the available computer power on the other. Small areas are conveniently covered with a Cartesian grid (or different projection maps); larger areas are necessarily covered with a geographic grid.

Nesting: if a sufficiently small grid size is not possible for the entire area of the model, we may have to resort to the nesting technique. For accurate results the nested grid has to extend well beyond the area of immediate interest.

Bathymetry: for a meaningful application, this must be available at least at the resolution of the grid. If significant depth variability is present at the subgrid scale, some processes (e.g. bottom refraction and bottom dissipation) will be poorly represented by the model.

Processes to consider: the choice will depend on the relevant processes in the area of interest and on the aim of the application. A smaller number of processes means a saving of computer time. A careful examination of the magnitude of the different terms in the energy balance is required before one can make an appropriate choice.

Spectral resolution: on a discrete representation, the two-dimensional spectrum is represented by a finite number of frequencies and directions. Their actual number is a trade-off between the resolution required for the specific application and the computer time required to run the model.

Integration step: this is established by the numerical stability of the advection procedure. In some cases, when very high gradients are present in the field (in hurricanes, for example), it may be necessary to choose a smaller time step to resolve the physics adequately (with correspondingly larger computer time).

Wind fields: the quality of the input winds is a very critical factor. Usually we must use what is available, but in some special cases a dedicated study can be carried out to improve the meteorological information. This is usually quite expensive and worthwhile only for very special purposes.

Table 4.1. *Hindcasts discussed in this chapter and the sensitivities and possible pitfalls they illustrate.*

name	basin	period	sensitivity to (impact on)	§ IV.*a.b*
SWADE	NW Atlantic	20-31/10 '90	wind accuracy	3.2
			depth refraction	7
			wave–current interaction	8.4
			grid nesting	9.2
LEWEX	NW Atlantic	9-19/3 '87	wind accuracy	3.2
Med87	Mediterranean	10-11/1 '87	wind (orographic effects)	3.3
			forecast wind	4.2
			grid resolution	9.1
			directional resolution	9.3
Gorbush	Mediterranean	2-3/12 '89	wind resolution	3.4
Adria86	Adriatic	1/2 '86	wave-dependent drag	5.1
Peru swell	Pacific	31/10 '89	wave-dependent drag	5.1
			frequency resolution	9.3
Hawaii storm	Pacific	5/11 '88	wind variability	5.2
Camille	Gulf of Mexico	16-17/8 '69	wind variability	5.2
Texel	North Sea	3/1 '76	bottom friction	6.1
Adria82	Adriatic	2/12 '82	bottom friction	6.1
Frederic	Gulf of Mexico	12-13/9 '79	bottom elasticity	6.2
North Sea87-88	North Sea	9 '87-1 '88	wave–current interaction	8.2
ERS-1 cal	NE Atlantic	21/11 '90	nesting	9.2
Irish surge	Irish Sea	26/2 '90	time varying depth	10.1
			impact on surge	10.1
North Sea surge	North Sea	11/2 '89	impact on surge	10.1
global	global	perpetual July	impact on atmosphere	10.2

While each parameter can be analysed separately, an overall consideration is required for the final choice. Some of the parameters are interdependent, and consistency is required in their choice to avoid unnecessary computation.

Overview of the chapter

We conclude the introduction of this chapter with an outline of what is to follow. The ordering into sections is based on physical arguments, but many different storms have been used to illustrate the points. Occasionally, the same storm has been used for the discussion of more than one physical aspect. Table 4.1 gives a summary of the different storm cases together with the area covered, the date of each storm and the main subject of discussion (wind, wave, model) and/or the process(es) considered. The last column refers to the corresponding (sub)section of this chapter. The table may be helpful in guiding the reader through the chapter, but can also be used as a directory on where to look for a specific storm or process.

In § IV.2 we address the wind problem. The aim is to give the reader a qualitative feeling of the physics involved, and of what to expect as input to a wave model.

In § IV.3 we consider the consequences of the accuracy of the wind field on the wave field. We will use the standard WAM model and drive it with different wind fields. The aim is to stress the sensitivity of the wave results to the wind input information, and to outline the best way to make use of it.

Next (§ IV.4) we consider the operational application of meteorological and wave models. We provide examples and statistics of results for both analysis and forecast. This section discusses the validation question on global and regional scales.

We move then to the sensitivity of the wave model results to the specifications and characteristics of the model itself.

First (§ IV.5) we analyse the relevance of the correct evaluation of the input source term as a function of the atmospheric and wave conditions. Here we will consider different ways of computing the wind input from a given wind field.

Next (§ IV.6) we analyse the role of the dissipative bottom processes; in particular we will show how different physical processes, bottom friction and bottom elasticity, may be dominant under different conditions.

Conservative processes that depend on the depth distribution, and refraction in particular, are discussed in § IV.7. The results give a feeling for their relevance in practical applications and illustrate the numerical accuracy required for their correct representation.

§ IV.8 addresses wave–current interactions. After outlining the basic principles, we discuss first the case of an extended uniform flow associated with open sea tidal currents. Then we consider the interaction with a current ring, typical of concentrated flows like the Gulf Stream. Finally, we quote an application from the North-West Atlantic Ocean.

In § IV.9 we analyse the sensitivity of the results to the numerics of the model. We discuss first the spatial resolution (grid and coast definition). This leads to the nesting technique and the conditions under which we can expect to find a substantial improvement of the results. Then we discuss the spectral resolution (the number of frequencies and directions used to represent the two-dimensional spectrum) and the implications for the results.

In § IV.10 we move to a broader subject by considering the coupling between meteorological, ocean circulation and wave models. We are here still in the early stages of development, so that further experience will be needed before definite conclusions can be drawn. We report on two sets of experiments, differing in the spatial and time scale of the simulation. The first experiments concern two limited basins, namely the Irish Sea and the North Sea, and span the duration of a single storm. The second experiments are a first attempt to quantify the effect of waves on the general circulation of the atmosphere.

IV.2 Surface wind fields

L. Cavaleri and V. Cardone

In this section we consider the characteristics of wind fields and we discuss the problem of their determination with the help of numerical models. The discussion is qualitative and focusses on those aspects that are relevant to the evolution and computation of the wave field. For a more technical discussion we refer the interested reader to § I.3 and to the extensive literature on the subject.

We begin (§ IV.2.1) by considering the scale of the marine wind field, the effect of the air/sea stability conditions and the characteristics of a gusty wind. Then (§ IV.2.2) we briefly review how wind fields are computed in meteorological models. We discuss their general structure and the scales of application. We end (§ IV.2.3) with a description of two important phenomena, namely extratropical storms and tropical cyclones.

IV.2.1 The marine wind field

Scales of motion

The physics of the lower layer in the atmosphere is described in § I.3. Starting from the upper layers, and for purposes of discussing the determination of surface wind fields over the ocean, a distinction can be made between geostrophic motion and flows in the Ekman layer and constant stress layer. In § II.2 it is argued that the relevant information for wave generation is the wind stress at the surface, but we need to begin our considerations from somewhere above the surface, because the dynamical behaviour of the different atmospheric layers is strictly interconnected.

The general structure of a wind field is characterized by its large overall horizontal scale in comparison with its vertical scale. This reflects the geometry of the atmosphere. In the vertical direction a storm extends up to 10 km, but horizontally it extends to a scale of 10^3 km, with a ratio horizontal/vertical scale of 10^2. The aspect ratio is smaller for tropical storms, for example hurricanes, but is still much larger than 10. The ratio for the horizontal and vertical wind components is even larger. Horizontal velocities vary between 1 and 50 m/s, vertical velocities are limited to $10^{-(2 \div 5)}$ m/s (notation: roughly between 10^{-2} and 10^{-5}) with a ratio $10^{4 \div 5}$. Although the atmosphere is dominated by its horizontal dimension, the vertical component of motion w is very important. At short time scales, the interaction of w with the horizontal components u and v leads to the vertical flux of momentum, and consequently to the surface stress; at longer time scales it is essential for the dynamical evolution of a storm.

Because of the geometry of the atmosphere, the values of the ratio of the gradients of u, v and w are roughly the inverse of those obtained for the velocities themselves. The extended horizontal dimension leads to a relative horizontal uniformity. Usual gradients are $10^{-(4 \div 5)}$ s^{-1} for the modulus and $10^{-(4 \div 5)}$ rad/m for the wind direction: the largest values being typical for frontal zones. The corresponding vertical values are much larger, 10^{-1} s^{-1} and 10^{-3} rad/m, respectively for modulus and direction.

Air/sea stability

For wave generation at a given location, the local vertical structure is important, because wind stress, the driving element in wind wave generation, depends on the vertical gradient of the horizontal wind components, particularly in the lowest few metres. For given surface roughness and for given geostrophic conditions, the wind shear near the surface and the corresponding friction velocity vary with the vertical exchange of momentum and with the density stratification of the atmosphere. The mathematical description of this effect has been given already in § I.3.7. For a given geostrophic wind U_g, say at 1000 m height, a more intense vertical mixing in the underlying layers causes a stronger wind close to the surface and therefore a larger velocity gradient, hence stress, at the surface. Both the physics of the process and the experimental evidence strongly suggest that in unstable conditions (water warmer than air) the stress is increased relative to the neutral situation. The opposite is true for stable atmospheric conditions.

As a result the expression for the surface drag coefficient C_D (defined in § I.3) depends explicitly on the air/sea temperature difference. In very qualitative terms, when water is warmer than air there is an upward heat flux. This implies turbulence with strong vertical mixing, and, following the previous arguments, a larger drag at the surface.

The practical effect of air/sea stability may be expressed by the concept of 'effective anemometer height wind' (Cardone, 1969, Cardone *et al*, 1990). Given a wind U at a reference height h and under, for example, unstable conditions, we define as U' the wind speed that under neutral conditions would produce the same stress as U. Table 4.2 (Cardone, 1969) demonstrates the influence of atmospheric stability. Under highly unstable conditions $U = 20$ kn (about 10 m/s; one knot is one nautical mile (= 1856 m) per hour) produces the same stress as a neutral wind of 22 kn. Because stress $\tau \sim U^2$, it follows that strong instability can increase the stress up to 20%. For stable conditions the stress is reduced. Note, however, that the influence of stability tends to decrease with increasing wind speed.

A different way of showing the effect of air/sea stability is to plot the variation of the drag coefficient C_D with the value of U, typically U_{10}, and with the stability. This is done in figure 4.1 for various possible conditions,

Table 4.2. *Effective neutral 20 m height winds as function of the air/sea tem-*
perature difference △*T. Winds in the top row are the winds actually observed.*
For each wind speed the lower rows give the winds that, for a given △*T, would*
produce the same stress in neutral conditions.

△T (°C)	wind speed (knots)		
0	20	40	60
-8	22.1	41.8	61.5
-4	21.4	41.0	60.8
+2	17.8	38.9	59.2

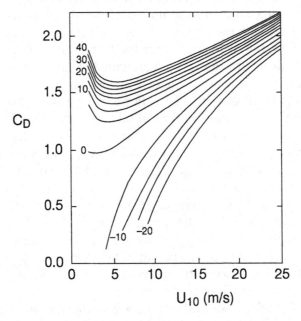

Fig. 4.1. Variation of 10^3 drag coefficient for sea-air potential virtual temperature
difference of $-20°C$ to $+40°C$ (after Smith, 1988).

including the range of interest for wave generation, $10 < U_{10} < 20$ m/s
and $-10°C < \Delta T < 5°C$. The drag coefficient C_D shows variations from
-10 to $+20\%$. The problem is further complicated by the approximation
present in the parametrization used for the heat exchange (Blanc, 1985) and
consequently in the estimate of the stability.

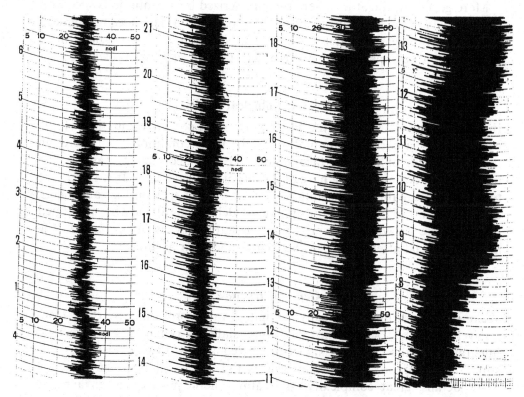

Fig. 4.2. Wind records with similar mean wind speeds, but different turbulent levels. Vertical units: hours; horizontal units: knots.

Gustiness

Another relevant effect connected to the stability conditions of the atmosphere, particularly close to the surface, is the space and time variability of the wind. Meteorological models, operating at both large and small spatial scales, always supply average conditions in time, that is within the time step of numerical integration. The wind so derived does not represent subgrid and subtime step processes. Some aspects of this variability, like gustiness, are relevant for wave modelling.

Figure 4.2 shows a series of wind records, from an oceanographic tower in the Adriatic Sea (Cavaleri *et al*, 1981), characterized by the same mean speed, about 30 kn, but with different level of gustiness. This is due to the turbulence of the atmosphere, and it increases with increasing air/sea instability. The turbulence level is conveniently characterized by the parameter $\tilde{\sigma} = \sigma_U/U$, with U the mean wind speed and σ_U the rms deviation from the mean. In highly unstable conditions $\tilde{\sigma}$ can reach values of up to $0.2 - 0.3$ (Monahan and Armendariz, 1971, Sethuraman, 1979).

More generally, the atmosphere is characterized by a continuous spectrum of variability from microturbulence to the space and time scales reproduced in the meteorological models. As we will see later in this chapter, these oscillations have consequences for the performance of wave models.

IV.2.2 Meteorological models

In § IV.1 we have stressed the importance of a correct evaluation of the wind field for a correct computation of the wave conditions. Any error in the distribution of these surface winds and any incorrect forecast will be reflected in a corresponding miscalculation of the wave field. It is important to know the capabilities and limitations of present meteorological models. The mathematical basis for atmospheric modelling has already been outlined in § I.3. Here we will give a brief and qualitative review of the state of the art, with special attention to surface winds and their accuracy.

General structure

Weather forecasting is basically an initial value problem. Starting from a given situation, a sequence of future states is obtained by numerical integration of the classical Navier-Stokes equations forward in time. This situation, apparently simple and well established, exhibits, when observed in its details, great complexity. The atmospheric processes span a range of more than ten orders of magnitude in space, from the most elementary microscopic processes to the thousands of kilometres of the largest weather systems. The range of time scales too, extends over at least seven orders of magnitude, from seconds to months. These scales interact with each other, making the problem unstable because the small details, which we do not know with sufficient accuracy, can affect and eventually determine the later large-scale structure of the field.

Numerical meteorological models have been in use for about forty years. In the last two decades particularly they have shown impressive improvements. The main reasons for this have been a better understanding of the physics of the atmosphere and a tremendous increase in computational capability. These are the same factors that, *mutatis mutandis*, have determined the development of wave models. But meteorological models have the advantage, and need, of a wide observational network with an enormous and continuous flux of data that has led, among other things, to the development of sophisticated assimilation techniques.

As discussed in § I.3 meteorological models integrate numerically in time a set of equations that describe the evolution of the state of the atmosphere. The equations concern momentum, thermodynamics of the system, conservation of mass and humidity, and the hydrostatic equilibrium (see, among others, Haltiner and Williams, 1980 and Holton, 1992 for a thorough discussion of the subject). A predictive equation for pressure is obtained by integrating the equation for mass along the vertical. The general formulation is done in three-dimensional space using spherical coordinates.

The classical method for the integration of geophysical differential equations is by finite differences. The evolution of the different variables that characterize the system is evaluated at the intersections of a grid covering the area of interest. The extension of this technique to the whole globe has required the proper treatment of the singularity at the poles, and also methods of filtering to avoid problems connected with the convergence of the meridians at the high latitudes.

An alternative to the classical method is provided by the spectral techniques first proposed by Eliasen *et al* (1970) and Orzag (1970). In this approach the horizontal distribution of the physical quantities is described using a two-dimensional spectral representation. This has led to the use of more efficient integration methods, such as the semi-implicit scheme of Robert *et al* (1972) that allows a separate treatment of waves with different propagation speed (Rossby and gravity waves). In a spectral model the variables are represented by means of truncated spherical harmonics. The linear terms in the equation are calculated in spectral form, while the nonlinear ones and the forcing are evaluated in the grid-point domain. This requires the use of efficient spectral transforms as the model has to switch back and forth between the two representations at each integration step.

Scale and resolution – global and local models

The model equations can only properly describe the phenomena that take place on a scale larger than the resolution of the model itself. Given also that the orography of the continents strongly influences the general meteorological patterns, and that the accuracy of its distribution is connected to the resolution of the model, it is not surprising that this resolution effectively controls the overall model performance. Figure 4.3 shows the improvement in time of the performance of the forecast produced at five different meteorological centres, and in particular at the European Centre for Medium-Range Weather Forecasts (ECMWF). The notable improvement centred at 1985 was associated with the implementation of the higher resolution models. Table 4.3 provides the basic characteristics of the present and previous models in use at ECMWF and at the UK Meteorological Office (UKMO).

Most of the global models in present use have a resolution of 70 – 100 km. On a geographical grid the distance between adjacent longitudinal coordinates decreases moving away from the equator. The resolution in a spectral model is more difficult to define. In a very simplified manner a model with n components can distribute $2n$ independent bits of information

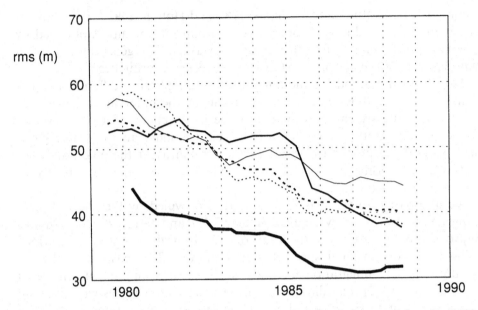

Fig. 4.3. Standard deviation scores for 72 h operational forecasts for the 1000 hPa height fields for five different meteorological centres. The thick line refers to ECMWF (after Bengtsson, 1991).

Table 4.3. *Characteristics of the previous and present ECMWF and UKMO meteorological models. See text for explanation.*

model	ECMWF	UKMO	UKMO
range	global	global	LAM
type	spectral	grid	grid
resolution until '91			
horizontal	T106	1.5° × 1.875°	0.7° × 0.9375°
vertical	17 levels	15 levels	15 levels
resolution after '91			
horizontal	T213	0.833° × 1.25°	0.422° × 0.422°
vertical	31 levels	20 levels	20 levels

over the circumference of the earth, so that the associated resolution in kilometres is $40000/2n$.

When implementing a model, a scale analysis of meteorological events is essential to establish the necessary model resolution. One can never resolve all processes, because some, such as heat convection, are active on a very small scale, but they can often be parametrized. Sometimes the evolution of an event is determined by phenomena occurring on a very small time and space scale where parametrization is not possible. In such cases one must make sure that the resolution is sufficiently small for the model results to be reliable. When the required resolution is not feasible in practice, the results of the model are likely to be approximate. This occurs in particular during the peak of a storm, or in frontal zones, in which case results are at best a smooth representation of reality, and thus an underestimate of peak conditions.

If we require a more detailed analysis or forecast for a given area, or if physical reasons suggest a higher resolution, the solution is to go to limited area models (LAMs). Usually these models are a reproduction of the global ones, with obvious rearrangements for the grid spacing and dimensions. If a starting condition is required, this is obtained by interpolation from the global results. The LAM then evolves in time, with inflow information at its boundaries determined by the contemporary integration of the global model. We see from table 4.3 that, besides a global model, UKMO also runs a limited area version of it. The UKMO LAM covers Europe and the surrounding basins. In its present version, to minimize the number of points, and hence its computational requirements, a virtual pole is used, located at 160° East, 30° North, that places the area of interest across its virtual equator.

A more sophisticated approach is the use of a grid with variable resolution, to be made denser in the area of interest. This technique is still under development, but it is likely to be extensively applied in the near future. A description of it is given by Courtier *et al* (1991).

A change of resolution, particularly when made both in the horizontal and vertical directions (the latter changing the number of layers considered in the model), is not only a matter of computational software or hardware. All the subgrid phenomena must be parametrized. Examples are cumulus convection, stratiform precipitation and dissipation of momentum (see Bengtsson, 1991). The parametrizations can be very sensitive to the scale/resolution ratio, and it is by no means certain that the same algorithms can be used when using a higher resolution. This is a serious problem because the tuning of a meteorological model requires considerable testing, and the rapidity at which computers are updated and models are upgraded carries the risk of leaving little time for their testing and optimization.

We have already mentioned that an advantage for meteorological models is the very large quantity of (real-time) observations at their disposal. Unfortunately this advantage is partially spoiled by their uneven distribution. The data are mostly concentrated over the continents and in the northern hemisphere. This is plainly reflected in the accuracy of the analysis, and consequently of the forecast. Figure 4.4 shows, respectively on the upper and lower panels, the forecast skill of the ECMWF model over the two hemispheres in the last decade. The number of days of predictive skill is a parameter used to describe the performance of a meteorological model. It can be derived from monthly means of daily averages of the anomaly correlation

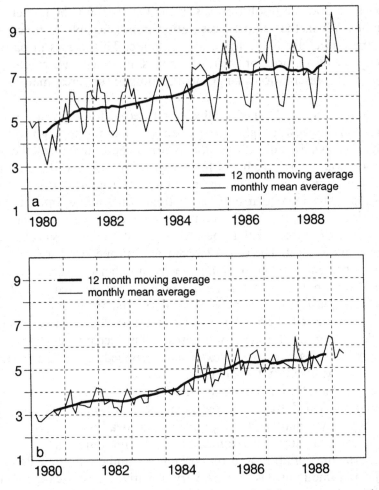

Fig. 4.4. Skill (in days) of the ECMWF extra-tropical forecasts: a) northern hemisphere: b) southern hemisphere (after Bengtsson, 1991). See text for explanation.

and standard deviation of the errors of geopotential height and temperature forecasts for the level 850 to 200 hPa. For the northern hemisphere the number of days of predictive skill is about one day and a half larger than for the southern hemisphere. Predictability studies (Lorenz, 1987) suggest that a halving of the analysis errors could extend the predictive skill by about two days. It is instructive that the 24 – 48 hour forecasts downward of continental areas are generally more accurate than the present satellite observations (Bengtsson, 1991).

Given the present resolution of meteorological models, the relative abundance of data over the northern continental areas may appear unnecessary in most situations, because the data are not independent, but reciprocally

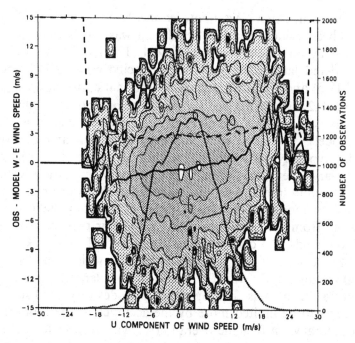

Fig. 4.5. Comparison between automatic ship reports and analysis winds at ECMWF. Period: December 1991 – February 1992 (Stoffelen, private communication).

connected by dynamical constraints. However, this is not true for tropical meteorological disturbances that, due to their reduced dimensions and strong gradients, require a very detailed description for their correct forecast. Also, as discussed in § IV.2.3, recent ideas about intense extratropical disturbances suggest that, at least in areas where rapid changes are taking place, more dense and detailed information is required, both for a better understanding of the physics involved and for the forecast of the storm evolution.

Of course, one would like to have a feeling for the quality of the computed sea surface winds by comparing them directly with wind observations over the sea. Here we face a difficult problem, because apart from some special and dedicated campaigns, the error present in wind measurements over sea is not so much less than the model error. Therefore, conclusions have to be drawn with care. Figure 4.5 shows the result of a three-month comparison for the east–west surface wind component between the T213 model of ECMWF and automatic ship reports (the technical details are given in Stoffelen and Anderson, 1992). The scatter is quite large, but 80% of the data show an average error less than 1 m/s. This can be interpreted as a basic agreement between the two sources, with a scatter due to intrinsic variability of the signal. Actually, whenever detailed and extensive measurements of wind have been available, as in ERICA (Wakimoto *et al*, 1991) and SWADE (Weller *et al*, 1991), it has been evident that the wind field was much more complicated than indicated by any meteorological model. The issue of wind validation will be considered later when we discuss specific wave model application (§§ IV.3 and IV.5) and when we discuss operational applications in § IV.4.

In section § IV.2.1 we mentioned the relevance of air/sea stability for wave generation. It is important therefore to see how this aspect is treated in numerical models. Once the wind has been estimated at the lowest level of the model, typically between 20 m and 30 m above mean sea level, the surface boundary layer is modelled with some parametrization, which takes atmospheric stability into account. It is important that wave modellers make consistent use of model information. Either they should take the friction velocity output from the boundary layer package or they should take 10 m or lowest level winds and convert them to friction velocities taking stability into account. Cardone (1969) describes a method which matches solutions of Monin-Obukhov theory with the results of the planetary boundary layer from meteorological models. This approach provides very good solutions. At present it seems to be limited more by the accuracy of the input data than by its own accuracy. In fact the overall error with respect to the measured data seems to decrease with the increasing quality of the data, suggesting that, if properly used, these methods can provide results whose quality is comparable, if not superior, to that of the measurements. An example of application of this technique has been given by the wind fields evaluated for the SWADE experiment (Weller *et al*, 1991), that are discussed later in § IV.3.2.

Gustiness, which we have discussed in § IV.2.1 and whose effects on wave growth we will describe in § IV.5.2, is not routinely predicted by meteorological models.

IV.2.3 Extratropical and tropical storms

In the previous sections we have outlined the present characteristics of the meteorological models, their basic structure, performance, problems, scale, skill, etc. We have also seen that the quality of the results is quite variable and strongly dependent on the kind of storm considered.

Meteorologists make a clear distinction between tropical and extratropical storms, a distinction arising from the general pattern of distribution of the stormy areas on the surface of the earth. The meteorological and oceanographic conditions vary with latitude and this leads to a different physical characterization of the storms developed at different latitudes. Therefore, we will now discuss the specific approaches to extratropical and tropical storms.

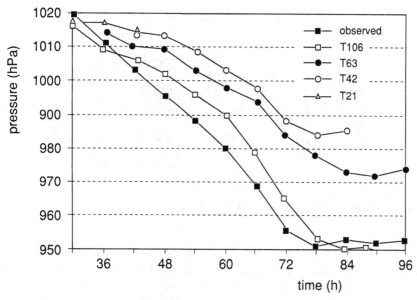

Fig. 4.6. Surface pressure as a function of time in the centre of a deepening low, forecasts from 12 UTC 28 January 1988. Forecasts for T21, T42, T63 and T106 are shown separately (after Bengtsson, 1991).

Extratropical storms

The midlatitude general circulation of the atmosphere, characterized by a strong permanent eastward flow, is baroclinically unstable. This leads to the development of cyclonic eddies through the year, whose intensity varies with the season, the strongest events taking place in the winter months. The large-scale events, with a spatial extension of 10^3 km or more, are well described by present meteorological models. Difficulties arise in areas with very strong spatial or temporal gradients. A good example is given by Bengtsson (1991) with the hindcast of a case of strong cyclogenesis in the North Atlantic Ocean which happened in January 1988. The evolution of the storm was modelled with different resolutions by the T21, T42, T63, T106 versions of the ECMWF spectral model. Each different resolution leads to a different rate of deepening of the minimum (shown in figure 4.6). The accuracy improves with the resolution of the model. T106 was the only one to produce a reasonable agreement with the observations. When the resolution was less than T106 there were also large displacements of the centre of the storm. Note, however, that even T106 does not succeed in reproducing the observed deepening rate of the minimum. As we have discussed in § IV.2.2, the problem lies in the gradients present in the field. Each model can handle up to a certain spatial gradient, beyond which its accuracy begins to deteriorate.

When we analyse a meteorological field in detail, we soon realize that

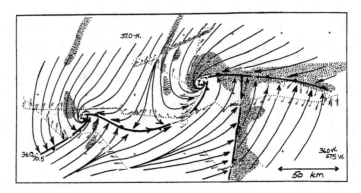

Fig. 4.7. Mesoscale vortices in the ERICA IOP-4 4 – 5 January 1989 cyclone bent-back warm-front (after Shapiro *et al*, 1991).

things are not as simple and smooth as in the models. This is even more evident in areas where rapid changes are taking place. The examples recorded during special campaigns like ERICA (Wakimoto *et al*, 1991) and SWADE (Weller *et al*, 1991) illustrate this very well. Shapiro *et al* (1991) and Cardone (1992), among others, have given a clear presentation of new ideas and evidences for extratropical storms, particularly in cases of explosive cyclogenesis. Figure 4.7 gives an idea of the possible complexity of the frontal area during the intense development stage. It is clear that the situation does not resemble the classical warm front–cold front system. The complexity is probably better introduced by the words of Shapiro *et al* (1991) who, on describing the details in the frontal area, state

These internal motions and associated physical processes include along-front variability and instabilities; cross-front moist- and dry-symmetric instability; vertical- and slantwise-convective precipitation systems; air/sea interactive processes, including sea-state (ocean-wave) atmosphere interactions; mesoscale moist baroclinic instability; pre- and post-frontal banded precipitation systems; cloud microphysical and radiative processes; inertia-gravity wave motions; turbulent mixing within elevated vertical shear layers; and isentropic potential vorticity concentration and dilution processes.

It is clear that a 100 km resolution model, even with the correct physics, cannot detect and reproduce such a complexity. We need here resolutions of order 10 – 20 km, together with an accurate representation of the three-dimensionality of the processes. However, even if we could deal with such resolutions in daily operations, the model initialization with observed data would still be a problem. The more detailed the model, the more dense and accurate must be the observations. Admittedly, if the model is sufficiently detailed in space, it will indeed develop the structure of the storm, independently of the availability of data, but to get details of the structure and its positioning and timing correct we would still need the data.

Tropical cyclones

Tropical cyclones are intense vortical storms that develop, through an instability mechanism, over the tropical oceans in regions of very warm surface water. The presence of the Coriolis force is essential. As a result tropical cyclones tend to originate in belts along the equator (Anthes, 1982), but not on the equator. These storms are given different names in different areas. So we find 'hurricanes' in the Gulf of Mexico, 'cyclones' in the Bay of Bengal and 'typhoons' in the China Sea. Incidentally, these names retain clear indication of their origin. The word 'hurricane' is derived from 'Huracan', the god of the storms of the preColombian Antilles population in the Caribbean Sea. The word 'typhoon' is the English transliteration of the Chinese word 'taai fun' meaning 'great wind' in Cantonese (but the presence of the ancient Greek god Typhon makes us wonder about the direction of flow of the information).

The problem of dealing with tropical cyclones (we will refer to them as hurricanes for simplicity) is their reduced dimension. This requires a high model resolution, as do cases of explosive cyclogenesis in extratropical storms. For this different possibilities exist. The simplest one is to cover the whole area of interest with a dense grid, 1/4 degree being the bare minimum necessary to lead to a sufficiently detailed description of the conditions around the eye. In the northern hemisphere, with counterclockwise rotating winds, the peak wind and wave conditions are found just to the right of the eye with respect to the direction of motion of the hurricane. Symmetrical conditions hold in the southern hemisphere. The gradients in this area of the hurricane are really extreme, peak values reaching 0.5 (m/s)/km and 2°/km for modulus and direction respectively. Examples are found in WAMDI (1988) and Dell'Osso (1990).

As we mentioned in § IV.2.2, if the area we need to consider becomes too large to be uniformly resolved at a high level, we can use a nested grid centred on the area of interest. In principle, a hierarchy of nestings is also possible. If the hurricane approaches the boundary of the high-resolution grid, one can shift that grid to a new position and continue the simulation. Making the shifts more frequent and smaller, we reach the solution of the grid centred on the eye and moving with it. This technique has been applied by Aberson and De Maria (1991).

As in intense extratropical storms, the modelling of a hurricane is very sensitive to the resolution of the model. A good example has been given by Dell'Osso (1990) hindcasting the development of hurricane Hugo (September 1989, in the Caribbean Sea) with three different resolutions, namely T106,

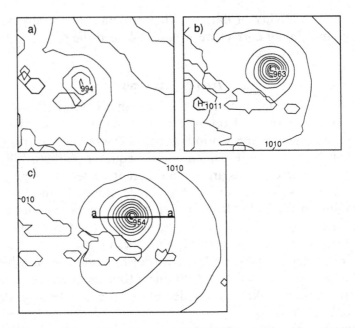

Fig. 4.8. Deepening of hurricane Hugo at 12 UTC 19 September 1989 after 48 h simulation with the a) T106, b)T260, c)T444 versions of the ECMWF meteorological spectral model (after Dell'Osso, 1990)

T260 and T444, of the ECMWF spectral model. The three results at T+48 h forecast are shown in figure 4.8. While all the three simulations have picked up the deepening of the hurricane, the level of deepening changes drastically with minimum pressure at 994, 963 and 954 mbar respectively. The wind speed changes accordingly, from 35 to 50 and 60 m/s, to the right of the eye.

That these approaches are effective in modelling the development of a hurricane is demonstrated by the hindcast of hurricane Camille (WAMDI, 1988). Wave results obtained for this storm with the WAM model will be discussed in § IV.5.2. We will also return there to the problem of wind variability.

IV.3 Wave sensitivity to the accuracy of the input wind fields

In this section we discuss the dependence of the accuracy of wave computations on the quality of the input wind fields. With this aim we analyse a number of cases, both in the open ocean and in a closed basin. The sensitivity of the wave results on the input wind is expressed by hindcasting the same storm with different wind fields and relating wind and wave differences.

First (§ IV.3.1) we discuss some general aspects of the relation between wind and wave fields. In § IV.3.2 we consider a severe storm, which occurred in the Atlantic Ocean during the Surface Wave Dynamics Experiment (SWADE). In this case, a careful and dedicated evaluation of the wind fields led to a drastic improvement of the wave results. We also quote results from the Labrador Extreme Waves Experiment (LEWEX), where the level of different spectral components was carefully related to the vector wind components in the corresponding source regions. A severe storm, which occurred in the Mediterranean Sea, is used in § IV.3.3 to show the influence of orography. A different storm in the same area, but with quite different characteristics, the Gorbush storm, is used in § IV.3.4 to demonstrate the effect of a higher resolution in the meteorological models, when strong gradients are present in the wind field. The problem of interpolating winds from one grid to another is commonly encountered. Its consequences for the computation of the wave fields are discussed in § IV.3.5.

IV.3.1 Wave models and input wind

L. Cavaleri

Surface waves are a direct effect of the wind blowing over the sea. Any error in the input wind field is reflected in an error in the computation of the wave conditions. The problem is acutely felt by interested users, because wind waves are also very sensitive to small variations of the input. In fully developed conditions $H_S \propto U^2$; in general, $H_S \propto U^{1 \div 2}$. A 10% bias in the estimate of the surface wind speed U is believed to be acceptable by a meteorologist. However, the consequent $10 - 20\%$ error in H_S and $20 - 50\%$ in wave energy is often not acceptable.

The spatial distribution of the errors is quite different in wind and wave fields. Wind errors are sometimes concentrated in certain areas, say for orographic reasons or close to a front. Even in these cases the consequences for waves spread throughout the basin, so that we never come across a localized incorrect peak value, but we find rather, for example, that a whole area shows a general over- or underestimate. In other words wave conditions at a given point are an integrated effect, in space and time, of the previous winds throughout the basin. Consequently, it is not an easy task to deduce where a miscalculation comes from.

In an ideal case it is possible to conceive the definition of a function of influence between a ΔU and the consequent ΔH_S at any point and time. Günther and Rosenthal (1989) have analysed the case of fetch-limited generation. They estimate the contribution of different sources of errors to the overall error of the model results. In the real world the variability of the phenomenon tends to hide the consequences of a single event, and a comparison between wave model results and measured data can only provide an overall indication of the generating wind field. However, often the whole storm is over- or underestimated and in this case the resulting waves provide an effective indication of its accuracy. Intuitively, this is more easily shown when the source of error is closer, which is the case for the wind sea part of the spectrum. Janssen *et al* (1989b) made use of this when they proposed their wave data assimilation technique, which was implemented at ECMWF by Lionello *et al* (1992). (See discussion in chapter VI.)

Sometimes, when there are strong gradients in the wind field, wind errors may be localized in space. An example is discussed in Cavaleri *et al* (1991), who considered the effect of a strong airflow through the Strait of Bonifacio. This flow is a local effect, enhanced by the orography of Corsica and Sardinia and is not resolved in the usual synoptic analysis. The effect on the wave conditions is significant, however, and could be followed all the way to the Italian coast.

Once a wave model has been defined, the quality of its predictions is controlled by the quality of the input wind. Whether the winds come from a hindcast, an analysis, a forecast or an idealized case, is irrelevant for the 'wave model machine'. Those who need the wave predictions for applications should realize that the quality of the model results depends both on the accuracy of the model itself and on the accuracy of the input winds. To get a proper judgement of the wave model itself, one should drive it with high quality winds, which is not easy in practice. However, one may get a feeling for the effect of the wind errors by estimating, for a given wind variation in a certain zone, the resulting wave variation, that will change from point to point and depend on the structure of the event.

Depending on their structure and the sophistication of their dynamics, wave models react differently to the same variations of the input wind fields. In the case of a third generation wave model, like WAM, the intrinsic errors, due to approximations associated with the formulation of the wave model, seem to be smaller than the errors commonly present in the wind fields, but this seems hard to prove in general, since it is difficult to separate both error sources. To isolate the effect of wind we will compare cases in which the WAM model has been forced with different wind fields, so that differences in wave response are necessarily due to differences in the input winds.

IV.3.2 The accuracy of wind field description

H. Graber, V. Cardone, R. Jensen, S. Hasselmann, H. L. Tolman and
L. Cavaleri

We consider the case of a storm in the open ocean, and the description of the associated wind and wave fields. This case occurred during the Surface Wave Dynamics Experiment (SWADE) and was particularly enlightening in several respects. SWADE was a carefully planned and extensive experiment, carried out from October 1990 until March 1991 off the east coast of the United States. The primary objectives were to understand the dynamics of the evolution of the surface wave field and to determine the effect of waves on the air/sea transfer of momentum. In order to achieve these objectives, concerted efforts were made 'to measure the surface meteorology with sufficient accuracy and spatial coverage such that the wind input to numerical models will not be the source of overwhelming uncertainty that it has been in most field experiments to date' (Weller *et al*, 1991). With this aim a dense array of meteorological and wave buoys was used throughout the experiment.

To minimize the consequences of wind inaccuracy on the evaluation of the wave conditions during SWADE, a hierarchy of nested model grids with different spatial and temporal scales (see figure 4.9) were set up. The prediction of the wave field was done in the north and south Atlantic on a basin scale (1° resolution), then in the north-west Atlantic on a regional scale (0.25° resolution), and finally within the experimental area on a fine-resolution scale (0.05° resolution) (Graber *et al*, 1991). While the overall Atlantic wind fields were supplied by ECMWF, six different wind field descriptions (ECMWF, FNOC, NMC, NASA, OW/AES, UKMO) were applied to drive the WAM model on the regional grid from 20 to 31 October. This period included three major storms, one of which has become known as the SWADE storm (Morris, 1991). From comparison with experimental data, none of the five meteorological models, namely ECMWF, FNOC, NMC, NASA and UKMO, succeeded in providing wind fields sufficiently accurate for the scope of the experiment. In contrast to the classical use of meteorological models, the winds provided by OW/AES are the result of a 'man–machine mix' procedure that takes maximum advantage of all the available tools and information (Cardone *et al*, 1980): numerical modelling, the know-how of the experienced meteorologist and the measured data. In the case of SWADE, high-resolution wind fields were produced for the focus period of the storm and for a limited domain coinciding approximately with the fine-resolution grid mentioned above. The manual kinematic analysis makes use of the available pressure, air and sea surface temperature fields, ship and buoy data. All the data are carefully screened for inconsistencies, and measured winds are adjusted in height and

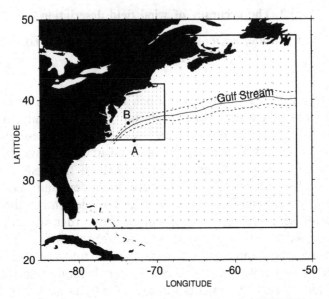

Fig. 4.9. Regional and fine-resolution grids area for the SWADE experiment. Resolutions are 0.25° and 0.05° respectively. The two dots show the position of the buoys used for comparison. A = 41001, B = 44015.

corrected with a stability-dependent surface layer model to effective neutral 20 m height hourly averages. The detailed hand analysis of the pressure and temperature fields carefully preserved temporal and spatial continuity of low pressure centres and frontal boundaries. Outside the limited domain and focus period, effective neutral winds were calculated with the marine boundary layer model of Cardone (1969). Time histories of measured and modelled winds and standard statistical measures of differences were used to evaluate the quality of the final wind fields.

It is obvious that the manual effort and the time required by such a direct analysis of the fields are not compatible with routine daily procedures. However, where and when needed for special reasons, the results are rewarding. The ECMWF and OW/AES wind fields at 18 UTC 26 October 1990, at the peak of the SWADE storm, are shown in figure 4.10. At first glance the differences in the flow pattern appear small although ECMWF locates the centre of the cyclonic low further north-east than OW/AES. ECMWF shows a stronger convergent flow offshore, while in OW/AES the convergence is more diffused. The main noticeable feature is the generally higher wind values in OW/AES, with a maximum speed difference of 5 m/s. However, the results of the careful manual analysis are better highlighted by the comparison, in figure 4.11, of the two winds with the values recorded at the two buoy positions, A and B, shown in figure 4.9. It is interesting to note, particularly

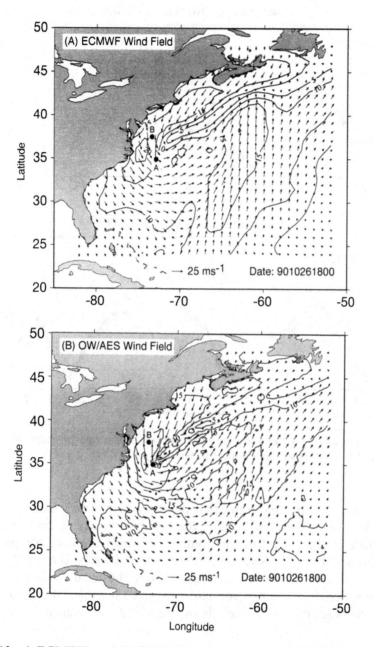

Fig. 4.10. a) ECMWF and b) OW/AES wind fields at 18 UTC 26 October 1990. Isotachs at 5 m/s intervals.

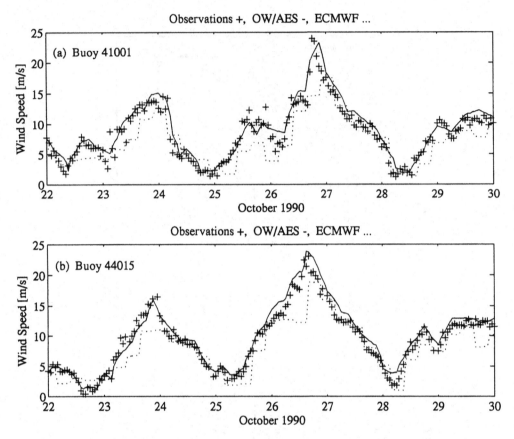

Fig. 4.11. Comparison of the wind speeds recorded at the two positions shown in figure 4.9 and the corresponding values provided by ECMWF and OW/AES.

for the air/sea stability conditions, that the two buoys were selected for their positions relative to the Gulf Stream, the first being situated south and the second north of the Gulf Stream. While the wind direction comparisons (not shown) are uniformly excellent for either wind fields, the ECMWF wind speed clearly shows significant deviations from the wind observed at the buoys. The statistics of the two comparisons, given in table 4.4, corroborates the above conclusion. Further details of this study are given in Cardone *et al* (1994).

We stress that the SWADE storm is representative of the average performance of global meteorological models in areas with strong gradients and/or intense activity. In the present case we could assess this performance because of the availability of the measured data and of the dedicated reconstruction of the wind fields. In general, this is quite difficult, because the data used for model validation are usually also available in real time on the global

Table 4.4. *Statistics for the wind speed comparisons in figure 4.11.*

| station | ECMWF | | | | | OW/AES | | | | |
	bias (m/s)	RMSE (m/s)	SI (%)	corr	slope	bias (m/s)	RMSE (m/s)	SI (%)	corr	slope
41001	-1.93	2.28	25	0.87	0.71	0.11	0.99	11	0.98	0.95
44015	-2.23	2.37	23	0.90	0.71	-0.21	0.93	9	0.99	0.97

telecommunication system (GTS) of the World Meteorological Organization, which makes it very likely that they have been assimilated in the meteorological models. Obviously, this biases any comparison between the measured wind data and the corresponding output of the model.

The differences between the input wind fields are further highlighted by the comparison with the corresponding wave fields. As we have mentioned, the quality of the wave results is a very good indicator of the correctness of the input wind. This is due to the reliability of the advanced wave models and to the accuracy of the wave measurements. Because the wave conditions at a given time and location represent an integrated effect, in space and time, of the previous wind fields, we can easily judge from their result the quality of the input wind, particularly for the locally active part of a storm.

Let us therefore look at the wave results of SWADE, obtained by driving the WAM model with the different wind fields. Wave–current interactions were considered (see § IV.8 for a more complete discussion of this point). Figure 4.12 shows the two wave fields corresponding to the winds in figure 4.10. As expected from the wind results, the OW/AES waves are much higher than the ECMWF ones. It is of particular interest to note how apparently limited differences in the structure of the wind fields can produce, as in this case, large differences in the wave height distribution. While the ECMWF wave field has different separate maxima, the absolute maximum being located far off the coast, the OW/AES field is much more concentrated around the position of the 'low' close to the coast. These differences are caused by the different strengths of the two wind fields. The presence of the ECMWF wave height maximum off the coast can be understood as a consquence of the strongly convergent wind field across the front shown in figure 4.10a.

The comparison between the measured wave heights at the two buoy positions in figure 4.9 and the corresponding results from the WAM model using the ECMWF and OW/AES wind fields is given in figure 4.13. The results follow clearly those for the wind. There is a quite remarkable fit for the OW/AES results, while the ECMWF ones exhibit a consistent negative bias throughout the hindcast period. Since errors in peak period are correlated

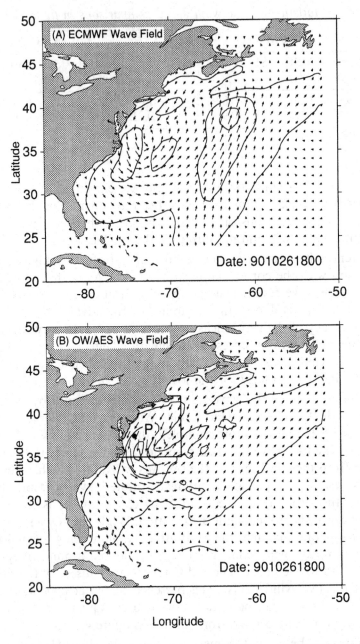

Fig. 4.12. Wave fields obtained with the a) ECMWF and b) OW/AES winds at 18 UTC 26 October 1990. Isolines at 2 m intervals.

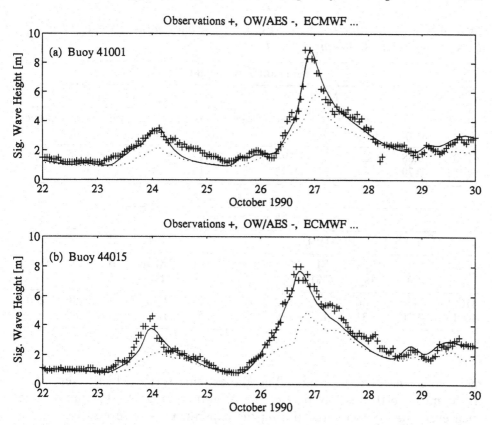

Fig. 4.13. Comparison between the significant wave height recorded at the two positions shown in figure 4.9 and the corresponding values obtained using the ECMWF and OW/AES wind.

with errors in wave height, a similar trend is also found in the peak wave period time series (not shown).

The overall statistics for the wave results are given in table 4.5. In general the OW/AES hindcast produced the smallest overall errors for all the buoys (Cardone *et al*, 1994).

The SWADE example clearly demonstrates the impact and importance of knowing the surface meteorology with sufficient accuracy and spatial coverage and the need for 'good' wind fields for ocean wave modelling.

In relatively simple meteorological conditions, as during SWADE, the significant wave height is a good parameter for judging the model performance. However, this is not always the case. A very good example is given by the hindcast carried out by Zambresky (1991) for the Labrador Extreme Wave Experiment (LEWEX). Like SWADE, the experiment involved a large number of modellers and a large diversity of instruments. LEWEX aimed at

Table 4.5. *Statistics for the wave results at the buoy positions marked in figure 4.9, and partly shown in figure 4.13.*

	significant wave height									
	ECMWF					OW/AES				
station	bias (m)	RMSE (m)	SI (%)	corr	slope	bias (m)	RMSE (m)	SI (%)	corr	slope
41001	-0.68	0.60	23	0.94	0.68	-0.12	0.35	13	0.98	1.05
44015	-0.79	0.84	32	0.92	0.54	-0.15	0.34	13	0.98	0.93

	peak wave period									
	ECMWF					OW/AES				
station	bias (s)	RMSE (s)	SI (%)	corr	slope	bias (s)	RMSE (s)	SI (%)	corr	slope
41001	-0.54	1.46	17	0.82	0.61	-0.22	1.36	15	0.86	0.80
44015	-1.20	1.31	14	0.79	0.70	-0.55	1.02	11	0.88	0.70

measuring, analysing and modelling the extreme waves which often occur in the most north-westerly part of the Atlantic Ocean. As it happened, no such event was encountered during the experiment, but several interesting situations were carefully analysed. One in particular (13 – 19 March 1987) is of interest for our present purposes. Zambresky (1991) reported for this storm a good fit between the significant wave height observed with a buoy and the corresponding estimate with the WAM model driven by LEWEX winds (evaluated by Oceanweather). However, when the comparison was extended to the two-dimensional spectra, a more complicated situation became evident. The spectra had several separate components, travelling in completely different directions. A careful analysis revealed evident biases in the model estimates of the individual components. The model had the overall energy level correct, but sometimes coming from the wrong direction with an incorrect frequency distribution. By comparing wind fields with available ship observations, Zambresky was able to show that biases observed in components of wave spectra were directly related to wind biases observed in generating areas. This clearly shows that the significant wave height does not necessarily provide a reliable measure of the skill of a wave model. It also proves the capability of a third generation wave model to handle very complicated situations and to act as a reliable tool to judge the quality of the input wind fields. This has implications for data assimilation, when observations of wave spectra become more generally available, as from the

SAR, which would allow, in principle, an 'observed' wind field to be derived from an observed wave field (§ VI.10).

IV.3.3 Orographic effects

L. Cavaleri and L. Bertotti

Obtaining accurate wind fields becomes even more of a problem when we approach the coast or when we want to model a storm in an enclosed basin with complex orography. This is illustrated by a severe storm that occurred in the Mediterranean Sea in January 1987 (Cavaleri *et al*, 1991). Figure 4.14 shows two wind fields corresponding to the peak of the storm, produced respectively by the T106 of ECMWF and by the LAM of UKMO (see table 4.3 for the characteristics of the models). Note that the ECMWF wind is available at 10 m height, while the UKMO wind is supplied at 0.997 σ level, that is at a height where the pressure is 0.997 of the atmospheric surface pressure. In practice this corresponds to 3 mbar, $\sim$ 24 m above the sea surface. Assuming neutral conditions, a 0.94 reduction coefficient has been used to reduce the UKMO wind to the 10 m height.

The different strengths of the two fields are quite evident. There is a 4 – 5 m/s difference both for the average and for the peak values. Because the T106 and the UKMO global models have similar performances on the open ocean (Günther and Holt, 1992), it is logical to associate the differences with the different resolutions. The explanation is easily found by looking at figure 4.15, which shows the orograhy used for the T106 and T213 models, the latter being also representative of that used in the UKMO LAM until 1991 (see table 4.3). For comparison, the detailed orography of the Mediterranean basin, with 1/12 degree resolution, is also given. The excessive smoothing of the orography in T106 is striking. This smoothing suppresses the effect of geographical features like the Alps, the Atlas mountains and the Apennines, which have a strong influence on local wind fields.

The WAM model was run with both the above winds and the resulting wave fields (peak conditions) are shown in figure 4.16. We see at once the remarkable difference between the two maxima, 7 m against 12 m! At this stage the sea is well developed, and the 30% difference in wind speed (20 m/s versus 26 m/s) is plainly reflected in the 70% difference for H_S. It is worthwhile to point out (Cavaleri *et al*, 1991) that the comparison with the available wave measurements is strongly in favour of the hindcast with the higher resolution UKMO wind. It is also remarkable that the difference between the two meteorological results is not due to any specific local effect, but only to the orographic structure of the basin, which is not sufficiently represented in the resolution of the T106 model of ECMWF.

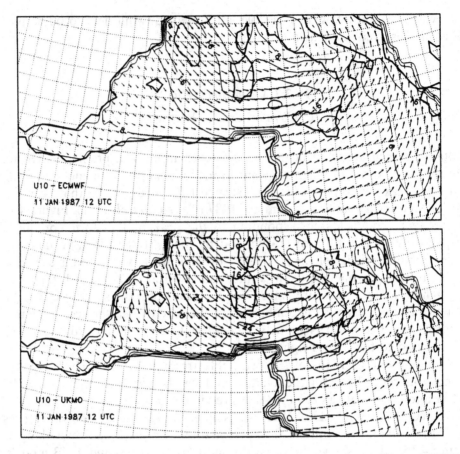

Fig. 4.14. ECMWF and UKMO wind fields in the Mediterranean Sea at 12 UTC 11 January 1987. The different strength of the two fields is mainly associated with a more detailed description of the orography of the basin in the UKMO model (see figure 4.15). Isotachs at 4 m/s intervals.

IV.3.4 Wind resolution

L. Cavaleri and L. Bertotti

The principal reason for using a LAM is the presence of very strong gradients in the meteorological field. A global model provides acceptable results as long as the field variations are not too large. An example is given in figure 4.17, which shows the most interesting moment of a storm that hit the Ionian Sea, south of Italy, on 2 and 3 December 1989 (Dell'Osso *et al*, 1992). This storm was very well documented, because it occurred during a summit meeting between US president Bush and Soviet leader Gorbachev (for this reason it has been nicknamed the Gorbush storm). Figure 4.17a shows the T106

(a)

(b)

(c)

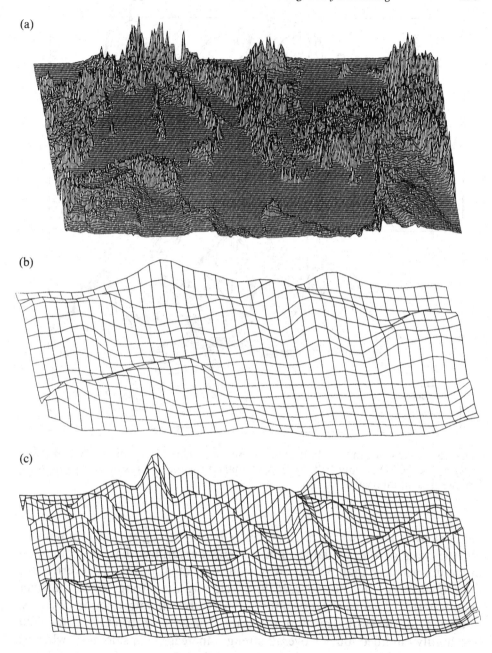

Fig. 4.15. a) Orography of the Mediterranean basin at 1/12 degree resolution and its representation in b) the T106 and c) the T213 spectral models at ECMWF.

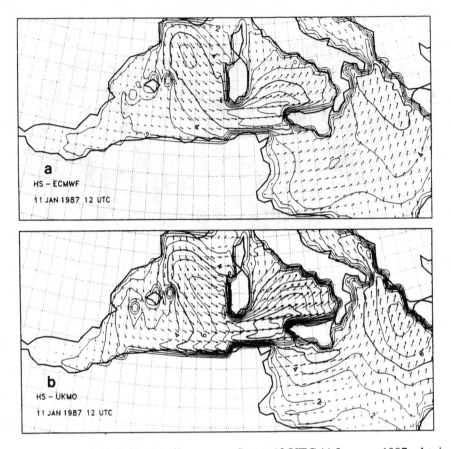

Fig. 4.16. Wave fields in the Mediterranean Sea at 12 UTC 11 January 1987, obtained using a) the ECMWF and b) the UKMO analysed wind fields shown in figure 4.14. Isolines at 1 m intervals.

analysis by ECMWF, which is characterized by a strong pressure minimum. On the basis of the available data it could be verified that the computed minimum was less than 50 km away from the actual position. Nevertheless, the global model did not succeed in developing the strong winds that had been reported. This was significantly improved by the introduction of a LAM in the area of the storm (resolution 40 km). The general structure of the field was hardly modified, but its overall strength increased considerably, whereas extreme gradients appeared in the area of the minimum. It is remarkable that this enhancement is not due to orographic effects, but to the physics of the process.

As indicated earlier, the following point, applicable to any limited area model, must be stressed. A LAM starts from a given situation and, controlled by the initial and boundary conditions that are interpolated from the global

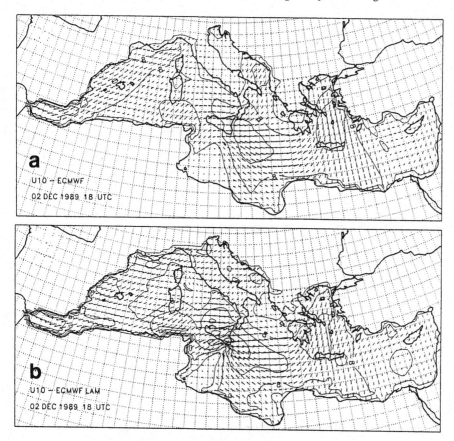

Fig. 4.17. a) Surface wind from ECMWF analysis at 18 UTC 2 December 1989. b) As a), but obtained with the 40 km resolution limited area model. The same orography has been used for both the cases. The higher resolution in b) allows a more effective modelling of the evolution of the wind field. Isotachs at 4 m/s intervals (after Dell'Osso *et al*, 1992).

model, integrates in time on its own. Its results will certainly be very detailed, but correct only as far as these initial and boundary conditions are correct. In other words a LAM believes in whatever we say and carries on with that. There is no way that a LAM can correct a wrong input from the global model. Resolution does not necessarily mean accuracy, and this is too easily forgotten in practical applications.

So far we have demonstrated that, in the case of very intense and small storms, an increased resolution of the meteorological model can lead to a significantly improved description of the local small-scale physical processes, and consequently to a better, usually higher, estimate of the wind. It is instructive now to look at the consequences for the wave field. To this end, the Gorbush storm has been hindcast with the WAM model at 0.5 degree

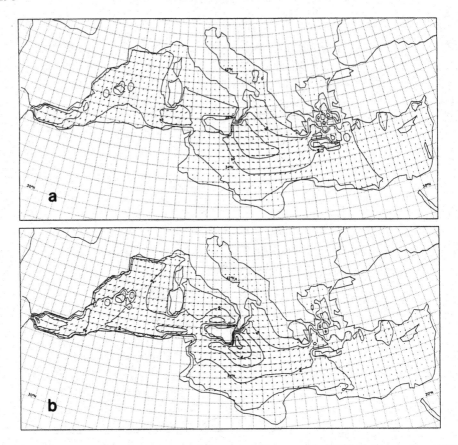

Fig. 4.18. Wave fields in the Mediterranean Sea at 18 UTC 2 December 1989, obtained using a) the ECMWF analysis and b) the LAM winds shown in figure 4.17. Isolines at 1 m intervals.

resolution (Dell'Osso *et al*, 1992). The two wave fields corresponding to the winds shown in figure 4.17 are given in figure 4.18. There is an evident increase of wave height from 3 to 5 m from the low to the high resolution, the latter figure being very close to the results of measurements made at Catania, on the east coast of Sicily.

The strong gradients of the high-resolution wind around the storm centre, estimated at 0.3 (m/s)/km and 5°/km for modulus and direction, have important consequences for the local sea conditions. This can be seen in figure 4.19. The main sea is coming from the east, but, with increasing frequency, the wave direction changes gradually, until it aligns with the local wind. This corresponds with the severe cross-sea conditions that have visually been observed by the different meteorological services (Malta and USA) present at the time.

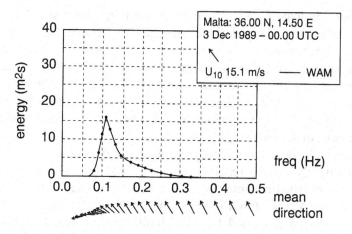

Fig. 4.19. One-dimensional spectrum at Malta position at 00 UTC 3 December 1989. Lower arrows show the mean direction for each frequency. The single arrow indicates the local wind. The directional dependence on frequency reflects the strong gradients in the wind and wave field.

IV.3.5 Wind interpolation

M. Holt and L. Cavaleri

In general, wind and wave models will have different resolutions. This is almost certainly so when the models are run at different operational centres. Because a wave model requires the wind to be specified at its grid points, the usual procedure is a more or less sophisticated interpolation from the wind to the wave grid points. The question arises of the consequences of the interpolation, and of its optimization for the best wave results. An interesting experiment has been done by Holt (1991) who has examined the impact of an increased resolution on the results of the global wave model run at UKMO (Ross, 1988). The wind was available on a geographical grid with resolution $1.5°$ latitude $\times$ $1.875°$ longitude.

Three different wave grids were used for the same model, A ($3° \times 3.75°$), B ($1.5° \times 1.875°$) and C ($0.833° \times 1.25°$). The B grid is identical to the original wind grid. Both A and B are staggered with respect to C. For grids A and C at each wave grid point the wind was evaluated by bilinear interpolation from the four surrounding wind grid points. The test covered a two week period in July 1989, including a somewhat atypical vigorous depression in the North Atlantic Ocean with more than 8 m significant wave height. The validation was done against buoy data.

The best results were obtained by B, with equal and colocated wind and wave grids. As expected, the coarser solution A missed the peak of the storm, with an underestimate of 3 m/s for U and up to 2 m for H_S. Also the high-resolution hindcast C missed the peak, although by a smaller amount, because the necessary interpolation smoothed out the peaks in the wind distribution. The conclusion is that, if for any reason we want to use a high-resolution grid, we must make sure that the grid points of the input wind grid are also points of the finer grid. However, Holt (1991) points out that even in this case the nonlinearity of the process, combined with the smaller area where the highest winds are assumed to be present, will cause an underestimate of C with respect to B.

The problems derived from interpolating the input wind are felt more acutely on relatively coarse resolutions. A partial way out would be the use of more sophisticated interpolation techniques that would limit the smoothing of the peaks. On the very fine grids of the local models, for example the 0.5° step size used for the Mediterranean Sea, the problem is hardly appreciable.

IV.4   Operational applications

The analysis of the analysis, the application of the application, the average of the averages, the forecast of the forecast (skill) ...

In operational centres meteorological models and wave models are run routinely, in real time, to provide forecasts to interested users. Naturally, these users want to know the reliability of the predictions, which can be given in statistical terms in a number of ways. In this section we discuss various statistics of the WAM model, the information they convey and how to interpret them.

We begin our discussion by looking first (§ IV.4.1) at the quality of the model analysis, which is obtained by forcing the wave model with analysed winds. This will give the background for a correct interpretation (§ IV.4.2) of the results obtained in forecast mode. The decrease of wave forecast skill with increasing forecast range reflects the corresponding decrease of atmospheric forecast skill. To illustrate this point we will discuss a specific example in some detail.

IV.4.1 Analysis
L. Cavaleri, G. Burgers, H. Günther, V. Makin and L. Zambresky

Wave model results are the product of a meteorological model and a wave model, which are usually integrated in sequence. The wave field analysis is obtained by running the wave model with the analysed wind. Starting from the latest analysis, operational centres produce a new forecast every day. The forecast period is typically ten days for global applications or less for local applications. After 24 hours, or whatever the hindcast cycle, when more recent analysed wind fields are available, the wave model is run again for this period. This provides the most recent wave analysis field, from which the next forecast is started. Recently, several centres have begun to assimilate wave data, but sufficient results are not yet available. Therefore, the wave analyses to be considered in this section have not assimilated wave observations. In chapter VI we will consider recent progress in this field.

Analysed wave fields are essential for model verification. They provide the background information for judging the performance of the forecast. The quality of the wave forecast usually deteriorates when the forecast period increases. When no wave data are assimilated this can only be due to the atmospheric component of the wind and wave model system. It does not say anything about the quality of the wave model.

It is important to realize what can be inferred from the statistics of operational applications. The careful analysis of specific events, as carried out in the previous section, aims at a better understanding of the model physics and the model performance under certain conditions and at ways of improving it. In contrast, the statistics of the long term performance of the wind and wave model system is the reply to the needs of users purely interested in the final output of the wind plus wave forecasting system. They simply want to know the reliability of a certain result and its expected accuracy, without going into details of the reasons for a possible error. Nevertheless, valuable information can still be obtained from integrated statistics, particularly when we stratify results or when we compare wind and wave statistics. It is worthwhile to stress that only winds measured in the open sea can be used for our purposes. Wind data from coastal stations are drastically affected by the presence of land and are not comparable with open sea data.

Table 4.6, extracted from Zambresky (1989), gives the verification statistics for wave height and peak period, and for wind modulus for four different areas and for the period December 1987 – November 1988. The global WAM model was run with $3° \times 3°$ resolution, using as input the analysed wind of ECMWF (see table 4.3 for its characteristics). The comparison has been made against data obtained from moored wave measuring buoys. The parameters used for the statistical comparison, here and later on in this chapter, are bias = mean error, rms error, and scatter index = rms error normalized by the mean observed value of the reference quantity.

In general, there is a clear underestimate of the wave characteristics in the model results, even if the bias is small enough to make them acceptable for most practical purposes. However, it is instructive to see how the biases

Table 4.6. *Statistics obtained by the WAM model driven by ECMWF wind for wave height H_S (m), peak period T_p (s), and wind speed U (m/s) for four different areas and for the period December 1987 – November 1988. Bias is with respect to measurements. SI is scatter index. In brackets are the numbers of observations used for the statistics.*

area (number of obs)	H_S		T_p		U (scalar)	
	bias	SI	bias	SI	bias	SI
Alaska (4657)	-0.22	0.22	-0.70	0.18	0.49	0.27
Hawaii (2061)	-0.28	0.22	-0.01	0.21	-0.38	0.18
East Coast USA (4284)	-0.38	0.37	-0.71	0.19	0.07	0.30
North UK (2825)	-0.40	0.29	–	–	-0.54	0.28

of both H_S and T_p have little relationship to the bias of the wind speed. Apart from possible errors in the measurements, this could simply reflect deficiencies in the wave model. Before one can draw this conclusion, it should be realized that the wave conditions at a given location depend not only on waves generated by local winds, but also on swell coming from long distances. Swell systems result from storms which can be generated far from a given location and at (much) earlier times. In fact, one cannot expect a simple relation between local winds and waves. The relation is further blurred by the fact that the wind data used for the comparison in table 4.6 have been assimilated into the ECMWF meteorological model. Therefore, the analysis wind fields were in a certain sense forced to fit these data, so that it would be incorrect to assume a comparable skill in areas without wind measurements.

It is important to realize that, as a result of the continuous upgrading of both the meteorological and wave models, the wave model performance is steadily improving. Table 4.7 shows results corresponding to part of those in table 4.6, but for the period September – December 1991. Relative to the 1987 – 1988 period there is a large reduction in the bias of H_S. The improvement in the scatter index is less dramatic; this appears to be connected with a fundamental problem of model validation that will be discussed later in this section. These reductions are probably significant, although, strictly speaking, one is not allowed to compare table 4.6 and table 4.7 directly, because they do not cover the same seasons and, even between successive years, there may be large differences, due, for example, to the different number of severe storms.

Due to their costly operation, the number of wave measuring buoys in the ocean is rather limited. They are mostly located close to the coast, and data are scarce for the open ocean. Fortunately, ocean satellite observations, to

Table 4.7. *As table 4.6, but for the period September – December 1991.*

area (number of obs)	H_S bias	SI	U (scalar) bias	SI
Alaska(1932)	-0.13	0.21	0.32	0.26
Hawaii(1478)	-0.05	0.17	-0.54	0.22
East Coast USA (3150)	-0.08	0.31	0.25	0.15
North UK (1395)	-0.18	0.35	0.61	0.26

be discussed in chapter V, are beginning to fill this gap.

To get a feeling for the effective performance of the wind plus wave forecast system, it is instructive to look at extended time series, comparing model and measured data. Figure 4.20 shows two one-month comparisons for December 1991, at two buoy locations, a) on the east coast of USA and b) close to Hawaii. At the former location a sequence of winter storms occurred. Two or three peaks are missed, but the general behaviour is quite satisfactory (the statistics are given in the caption). Note that the measured data are provided only as rounded off, discrete values, which leads to an increase of the rms error and of the scatter index. The comparison at Hawaii (figure 4.20b) is less spectacular, with the wave height varying around a mean value close to 3 m. However, the results are perhaps more significant, in terms of model performance, than those on the east coast of USA. The peaks in figure 4.20a, associated with winter storms, are likely to be of more or less local origin, while the waves at the Hawaiian Islands are mainly swell. The correct evaluation of swell requires both a correct estimate of the wind field on a very large scale and an effective propagation scheme in the wave model. The results of figure 4.20b are very positive in this sense.

An example of a regional application is given by the operational NEDWAM (*Nederlands* WAM) system for the North Sea (Burgers, 1990). In this system, wave model predictions are obtained by WAM (cycle 3) driven by the atmospheric fine-mesh LAM. Figure 4.21 (from Van Moerkerken, 1991) shows a one month comparison between analysis, 24 hour forecast (discussed in the next section) and recorded data, for wind direction and speed, significant wave height and period and low-frequency wave height, at station AUK in the central North Sea. The statistics for the corresponding storm season (October 90 – April 91) for AUK and two more stations are given in table 4.8. The closest grid points were used for comparison.

It is interesting to note that, in general, in areas like the North Sea and the Low Countries, where virtually no orographic effects are present, high-resolution atmospheric models do not necessarily perform better than global models. In such cases it is the resolution of the wave model, rather than the

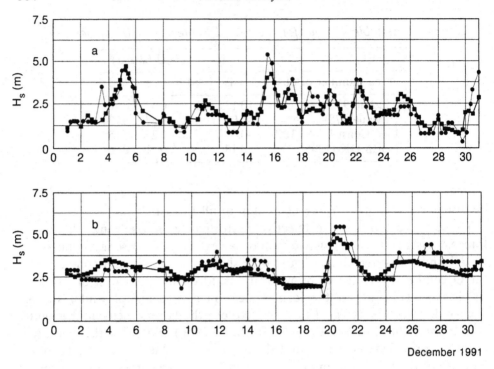

Fig. 4.20. a) Intercomparison between WAM model results (continuous line) and measured data (dotted line). Location: east coast of USA (40.5°N, 69.4°W). Time in days. The model mean wave height is 2.3 m; the observed wave height is 2.2 m. Rms error, bias and SI are 0.53, 0.07 and 0.24 respectively. b) Same as a) at a location close to Hawaii (19.3°N, 160.8°W). Here the model mean wave height was 3.0 m; the observed wave height is 3.1 m. Rms error, bias and SI are 0.46, - 0.05 and 0.15 (after Günther *et al*, 1992).

Table 4.8. *Statistics obtained by the WAM model in KNMI's operational NED-WAM system for wave height H_S (m), mean period ($T_{m0,-1}$, in s) and wind speed U (m/s) for three different stations in the North Sea and for the period October 1990 – April 1991.*

| | analysis | | | | | | + 24 | | | | | |
| | H_S | | T_m | | U (scalar) | | H_S | | T_m | | U (scalar) | |
station	bias	SI	bias	SI	bias	SI	bias	SI	bias	SI	bias	SI
EPF	-0.27	0.25	-0.3	0.17	-0.9	0.16	-0.19	0.36	-0.2	0.19	-0.8	0.26
K-13	-0.12	0.24	-0.4	0.20	0.2	0.13	-0.02	0.35	-0.2	0.21	0.2	0.24
AUK	-0.14	0.24	0.1	0.24	0.6	0.13	-0.18	0.33	0.1	0.25	0.4	0.26

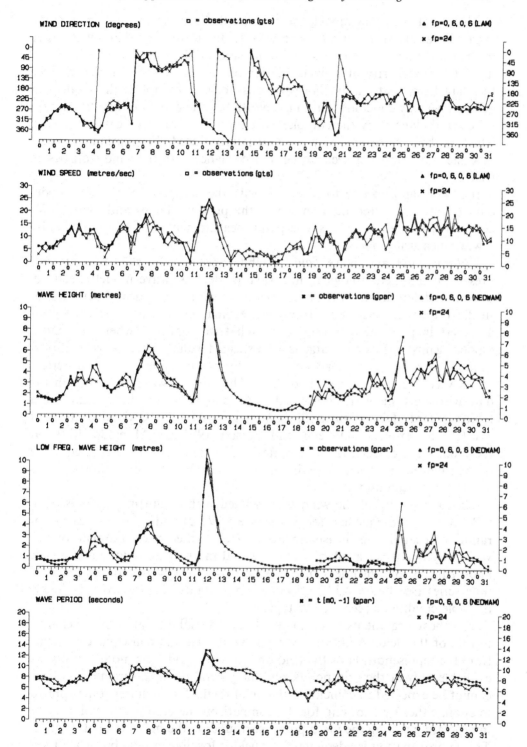

Fig. 4.21. Comparison between analysis (△), 24 h forecast (×) and measured data (□) at station AUK in the North Sea for December 1990.

resolution of the wind model, that can be the limiting factor. The relative large bias at EPF in H_s and U is very likely due to the fact that EPF is closer to the coast than K-13 and AUK and more susceptible to resolution effects. To judge the model error at a given location one needs also to consider whether the data from that station have been assimilated or not in the model. In general, we may conclude that in the North Sea the benefit of high resolution is counterbalanced by the complex structure of prevailing weather systems, which is the very reason why the LAM is introduced. For stations K-13 and AUK the wave statistics seem to be in accordance with the wind statistics as well.

It is now appropriate to discuss briefly the comparison of wave model data against measured data shown in the previous tables and figures. We consider here the case of a single point measurement, typically provided by a wave measuring buoy.

Natural wind and wave fields exhibit random properties that are not present in the meteorological and wave models. A wave model effectively deals in a deterministic way (with equations) with a statistical description of the field (the wave spectrum). It provides estimates of the 'average' conditions at a certain point at a certain time. What we compare, when verifying a model against measured data, is an expected value versus a sample of a random process. It is, therefore, physically not possible to reach a perfect fit between model and measured data. There will always be scatter in the data with consequent rms error and scatter index. For wind waves and for a typical record of 20 minutes, the sampling variability is estimated at about 8% (see, for example, Donelan and Pierson, 1983 and Monaldo, 1988). If the models are correct, their estimate is coincident with the expected value of the random process and we find no bias in sufficiently long statistics. This is a result we can aim at.

The randomness of the wind is associated with its gustiness, discussed in § IV.2.1. The consequences for waves (see § IV.5.2) add a further reason for random variability in measurements of H_S. A theoretical estimate of this variability is not yet available, and it would in any case require a complete knowledge of the gustiness of the wind. By direct inspection of recorded time series (see figure 4.38), a sampling variability close to 10% in highly unstable conditions seems to be typical.

The above arguments result in geophysical variability, an intrinsic characteristic of the field. Additionally, we have to face the consequences arising from the representation of the field on a discrete grid. The model results are available only at the grid points, and, in general, the measurement location cannot be expected to coincide with one of them. This requires some approximation, either by choosing for the comparison the closest grid point, or by interpolating among the mesh points surrounding the measurement location. This approximation is adequate if the field is locally smooth, but when there are strong spatial gradients the error may be significant. The errors are increased by the fact that most buoys report wave heights rounded to 0.5 m. Again, because the error has no preferential tendency, the resulting bias is small. This is not so for the scatter. As a consequence, there is an additional 'sampling variability', of different origin from the previous ones. Note that this variability depends on the step size of the grid. The smaller the step size, the more limited are the consequences of the above approximation.

In conclusion, we see that we can aim at zero bias in the comparison of estimated and measured significant wave heights. There is, however, a minimum rms error or scatter index, below which it is impossible to go. This limit varies with the specific conditions in the field and with the resolution of the grid. On average we estimate its value to lie between 10 and 20%. With this in mind we can appreciate better the results shown in tables 4.6 and 4.7. The improvement in modelling between 1988 and 1991 has led to a marked decrease of the bias of the results, and a marginal decrease of the scatter index. Over the same period the reduction of the scatter index was only marginal. This is in agreement with the above arguments, and indicates that we are close to the intrinsic limit of the comparison, either because of geophysical variability or due to the resolution of the wave model.

The discussion of geophysical variability can be extended to the comparison of one- and two-dimensional wave spectra. The problem here is more complicated because we are not dealing with a single value, the significant wave height H_S, but with one-dimensional or two-dimensional matrices. The statistical variability of the estimate of the different wave components is much larger than for the single H_S values (see, for example, Borgmann, 1972 and Donelan and Pierson, 1983). Besides, most measurement techniques provide only filtered estimates of the actual wave spectrum. This is still an open problem and, while some attempts have been made to address it (see, for example, Guillaume, 1990), no standard technique is yet available.

IV.4.2 Forecast

L. Cavaleri, G. Burgers, H. Günther, V. Makin and L. Zambresky

The ability of meteorological models to analyse in detail a given situation has been steadily improving with time. This is associated with an increase of spatial resolution, with improvements of data assimilation techniques, with the growing availability of atmospheric observations in real time and with improvements of model physics and numerics. Analysis of data from different sources, such as the ERS-1 scatterometer and altimeter data, suggests that the rms error of the vector wind analysis is presently between 2 and 3 m/s (Hollingsworth, personal communication), with specific values varying

with the geographic area and the season. This is a good figure, especially
in view of the assumed ship reports error of 2.5 m/s. The reasons for this
achievement are the use of other data (upper air data, satellite information) in
the assimilation system. However, the error is present and must be expected
to grow in the forecast. A comparison between analysis and forecast fields
suggests that the rms vector wind forecast error for the North Atlantic and
North Pacific is about 3.5 m/s at day 1 and 5 m/s at day 2. The seasonal
variation is ± 0.5 m/s and ± 1 m/s, respectively. These figures agree with the
rule of thumb that the forecast error has a two-day doubling time. Larger
errors must be expected in areas with sparse data coverage. Figure 4.22 shows
the time history of the rms error and forecast skill over a representative area
of the North Atlantic Ocean for the day 1 and day 2 forecasts of the 1000
hPa wind. The skill score, different from the one quoted in § IV.2.2, is here
defined as

$$\text{skill score} = 100 \left[1 - \frac{\text{rms}_{fc}^2}{\text{rms}_{pers}^2} \right], \tag{4.1}$$

where the mean square difference between forecast and verifying analysis is
normalized by the mean square difference between the verifying analysis and
the analysis from which the forecast started. This weights the forecast error
by the magnitude of the observed change.

Figure 4.22 shows that the largest absolute errors take place during the
active winter months. The maxima in the score are reached during the same
period, which suggests that the model behaves better in active situations,
when the general pattern is more clearly defined by synoptic scale systems.
In calm conditions, the onset of an event may be more difficult to anticipate.

Considering the results of the last decade in figures 4.4 and 4.22, one
gets the feeling that the general improvement in forecast skill has not
been parallelled by an analogous result for the surface wind field over
the oceans. The marked improvement until 1986 is followed by an apparent
plateau in forecast skill with pronounced seasonal and interannual variations.
Even more recent results (Hollingsworth, personal communication) show
only minor improvements in this respect. The point, to the credit of the
meteorologists, is that the analysis is already very good, given the amount
and quality of the data available at sea. The figures given above, describe
the present situation, which we have to accept, whether we like it or not.
Further progress requires a marked improvement in the analysis skill (like
better data or the use of variational analysis) and substantial improvements
in forecasting skill (from better atmospheric models).

It is essential to be aware that the values plotted in figures 4.4 and 4.22 are
averages in space and time. The forecast skill shows an extreme variability,
and we meet major difficulties in areas with strong gradients, that often
deserve special interest. Some areas are more difficult to deal with because

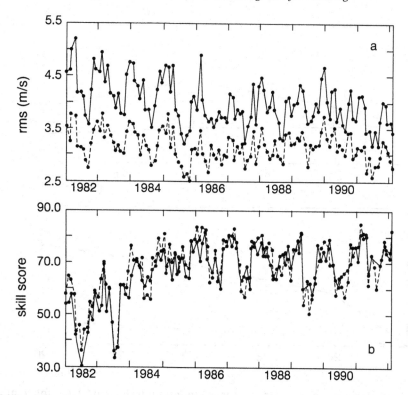

Fig. 4.22. a) Root mean square error and b) skill score for the forecast of the 1000 hPa vector wind of the ECMWF meterological model over a representative area of the North Atlantic Ocean. Dotted line = +24 h; continuous line = +48 h.

of surrounding complicated orography, as we have seen in § IV.3.3 to be the case for the Mediterranean Sea.

The problems and limitations of meteorological forecasting are reflected into analogous characteristics of wave model forecasts. Since the same wave model is used to compute the analysis and forecast fields, a downgrading in time of the meteorological forecast is normally reflected in a deterioration of the wave forecast. Figure 4.23 summarizes the results of one month's statistics for the meteorological model of ECMWF and the associated WAM results (Carretero Albiach and Günther, 1992). The figure provides the average global values of bias and rms error with respect to the verifying analysis for wind speed U and significant wave height H_S, each value complemented by a vertical bar showing the variability of the statistics in the three large areas considered: upper northern hemisphere, tropics, lower southern hemisphere. We note the surprisingly small mean bias, both for wind and waves, with no obvious tendency to grow in time, as the forecast progresses. In contrast, such a tendency is very clear for the rms error,

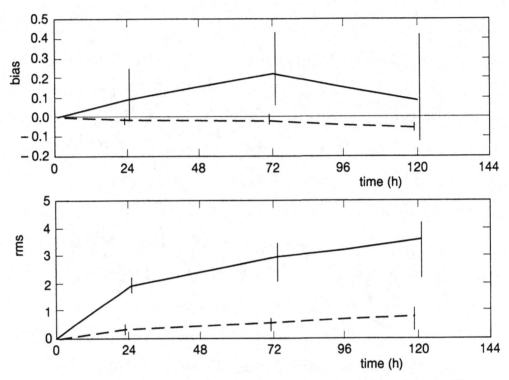

Fig. 4.23. Bias and rms error for wind speed (continuous line, m/s) and for significant wave height (dashed line, m) for different forecast periods relative to the ECMWF analysis. One month's statistics, 1 – 30 November 1988. The lines show the global averaged values, the vertical bars their geographical variability.

suggesting that storms do exist both in the analysis and in the forecast, but that the error in their positioning grows in time.

There is a striking variability with geographical position, a consequence of the differing seasons in the two hemispheres and of the different densities of available meteorological information, and therefore of the accuracy of the analysis. Figure 4.23 also warns against the straightforward use of mean values in the statistical interpretation of model results.

It is quite remarkable that often the 12 – 24 hour forecasts are of the same quality, or even better, than the analysis. The explanation comes once more from the wind. The analysis is always a mixture of first-guess and measured data. The following integration in time is likely to shape better the general situation, adapting to both the physics of the event and to the possible orographic constraints. At the same time, because of the limited forecast period, the associated error is still kept small. As a result the forecast wind fields are, on the average, comparable in quality to the analysis.

A good example of this is given by the comparison between the analysis and forecast results shown in table 4.8. When compared with measured data, H_S and U for the 24 hour forecast have a bias lower than the corresponding analysis results. The opposite is true for the scatter index of both quantities. This can be for different reasons. One could be the error in position that we have just discussed. This would result in a variable bias, for example positive at a certain time and negative afterwards, which would average out in the long term statistics. In this case the scatter index, which is insensitive to the sign of the individual error, would be enhanced. A more likely reason is large- and small-scale variability intrinsic to a wind field. This is discussed more extensively in the next section. For now it suffices to say that, given such variability, the analysis tends to fit it, whatever the distribution, while the forecast follows from a numerical model where variability cannot be modelled. The result is an increase not of the bias, because the field is correct on average, but of the standard deviation and scatter index of the error, because of the spatial variability of the wind and consequently the wave field.

It would clearly be of interest to be able to anticipate the forecast skill. Molteni and Palmer (1991) have attacked this problem with statistical methods, but their results are not yet conclusive. This is an area where improvements are expected in the near future. The concepts are, in principle, also applicable to waves. The sequence of deductions that would lead to the result (meteorological forecast → wave forecast, prediction of meteorological skill → prediction of wave skill) could be quantified in Monte Carlo simulations, but this would be rather expensive and has not yet been attempted.

It should be noted that the skill in wave forecasting depends heavily on the skill of predicting surface winds. As noted above, this skill deteriorates rapidly with increasing forecast period. This is directly reflected in deterioration of the wave forecast. For the 24 hour forecast especially, it is hard to see an improvement over the North Atlantic in recent years.

We end this section with a representative example, which is derived from the Med87 storm in the Mediterranean Sea (Cavaleri *et al*, 1991), already encountered in § IV.3.3. Figure 4.24 shows the analysed wind at 12 UTC 11 January 1987, and the corresponding forecast at 72 hours. There is an evident dramatic deterioration. The general structure of the field is correct, with the mistral wind blowing from the Spanish–French border and a front in the Ionian Sea south of Italy, but the strength of the storm is largely underestimated.

The wave forecast reacts accordingly. To stress the point, a procedure different from the standard one has been followed. The usual procedure makes use of the analysis wind fields until time 0, after which the wave forecast is produced using the 6, 12, 18, 24, 30, 36, etc. hour wind forecast (when the wind fields are available at 6 hour intervals, as is the case at ECMWF). Because waves are an integrated effect in space and time of the

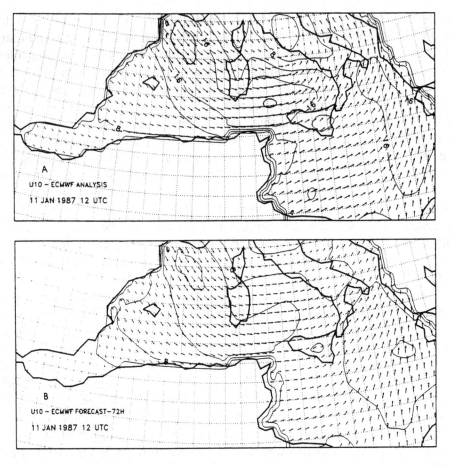

Fig. 4.24. Surface wind field on the Mediterranean Sea at 12 UTC 11 January 1987 from the a) ECMWF analysis, and b) the corresponding forecast at 72 h. Isotachs at 4 m/s intervals (after Cavaleri *et al*, 1991).

input wind fields, this approach does not stress the quality of the forecast wind. With this aim, the wave forecasts, such as the one shown in figure 4.25b, have been produced making use only of the forecast winds. In other words for the '24 hour' wave forecast the input wind for day N was given by the wind produced as forecast at day $N - 1$; for the '48 hour' wave forecast the wind produced at day $N - 2$ was used; for the '72 hour' wave forecast (shown in figure 4.25b) the wind produced as forecast at day $N - 3$ was used. In this way a permanent bias in the wind forecast is immediately apparent in the wave forecast. Note that we have previously pointed out the inadequacy of the ECMWF wind in this case, but the general argument keeps its validity.

There is a dramatic loss of skill at 72 hours. We can hardly speak of a forecast storm at all. Such a strong change of wave height conditions is also

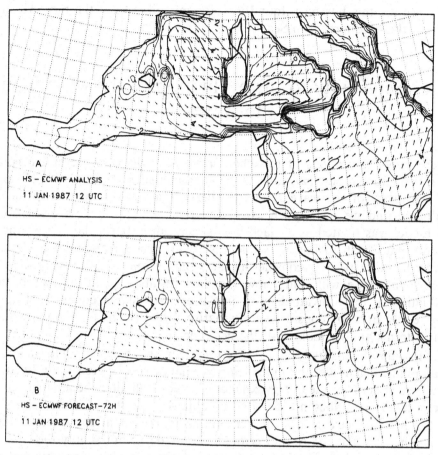

Fig. 4.25. Wave fields in the Mediterranean Sea at 12 UTC 11 January 1987 obtained using a) the analysis and b) the 72 h forecast winds of ECMWF. See figure 4.24. Isolines at 1 m intervals (after Cavaleri *et al*, 1991).

related to the geometry of the basin. With complicated geometry, a limited shift of the storm can lead to drastic changes due to the variation of fetch, to shadowing or to a different angle of the wind with respect to the coast. This is the case in the Tyrrhenian Sea, between Italy and Sardinia. In more open seas, such as in this example in the Ionian Sea, or in general the open oceans, a shift of position of the storm is likely to lead to much less dramatic consequences, and one is mainly concerned with the strength of the event.

The above example serves to show how, in enclosed basins, the dependence of waves on wind is more critical than in the open ocean. It is also worthwhile to stress that, according to the characteristics of the meteorological models shown in table 4.3, the transition to high-resolution models has dramatically improved the average quality of the forecast, including the surface wind.

Evidence for this has been found also for mistral storms, the most violent wind in the Mediterranean Sea.

IV.5 Sensitivity to the evaluation of the wind input to waves

In the previous sections of this chapter we have discussed several general aspects of wave modelling, such as the characteristics of the wind fields and the sensitivity of the wave results to the accuracy with which the wind is evaluated and we have given extensive applications. In these discussions the model was treated as a black box, generating a particular output for a given input. In the following sections we will turn our attention to details inside the black box, by considering the separate role of different processes. In particular, we will study how the final results depend on the correct evaluation of the individual terms in the energy balance equation. This is most easily done by considering alternative formulations for the individual source terms and by looking at the net effect.

This type of sensitivity analysis is useful in understanding what are the weak points in a model and in realizing to what extent the eventual correction of a specific term of the energy balance equation will alter the model results. Such an analysis does not tell one anything about the absolute reliablity of the model, because the different terms in the equations interact with each other, so that the overall error of the model is not simply the sum of those due to the individual terms. Overall statistics can only be obtained by comparison with measured data, as has already been discussed. A sensitivity analysis will tell us where it is more convenient to look for improvements, following the principle 'larger corrections first'.

Wave–current interactions and typical shallow water effects will be considered in the following three sections. In this section we begin by discussing the normal, deep water situation for which there is an interaction between the wind input term S_{in} and the other source terms (§§ III.3.1, III.3.2). The wind is the source of practically all energy in sea surface motion at supertidal frequencies. Wave generation by the wind is extremely common in nature. We can observe the process daily when we observe the effect of wind on a water surface. The physics of the process is relatively well known (as explained in § II.2). In this apparently happy situation it is disappointing that so little experimental evidence exists on the correct formulation of the wind input. The reason for this is the strong continuous interaction between wind input, nonlinear interactions and whitecapping. Wave generation by the wind is intrinsically associated with the presence of all three processes.

They have comparable magnitude, and wave growth depends on relatively small differences in their balance. Furthermore, generation is not localized in space and time. This makes any attempt at obtaining experimental information about the magnitude of the individual source terms a formidable task. In several dedicated field experiments the wind input source term was measured at a particular location. The uncertainty is still significant. In addition, one only knows the result of the overall process in terms of the evolution of the wave spectrum (see, for instance, Hasselmann *et al*, 1973, Snyder *et al*, 1981, Kahma, 1981a and § II.8).

To verify the theory, the ideal situation would be a case where one or two of the above terms have zero or very low values. A typical example is a sudden drop of wind after a period of active generation, with a consequent vanishing of S_{in}, but even then it would be hard to separate the effect of S_{nl} and S_{ds}. This idealized situation is approximated in nature by waves leaving the area of a highly concentrated storm, with a consequent rapid transition from wind waves to swell. There is little chance of finding extensive conventional measurements in such a situation, but satellites offer new possibilities, because they combine continuous, weather-independent operation and extensive spatial coverage. The conclusion is that at present we have very little, if any, quantitative information on S_{nl} and S_{ds} from field experiments.

Systematic tests with different dissipation constants have been made by Komen *et al* (1984), who determined the dissipation coefficients by matching growth with observations. In this section we focus on the sensitivity to different formulations of the wind input. The wind input into waves has been thoroughly discussed in previous chapters. The problem was approached mainly from the following viewpoint: given u_* and the wave spectrum, to compute the corresponding energy input into waves. However, in practice, u_* is not given (or not reliably), so that we have to determine the wind input from whatever parameters are provided by the meteorological model. These are, in general, wind speed, temperature, humidity, etc. at different levels, well above the sea surface, with the additional complication that small-scale variations are not resolved, so that the atmospheric parameters are actually smoothed values, averaged over a time step (usually 20 minutes). The computation of the wind input can therefore be considered as consisting of two steps:

a) the transition from the (smoothed) U_{10} to the (smoothed) u_*, which may depend on parameters, such as gustiness, air/sea temperature difference and (wave-dependent) surface roughness,

b) the computation of the actual energy flux into the waves (expressed by source term expressions like (2.43), (2.56) or (3.39)), depending on

u_*, but also on other characteristics of the atmosphere, such as, again, the air/sea temperature difference and the variability of the wind.

We will study the sensitivity of the wind input to details of both steps. First (§ IV.5.1), we analyse the effect of the surface wave conditions on the evaluation of u_* and we see that this has remarkable consequences for the distribution of energy in the spectrum. Next (§ IV.5.2), we address the general problem of wind field variability and finally, in § IV.5.3, we will analyse the role of air/sea stability. Where possible we illustrate the points with specific case studies.

IV.5.1 Wave-dependent drag coefficient

L. Cavaleri, L. Bertotti and P. A. E. M. Janssen

Most wave and surge models assume the drag coefficient C_D to be dependent only on the local wind speed (and possibly stability, see § IV.5.3). Expressions for C_D as a function of U_{10} have been given by Wu (1982), and Smith and Banke (1975). These expressions are based on empirical evidence and implicitly on the hypothesis that the sea conditions are a function only of the local wind. The physical background for this is that the sea surface roughness felt by the atmosphere depends on the energy content of the high frequency range of the spectrum, and these frequencies respond almost immediately to any variation of the local wind. However, the energy content of the highest frequencies depends on the overall energy distribution in the spectrum and is formally represented by the variability of the Phillips constant in the spectral tail. In the typical evolution of a storm, the peak period of the spectrum is initially rather short, with high values for the Phillips constant and the drag coefficient. When the peak then gradually shifts to lower frequencies the value of C_D decreases rapidly. It is clear, therefore, that its dependence should be both on U_{10} and on the wave conditions. Different ideas and proposals have been advanced on the explicit formulation of this dependence, the most recent contributions being by Janssen (1991a) and Donelan (1992), described in § II.2. The standard WAM model (cycle 4) includes a description of this process. Earlier cycles did not. It is therefore interesting to look at the effect by comparing results of the standard WAM model with runs in which the coupling is ignored.

One such comparison was made in figure 3.17. In comparison with the uncoupled case there is a 50% increase in the initial value of C_D, that rapidly disappears within a few hours and then stabilizes at a lower value. As explained in chapter II, the growth rate depends directly on C_D, so that this would lead to higher H_S values in the first period of a storm. In the standard version of the WAM model this has been compensated by a

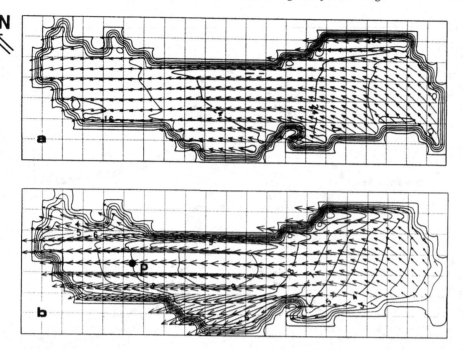

Fig. 4.26. Wind and wave fields in the Adriatic Sea at 00 UTC 1 February 1986: a) wind; isotachs at 4 m/s intervals; b) significant wave height obtained with the uncoupled WAM model. Isolines at 1 m intervals. Copyright: Società Italiana di Fisica.

slightly different formulation of the whitecapping source function, so that the original evolution of the significant wave height was recovered.

Sirocco in the Adriatic Sea (Adria86)

There are, however, more subtle differences that are worth exploring. To this end we consider a strong sirocco storm that occurred in the Adriatic Sea (figure 4.40) in January – February 1986. The wind fields were computed with the wind model of Bergamasco *et al* (1986) and verified against the data obtained from the oceanographic tower present off the coast of Venice (Cavaleri, 1979). The wind field at the peak of the storm is shown in figure 4.26a. The wind is blowing strongly along the main axis of the basin. This is the typical condition for a surge in the northern part (to the left in the figure), and indeed Venice experienced in this case one of the highest recorded floods. The storm has been hindcast with and without coupling. The results of the latter at peak conditions are shown in figure 4.26b.

As explained, the introduction of coupling does not lead to substantial differences in the distribution of H_S. There is a definite but very limited

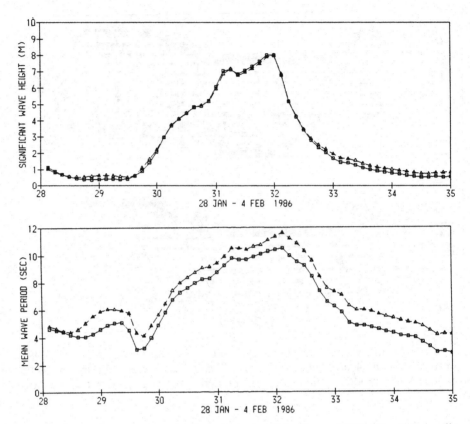

Fig. 4.27. Time series of H_S and T_m at point P in figure 4.26b. Continuous line – uncoupled; broken line – coupled hindcast. Copyright: Società Italiana di Fisica.

increase of the wave height, particularly in the central part of the basin, that is in the area of maximum wave height. It is of interest to analyse in figure 4.27 the time series of H_S and T_m at a single point, P in figure 4.26b. While the two time series for H_S do not differ appreciably, there is a significant difference between the mean periods T_m throughout the storm, the coupled case being higher by more than one second. The explanation comes from figure 4.28 where we compare the one-dimensional spectra at P at the same time as in figure 4.26. The two spectra have similar peak frequencies, but in the coupled case the peak is more enhanced, providing more energy in this frequency range. This is compensated by a lower tail, that is a lower Phillips constant, with the result that the two values of H_S equal each other.

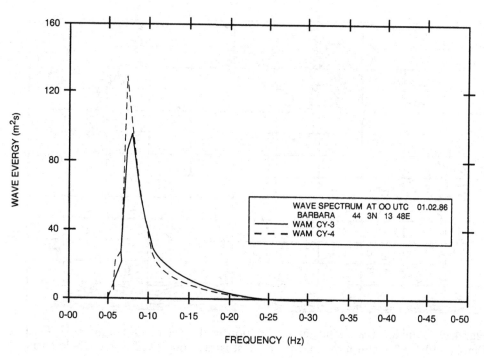

Fig. 4.28. One-dimensional wave spectrum at point P, at the same time as in figure 4.26b. Continuous line – uncoupled, broken line – coupled hindcast. Copyright: Società Italiana di Fisica.

Pacific Ocean (Peru swell)

As explained in more detail in chapter III, the results of the Adriatic storm reflect the characteristics of the coupled and uncoupled models, the different wind input and whitecapping source functions leading in the former case to a higher concentration of energy towards the lower frequencies. As an interesting consequence, particularly for large scale applications, there is a higher percentage of swell in the basin, whose relative absence was a characteristic of many of the previous applications. This is well illustrated with the storm shown in figure 4.29, in which the significant wave height reached almost 14 m to the north-east of Japan. Swell was radiated throughout the Pacific basin. Eight days later this swell was recorded off the coast of Peru. The relative comparison between model and measured data is shown in figure 4.30. We see that the coupled hindcast produces more swell energy, with a definite difference in its period. It also leads to a better fit with the measured data.

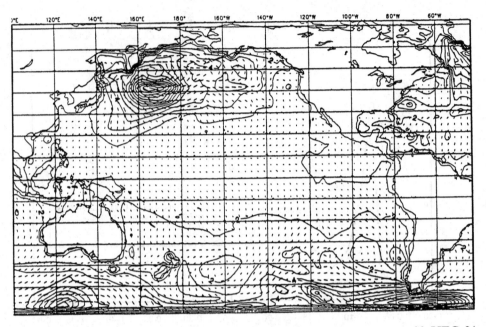

Fig. 4.29. Significant wave height distribution in the Pacific Ocean at 00 UTC 31 October 1989. The storm off the Asian coast is radiating swell in the whole basin. Isolines at 1 m intervals.

IV.5.2 Wind variability

L. Cavaleri

In § IV.2.1, while discussing the general characteristics of a wind field, we noted the presence of random oscillations superimposed on the long term evolution of the field. These oscillations are present with a continuous spectrum, from microturbulence to the scales represented in the meteorological models. In effect, there is no discontinuity between what we consider *random motion*, for instance the fast oscillations shown in figure 4.2, and the *deterministic* variability that we can anticipate by integrating the equations of a model. The distinction arises only from our will or ability to resolve these oscillations. Nevertheless, for practical purposes a distinction between subgrid scale phenomena and resolvable oscillations is unavoidable. The smallest spatial and temporal scale that we can resolve in a deterministic way may depend not only on the resolution of the meteorological model, but also on the density and distribution of the available data. When data are scarce and information on the initial condition is lacking, the model will indeed develop oscillations, which are correct in a

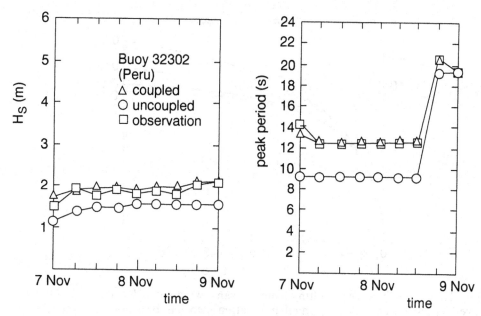

Fig. 4.30. Swell at the coast of Peru seven days after the situation shown in figure 4.29. The coupled hindcast leads to more swell energy and to a better fit with the measured data than the uncoupled case.

statistical sense but wrong in time and place. An estimate of the subgrid variability of the wind field is not normally given by meteorological models. This is unsatisfactory because the wind variability can be quite important, affecting both the rate of growth of waves and the verification of model results.

We will show the importance of this by considering oscillations of different characteristic periods together with their consequences for the evolution of the wave field. First, we analyse the effect of gustiness, defined, as usual, as random wind oscillations with periods up to 20 – 30 minutes, which is of the order of the integration time step of the wave model. We move then to longer period oscillations ('long term variability'), by considering the hindcast of a Pacific storm, in which the oscillations lead to enhanced growth. We also show how one could deal with wind variability in daily applications. We conclude by considering the very slow oscillations (12 – 24 hours) that occurred during hurricane Camille in the Gulf of Mexico. These oscillations are clearly visible in recorded wind and wave data, but, in as far as they are not resolved in meteorological models, we have no way of treating them correctly in wave modelling. The best we can do is to be aware of their existence.

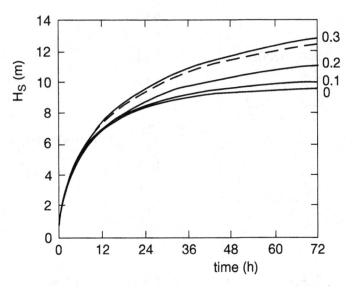

Fig. 4.31. Growth curves resulting from the same wind speed U_{10} but with different levels of turbulence. A very small integration step has been used to smooth the curves. The turbulence is taken as Gaussian with normalized standard deviation $\tilde{\sigma}$. The dashed line for the case $\tilde{\sigma} = 0.30$ shows the consequence of a Gaussian distribution of u_* instead of U_{10} (after Cavaleri and Burgers, 1992).

Gustiness – a synthetic wind case

In § IV.2.1 we have shown that the wind oscillates around its mean value, and that the normalized standard deviation of these oscillations, defined as $\tilde{\sigma} = \sigma_U/U$, varies with the air/sea stability conditions, up to values of 0.3. As the wave conditions are directly related to the generating wind, it is obvious that the oscillations of the wind affect the response of the sea.

The problem is due to the asymmetry of the generation mechanism described by Miles (1957) and explained in § II.2.1. According to Miles, wind can pump energy to a slower wave, while a faster wave remains unaffected. High wind speed fluctuations transfer extra energy to the waves, which is not compensated during phases of lower speed. As a consequence of this 'rectification' mechanism the net energy transfer to waves is increased. (There are suggestions, see, for example, Burgers and Makin (1992), that, when running faster than the wind, wind waves can in effect pump energy into the atmosphere. Even if true, this effect is much smaller than the main Miles mechanism, and its consideration would not alter the conclusions of the present argument.) The process has been studied by Janssen (1986) and Cavaleri and Burgers (1992). The theory is given is § II.2.4.

The overall effect is well represented in figure 4.31 by the growth curves for different $\tilde{\sigma}$ values. These have been obtained in a 'Monte Carlo approach' by running the WAM model with a time step of one minute, forcing it at each integration step with a wind that was randomly chosen from a Gaussian distribution centred around U_{10}, but with different $\tilde{\sigma}$. We see that, with respect to the $\tilde{\sigma} = 0$ case, a $\tilde{\sigma} = 0.10$ has very little effect, but $\tilde{\sigma} = 0.30$ leads to an increase of the final H_S of more than 30%.

The theoretical approach by Janssen (1986) assumes a Gaussian distribution for the $u_\star$ values. Experimental evidence (see, for example, Smith *et al*, 1990) suggests that U_{10} rather then $u_\star$ has a Gaussian distribution. If this is indeed the case, as Cavaleri and Burgers (1992) have assumed, then there is another, second order, mechanism by which gustiness enhances the wave growth, because a symmetrical oscillation of U_{10} leads to a skew distribution of $u_\star$, with $\bar{u}_\star > u_\star(\bar{U})$. In the presence of gustiness this implies a net extra input of energy with respect to $\tilde{\sigma} = 0$ and an enhanced growth rate. The effect is illustrated in figure 4.31 for $\tilde{\sigma} = 0.30$ by the difference between the continuous and dotted lines, the dotted one representing the growth for a Gaussian distribution of $u_\star$.

One may wonder whether the Monte Carlo results obtained so far can be reproduced with the subgrid parametrization given in (2.49). To this end the energy balance equation was solved using this parametrization. Excellent agreement was obtained. In fact, the results for $\tilde{\sigma} = 0.3$ coincide with the dotted line in figure 4.31. This is an important result, because it opens up the possibility of parametrizing the subgrid scale variability in operational applications.

Unfortunately, the overall picture becomes slightly more complicated when we consider in detail the characteristics of a turbulent wind. When analysing records of wind speed U, such as the ones shown in figure 4.2, we soon realize that a Gaussian distribution with mean value $\bar{U}$ and variance σ_U does not fully characterize the time series. A detailed analysis reveals that the sequential values show a strong correlation α over a short period of time. This is still consistent with a Gaussian distribution of the moduli, but it implies that U oscillates around U_{mean} rather slowly, the more slowly the higher the correlation between the sequential values, showing a tendency to remain for a while at values higher or lower than U_{mean}. Figure 4.32 shows four synthetic time series, obtained from the same sequence of random numbers, but with different α values. These oscillations of the wind are reflected in oscillations of the wave field. The actual growth curve of H_S is not as smooth as shown in figure 4.31, but it varies in a random, but coherent way around the smooth curve, following the variability of the forcing wind. Figure 4.33 shows the actual growth curves for $\tilde{\sigma} = 0.25$ and for different correlations α. A time step Δt of one minute has been used. The smooth growth curves have been obtained with $\alpha = 0$. Considering the case $\alpha = 0.90$, we see that oscillations of H_S of the order of 10% are very frequent. We stress that these oscillations are not connected with the sampling variability

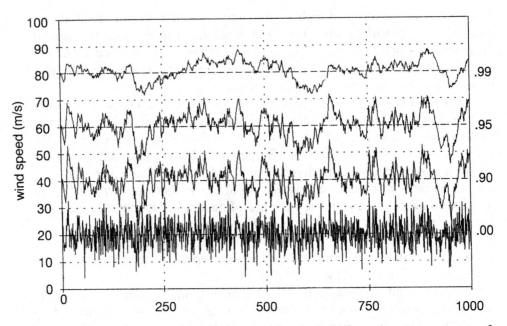

Fig. 4.32. Four synthetic wind speed time series obtained from the same sequence of random numbers, but with different correlations (shown on the right hand vertical scale) between sequential values.

of the measurement of the wave spectrum and H_S, typically lasting 10^3 seconds. The oscillations shown in figure 4.33 represent actual oscillations of the mean local conditions and, as pointed out in § IV.4.1, they must therefore be taken into account when comparing model outputs and measured data.

Both in the Monte Carlo approach used above and in the parametrization based on (2.49) it is assumed that the wind is spatially homogeneous and that the wind fluctuations are coherent. In reality this is not correct, because the oscillations have a correlation that rapidly decreases with distance and becomes negligible already at one spatial step size. Consequently, oscillations at neighbouring grid points are completely independent. In a simulation with the WAM model it was found that the oscillations in H_S are still relevant, but somewhat smaller than suggested by figure 4.33.

The examples of figure 4.33 have been obtained with $\Delta t = 1$ min. In practical applications we have to deal with the time step of the wave model, namely 20 min for the standard WAM model. Because we cannot use the Monte Carlo approach for the subtime step fluctuations, the best we can do is to parametrize these fluctuations by using (2.49) and to simulate the effect of the slower fluctuations in a Monte Carlo approach. This implies a distinction between turbulence with characteristic period T_g larger or smaller than the integration time step Δt of the wave model. Cavaleri and Burgers

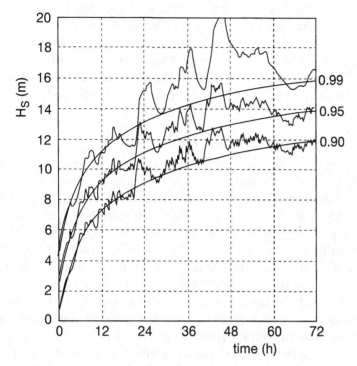

Fig. 4.33. Growth curves resulting from wind with a 0.25 turbulence level. Different correlation values α between sequential wind values have been used, the smooth reference lines corresponding to $\alpha = 0$. The large oscillations result from those of the input turbulent wind (see figure 4.32). There is an up shift (of 2 m) between the diagrams for better visualization (after Cavaleri and Burgers, 1992).

(1992) estimate that most of the variance (80%) is in fluctuations on a time scale above 20 min. Therefore, at each time step, we use a wind value U' derived from a Gaussian distribution centred at the value provided by the meteorological model, with a $\tilde{\sigma}_{20} = 0.8\tilde{\sigma}$ and taking correlations between sequential U' values properly into account. The wind input term is then evaluated with (2.49) with a variance $\tilde{\sigma}' = 0.6\tilde{\sigma}$. Note that the correlation $\hat{\alpha}$ between wind values used in sequential time steps at a given point is higher than α. The U values represent averages over the model time step of 20 min. Cavaleri and Burgers (1992) show that in this case the correlation is given by

$$\bar{\alpha} = \frac{\alpha}{n^2} \left(\frac{1 - \alpha^n}{1 - \alpha} \right)^2 , \qquad (4.2)$$

with n the number of points in the averaged period. For $\Delta t = 20$ min, $n = 20$. Of course, in this Monte Carlo approach, the wind oscillations are only indicative of the oscillations that are actually present at sea. Nevertheless, the approach is promising, because it offers a method for the determination of the statistics of the oscillations. Using different, but statistically equivalent series of random numbers, we could run the model several times to generate an ensemble of hindcasts and forecasts. These would then provide both the mean smoothed wave height curve and the statistical characteristics of its oscillations. So far this has not been done in a systematic fashion.

Pacific hindcast (Hawaiian storm)

A good practical example of the effect of gustiness on the evolution of a wave field is given by a severe storm that occurred in the northern Pacific Ocean in November 1988, which was analysed by Cavaleri and Burgers (1992). A pressure minimum in the area of the Aleutian Islands led to a first storm and to waves propagating towards the south-east. North of the Hawaiian Islands the cold front, moving towards the south, led to the development of a second storm that, acting on the existing background, led to significant wave heights of more than 10 m, as recorded by a waverider buoy north-west of Kauai. Figure 4.34 shows the meteorological situation at 12 UTC 5 November 1988, as seen by the ECMWF analysis. Isolines of wind speed are traced at 4 m/s intervals. The maximum speed is above 28 m/s. The wave hindcast obtained with the standard WAM model is given in figure 4.35a. The large wave area, west of Hawaii, is quite evident in the lower central part of the figure. A comparison between the measured and modelled H_S data is shown in figure 4.36a. The closest grid point has been taken for comparison. There is an evident underestimate of the significant wave height throughout the storm (H_S bias $= -0.39$ m, rms error $= 0.90$ m).

The cold air from the north and the relatively warm water of the Pacific Ocean (the $\Delta T_{air-water}$ was estimated at -10 °C) suggest the presence of a strong instability of the lower atmospheric levels, hence of a strong gustiness of the wind. By comparison with other similar situations the turbulence level was estimated at $\tilde{\sigma} = 0.25$. Given the large distance corresponding to one grid step (180 miles in latitude, 120 – 170 miles in longitude), no correlation in space was assumed for the wind values. A five minute time step was used, which was short enough to ensure that most energy was in oscillations with a period larger than the time step. At each time step and at each grid point the actual wind value was randomly chosen from a Gaussian distribution centred on $\bar{U}$, with $\tilde{\sigma} = 0.25$. The gustiness was uniformly applied to the whole grid throughout the storm. The resulting wave field is shown in figure 4.35b. There is a substantial increase of the whole field, with the peak value from 11 m up to 13 m.

The comparison with the measured data in figure 4.36b is enlightening. The model now reproduces well the passage of the storm, with the bias in H_S reduced to 0.26 m, the rms error to 0.80 m. This is already an improvement

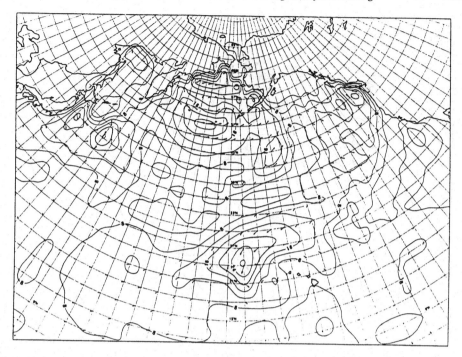

Fig. 4.34. Surface wind on the North Pacific Ocean at 12 UTC 5 November 1988. Hawaii is in the lower centre of the figure. Isotachs at 4 m/s intervals.

over the original values. But even better results can be obtained, when details of the evolution of the storm are considered, because then it becomes clear that the wave conditions in the Hawaiian area were not affected by the air turbulence until the arrival of the cold front, or better until that of the 'forerunners': the long waves anticipating the storm. The comparison with the measured data has therefore been repeated using the no-gustiness results until 00 UTC 4 November 1988, and only subsequently the 'gusty' ones. The bias is thus reduced to 0.09 m, the rms error to 0.70 m.

It is worthwhile to point out that other factors could potentially be called upon to justify the discrepancy between model and measured data. The same air/sea instability that leads to gustiness leads also to an increased wind stress at the sea surface. Or, given the large spatial gradients present in the field, a small error in the location of the storm could easily justify the difference. Whichever the case, the main message here is that wind gustiness affects the wave growth, and can lead to substantial differences with respect to the not gusty case.

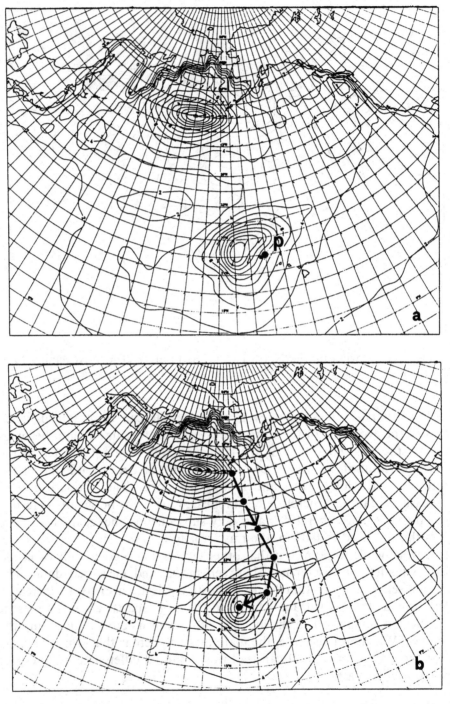

Fig. 4.35. Wave field in the North Pacific Ocean at 12 UTC 5 November 1988, a) without and b) with wind gustiness taken into account. The dots show the track of the storm at 12 h intervals. Point P in a) shows the position of the measurements. Isolines at 1 m intervals.

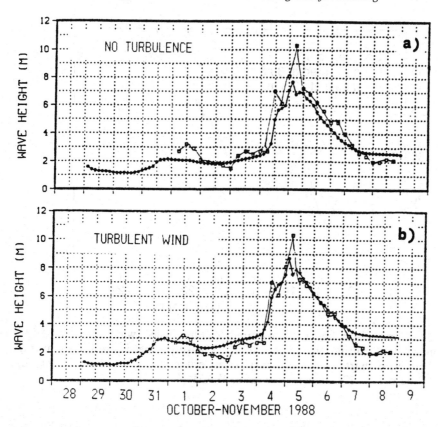

Fig. 4.36. Comparison between measured (thin line) and hindcast (heavy line) significant wave heights west of Hawaii a) without and b) with wind gustiness taken into account.

Larger scale variability (Camille)

If we analyse the results of a meteorological model, we find that, off the coast and away from the peaks and the fronts of the field, the distribution of the various parameters is usually rather smooth, as is their time evolution. In particular the wind at a given location increases steadily as a storm approaches, and then decreases steadily to lower values once the storm has passed.

This is not what we experience in the field. Measured data show large variations around the general trend of growth or decay, on a scale that varies from the time interval of the measurements up to half a day or so. Longer oscillations are considered as part of the evolution of the field. The wind recorded on a rig during hurricane Camille (figure 4.37) provides a good example. Note that the amplitude of the oscillations reaches 30 – 40% of the mean value. The dynamics of these large scale oscillations (with periods

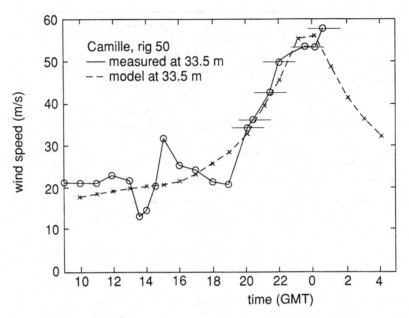

Fig. 4.37. Comparison between measured and modelled wind speed at Rig 50 anemometer level during hurricane Camille. Measured speeds are hourly averages taken from strip chart. Horizontal bars estimate uncertainty of timing in the chart record. Date is 16 – 17 August, 1969 (after WAMDI, 1988).

of a few hours, corresponding to a spatial scale of the order of 10^2 km or more) is not completely clear and is not, at present, modelled in numerical simulations of the atmosphere. A meteorological model acts in this respect as a low pass filter. However, these oscillations have a noticeable effect on the evolution of the wave field. To prove this we consider the corresponding wave hindcast and measured values.

Camille (16 – 17 August, 1969) was one of the most severe hurricanes recorded in the Gulf of Mexico. For wave modelling it was fortunate that Camille passed close to observing stations, which recorded exceptionally high wave conditions until the system failed at the peak of the storm (this is the reason why the measured data in figure 4.37 stop soon after 00 UTC and slightly earlier in figure 4.38). This provided excellent data to verify the wind and wave model results. The hurricane was hindcast with the WAM model (WAMDI, 1988), and the results for two stations, located close to the Louisiana coast on the left of the eye track, are shown in figure 4.38. The vertical bars indicate the 90% confidence limits of the measurements. The agreement is excellent. Even the peak value at station 1 is reproduced well.

In a hurricane, wind and waves are highly variable. The Camille results indicate well how a third generation model can handle such very complicated

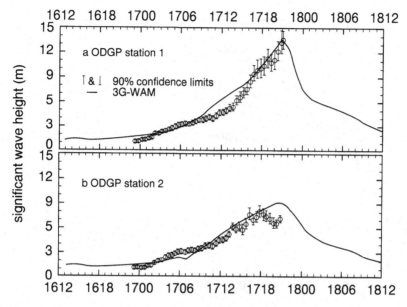

Fig. 4.38. Comparison of measured and hindcast significant wave height at ODGP deep-water measurement stations 1 and 2 in hurricane Camille (after WAMDI, 1988). On the *x*-axis time is labelled by date and hour. The vertical axis gives the significant wave height.

situations. The evolution of the average wave height is computed correctly, but the oscillations of the recorded H_S deviate from the smooth growth in the model. The observed H_S oscillations follow the observed wind oscillations to some extent (Rig 50 is station 1 in figure 4.38), but are proportionally smaller than those of U, which is in agreement with the arguments expressed above on the spatial coherence of the oscillations. As discussed in § IV.4.1, these oscillations make it impossible to reach a perfect fit, with zero rms error, between model and measured data. Also, extremes can be higher than computed. This is important and must be taken into account when we estimate the likely extremes at a given location.

IV.5.3 Atmospheric stability

L. Cavaleri

The effect of air/sea temperature difference has been discussed in various places. A temperature difference may enhance or reduce gustiness, and hence wave growth, but in addition, it has a direct effect on wave generation, for the following reasons:

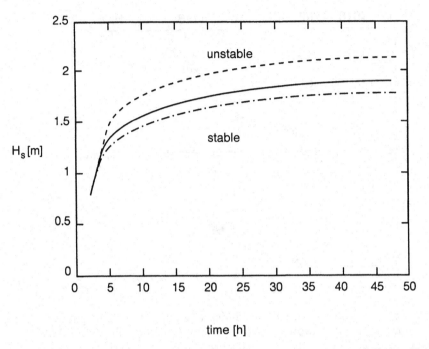

Fig. 4.39. Wave growth for different air/sea temperature differences (after Voorrips *et al*, 1994a).

a) when one computes u_* from U_{10}, the stability of the boundary layer (and hence the air/sea temperature difference) enters through the Obukhov length L (see § I.3.7),

b) theory suggests that the wind input source term depends not only on u_*, but also explicitly on the dimensionless Obukhov length L_*.

Voorrips *et al* (1994a) found that in practical cases effect a) is normally dominant. Given this fact, they could quantify the effect by simply rescaling the standard WAM model growth curve with the appropriate stability-dependent relation between U_{10} and u_*. The result is shown in figure 4.39, where the wave height is given as a function of fetch for a given U_{10} and for different stability conditions. The effect is about 10% of the wave height.

It is rather remarkable that we are not able to illustrate the model sensitivity to air/sea stability with hindcast results. In the ideal case one would expect to have two hindcasts, one with u_* derived from ten metre winds using neutrally stable drag, and one using the appropriate stability-dependent wind input source term. Then one would like to demonstrate the superiority of the latter by comparing wave height predictions with observations. So far this has not been done, because the air temperature over sea is still poorly computed in meteorological models, which inhibits

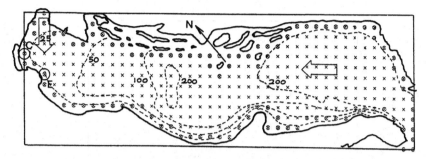

Fig. 4.40. WAM grid on the Adriatic Sea. The grid step is 20 km. Isobaths are in metre. Croatia is at the top of the figure, Italy at the bottom. The north is towards the upper left corner. The large arrows show the two main winds in the basin.

useful applications. That the effect may be important, nevertheless, is illustrated by an example considered by Voorrips *et al* (1992, 1994b). They selected situations characterized by strong stable or unstable conditions in the North Sea, for which measurements were available from a tower. For these situations the NEDWAM model (Burgers, 1990) gave an underprediction of wave height, when driven with winds from the operational atmospheric LAM. In that model, wind observations from the top of the tower had been assimilated after conversion to 10 m height without the appropriate stability-dependent corrections. By taking air/sea stability properly into account, the consequent correction of the values of the surface wind U_{10} was sufficient to justify the underestimate of the wave height.

We end this discussion with 'circumstantial evidence' for the importance of air/sea stability. Some of this evidence comes from the northern Adriatic Sea (figure 4.40). This area is characterized by two main winds (see Cavaleri *et al*, 1989). The bora, a cold and dry wind, blows from the north-east; the sirocco, usually warm and humid, comes from the south-east, along the main axis of the basin. The different stabilities associated with the two winds are evident from their analogue chart records (figure 4.2) by the amplitude of the oscillations around the mean wind speed. In figure 4.2 the left record is of sirocco, all the other ones are of bora. Unstable conditions imply a large vertical exchange of mass, and consequently strong turbulence. The local fishermen claim that the bora 'grasps' the waves, while the sirocco 'slides' over them. The two different levels of interaction are reflected in the different growth rates and, at a given instant, in a different wave steepness. Analysis of the long term time series of H_S available on the local oceanographic tower (Cavaleri *et al*, 1981) has shown that the waves produced by the bora are on the average 20% steeper than those produced by the sirocco. This is true also when the sirocco is blowing only over the northern part of the basin, so excluding the different fetches as the only reason for the different steepness.

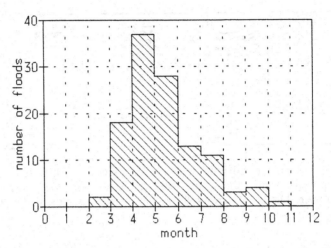

Fig. 4.41. Number of floods in Venice throughout the year (beginning in July). Period from 1923 until 1989. After Camuffo (1984). Reprinted by permission of Kluwer Academic Publishers.

This fits in well with our theoretical understanding: for given fetch and U_{10}, unstable situations give larger u_* and hence younger wave age and thus steeper waves.

Other suggestive statistical evidence on the effect of air/sea stability is given by the distribution of floods in Venice (located on the far left in figure 4.40). These floods are associated with the surge produced by the sirocco. Similarly to wind wave generation, a storm surge depends on the stress on the surface, hence on the air/sea stability conditions. The statistical distribution of the forcing function, that is of the wind speed, and in general of the weather patterns, is characterized by two strong peaks, in the autumn and in the spring (see Camuffo, 1984). However, this distribution is not reflected at all in that of the floods that, as seen in figure 4.41, show a dominant concentration in the autumn. Autumn and spring are characterized by quite different stability conditions: stable in the spring, because of warm air and cold water, with $\Delta T = 3$ to $5\ °C$ and highly unstable in the autumn, because of cold air and warm water, with $\Delta T = -10$ to $-12\ °C$. While other aspects of the overall phenomenon are probably playing a role in the asymmetry of the distribution of the floods, the correlation with the variation of the air/sea stability in the year is suggestive of its effective role in determining the level of a flood, hence on the forcing function of the wind. Because of the similarity of the storm surge and wave generation processes, it is natural to extend the same conclusion also to wind waves.

IV.6 Effects of bottom dissipation

In shallow water, wind waves interact with the bottom and a number of phenomena take place. Shallow water areas form a minor fraction of the total ocean surface. They are limited to the continental shelf and not normally represented in global wave models. However, they are also the areas where human activity is concentrated. It is therefore of interest to explore the sensitivity of the results of a wave model to the various formulations used to represent shallow water phenomena.

In this section we deal with the shallow water processes that lead to dissipation of energy by bottom friction, percolation and viscous bottom motion. Of the three, percolation, discussed in § II.5, is associated with the presence of coarse sand or shingle on the sloping bottom close to the coast. The presence of relatively large material makes these slopes very steep, and therefore of limited extent in the direction perpendicular to the shore. It is therefore obvious that percolation is potentially relevant only for limited area wave models. Its importance in comparison with other processes can be estimated from the diagrams of Hsiao and Shemdin (1978). Usually percolation is much less important than bottom friction. Occasionally, it can be of the same order of magnitude. A precise quantification is hampered by the lack of sufficient experimental data. In any case, the actual energy loss would be highly dependent on the size and shape of the bottom material. The wide range of possibilities makes it difficult to provide a general rule. In practical cases one has to resort to specific tests on the permeability of the material one is dealing with. We will therefore limit our further discussion to the other two processes mentioned above. In § IV.6.1 we analyse the effect of different formulations for bottom friction on the wave model results. In § IV.6.2 we explore the potentially dominant role of the viscous dissipation at the bottom. The results for a viscous bottom have not been obtained with a third generation model, but we include them, nevertheless, because in this case the dissipation is so large that it overwhelms all other terms in the energy balance.

IV.6.1 Bottom friction

L. Cavaleri and S. L. Weber

The energy loss by bottom friction arises from the relative motion between the bottom and the horizontal alternating free stream velocity close to it. A thorough description of the physics of the phenomenon and of the related theoretical approaches is given in § II.5. Bottom friction and, in general, bottom processes are usually neglected in a global or large scale wave model. At resolutions between one and three degrees the continental shelf is very limited in space, and in any case its description is too approximate to allow a proper evaluation of the energy loss due to the limited depth. This is usually dealt with by local models, the necessary resolution being of the same order as the spatial scale of variability of the characteristics of the bottom.

In this section we consider two different cases, one in the North Sea, the other in the Adriatic Sea, both representative of the effect of bottom friction on the results and of the differences arising from the different approaches.

North Sea (Texel storm)

In § II.5 we have discussed three approaches to the evaluation of the energy loss by bottom friction: the linear expression suggested by JONSWAP (Hasselmann *et al*, 1973); the nonlinear formulation with a drag law by Hasselmann and Collins (1968); and the eddy viscosity approach by Weber (1989), henceforth referred to in this section as J, H-C and E-V. All of these have been applied by Weber (1991a) to a very severe storm that occurred in the southern North Sea on 2 – 3 January 1976. An extensive description of the Texel storm, so named from the name of the West Frisian island where the storm reached its peak, is given by Harding and Binding (1978). The storm was characterized by a depression that, while moving from Scotland towards the eastern side of the North Sea, produced extended and prolonged strong northerly winds blowing directly towards the Dutch coast with speeds between 25 m/s and 30 m/s. At Texel, close to the island, significant wave heights between 6.5 m and 7 m were recorded for more than 15 hours. The storm was hindcast with the standard shallow water version of the WAM model, that is with the bottom friction evaluated with expression J. Figure 4.42 shows the 75 km resolution grid used for the hindcast. Measurements were available at Eurochannel and Texel positions, respectively in 25 m and 30 m of water depth. Their locations are shown at the bottom of the figure. The input wind was obtained by careful manual analysis of the storm, taking into account all the available data. Wind data for this storm must be considered of high quality, providing a very good chance to verify different approaches in a wave model. No consideration was given to refraction because the 75 km step size was not considered sufficiently detailed to resolve the spatial variability of the bottom topography.

The comparison between the hindcast results and the measured H_S is given in figure 4.43a,b, respectively at Texel and Eurochannel locations. There is an evident overestimate by the model results, indicated by J in the figure, particularly at Texel.

The hindcast has been repeated using both the H-C and the E-V approach. The two results, shown again in figure 4.43, do not differ much from each

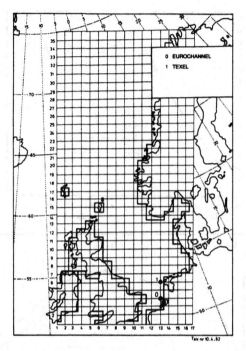

Fig. 4.42. The grid covering the North Sea for the hindcast of the Texel storm. The grid step is 75 km.

other and are much closer to the measured data. As all the other parts of the model were left untouched, the differences are clearly due to the different formulations of the bottom friction.

The J expression seems to underestimate the energy loss considerably. However, this is not always the case. Weber (1991a) has also analysed the case of low-frequency swell ($H_S = 1 - 2$ m, peak frequency 0.06 Hz), arriving at the Dutch coast from the Norwegian Sea. There was no wind in the southern North Sea, and the low average wave slope made the whitecapping and nonlinear interaction terms virtually negligible. In practice, the only active term was bottom friction. In that case, all three approaches gave good and similar results.

The explanation lies in the different bottom (horizontal) velocity present during the two storms. Cavaleri and Lionello (1990) have analysed the loss evaluated according to J and H-C in two ideal cases, a pure swell and a JONSWAP spectrum, propagating towards the shore with a uniformly sloping bottom. They found the difference between the two approaches to depend on the average $<U>$ of the modulus U of the bottom horizontal velocity. If $<U>$ is larger than a given critical value U_{cr}, the estimate of the loss by J is smaller with respect to H-C; the opposite is true when $<U>$ is smaller

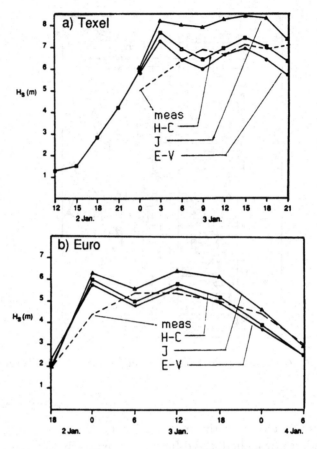

Fig. 4.43. Comparison between measured and hindcast significant wave height for the Texel storm in January 1976 at a) Texel (1) and b) Eurochannel (0) locations in figure 4.42. Different expressions for bottom friction have been used: J = JONSWAP, H-C = Hasselmann and Collins, E-V = eddy viscosity coefficient (after Weber, 1991a).

than U_{cr}. The exact value of U_{cr} is actually dependent on the directional distribution of energy in the spectrum and varies between 0.14 m/s and 0.18 m/s. Note that this holds in the case of no current. If a current with speed comparable to or larger than the above values is present, conditions are quite different. This possibility is not considered here. Applications with interactions between waves and current are described in §§ IV.8 and IV.10.1.

Quite different conditions were present during the Texel storm and the swell case discussed by Weber (1991a). For the former she estimates $<U>$ at about 0.7 m/s (actually the estimate was done on U_{rms}; we have here assumed an approximate ratio $<U>/U_{rms} = 0.9$). On the basis of the results of Cavaleri and Lionello (1990) quoted above, $<U>$ being much larger

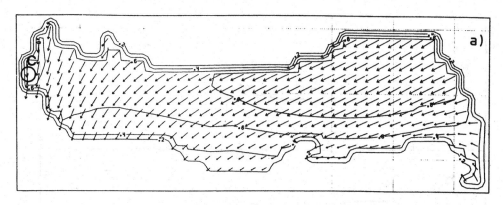

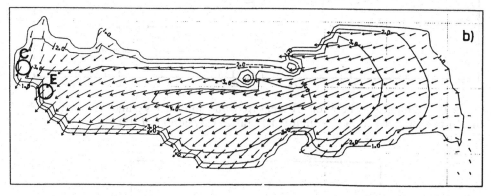

Fig. 4.44. a) Wind and b) wave fields in the Adriatic Sea at 00 UTC 2 December 1982. Wind friction velocity is shown. Isolines at 0.2 m/s and 1 m intervals respectively (after Cavaleri *et al*, 1989).

than U_{cr}, the J approach leads to an underestimate of the loss by bottom friction, hence to an overestimate of H_S. This is clearly the case in figure 4.43. In contrast, in the swell case, $<U>$ is reported to have values between 0.1 m/s and 0.2 m/s. This implies similar estimates of the loss by all three approaches, and therefore, consistently with the results of Weber (1991a), similar estimates of the significant wave height.

Adriatic Sea (Adria82)

Another clear example of the lack of dissipation in J is given by Cavaleri *et al* (1989) with the hindcast of a sirocco storm in the Adriatic Sea. The storm was not very intense, but prolonged winds blowing along the axis of the basin produced a large swell that in the northern part (to the left in figure 4.44) came across an active transverse wind. Figure 4.44 shows the wind and wave fields at 00 UTC 2 December 1982. Measurements were available at points C (wind) and E (wave). The storm was hindcast with the standard WAM

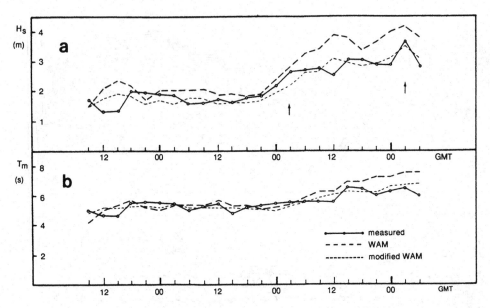

Fig. 4.45. Comparison between a) measured and hindcast significant wave height and b) period at position E in figure 4.44 (after Cavaleri *et al*, 1989).

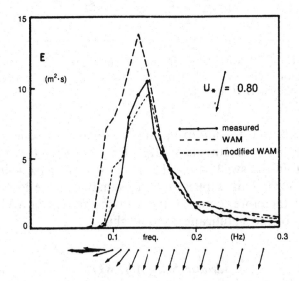

Fig. 4.46. One-dimensional spectrum at position E in figure 4.44 at 00 UTC 2 December 1982. The lower arrows show the mean direction for each frequency. The single arrow indicates the local wind (after Cavaleri *et al*, 1989).

model, that is with J, and then repeated with H-C. The comparison with the data measured at position E is shown in figure 4.45. The overestimate by the J approach is evident, both in the significant wave height and in the period. The results are even more convincing when we consider in figure 4.46 the one-dimensional spectrum at the time of figure 4.44. The variation of the mean direction with frequency indicates the presence of severe cross sea conditions. In the low-frequency range the waves are along the axis of the basin, while at high frequencies they are in the local wind direction. There is an evident overestimate by J in the range where the bottom friction is actively absorbing energy. The two hindcasts almost coincide for $f > 0.14$ Hz, where in 20 m depth, bottom friction has negligible effects.

For practical applications it is worthwhile to point out that for $<U> < U_{cr}$, and particularly for short distances, the energy loss by bottom friction is very limited, and the overestimate by J has little effect on the final results. Cases for which the effect becomes relevant are all characterized by $<U> \gg U_{cr}$ and extended distances. In such situations the use of H-C or E-V is strongly recommended.

IV.6.2 Bottom elasticity

L. Cavaleri

In the introduction of § II.6 we hinted at the possibility of strong energy dissipation at the bottom when viscous material is present. The usual bottoms consisting of sand or stiff clay have very limited elasticity, which admits small vertical oscillations only. These oscillations are in phase with the forcing function, that is with the surface water wave, so there is no or very little energy absorption. In contrast, in the presence of elastic viscous material, the action of waves generates large out of phase bottom oscillations (mud waves) and a strong absorption of energy results. Experimental evidence, which is scarce, comes from the consequences of the phenomenon rather than from its direct verification.

In a study over several years in the Mississippi delta, Forristall and Reece (1985) have obtained quite remarkable data. Their measurements were made at two platforms, Cognac, located offshore, in a depth of 300 m, and VV, closer to land in a depth of 19 m. The distance between the two platforms is 30 km. The most spectacular case was hurricane Frederic (11 – 12 September 1979). While significant wave heights up to 8.6 m were experienced at Cognac, the maximum H_S at VV was 2.45 m. The analysis by Forristall and Reece (1985), with the inclusion of shoaling and refraction, shows that the usual bottom dissipation processes, friction, percolation and vertical oscillations, cannot account for the measured attenuation. This case

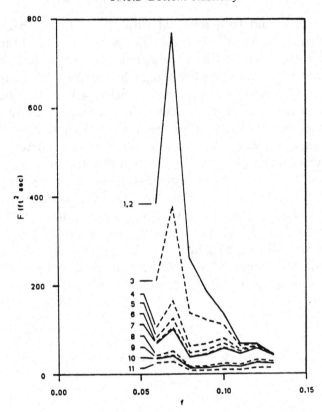

Fig. 4.47. Attenuation of the wave spectrum during hurricane Frederic at platforms Cognac (curve labelled 1,2) and VV (labelled 11). The other curves represent the spectrum at intermediate locations. Reproduced, with permission, from Forristall *et al* (1990).

was parallelled by other events of lower wave heights, where the decay of energy from Cognac to VV was well described by shoaling and refraction. This suggested that the efficiency of the process increases when the significant wave height exceeds a certain level. In the area between the two platforms there are large deposits of soft mud (mud flow gullies and lobes) up to 30 m deep. Therefore, Forristall and Reece (1985) conclude that these deposits are responsible for the attenuation.

A quantitative explanation has been given by Forristall *et al* (1990) by modelling the interaction of water waves with a soft bottom composed of visco-elastic layers. As the bottom properties were known only at the platform VV, the characteristics of the soft mud in the gullies between the two platforms were deduced by tuning the model to reproduce the results of hurricane Frederic. The model was then verified against the other available measurements with highly satisfactory results, with errors at VV of the order of 10% for the energy loss at a given frequency. The spectacular attenuation between Cognac and VV is clearly shown in figure 4.47 by the variation of the wave spectrum along the eleven steps into which the distance between the two towers has been divided. Note that, in this case, by taking only shoaling and refraction into account, Forristall *et al* (1990) had been able to justify only a minor fraction of the attenuation between Cognac and VV.

In cases like Frederic the wave height attenuation on the soft mud areas reaches 30% per km. This contrasts with the cases of low wave height (1 m or 2 m at Cognac), when very little attenuation was observed. In the model of Forristall *et al* (1990) this is related with the nonlinear behaviour of the soil. Large wave pressures lead to large strains, which reduce the shear modulus leading to even larger strains, with an extremely high damping of waves.

We may conclude that dissipation by mud waves may be very important and should be considered when mud is present. However, its importance depends on very special conditions. There should be unconsolidated layers of soft mud and these are found only in certain areas, the south-west coast of India being another notable example. When the geographical extent of the mud is very limited, as in the Mississippi Delta, the problem can be dealt with on a local basis. Therefore, we can neglect the process in standard wave modelling, but we need to keep in mind that we must consider it when running local models in some special areas.

IV.7 Shoaling and depth refraction

L. Cavaleri, R. Flather, S. Hasselmann and X. Wu

In the previous section we have considered processes that dissipate energy in shallow water. In this section we consider shoaling and depth refraction, two processes which affect the spectral and spatial energy distribution of the wave field without changing the overall energy budget.

Another conservative bottom interaction process, scattering by random bottom irregularities, was discussed in § II.6. There it was pointed out that this potentially relevant process has not received much attention from wave modellers because of its heavy computational demands and the difficulty of acquiring the necessary input data in the form of the two-dimensional bottom topography spectrum. Moreover, its verification requires very careful and detailed directional wave measurements. We shall therefore consider only shoaling and refraction in the following.

In numerical wave modelling the processes of shoaling and refraction are often referred to by the single name of *refraction*, which then comprises both *shoaling* and *angular refraction*. Taking for simplicity a Cartesian rather than a spherical grid, the kinematic part of the energy balance equation (3.11) for the propagation of a two-dimensional wave spectrum $F(f, \theta)$ in water of spatially variable but time-independent depth (for which the frequency remains constant) reduces to

$$\frac{\mathrm{d}}{\mathrm{d}t}F = \frac{\partial}{\partial t}F + \mathbf{c}_g \nabla F = \frac{\partial}{\partial t}F + (\nabla \mathbf{c}_g)F + \frac{\partial}{\partial \theta}(\dot{\theta}F), \qquad (4.3)$$

where $\mathbf{c}_g$ is the group velocity, $\dot{\theta}$ denotes the rate of change of the direction of a wave component following the path of the wave component and we have excluded the presence of currents. The first term on the right hand side of (4.3) represents the change in the spectrum due to the divergence $\nabla \mathbf{c}_g$ of the group velocity and is termed the *shoaling* contribution. The second term is the *angular refraction* contribution. On a spherical grid, angular refraction terms also arise in deep water through the change in wave propagation direction along a great circle path.

For quantitative studies of refraction in the general case of a two-dimensional bathymetry, the distinction between shoaling and angular refraction is not particularly useful. It is more convenient to regard refraction as an integral process involving the propagation of wave packets in four-dimensional $\mathbf{x}, \mathbf{k}$ space, as discussed in § III.2. Thus the concepts of shoaling and angular refraction are not usually invoked in numerical refraction studies, which typically involve complex two-dimensional bathymetries. We shall similarly not distinguish between these processes in the numerical examples discussed below and elsewhere in this chapter.

However, shoaling and angular refraction are useful concepts for the simple case of a one-dimensional geometry, i.e. when the depth contours are parallel and the wave field is time-independent and homogeneous in the direction parallel to the isobaths; the wave spectrum itself can be two-dimensional. In this case the refraction equations can be integrated explicitly and the interpretation of the shoaling and angular refraction terms becomes particularly simple. Even when this simplified geometry does not strictly apply, however, the concepts of shoaling and angular refraction can still be helpful in qualitative discussions of refraction phenomena. For this reason shoaling, in particular, is a much used concept in coastal engineering.

In the one-dimensional and time-independent case, (4.3) becomes

$$\frac{\mathrm{d}}{\mathrm{d}x}(c_g \cos \theta F) = -\frac{\partial}{\partial \theta}(\dot{\theta}F), \qquad (4.4)$$

where x is the coordinate orthogonal to the depth contours ('normal to shore') and θ is measured relative to the x direction.

We consider first the effect of shoaling. To exclude the angular refraction term we can consider the case of propagation in the normal-to-shore direction. In this case (4.3) can be immediately integrated to yield (writing F to denote either $F(f,0)$ or $F_1(f)$),

$$Fc_g = \text{energy flux normal to shore} = \text{const} \tag{4.5}$$

corresponding to (3.26), but using F instead of the action density N.

The effect of shoaling is easy to understand. As the waves travel from, say, deep water into shallower water, the group velocity, defined by equation (1.71), and explicitly given by

$$c_g = \frac{c}{2}\left(1 + \frac{2kh}{\sinh 2kh}\right), \tag{4.6}$$

initially increases slightly and then decreases, approaching zero for zero depth. Since the energy flux remains constant, this implies that after a minor initial decrease, shoaling increases the wave energy density and the wave steepness, as the waves propagate towards shore. Close to shore, the wave steepness rises very sharply, and nonlinear effects can no longer be regarded as weak. Consequently, the linear superposition principle – the foundation of wave spectral modelling – is no longer a valid first approximation and wave modelling as discussed here breaks down. Although important for coastal engineering, the excluded near coastal region is very narrow – essentially the spilling-breaker or surf zone – and has negligible impact on the region in which standard spectral wave models are applicable. In the surf zone quite different (normally deterministic) wave models are needed. These can be driven by the output of wave spectral models, which yield the input boundary conditions at the edge of the strongly nonlinear region, but are otherwise independent of spectral models.

The pure effect of angular refraction, or simply refraction, as it is commonly referred to in coastal engineering, is associated with the redistribution on a different width of the energy flowing between two adjacent rays, following a change of the wave propagation direction. The classical case is that of a δ-function spectrum propagating at an angle θ to the normal-to-shore direction. As the waves travel into shallow water, refraction bends the rays towards the normal-to-shore direction in accordance with Snel's law (usually called Snell's law, but the name of the Dutch mathematician was Snel or Snellius)

$$\frac{c}{\sin\theta} = \text{const}, \tag{4.7}$$

where c is the phase velocity. The bending into the normal incidence direction leads to a widening of the distance between adjacent rays, hence to a decrease of the overall energy level.

Hubbert and Wolf (1991) have tested whether the WAM model correctly reproduces Snel's law. To this end they studied the idealized case of a one-dimensional sloping bottom, rising from 80 to 15 m depth within 6.5 degrees

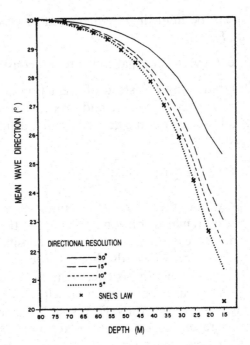

Fig. 4.48. Refraction on a sloping bottom as a function of directional resolution (after Hubbert and Wolf, 1991).

of latitude, approximately 720 km. This represents a very mild slope of $1 : 10^4$. They used a grid step size of 0.5 degrees and the standard frequency resolution. The input was given by a hump of swell with a mean period of 10 s, coming from deep water under an angle of 60° with the isobaths. In figure 4.48 the results of the test are compared with the exact solution of Snel's law. Excellent agreement was obtained with a directional resolution of $\Delta\theta = 5°$. Coarser resolutions introduced errors, but they remain very limited for $\Delta\theta < 15°$. As expected, waves are refracted mostly in the shallowest water, where the spatial gradient of the phase speed is larger, with only small change of direction in the deeper water (50 – 80 m). The slope used in this test is actually quite small and characteristic for a large shallow basin. Close to shore we must expect slopes at least one order of magnitude larger. This is likely to decrease the accuracy of the numerical evaluation of refraction, or, in other words, if we want to keep the same accuracy, we need to increase the grid resolution.

In general, we have to deal with a complicated bottom topography and with a continuous energy spectrum. In this case the uneven distribution of depth in shallow areas leads to local convergence and divergence of the wave rays and a corresponding local increase and decrease of energy. In some particular cases this can lead to spectacular effects, an historical example being the high waves experienced in 1930 by a short segment of the Long Beach, California, breakwater. These high waves, in an otherwise calm environment, were caused by a local hill on the bottom, which happened to focus long swell from a remote storm in the southern hemisphere occurring many days earlier (Bascom, 1980, p84). However, such events are very rare and happen only for nearly monochromatic waves. In general, different frequencies tend to converge in different areas, while nonlinear interactions among different wave components further blur the effect.

In a theoretical approach the depth of the basin is sometimes taken as a random distribution. Pure ('optical') refraction then leads to a random distribution of rays. This is comparable with the scintillation of stars, when random fluctuations in the atmosphere create fluctuations in the light received at a particular point of observation. Theoretically, in the case of ocean-wave refraction, small fluctuations in direction and frequency of incoming waves, combined with spatial bottom variations, can lead to large, chaotic fluctuations in swell intensity and direction at a particular location (Graber *et al*, 1991). This scintillation phenomenon is not yet completely understood.

It should be stressed that it is difficult, in practice, to resolve refraction effects in sufficient detail. The characteristic length scale of bottom variations may be as small as 100 m. For this reason operational models sometimes prefer to ignore refraction, simply because their resolution, typically from 50 km upwards, is too coarse.

What in practice can be achieved with the WAM model is well illustrated by figure 4.49, comparing two spectra evaluated with and without refraction at a point at a depth of 20 m near the North American Atlantic coast, just south of Cape Hatteras, in an area with steep depth gradients. These spectra were obtained in a hindcast of the SWADE storm by running the WAM model on a hierarchy of nested grids (discussed in more detail in IV.3.2). In the figure, swell is running in south-westerly direction, while local wind waves are generated by a northerly wind. In this situation the most striking effect is the large change in the swell direction of 45° towards the coast. On closer inspection it can be seen that the amount of turning decreases with increasing frequency and decreasing wavelength. When we consider the local wind wave part of the spectrum, it is clear that there is no large effect of refraction. There are minor differences, but these may be explained by details in the interactions (such as whitecapping) between wind-sea and swell.

Onvlee (private communication) has made a systematic comparison of a run of the WAM model with and without depth refraction. The run was made for the North Sea on a fine mesh grid and covered the first four months of 1993. Results were compared systematically with observations. The effect on the statistics of the model performance was small, but consistently positive. On the average, the rms error was one centimetre smaller when refraction

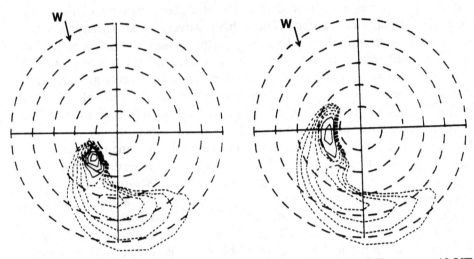

Fig. 4.49. Two-dimensional spectra computed during the SWADE storm at 18 UTC 27 October 1990, with (right) and without (left) refraction included. The location is slightly south-east of Cape Hatteras, near the lower boundary of the high-resolution grid in figure 4.12. The arrow indicates the local wind direction.

effects were incorporated. A detailed analysis of time series showed that most of the time refraction had no effect at all, but when there was an effect it was of the order of decimetres and it always brought the model results in closer agreement with the observations.

The problem of depth refraction may become complicated when the bathymetry changes with time. The largest features of the seabed, on the scale of tens of kilometres or more, have a typical lifetime of the order of several years (see, for example, de Moor *et al*, 1989), but smaller scale features can change in a shorter time, invalidating the results of earlier refraction calculations. If the actual topography is not known with sufficient accuracy, then the results of refraction calculation will at best be indicative of the variability in space of the wave characteristics. This should be taken into consideration when analysing model results.

Another complication arises in tidal basins, when the effective depth is a function of tidal phase. This situation will be considered in more detail in § IV.10, as part of our discussion of the interaction of wave and tidal models. There we will show how this can have significant consequences for the computed wave height.

IV.8 Wave–current interactions

When waves propagate on a variable current several mutual interactions occur. This can be described with a combined wave–current–surge model. A description of the theoretical aspects can be found in Jonsson (1990). The numerics was discussed in § III.2. Some related results will be given in § IV.10. In this section we discuss the consequences for the wave field from the point of view of wave modelling and under the assumption of slowly varying and limited current speed. This condition, $U < c_g$ in practice, is satisfied in almost all practical cases. Exceptions occur when the current reaches values of the order of several knots, which may happen in the estuary of large rivers like the Columbia river. In that case the interactions can be much more intense. Wave components in the high frequency range may break or be blocked by a strong adverse current. This possibility is not considered here, although an application of the WAM model also in this range is favoured by the transformation to a system moving with the current.

We begin (§ IV.8.1) by analysing how the parameters that characterize a wave field change when the waves interact with a current field. Then (§ IV.8.2) we address the case of a large scale unsteady current, such as tidal flow in a large basin. In the final two subsections we consider the opposite situation of a steady current with strong spatial variation; in § IV.8.3 we treat the effects of an idealized Gulf Stream eddy on idealized swell and wind seas and we end in (§ IV.8.4) by discussing a real case taken from SWADE.

IV.8.1 Basic concepts

L. Cavaleri and H. L. Tolman

When considering waves on currents it is important to distinguish between two different reference systems, an absolute one (fixed in space) and a relative one (moving with the current). The first system is relevant for fixed structures in the sea, the second for free floating vessels, for example ships. The wave period changes from one system to the other. We will refer to the absolute period T_a and the relative period T_r in the fixed and moving system, respectively. These correspond to the absolute and intrinsic frequencies ω and σ, which appeared in equations (1.60) and (3.4) describing the Doppler shift. Wave height and wavelength are unaffected by the choice of reference system. The wave period changes from one system to the other.

Currents affect waves through their spacial and temporal variability. Otherwise the effect could be removed by a simple uniform translation to a coordinate system moving with the current. Inhomogeneity and unsteadi-

Table 4.9. *Table indicates whether wave characteristics remain constant (c) or vary (v) in different current conditions.*

	T_a	T_r	L	H_S
inhomogeneity	c	v	v	v
unsteadiness	v	c	c	c

ness have rather different effects on the wave field. Let us first consider a steady one-dimensional current with a monochromatic wave moving from an area at rest into a (gradually increasing) adverse current (see, for example, Phillips, 1977, p74 – 76). The absolute period T_a and frequency ω then are constant, because the number of waves passing by each location has to be the same. This is described by (3.34d). Through the Doppler shift the intrinsic frequency σ increases so that the relative wave period T_r decreases. The wavelength L, which is related to T_r through the dispersion relation, also decreases. Apart from these 'kinematic' effects, there are also 'dynamical' effects resulting in changes of the wave height. In the present case this results in an increase of H_S (Phillips, 1977, figure 3.6). If the current is not adverse but following, all above modulations are reversed.

Another idealized case concerns a spatially uniform wave field, which initially has no underlying current. At a given time the whole field begins to move with velocity U. This is equivalent to a stationary wave field and an observer who begins to move with speed U along the wave direction. In the moving frame all wave properties remain unchanged, except for the period measured by the observer, T_a, which is Doppler shifted. The above considerations are summarized in table 4.9 (c = constant, v = variable).

In practice, the situation is more complicated for several reasons. First, currents are neither steady nor homogeneous. As a result all parameters in table 4.9 are affected. Secondly, real spectra contain wave components with many frequencies and directions, which may obscure the above monochromatic behaviour. Finally, the above mentioned 'conservative' interactions can change the wave steepness significantly, which can have a significant impact on all source terms in the wave energy balance equation.

IV.8.2 Large scale tidal flows

H. L. Tolman and L. Cavaleri

In a study with a third generation wave model, Tolman (1991a,b) has analysed the effect of a tidal current on the characteristics of wind waves during several storms in the North Sea. Figure 4.50 shows the variation of

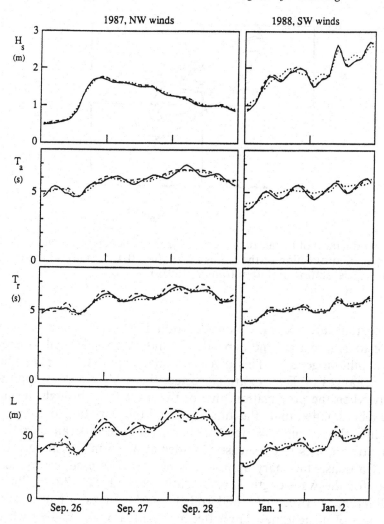

Fig. 4.50. Significant wave height H_S, mean absolute and relative periods T_a and T_r, and mean wavelength L at Euro-0 for two cases in the storm season 1987 – 1988. Dotted lines: without tides. Solid lines: with tides. Dashed lines: with tides, quasi-stationary approach ($\dot{\omega} \equiv 0$).

H_S, T_a, T_r and L at station Euro-0 in the southern North Sea, near the Dutch coast. The two cases are characterized mainly by swell and wind waves, respectively. The effect of the tide is shown by the difference between the results obtained considering the full interaction (continuous line) and results obtained by neglecting the tide (dotted line). The modulation of both T_a and T_r indicates that both space and time variations of the current have a distinct effect on the wave field.

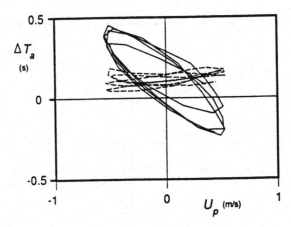

Fig. 4.51. Modulation of the absolute period T_a, as a function of the current velocity in the propagation direction of the waves U_p at Euro-0. Continuous line: unsteady; dashed line: quasi stationary (after Tolman, 1991a).

When currents are taken into consideration, it is common practice in wave modelling to ignore the time dependence and to consider only the effects of current inhomogeneity. This quasi-stationary approach implies that the kinematic changes of T_a or ω are neglected, that is $\dot{\omega} \equiv 0$. This approach is valid only when the propagation time of the wave field through the current field is much smaller than the time scale of the variation of the current field. While this is the case for waves crossing deep-ocean currents like the Gulf Stream, it is not the case for tides away from the shore (Tolman, 1990b). If a quasi-stationary approach is nevertheless used, errors occur in particular for the wavelength L and relative period T_r. This is illustrated by the difference between the solid and the dashed lines in figure 4.50. The consequences of unsteadiness are better highlighted by figure 4.51, where the differences of T_a between the full model and the quasi-stationary approach are plotted against $U_p = \mathbf{k} \cdot \mathbf{U}$. The strong modulation of T_a, which occurs only if temporal variations of currents are accounted for, shows that the modulation is mainly a consequence of the temporal variability of the current field. Tolman (1991a) furthermore shows that the relative importance of temporal and spatial variability of the current varies with location and from storm to storm. Consequently, local effects of wave–current interactions cannot be estimated from the local currents only, that is by simply computing local Doppler shifts.

The modulation of the mean wave characteristics due to tides is not very strong. Tolman (1990a, 1991a) shows them to be typically around five to ten per cent for most of the southern North Sea. Given the relatively strong current in the area, these percentages can be considered as characteristic of the wave–current interaction due to tidal currents in shelf seas away from the shore.

The modulations due to tidal current can be much more relevant in individual spectral bands, with variation up to 50% in the high frequency range. Because the amplitude of the modulation depends on the ratio U/c, it is usually assumed that effects of interactions decrease with decreasing frequency. However, due to the sharp spectral shape of wind waves, small modulations of the spectral peak can result in large modulations of energy in narrow frequency bands throughout the spectrum, in particular for frequencies in the steep, low-frequency flank of the spectrum.

IV.8.3 Current ring

H. L. Tolman, L. Cavaleri and L. H. Holthuijsen

It is of interest to analyse the effect of complex current distributions on the wave field, including the dynamics of wave growth and decay. Holthuijsen and Tolman (1991) have analysed the consequences of the presence of a clockwise turning ring on a uniform wave field propagating through it. This ring is representative for a warm-core eddy of the Gulf Stream.

Two cases were considered. The first was a swell of $H_S = 1.99$ m with a Gaussian spectral frequency distribution centered at $f_p = 0.071$ Hz and a $\cos^{20}(\theta - \bar{\theta})$ directional spread. The spectrum was represented by 13 frequencies ($f_1 = 0.042$ Hz, $\Delta f = 0.1 \cdot f$) and 48 directions ($\Delta\theta = 7.5°$). In the second case a wind sea was considered with a JONSWAP spectrum near full development, with $U_{10} = 20$ m/s, for which 26 frequencies with the usual geometric distribution and 24 directions ($\Delta\theta = 15°$) were used.

Figure 4.52a shows the resulting wave heights for the swell case. Because of the low steepness of the swell and the absence of wind, the effects of source terms can be neglected (as was confirmed by the model). Also, since the current is steady, the absolute period remains constant and the change in wavelength is a function of the local current only. Modulations of the wave height, however, appear outside the current field due to the cumulative effects of refraction. This is illustrated in figure 4.53, which shows the optical ray paths for a comparable 14 s wave. Increased wave heights in figure 4.52a correspond to focussing areas with converging rays in figure 4.53. Similarly, decreased wave heights correspond to diverging rays. Far away from the ring, effects on the wave heights tend to disappear due to the divergence of the rays and the angular dispersion associated with the large but finite scale of the wave crests of the swell.

Wave–current interactions for wind seas are more complex because of the additional effects of the modified source terms on the wave evolution. In

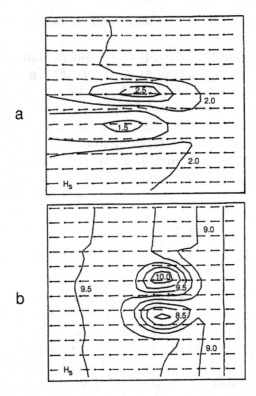

Fig. 4.52. Contour line plots of a wave system crossing a clockwise turning current ring from right to left. a) swell, b) wind waves. Isolines of significant wave height in metres (after Holthuijsen and Tolman, 1991).

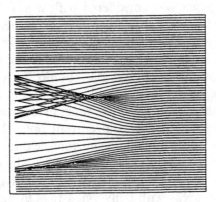

Fig. 4.53. Wave ray pattern for a 14 s period wave passing across a clockwise turning current ring (after Holthuijsen and Tolman, 1991).

the northern part of the ring, where the current flows against the waves, Holthuijsen and Tolman (1991) find (figure 4.52b), as in the swell case, a decrease in wavelength and an increased wave height. Both effects lead to an increase in steepness of the already steep waves. This strongly enhances wind input, whitecapping and nonlinear interactions. Holthuijsen and Tolman (1991) have estimated a 45% increase in S_{in}, and a 90% increase in S_{ds} and S_{nl}. There is an additional shift of energy to lower frequencies, which compensates the local current-induced change of wavelength. Because the dissipation is enhanced more than the wind input, the dynamics reduces the wave height modulation compared to the case without source terms. Since the energy is distributed over many frequency and directional bands with different focussing points, no distinct focussing can be expected. The continuous action of the source terms furthermore tends to mask the effects of interactions beyond the ring, so that current-induced modulations of the wave height are mainly confined to the ring itself.

In a similar study, Hubbert and Wolf (1991) have analysed in detail the deformation of the two-dimensional wave spectrum at various points of the ring. As expected from the previous arguments, there are strong deformations of the directional distribution, mainly associated with different response times of different wave components to the straining by spatial gradients of the current. This is in agreement with Holthuijsen and Tolman (1991) and Tolman (1990a, 1991a). Hubbert and Wolf conclude that wave–current interactions should not be neglected in wave models.

IV.8.4 Application to the open ocean

H. L. Tolman, S. H. Hasselmann, H. Graber, R. E. Jensen and L. Cavaleri

To model the effect of currents on waves one needs to know the space and time distribution of currents in the area covered by the wave model. In general, this is not difficult in shelf seas, where existing circulation models can supply a rather detailed description of the currents near the coast and in tidal inlets or estuaries, but it can be a problem in the open ocean. Complex open-ocean current fields, as measured, for example, in the POLYMODE experiment (McWilliams *et al*, 1986), or observed in satellite images, are usually not modelled. For this reason wave modellers tend to neglect current-induced effects. Thus, in areas with strong currents, such as the Gulf Stream, the Kuroshio and the Agulhas current, wave model results without wave–current interaction should be used with care. A detailed computation, using a nested and fully coupled wave–current model, will generally be needed to obtain reliable results.

The effect of currents on waves is well illustrated by the SWADE storm,

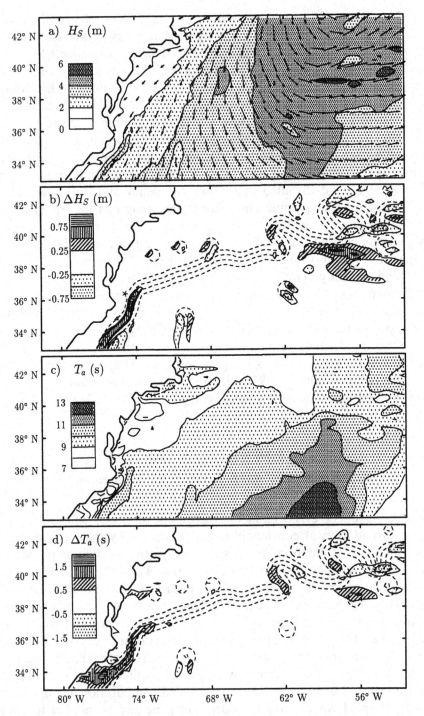

Fig. 4.54. Model results for 0300 UTC 28 October 1990 from the regional SWADE grid including the Gulf Stream. a) H_S distribution, b) current-induced modulation of H_S (i.e. the difference between model results obtained with and without currents), c) T_a distribution and d) current-induced modulation of T_a. Dashed lines in b) and d) denote 0.5 m/s and 1.5 m/s contours of the current velocity. ($\star$) in b) indicates the location of the spectrum of figure 4.55.

which was discussed in § IV.3. For this period reliable current estimates were available. Repeated runs of different wave models, with and without wave–current interactions, provided extensive data highlighting the effects of currents on the wave field.

The surface current distribution was estimated from the quasi-geostrophic Harvard open ocean model (Robinson and Walstad, 1987) and routine sea surface temperature analyses from NOAA/NMC. Input data for the ocean model were provided by the US Naval Oceanographic Office. Daily maps of surface velocity fields were supplied with a spatial resolution of 15 km, and an estimated accuracy of the Gulf Stream position of ± 30 km.

Figure 4.54a, computed with the WAVEWATCH model (Tolman, 1991b, 1992) including source terms identical to those of the WAM model and wave–current interactions, shows the distribution of the wave height H_S and the mean absolute period T_a in the western part of the North Atlantic at 0300 UTC 28 October 1990 (about 36 hours later than figure 4.12, after the storm has moved east). The location of the Gulf Stream is indicated in figure 4.54b. The Gulf Stream flows eastward and exhibits several large meanders in the eastern part of the figure. 'Warm core' eddies to the north of the Gulf stream turn clockwise, 'cold core' eddies to the south turn counterclockwise. Figures 4.54b and d show the current-induced modulation of H_S and T_a, namely the difference between model results obtained with and without currents.

Figure 4.54, which is fairly representative for interactions in a storm, illustrates the magnitude and scales of current-induced modulations of integrated parameters like the wave height and the mean period. Significant modulations of H_S and T_a in figures 4.54b and d are generally confined to an area near the Gulf Stream and its rings. Consequently, the spatial scales of these modulations are determined by the scales of the Gulf Stream. Temporal scales depend on the travel time of waves through the Gulf Stream and on the temporal scales of the wind field. Except for conditions where trapped waves travel within the Gulf Stream over long distances, the temporal scales of the wind field generally determine the interaction scales. Current-induced modulations of H_S and T_a are sufficiently large to produce the small-scale structure in the wave height in figure 4.54a. It is obvious from this figure, however, that the location of the Gulf Stream cannot easily be deduced only from modulations of the wave height.

The smallest interactions are found in the central part of figure 4.54b. The wave field in this area is dominated by swell, crossing a fairly straight section of the Gulf Stream at large incidence angles. In such conditions, current-induced modulations of the wave height are generally small (Holthuijsen and Tolman, 1991, figure 1). The large number of eddies in this area result in interaction patterns of the form discussed in § IV.8.3 (see figure 4.52a). Unlike for the pure swell in § IV.8.3, however, the eddies do influence the mean absolute period outside the current field. This is not because the swell period is modified, but is due to the spectral averaging of the period over a wave field consisting of swell and a wind sea, where the swell energy is modified by the current.

The largest interactions are found in the western part of figure 4.54, where the wave height is increased by the Gulf Stream by as much as 1 m or 50% and the period by 2 s or 25%. The wave field in this area is dominated by swell travelling in southerly directions. Swell trapping occurs at approximately 37° N, 74° W. Trapped swell energy follows the Gulf Stream to the coast of Florida, outside the figure. Note that the trapping of swell is accompanied by a 'shadow' area with lower wave heights around 33° N, 77° W. Figures 4.54c and d furthermore indicate that the Gulf Stream redistributes swell energy along the coast on fairly small scales.

The most complicated current effects are observed in the eastern part of figure 4.54, where large meanders interact with the centre of the storm. Because the storm is losing intensity, the wave field consists of a mixture of swell and wind sea. It is clear from figure 4.54b that the resulting interactions are very complex. In this area both 'conservative' interactions like focussing, trapping and reflection, and their interplay with the dynamics of growth and decay are important. A complete discussion of the details of the interactions in this area is outside the scope of this book. However, we call attention to the decrease in wave height produced by the interactions on and east of the southern turn of the meanders around 40° N, 56° W. The wind is here blowing eastward, in the direction of the stream. The reduced relative speed difference between wind and waves (because of advection by current) leads to a lower H_S. This condition propagates with the waves out of the current, till when the local active generation makes the two fields (with and without interactions) converge again. A similar pattern is present in the period T_a in figure 4.54d.

Even in areas where the effect of the current on the mean wave parameters is small, currents can have a distinct effect on the spectrum. This is illustrated in figure 4.55 (obtained with the WAM model), which shows spectra calculated with and without wave–current interactions at 36°35′ N, 74°50′ W, 15 hours later than figure 4.54. This location (marked ⋆ in figure 4.54b) is just inside the Gulf Stream. The wave field consists of swell travelling against the current (similar to the swell in figure 4.54), and a newly formed wind sea travelling with the Gulf Stream. Whereas the overall wave heights of the spectra differ by only 0.2 m (less than 10%), the spectra are clearly quite different. The current significantly widens the angular distribution of swell energy. Swell entering the Gulf Stream from northerly directions near

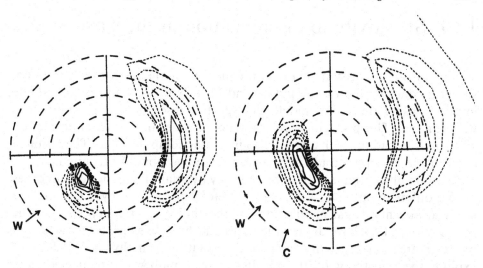

Fig. 4.55. Two-dimensional wave spectra for 1800 UTC 28 October 1990 at 36°35′ N, 74°50′ W (location marked ⋆ in figure 4.54b) with (right) and without (left) wave–current interactions. The arrows labelled w indicate the wind direction; the arrow labelled c gives the direction in which the current is flowing.

the location considered is refracted locally, shifting incoming swell energy northward. Swell entering the Gulf Stream more to the north crosses the centerline of the current twice (i.e., in the bend at 74° W 37° N), adding easterly components to the swell field at the location considered.

The current furthermore shifts the wind sea to higher frequencies, while reducing the corresponding wave height. This could be due to the reduced wind velocity relative to waves moving with the current. However, effects of the Gulf Stream on wind seas are generally difficult to estimate *a priori*, due to the complex interplay between 'conservative' interactions and active wave growth.

In summary, the SWADE experiment together with the more idealized examples considered earlier, demonstrate the wide variety of complex modifications of the wave field which can be produced by interaction with spatially and temporally varying currents. Although wave–current interactions do not always produce modifications in mean spectral parameters such as the significant wave height and mean wave period, they normally have a strong influence on the spectral distribution. Therefore, wave–current interactions should be included in wave models (the computational overhead is minimal), whenever strong tidal or ocean-circulation currents occur.

IV.9 Sensitivity to discretization in the wave model

So far, in our discussion of the consequences of the choice of resolution, we have concentrated on the input wind field. Now we want to analyse the implications of the finite spatial and spectral resolution of the numerical wave model. The problem is relevant because usually a higher resolution gives more accurate results, but it also increases computer time. The ultimate choice is necessarily a trade off between these two opposite requirements.

An optimal solution can be achieved only by quantifying the consequences of the different options. We begin therefore in § IV.9.1 by analysing the physical and numerical implications of the choice of grid step size. In the next section, § IV.9.2, we discuss the nesting technique, which is characterized by large area coverage and high local resolution. Finally, in § IV.9.3, we explore the sensitivity of the results to the number of frequencies and directions used to represent the wave spectrum.

IV.9.1 Spatial resolution

L. Cavaleri

The size of the grid step, the smallest distance we can consider in a wave model, affects the results in two different ways. On one hand it establishes the accuracy with which we can reproduce intense local events and refraction and dissipation in shallow water. On the other hand it is connected with the accuracy we can achieve in describing the bathymetry and the shape of the coasts. We will consider separately these two aspects of the problem.

Grid resolution

As discussed in the introduction of this chapter, the grid resolution Δx and the related integration time step Δt have to be sufficiently small to resolve relevant small-scale phenomena. Often, in wave modelling these small-scale phenomena do not result from the wave dynamics, but rather from the strong gradients that may be present in the forcing wind field, for example in the centre of a hurricane, in the frontal zone or at the maximum of a storm. In these areas an insufficient resolution will lead to a poor representation of the event and to the gradual divergence of the model results from the reality.

If the wave fields we want to reproduce derive from a very extended storm, with scales of several thousands of kilometres and several days, a global representation with three degree resolution is likely to be sufficient. A 20 minute time step is sufficient in such cases. (Larger time steps are excluded for the numerical reasons discussed in chapter III.) In contrast, if the storm is characterized by strong intensity and limited dimensions, a three degree resolution can easily miss the peak and lead to an underestimation of the strength of the storm.

The above considerations are confirmed by Zambresky (1989), who has hindcast one year of wave conditions on the globe, using the WAM model and the analysed winds of ECMWF. Zambresky reports that, for a very strong but extended storm in the North Atlantic Ocean (max $H_S > 18$ m), an increase of resolution of the WAM model up to $1° \times 1°$ did not change the results appreciably. On the other hand, for a less severe but compact storm the increased resolution led to an increase of H_S from 5 to 7 m in the central part of the storm. In general, the accuracy decreases whenever we find strong and rapid changes of energy. On a wave height map this corresponds to the area with the highest H_S gradients, usually near the frontal part of the storm, often just before the region with maximum wave height.

The above considerations implicitly assume no limitation of accuracy by the wind, that is the wind resolution is assumed to be equal or better than that of the wave model. If this is not the case, the wind becomes the limiting factor. However, when the physics of the meteorological event requires a high resolution, the wave field usually exhibits rapid changes, both in time and space. The examples given in § IV.3.4 for the Gorbush storm and hurricane Hugo are very illustrative in this sense.

On a more local scale the adequacy of the 0.5 degree resolution in the Mediterranean Sea (figure 4.56) has been tested by Cavaleri *et al* (1991) by repeating with 0.25 degree resolution the hindcast of the storm shown in figure 4.16. The results are very similar to the 0.5 degree case, but it is instructive to analyse the difference between the two hindcasts, shown in figure 4.57. We note the large areas of difference off the coast to the north of Sicily, and to the west of Sicily, between Africa and Sardinia. In zones not affected by coasts and boundaries the difference between the two hindcasts reflects the different accuracy reached in the two runs in the most intense area of the storm. Note that the 0.25 degree hindcast requires $\Delta t = 15$ minutes rather than $\Delta t = 20$ minutes for numerical stability reasons. In all the other zones, away from the coast and the centre of the storm, there is hardly any difference. Cavaleri *et al* (1991) quote the overall mean difference in the field as less than 2%, and less than 0.5% if we neglect areas near sharp coastal features, to be discussed below.

When the interaction between waves and currents (§ IV.8) is modelled, the scale of the current may determine the required resolution. This happens in particular close to coasts, when the currents have a strong spatial variability. If the current pattern is known sufficiently well a nested local model is a possible solution.

A similar argument holds for the refraction associated with bottom to-

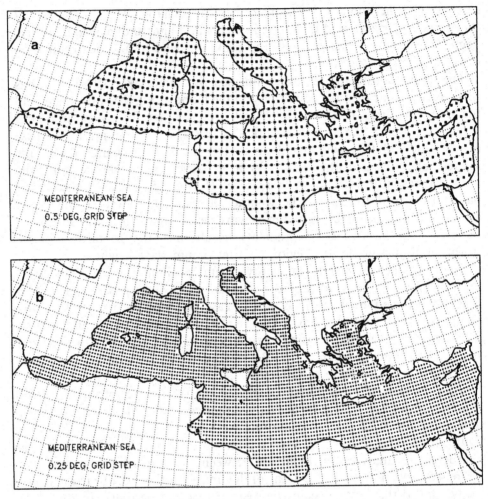

Fig. 4.56. Wave grids used for the Mediterranean Sea: a) 0.5 degree, b) 0.25 degree resolution.

pography (discussed in § IV.7). In this case the scale is usually even smaller, of order $10^{2 \div 4}$ metres, and for this reason it is often neglected in wave modelling. If relevant, a local model is again the obvious solution.

Definition of the coast

The resolution of the grid affects the accuracy with which we describe the boundaries of a basin. This affects, in turn, the accuracy of the wave fields in coastal areas. This is well illustrated in figure 4.57, which shows the difference between two hindcasts, performed, respectively, with 0.5 and 0.25 degree resolution, using the same input wind. As the resolution of the wind

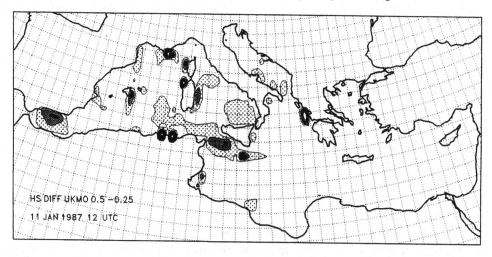

Fig. 4.57. Modulus of H_S vector difference between wave fields evaluated on the grids with 0.5 degree and 0.25 degree resolution respectively. Shading interval is 0.5 m.

field is 70 km, larger than that of the two wave grids, the wind cannot be the cause of the small but evident differences.

We consider first the large deviations, all close to coasts, two near Algeria and the others near Corsica, Sardinia, France and Albania. The example of Algeria is the clearest one. On the 0.5 degree resolution (figure 4.56) the coast appears straight, so that the waves are not hindered in their propagation (see figure 4.16). On the 0.25 degree grid two small peninsulas protrude into the sea for the length of a grid step, about 28 km. They shelter almost completely the grid points just to their east. The local difference between the two hindcasts is nearly equal to the entire wave height found with the 0.5 degree grid. Similar explanations hold for the other error maxima. We conclude that results close to coasts need to be used with caution, and that, generally, coastal details can be taken into proper consideration only by a high-resolution local model.

The difference field protruding out of the Strait of Bonifacio between Corsica and Sardinia, towards the Italian peninsula, is another good example of the impact of coastal resolution, associated in this case with the nonlinearity of wave growth. The strait is closed in the 0.5 degree grid, open in the 0.25 degree grid. In the finer grid there is some energy flowing through the strait. Although there is less than half a metre difference in H_S between the two cases, this is sufficient to trigger a more intense and faster wave growth through local wind, generating the difference seen to the east of the strait.

It is instructive to analyse the reason for the ΔH_S present on the east coast of Sardinia. Apart from some energy flowing around the southern end of the island, we have here wind seas generated by a wind blowing offshore along a slanting fetch. The two grid resolutions effectively locate the coast at two positions, separated by 25 km. Therefore, the corresponding sea points are subject to a different fetch. Cavaleri *et al* (1991) show that, according to the JONSWAP results, a fetch variation ΔX causes a wave height variation

$$\Delta H_S = 2.5 \times 10^{-4} \, U_{10} \times X^{-0.5} \Delta X, \tag{4.8}$$

with U_{10} the wind speed in m/s, and the fetch X expressed in m. Use of this equation reveals that the different fetches of corresponding points on the two grids explain the difference in wave height in the shaded areas in figure 4.57, not only close to Sardinia, but also on the French and Spanish coasts and on the Adriatic Sea.

The large differences on the lee of the Balearic Islands and of Malta, south of Sicily, are obviously due to the absence of these islands, or to their different size, on the 0.5 degree grid.

A final point to note is the large difference found in the area between Tunisia and Sicily. We are here close to the peak of the storm, where the presence of very strong gradients makes the results sensitive to the resolution. Differences may also arise through the presence of the Tunisian peninsula. Sharp features like this produce some unavoidable numerical noise, dependent on the accuracy of the advection scheme and on the directional resolution of the model. This noise is usually visible several grid points 'downwave' of the obstruction. The wake is not connected with the distance but with the number of points. Therefore, different grid resolutions may be expected to produce different wave fields in the wake of sharp coastal boundaries.

The distance over which the noise can persist also depends on the local sea conditions. If there is active generation, the rapidly growing sea and the interaction among the various components will efficiently absorb the difference. If the sea is decaying, and the wave–wave interactions are very weak, the difference will remain for a much longer time and it will be advected down stream. Depending on the situation, the time scale can vary from a few hours to days (Cavaleri *et al*, 1988).

IV.9.2 Grid nesting

S. Hasselmann, L. Cavaleri and G.J. Komen

In many of the examples discussed in this chapter, the use of high-resolution local models was essential for obtaining the required accuracy. This applied equally well to the meteorological models discussed in § IV.2 and to the wave model applications in subsequent sections. However, focussing on an area of particular interest is in conflict with the need to model wave generation over large areas of the ocean and to follow swell propagating over large distances. A compromise may be obtained by using a fine grid of limited extent embedded in a more extended and coarser grid. This numerical technique, which enhances the resolution locally, is known as the nesting technique. An alternative approach uses a variable resolution, which increases continuously towards the area of interest. This approach is still being developed and will not be pursued here.

The underlying reasons for using a fine grid have been just formulated in our discussion of spatial resolution. They are: 1) the presence of strong gradients in the wave field; 2) strong gradients in the underlying current or in the forcing wind; and 3) a complicated geometry or bathymetry. Whether one should use nesting or whether it is more convenient simply to use an extended high-resolution grid is often a compromise between total computing requirements and the amount of work actually involved in setting up the nested grids. Nesting is not necessary when the area of interest is of limited extent and bounded by coasts. However, when the area is open at the boundaries and part of a larger basin, information from the coarse grid is needed to provide boundary conditions for the fine grid. In general these boundary conditions vary in time with the evolution of the storm.

The effort required to run wave models on nested grids is only justified when the fine mesh grid provides information that is not already obtainable with the coarse grid. Whether this is the case depends on geophysical conditions. For example, when an extended storm generates a wave field with no fine-scale structure, in which the waves travel in deep water towards a straight coast, nesting will only increase the spatial resolution of the results, without adding any information. In practice, the need for nesting will also depend on the operational requirements. An example is provided by the case of a wind blowing offshore. In this situation a coarse model usually performs poorly near the coast, because the effective fetch at the first grid point is not very accurate and the rapid growth with fetch from the coast to the first model grid point is not resolved. If the point of interest is close to the shore, it may be worthwhile to run a nested model, but if it is further away this could be pointless, because the results of the coarse and fine grids converge rapidly when the distance from the coast increases.

Because of this convergence the flow of information between the grids is usually taken as one-way only, from the coarse grid to the nested grid, and not in the other direction. This has no consequences for an incoming swell or storm, but in the case of the offshore blowing wind the system behaves as if the wave model were run on two uncoupled grids. Only at a sufficient distance away from the coast will the results of the coarse grid be equally as accurate as those of the fine grid. The nesting must extend at least that far to avoid discontinuities in the description of the wave field.

The size of the nested grid must be chosen with care. The accuracy of the boundary information is limited by the resolution of the coarse grid. Therefore, the boundaries must be sufficiently far away from the region that needs to be modelled with high resolution, and the nested grid must extend beyond the area of interest. The actual extent depends on the complexity of the situation. If L is the linear dimension of the high-resolution area of interest, then it would be a reasonable guess to extend the nested grid by a length L in each direction.

The boundary conditions are specified by the values of the two-dimensional spectra at the grid points along the boundary of the fine grid. In general, these points do not coincide with grid points of the coarse grid, so that an interpolation has to be made. This interpolation requires some care. In principle, one could interpolate individual spectral components, but in certain situations this would lead to unrealistic spectra. Therefore, in § III.4.3, another method has been chosen in which, in essence, spectral parameters are interpolated and the two-dimensional spectrum is adjusted. This solution is adequate in most situations, but difficulties can arise, when the spectrum is changing more rapidly in space than the coarse resolution can resolve, for example when a wind sea with strong spatial gradients in direction is superimposed on a uniform background swell. In such situations it might be useful to consider implementing an improved interpolation scheme, in which the spectrum is first partitioned (§ V.4.3) into different wave systems. However, the basic problem is general and not related to a particular choice of interpolation scheme, but rather to the presence of nonresolvable gradients at the boundary of the fine grid. The only way to solve this is to extend the area of the high-resolution grid so that strong gradients at the boundary are avoided.

One of the first successful applications of grid nesting with the WAM model was made during the calibration/validation (Attema, 1992) of the instruments on board of the ERS-1 satellite, which was carried out in the Norwegian Sea. In this application the model was run on three grids, nested sequentially. Each grid provided the input information for the next finer grid. The resolutions used were $1° \times 1°$, $0.5° \times 0.25°$ and $0.125° \times 0.0833°$, respectively. This allowed for a correct description of incoming swell, while the local wind sea was resolved with sufficient resolution for the accurate calibration of the satellite instruments.

In another study with the WAM model Bertotti and Cavaleri (unpublished) nested a $0.18°$ resolution Adriatic version into a $0.25°$ resolution version of the full Mediterranean basin. They hindcasted a storm that occurred on 6 December 1992, and they compared results from runs with and without nesting with observations. They found that the nesting affected the model results in the northern Adriatic because of resolution effects, and in the southern part because of swell that entered from the Ionian Sea through

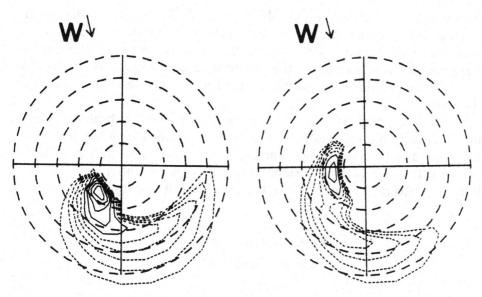

Fig. 4.58. Computed spectra at 18 UTC 27 October 1990, for a location in the North Atlantic Ocean south-east of Cape Hatteras (as in figure 4.49). The spectrum on the left was obtained on a coarse resolution grid; the spectrum on the right was obtained in a nested grid. Refraction was included in both cases.

the Strait of Otranto. In both areas the magnitude of the corrections was comparable with the actual wave height.

Yet another example showing the beneficial impact of grid nesting has already been encountered in our discussion of depth refraction during the SWADE storm (§ IV.3.2). There (figure 4.49) we illustrated the importance of depth refraction with the help of results from a nested grid. We compared two-dimensional spectra at 18 UTC 27 October 1990, for a location in the North Atlantic Ocean south-east of Cape Hatteras, obtained with and without depth refraction, and we found that refraction had a significant influence on the spectral shape. Most notable was the turning of swell towards the coast, by as much as 45°. Now we use this same situation to illustrate the importance of grid nesting. To this end we compare results (with depth refraction included in both cases) from the coarse grid (figure 4.58 left) and the fine grid (figure 4.58 right). The difference is striking. In particular, the pronounced turning of swell towards the coast is not reproduced in the coarser grid. Apparently, the crude description of the bottom topography in the coarse grid does not allow for a proper treatment of this turning. In fact, refraction is virtually absent in this case, as is evident from the comparison with figure 4.49, which shows the spectrum at the same location without consideration of refraction. In addition to differences in the treatment of swell one may also note that, in the fine grid, the wind sea is better developed than in the coarse grid. This is easily understood when we realize that we are dealing with fetch-limited conditions, which make the wave energy rather sensitive to the actual value of the fetch. For a correct description of the wave conditions in such a situation we must have an accurate determination of the fetch and this can only be achieved with a fine-resolution grid near the coast.

In summary, grid nesting is a numerical technique which allows one to increase the resolution locally. It should be used when both the large scale evolution and the small-scale structure are relevant. The grids have to be set up with some care concerning grid extension and interpolation at the boundary. Recent results illustrate the usefulness of nesting.

IV.9.3 Spectral resolution

H. Günther and L. Cavaleri

The representation of the spectrum by its values at a finite number of frequencies and directions unavoidably involves some approximation. We will now explore the effect of this on the model results.

Distribution of frequencies

The number of frequencies that we are allowed to use depends on the available computer power. The question is how to distribute them to optimize the calculations. In the past a uniform distribution, with a constant Δf, has often been used. However, what we must aim at is to concentrate the frequency values in the range where most of the action is taking place. For a wind wave spectrum this is around the peak. The problem is that the peak frequency exhibits drastic changes during the evolution of a storm. These changes are very rapid in the early stages, when the peak is in the high frequency range, and they slow down as the peak moves towards lower frequencies. Therefore, it is convenient to concentrate the frequency values in the latter range, where the peak dwells most of the time during storm conditions. This has been obtained in WAM (see § III.4) by distributing 25 frequencies in a geometric progression with a ratio of 1.1. This corresponds to a ratio of about 10 between the last frequency and the first one. The lowest value can be adjusted to the environment in which the model is to be applied. The standard global version of the WAM model has a prognostic range starting at 0.042 Hz and ending at about 0.4 Hz. One of the consequences of having a limited number of frequencies is the need to add a diagnostic tail (discussed in § III.4) to the high frequency part of the spectrum, to be able to evaluate the nonlinear transfer in the prognostic frequency range.

In a basin of limited dimensions, such as the Mediterranean Sea, the Baltic Sea or the Great Lakes of North America, we expect higher frequencies than in the open ocean. The frequencies in the model must be chosen accordingly. This goes hand in hand with the choice of a smaller grid step, because higher frequencies will be found at the first grid point off the coast in the case of an offshore blowing wind. The smaller the grid step, the higher the possible frequencies. This argument becomes relevant in nested modelling, when used in an area with an offshore blowing wind. In principle a different set of frequencies should be used for each nested grid. However, this would not be the case for an incoming storm or swell. Clearly the solution must be considered case by case, depending on the specific interests.

The minimum frequency 0.042 Hz used for the global WAM model is well below the wind wave range in the vast majority of cases. However, a very low-frequency swell is, in the oceans, a practical possibility. Such a swell is usually associated with a very intense distant storm. In these cases it would be appropriate to extend the frequency range towards the low side, well below the swell frequency (still keeping the numerical stability conditions under control). Such an experiment has been done by de las Heras (1990) for a case of swell on the coast of Peru (described in § IV.5.1), originated by an extreme storm on the north-west Pacific Ocean seven days earlier.

Concluding this short discussion, the frequency range on which to run a wave model must be chosen so as to include the energetic frequencies present during a storm. The smaller the basin and the smaller the grid step, the higher the frequencies to be considered. An extension towards lower values can be necessary to deal properly with some very low-frequency swell.

Directional resolution

The standard directional resolution, $30°$, used in many wave models, including WAM, may seem too coarse at first. After all, a pure wind sea, distributed on the $180°$ around the wind direction, is represented by only five directions. Therefore, it is of interest to check the eventual improvements arising from a better directional resolution. These can be expected in wind waves for a more detailed description of the physical processes represented in the source function (generation, dissipation, nonlinear exchanges). The swell propagation should improve, particularly over long distances, with a reduction of the garden-sprinkler effect discussed in § III.4.2. Close to coasts and islands a better description of the wave propagation around sharp coastal features is to be expected.

The accuracy of the discrete distribution can be checked against that of the continuous one in a theoretical case. The result is quite reassuring. For a $\cos^2$ distribution, taken as representative of generation conditions, the overall energy evaluated with a 12 direction distribution ($30°$ resolution) is

Table 4.10. *Intercomparison statistics with buoy wave height measurements, obtained with the WAM model with 15 degree angular resolution. In brackets are the numbers for the standard 30 degree model. The period is November 1988 (after Günther et al, 1992).*

	Alaska	Hawaii	US East Coast
number of data	356	356	367
mean H_S buoy (m)	4.10	2.57	2.18
bias (m)	-0.36 (-0.49)	-0.32 (-0.40)	-0.30 (-0.35)
STD (m)	0.79 (0.87)	0.47 (0.45)	0.45 (0.44)
scatter (%)	19 (21)	18 (18)	21 (20)

correct to within several digits. A similar conclusion holds for the wind input, considered as proportional to $\cos^3$ (the product of the energy density present in each direction times the cos arising from the angle between the wind and wave directions). However, it is clear that the conditions can be quite different in real cases, with asymmetrical, often peaked, distributions.

A thorough test has been carried out by Günther *et al* (1992), who have rerun the November 1988 period, already used in § IV.4.2 to asses the performance of WAM in analysis and forecast, with an angular resolution of 15°. The result of the test has been an increase of 0.11 m in the mean global wave height. The increase occurred not only in swell areas; the wind waves were affected in the same way. Changes of up to half a metre could be observed at individual grid points and times.

Table 4.10 compares the statistics of the test with that of the standard 30° results. Apart from the standard deviation (STD) in Alaska, the only improvement was in the bias, with a reduction of about 3% relative to the mean H_S value at the individual stations.

A similar test was carried out by Bertotti and Cavaleri (1993) in the Mediterranean Sea, in their hindcast of the Med87 storm (§ IV.3) with 15° resolution. The result was an increase of the mean H_S in the basin of 0.16 m. The relevant point is that there was a very small increase of the peak values, the changes being mostly located in the lee of the sharp coastal features.

The conclusion we can draw from these results is that, as expected, an increase of directional resolution from 30° to 15° leads to an improvement of the general performance of the wave model. However, in an extended basin the improvement is not enough to imply a firm necessity for the use of the 15° resolution. Clearly, substantial improvement is to be expected in the case of large spatial gradients in direction, as with a rapidly turning wind or a complicated geometry of the coastal features.

IV.10 Waves as part of the coupled atmosphere ocean system

From a physical point of view the evolution of the atmosphere/ocean system should be considered as a whole. So far, this has not been done. Instead, the interaction between different subsystems has been considered. For example, in § IV.8 we have seen how currents may affect the waves and in § IV.5.1 we have discussed the dependence of the evolution of the sea state on the dynamical coupling between the atmospheric boundary layer and the surface waves. This goes beyond the traditional picture in which there is a one-way flow of information from the atmospheric model to ocean wave and ocean circulation models. In fact, in reality, there are many mutual interactions, and both the interaction between ocean circulation and waves and the interaction between these and the atmosphere should be two-way. A complete description of the overall evolution of the ocean and atmosphere, both in the short and long term, can only be achieved with a single unified model. If run in a real time mode, such a model would also be ideally suited for the consistent assimilation of all ocean and atmosphere observations.

While we are still far from such a complete description, preliminary experiments are presently being carried out to investigate the sensitivity of the subsystems with respect to their mutual interaction. In this section we describe two sets of experiments, the first (§ IV.10.1) with special emphasis on the ocean, the second (§ IV.10.2) more concerned with the atmosphere.

IV.10.1 Combining waves and storm surge modelling

G. J. H. Burgers, R. Flather, P. A. E. M. Janssen, C. Mastenbroek, X. Wu and L. Cavaleri

The coupling between the atmospheric boundary layer and the wind waves, described in § II.2 and then discussed in § IV.5.1, leads to an increase of the surface drag coefficient C_D in the initial part of a storm. The increase is not only relevant for wind waves, but also for other phenomena such as the current distribution and the surge. As, in turn, these may affect to some extent the evolution of the wave field, it is obvious that the proper way to describe their evolution under a given atmospheric forcing is by means of a coupled wave–surge model.

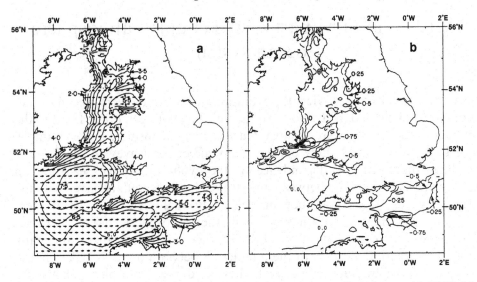

Fig. 4.59. (a) H_S (contour interval 0.5 m) and mean wave direction at 12 UTC 26 February 1990 in the Irish Sea. b) Change in H_S due to tide and surge (contour interval 0.25 m).

Irish Sea surge

A rather complete example is given by Wu and Flather (1992) with the hindcast of the severe storm in the Irish Sea that hit North Wales on 26 – 27 February 1990. The storm caused extensive flooding in North Wales with surges up to 2.5 m.

In the decoupled case the WAM model was run with constant depth. The standard surge-tide model used the empirical Smith and Banke (1975) formulation for the surface drag coefficient,

$$C_D = (0.63 + 0.066\bar{U}_{10})10^{-3}, \qquad (4.9)$$

while a quadratic bottom stress was considered, with a constant drag co-efficient $C_f = 0.0025$. A preliminary study by Wolf *et al* (1988) suggested that surface and bottom stresses are the source terms most affected by the interaction of waves with the surge. To test the impact of coupling on surge and wave results Wu and Flather (1992) considered several interactions. The roughness length of the sea surface was related to the inverse wave age using the formulation of Donelan (1992). For the bottom stress formulation the model of Christoffersen and Jonsson (1985) was used. The time varying water depth was provided to the WAM model by the circulation model.

We consider first the consequences for the wave height field. For the decoupled case H_S is shown in figure 4.59a, while part b of the figure shows the change in H_S due to tide and surge. We see that the changes in local

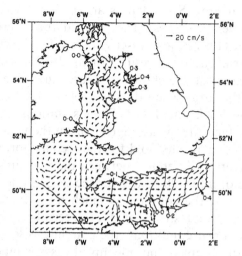

Fig. 4.60. Change in sea level and depth mean current at the same time as for figure 4.59, due to wave-dependent surface and bottom stress (contour interval 0.1 m): cgs units are used.

wave height are significant, particularly in coastal areas where they reach ± 1 m, which is $10 - 20\%$ of the local H_S.

We consider then the difference between the surge and current fields obtained with the two runs. These are shown in figure 4.60, which is for the same time as figure 4.59. The effects are generally small (~ 10 cm in surge elevation) on the North Wales coast, but in shallow areas, such as the Bristol Channel and the Gulf of St.Malo, the wave-dependent stresses cause significant changes in surge elevation of up to one metre during the storm.

North Sea surges

The wave age dependence of the surface drag coefficient C_D, with higher values in the early stages of a storm, leads to different effects at the different stages of development. It is instructive to analyse the hindcast of two storms of opposite characteristics, present in the North Sea on February 1989 and December 1990 (Mastenbroek *et al*, 1993). Both storms resulted in a considerable surge and high waves off the coast of the Netherlands, but while the first was characterized by a fast passage and therefore relatively young waves, the second storm moved slowly producing almost fully developed conditions. The reference runs were done with a decoupled barotropic storm surge model, similar to that used by Wu and Flather (1992) in their hindcast in the Irish Sea.

Mastenbroek *et al* (1993) used a bottom stress coefficient which depends on depth, but equals that of Wu and Flather (1992) in shallow waters with depths less than 42 metres. Such depths are, in practice, the most relevant ones for shallow water processes. The coupled runs were done using the approach of Janssen (1991a), also discussed in § II.2.7, with C_D depending on the state of the sea. The surge results for the 1989 storm at eight stations along the North Sea coast are shown in figure 4.61. There is a systematic underprediction of the surge when the uncoupled model is used, while, with the exception of the Wick and North Shields stations, the coupled model performs rather satisfactorily, its average error on the peak values being less than 0.11 metres. Similar results hold for the second storm. It is instructive, however, that the average difference between the two runs drops from 0.45 metres for the first storm to 0.28 metres for the second one. Being dominated by relatively young sea, the coupling is much more important in the first storm, while in the second storm, that of December 1990, characterized by almost fully developed conditions, the coupling is less effective and the difference between the coupled and uncoupled cases is much smaller.

The above results, obtained without additional tuning of the coupled wave–surge model, give indirect evidence for the effect of waves on the momentum flux. From a practical point of view it could be argued that similar results could be achieved by retuning the Charnock relation (see § II.2.3). Tuning on the observations of February 1989 resulted in a dimensionless roughness length $\alpha = 0.032$, which led also for the case of December 1990 to results comparable to those of the coupled run. It would be tempting to deduce from this that most of the improvement found using a wave-dependent stress can be reproduced if the stress relation is properly tuned. However, the value 0.032, or any similar one found in a different experiment, is not universal, but strongly depends on the characteristics, in particular on the geometry and bathymetry, of the basin under consideration. On the contrary, a sound approach is built on physics and as such is capable of providing good results in all the conditions. It is therefore highly desirable in all practical applications.

IV.10.2 Effect of waves on the atmospheric general circulation

L. Cavaleri and G. J. Komen

The classical approach of wave modelling is to force the wave model with given winds produced by an atmosperic model which has evolved according to its inner physics, with no feedback from the sea surface. The exchange of momentum, heat and humidity with the continents and the oceans is parametrized, but – over the ocean – this parametrization does not depend on the sea state.

The interaction between surface waves and the atmospheric boundary

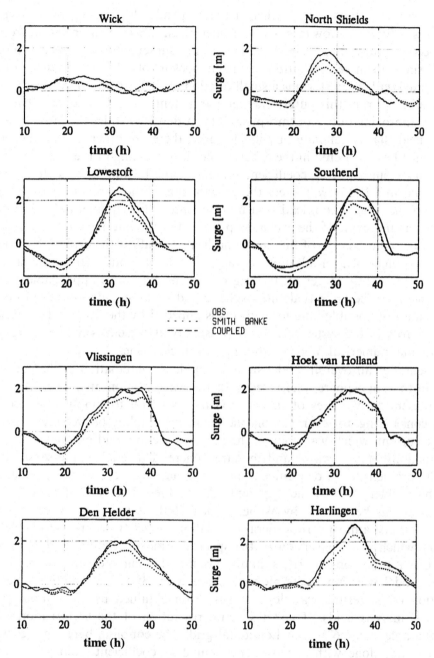

Fig. 4.61. Surge at eight stations along the North Sea coast. Start time is 00 UTC 11 February 1989. The results marked – – – are obtained by taking into account the effect of waves on the wind stress.

layer, with a surface drag coefficient that depends also on the wave conditions, is a first step towards also considering the flow of information from the ocean to the atmosphere. The impact of this interaction is expected to be small since a simple evaluation of the energies involved shows that the one associated with a sea storm is a small fraction of that present in the overlying atmosphere. From this purely energetic argument it can be deduced that to a first approximation an atmospheric storm develops independently of the wave field that is generated by it. However, there are other induced effects that must be considered in the effective physical coupling of the two systems. The larger surface drag coefficient present at the beginning of a storm (as discussed in § II.2.7) will affect the growth rate of the baroclinic instability and, on the other hand, lead to a more intense whitecapping and therefore to the enhancement of the exchange processes at the interface. For example, the increased exchange of humidity and heat may affect the evolution of the storm, appreciably. In addition, on a time scale of months, the cumulative effect of a slightly larger mean drag coefficient may have appreciable consequences for the large scale atmospheric and ocean circulation. The direct evaluation of possible long term effects is impeded by the deep and intrinsic nonlinearity of the system. The only possibility at present is the coupling of an oceanic numerical model with a meteorological one.

Essentially, an overall model should include the circulation in the oceans and in the atmosphere, interfaced with the wave motion at the surface. In the present early stages of this approach it is useful to simplify the problem, considering, for example, on one hand the coupling between wave and surge-current, which was done in the previous section, and on the other the interaction between waves and the atmosphere. This enables the sensitivity of the problem to be determined and the influence of the various elements of the problem to become apparent. Along these lines an interesting experiment has been done by Weber *et al* (1993), coupling the WAM model to the ECHAM atmospheric general circulation model (Roeckner *et al*, 1989, 1992), which is a T21 version of the spectral atmospheric model developed at ECMWF (Simmons, 1991), suitably modified for climate simulation. The WAM model was run in its global version, with $3° \times 3°$ resolution. Verification of its results was done versus those obtained using as input the ECMWF analysis fields. The T21 spectral resolution of the ECHAM model is comparable to a $5.6° \times 5.6°$ horizontal grid. The coupling between the two models was done with a wave-dependent drag coefficient from a preliminary parameterization of quasi-linear theory (Janssen, 1991a). Similarly the heat and humidity flux coefficients have been parametrized as functions of the sea state. The models were integrated under steady July conditions, so that the incoming solar radiation and the sea surface temperature did not vary. A fixed-month mode was chosen to get time series long and statisti-

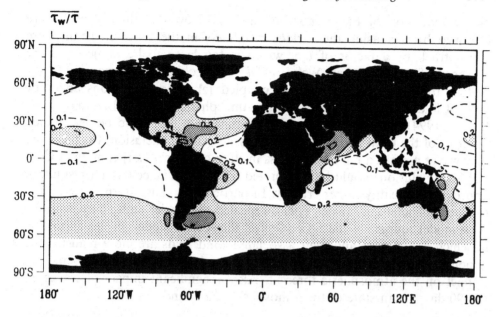

Fig. 4.62. Mean ratio between wave induced stress τ_w and total stress τ in the coupled wave-atmosphere experiment.

cally stationary enough to allow their meaningful statistical analysis. The results were compared against those obtained from the T21 model without coupling.

A good idea of the effect of the coupling is given by the spatial distribution of the ratio τ_w/τ between the wave-induced and the total stress. Due to the coarse resolution of ECHAM, the details of the distribution cannot be deduced from the experiment. However, some general characteristics are evident. There is a high spatial and temporal variability of the stress ratio, with young wind sea ($\tau_w/\tau > 0.4$) present only 10 – 20% of the time. The distribution under an individual cyclone is inhomogeneous: stress ratio maxima, young wind sea, occur on the equatorward 'frontal' side of the cyclone.

The average (over the model run) spatial distribution of the τ_w/τ ratio in figure 4.62 indicates the area where the impact of the coupling on the atmospheric flow is expected to be most important. Because of the period chosen for the simulation, a steady July condition, the major impact is, as expected, located in the southern ocean storm belt.

Weber *et al* (1993) speculate that the immediate consequences of the coupling for the atmospheric circulation would be a variation of the average storm intensity and a shift in latitude of the storm belt. And indeed, in an early experiment, Ulbrich *et al* (1993) obtained weakened westerlies and a reduction of the number of storms. However, these results were obtained by increasing the surface roughness in the southern hemisphere storm track by a factor of ten, which is too large to be induced by the presence of waves. Later, Weber *et al* (1993) obtained a much lower sea surface roughness enhancement in coupled T21/WAM simulations and they did not find a significant effect on the atmospheric flow. Further analysis by Weber (1994) showed that the air/sea momentum flux could be determined in terms of the local and instantaneous wind. Their conclusion was therefore that waves are not a relevant process in state-of-the-art (T21) climate models.

This conclusion should be interpreted with care. The coarse T21 resolution approximates reality rather crudely. In particular, intense storms are usually too weak. Therefore, Janssen (1994) repeated the model simulations with the WAM model coupled to the T63 version of the ECMWF model. In this approach, in which depressions and their associated wind fields are more realistic, the dependence of the drag on the sea state resulted in a small but beneficial impact on the atmospheric forecast. Also a significant impact of waves on the 90 days mean state of the atmosphere was found.

Chapter V

Global satellite wave measurements

K. Hasselmann *et al*

V.1 Impact of satellite wave measurements on wave modelling

K. Hasselmann and S. Hasselmann

Through the launch of ocean observing satellites, wave modellers are now for the first time receiving detailed wave data on a global, continuous basis. This can be expected to have a profound impact on wave modelling. The first US ocean satellite SEASAT demonstrated in 1978 that wave heights could be accurately measured with a radar altimeter and that a SAR (synthetic aperture radar) was capable of imaging ocean waves. Unfortunately, SEASAT failed after three months, and further satellite wave measurements were not made until the radar altimeter aboard GEOSAT was put into orbit in 1985. This changed with the launch of the first European Remote Sensing Satellite ERS-1 in July 1991. Since then, both radar altimeter and SAR wave data have been produced again globally in a continuous, near-real-time mode (cf. table 1.1).

Even before satellite wave data became available on a quasi-operational basis, the recognized potential of these data had a strong influence on wave modelling. One of the principal motivations for developing the third generation wave model WAM was to provide a state-of-the-art model for the assimilation of global wind and wave data from satellites for improved wind and wave field analysis and forecasting. Prior to the development of the WAM model, wave modellers had available only first and second generation wave models. The former were known to be based on incorrect physics, while the latter contained essentially correct physics but were restricted numerically through an artificial separation of the spectrum into a wind sea component of prescribed spectral shape and a swell spectrum, with rather arbitrary parametrizations of the coupling between the two spectral regimes. Both first and second generation models were unable to predict two-dimensional wave spectra reliably and yielded widely divergent results for complex sea states (SWAMP, 1985). We shall show in § V.4.3 that complex wave systems, consisting of mixtures of multiple swell systems and nonequilibrium wind seas, are the rule rather than the exception in the open ocean.

A novelty and major asset of ERS-1, which was being planned at the time the WAM group was formed, was that it would be able to provide quantitative global measurements not only of the significant wave height, through the altimeter, but also of the two-dimensional wave spectrum, through the global intermittent sampling SAR wave mode of the Active Microwave Instrument (AMI). These data could be meaningfully assimilated only in new third generation wave models based on correct physics with freely adjustable two-dimensional spectra. Moreover, to retrieve two-dimensional spectra from the

SAR image spectra, acceptable first-guess wave spectra were required as input to remove the 180° directional ambiguities inherent in all frozen surface imaging systems (and also to augment the SAR measurements beyond the 100 m short wavelength azimuthal cut-off due to motion effects).

Another motivation for developing a reliable wave model was the need for first-guess wave fields for the application of improved scatterometer algorithms using more sophisticated backscattering models including short-wave/long-wave interactions. This has already been discussed in § I.4, where it was also pointed out that reliable wave model data are required to compute sea-state corrections to altimeter sea level data for the derivation of near-surface geostrophic currents. These errors arise through differences in backscatter returns from wave crests and wave troughs, which introduce a bias in the mean pulse shape and thereby distort the measurement of the round trip pulse travel time. To a lesser extent biases are also introduced through weakly non-Gaussian asymmetries in the statistical distribution of the backscattering wave heights.

Work on the analysis and assimilation of satellite wave data was first stimulated by the availability of SEASAT and GEOSAT off-line altimeter data, but has received a major impetus since the provision of near-real-time ERS-1 data. In the following we give a summary of altimeter and SAR measurements of ocean waves, discuss the algorithms used for wave data retrieval and present examples of the validation of the satellite wave data against the WAM model and buoy data. Methods of assimilating satellite wave data in wave models are discussed in chapter VI.

V.2 Radar altimeter measurements of wave height

S. Hasselmann, K. Hasselmann, E. Bauer and B. Hansen

In addition to information on the sea level, through the pulse travel time, and the wind speed, through the backscattered energy (cf. § I.4), the radar altimeter provides a measure of the significant wave height through the distortion of the mean shape of the return pulse. The earlier return from the wave crests and the retarded return from the wave troughs leads to a broadening of the return pulse which can be directly related to the significant wave height. To determine the mean pulse shape, several hundred pulses need to be averaged, yielding one significant wave height measurement about every 7 km along the satellite track. For a Gaussian sea surface, the relation between the pulse shape and the rms sea surface displacement can be computed theoretically (small corrections due to short-wave/long-wave interactions, discussed above in relation to sea level measurements, can be

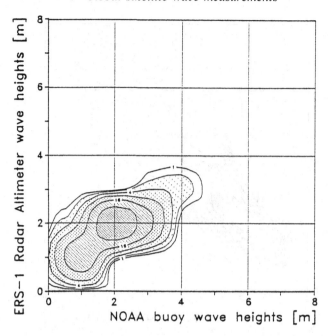

Fig. 5.1. Intercomparison of significant wave heights measured by the ERS-1 radar altimeter and by NOAA pitch–roll buoys in the North Atlantic and Northern Pacific for July 1992.

ignored to first order in this context). The model has been confirmed by numerous comparisons with *in situ* measurements. The typical accuracy of radar altimeter significant wave height measurements is of the order of 10% for wave heights between 1 and 20 m.

Comparisons between altimeter and buoy wave heights (Graber *et al*, 1994) for the SEASAT altimeter showed good overall agreement. Two recent studies for the ERS-1 altimeter (Goodberlet *et al*, 1992, Günther *et al*, 1993) also show satisfactory agreement for wave heights up to 4 m, although high waves tend to be underestimated by the altimeter relative to the buoy measurements (figure 5.1). The comparison was based on buoy data for a one month period from all USA NOAA ocean buoys in the western Atlantic and northern Pacific. Mastenbroek *et al* (1994) reached a similar conclusion from a comparison of ERS-1 data with North Sea observations.

The accuracy of altimeter wave height measurements (and the validity of the WAM model) is confirmed also by the global intercomparison of ERS-1 altimeter and WAM model wave heights for the month of July 1992 (figure 5.2). These data do not support the underestimation of high wave heights found in the intercomparison with buoy data. A regional decomposition of the data shows that the best agreement between model and altimeter data is found

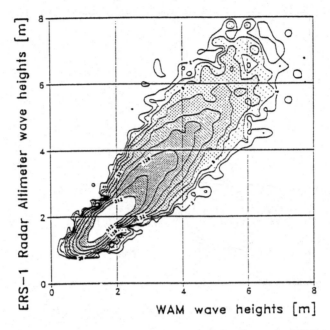

Fig. 5.2. Intercomparison of significant wave heights from the ERS-1 radar altimeter and computed by the WAM model for July, 1992, averaged over the globe.

in the northern hemisphere, while the WAM model wave heights tend to be somewhat lower than the altimeter wave heights in the southern hemisphere, where the wind fields are often underpredicted due to lack of data. This also leads to a model underprediction of swell in the tropics.

A more detailed regional intercomparison between the WAM model and altimeter wave heights is provided by the global mean wave height fields for January 1993 (figure 5.3). The figure shows the wave height data from two altimeters, ERS-1 and TOPEX/POSEIDON. The good overall agreement is again very encouraging for both the WAM model and the altimeter instruments. The strongest discrepancies are also seen here to occur in the regions of high winds in the Southern Hemisphere. A similar analysis with comparable results has been carried out for GEOSAT data by Romeiser (1993).

V.3 The Synthetic Aperture Radar (SAR)

S. Hasselmann, K. Hasselmann and C. Brüning

In contrast to the scatterometer, which is a low resolution, real aperture radar (RAR, cf. § I.4), the SAR is a high resolution imaging system. In the range direction, high resolution is achieved by measuring the travel time of short emitted pulses, while comparable resolution is achieved in the azimuthal (flight) direction by collecting the amplitude and phase histories of the returned signals from a large number of individual pulses to reconstruct the signal of a large virtual antenna. Spaceborne SARs on polar orbiting satellites flying at altitudes of approximately 800 km typically scan a swath of the order of 100 km with a resolution of 20 m × 20 m at incidence angles of 20° to 25° (SARs on lower flying space platforms can also operate at higher incidence angles). At these incidence angles the dominant backscattering mechanism is Bragg scattering: the resonant interaction of the incident electromagnetic microwaves with short ocean ripple waves which satisfy the Bragg resonance condition

$$\mathbf{k}_B = (0, 2k_r \sin \phi), \tag{5.1}$$

where $\mathbf{k}_B$ denotes the horizontal wavenumber vector of the short ocean ripples, the x and y components pointing in the azimuthal and radar look directions, respectively, k_r is the wavenumber of the radar waves and ϕ is the angle of incidence. The backscattered energy per unit surface area and unit incident radiation (specific radar cross-section σ_o) is proportional to the spectral density $F_r(\mathbf{k}_B)$ of the short (typically 5 to 30 cm) backscattering Bragg waves,

$$\sigma_0 = \gamma[F_r(\mathbf{k}_B) + F_r(-\mathbf{k}_B)], \tag{5.2}$$

where γ is a factor which depends on radar frequency, polarization, incidence angle and the dielectric constant of water. The energy density $F_r(\mathbf{k}_B)$ of the resonant short ripple waves depends on the wind speed, the azimuthal angle relative to the wind direction and the interactions with the long waves (Wright, 1968, Valenzuela, 1978, Hasselmann *et al*, 1990).

SARs are capable of detecting a variety of larger scale, oceanic phenomena which modulate the short (Bragg) ocean ripple spectrum, such as fronts, internal waves, natural surface films or man made oil slicks, bottom topography and ocean gravity waves.

Of particular interest to wave researchers has always been the SAR imaging of ocean waves, as two-dimensional instantaneous ocean-wave images clearly contain – if properly understood – valuable information for determining the full two-dimensional wave spectrum, which completely characterizes the local sea state. In the past, however, the very high data rates of the instrument have precluded the storage on board and subsequent transmission to a ground station of satellite SAR data, so that SAR images could be acquired only for areas within the reception zone of a ground station. High power requirements have furthermore limited the operational use of most SARs. On ERS-1, for example, the standard full-swath SAR operation is restricted to about 10 minutes per 100 minute orbit. A novel feature on the ERS-1 SAR, however, is

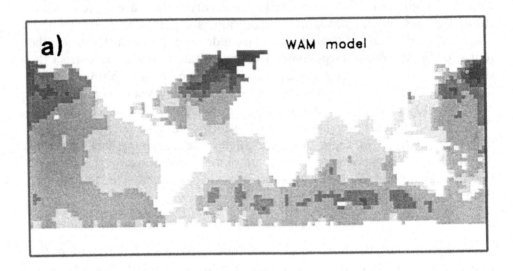

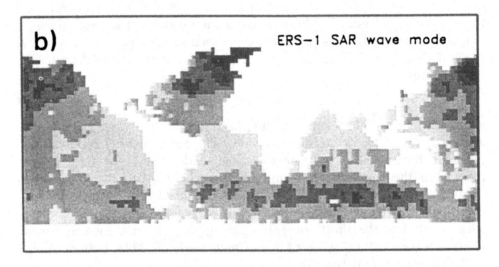

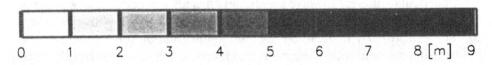

Fig. 5.3. Intercomparison of mean significant wave heights for January 1993 derived from different sources. Panel a: WAM model; panel b: the ERS-1 SAR wave mode. (*Continued on next page.*)

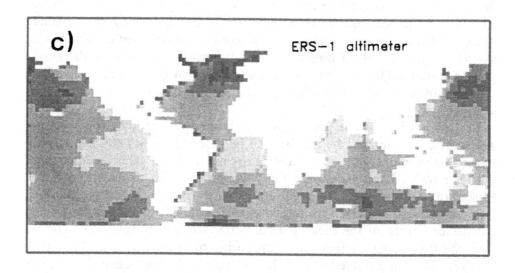

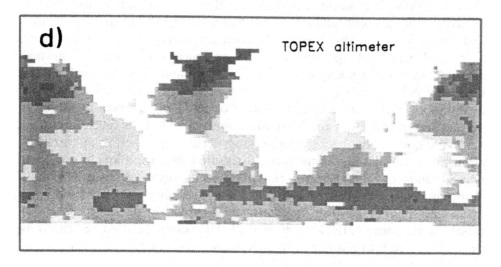

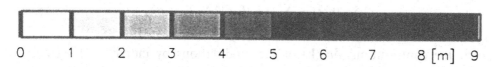

Fig. 5.3. (*Continued.*) Intercomparison of mean significant wave heights for January 1993 derived from different sources. Panel c: the ERS-1 altimeter; panel d: the Topex/Poseidon altimeter.

the incorporation of an intermittent sampling mode, which provides small 10 km x 5 km SAR images (imagettes) every 200 km along the satellite track. In this low power wave mode, the SAR can be operated continuously, the data being stored on board for later transmission to one of the ERS-1 ground stations. The ERS-1 SAR wave mode thus provides continuous, global sampling of ocean waves, yielding about 1500 two-dimensional ocean-wave images per day.

The modulation of the backscattering ripple waves by the longer ocean waves which are imaged by the SAR can be described by a two-scale model (Hasselmann *et al*, 1985, Alpers and Hasselmann, 1978). This assumes that the long waves have scales very much larger than the wavelength of the backscattering ripple waves – a condition which is well satisfied in practice. According to Bragg theory, the return signals from individual backscattering patches or 'facets' of the ocean surface arise through resonant interactions with the two Bragg components propagating towards and away from the antenna. These are filtered out of the full ripple spectrum defined over the small but finite backscattering patch. In the two-scale model the facet is regarded as large compared with the Bragg wave length but small compared with the scale of the long waves. The facet can therefore be regarded as an element of the local tangent plane of the long-wave sea surface. The normal direction and local velocity of the facet are determined by the local surface slope and orbital velocity of the long waves. Bragg theory is then applied to the local facet in a coordinate system oriented with respect to the facet normal and moving with the local advection velocity of the facet.

Modulations of the backscattering cross-section σ_0 arise through a) variations in the local angle of incidence associated with variations in the facet normal (*tilt modulation*) and b) variations in the energy of the Bragg backscattering ripples caused by hydrodynamic interactions between the short ripple waves and the longer waves (*hydrodynamic modulation*). To first order, both processes are linear in the long-wave field. The net modulation $\delta\sigma_o$ can thus be decomposed into Fourier components and represented as the sum

$$\delta\sigma_o(\mathbf{x}, t) = \sum_{\mathbf{k}} \delta\sigma_o(\mathbf{k}) e^{i(\mathbf{k}\cdot\mathbf{x}-\omega t)} + \text{c.c.} \qquad (5.3)$$

over the contributions $\delta\sigma_o(\mathbf{k})$ of the modulations by individual long-wave components $\mathbf{k}$, with

$$\delta\sigma_o(\mathbf{k}) = T(\mathbf{k})\eta(\mathbf{k})k, \qquad (5.4)$$

where $T(\mathbf{k})$ denotes the modulation transfer function (MTF), $\eta(\mathbf{k})$ is the local surface elevation, and k is the modulus of the wavenumber vector $\mathbf{k}$. The modulation transfer function is defined here, following common usage, as a dimensionless quantity referred to the wave slope $\eta(\mathbf{k})k$ rather than the wave height $\eta(\mathbf{k})$ (cf. Alpers and Hasselmann, 1978, Alpers *et al*, 1981, Monaldo and Lyzenga, 1986).

The tilt modulation transfer function can be computed from purely geometrical considerations, given the angle-of-incidence dependence of the Bragg backscattering cross-section and the form of the ripple wave spectrum. For a two-dimensional k^{-4} power-law ripple-wave spectrum one finds (cf. Alpers and Hasselmann, 1978)

$$T^{tilt} = 4ik_r \cot(\phi)(1 \pm \sin(\phi)^2)^{-1}, \tag{5.5}$$

where k_r and ϕ are defined as in equation (5.1) and the $\pm$ signs refer to vv and hh polarization, respectively. For small incidence angles (less than 30°) and horizontal polarization the formula reduces to

$$T^{tilt} = 8ik_r(\sin(2\phi))^{-1}. \tag{5.6}$$

The hydrodynamic mtf is less well known. Theoretical expressions containing some free empirical parameters have been derived using wkb hydrodynamic interaction theory by Wright (1968), Feindt *et al* (1986) and others. The ripple spectrum is described in the wkb two-scale theory by a spectral transport equation in which the long waves act on the ripple waves through refraction caused by the variable long-wave orbital velocity field. In addition, the long waves can modulate the wind input to the ripple waves. The resulting transfer function, as proposed by Feindt *et al* (1986), has the form

$$T^{hydr} = \frac{\omega - i\rho}{\omega^2 + \rho^2} k\,\omega \left(\frac{k_y^2}{k^2} + Y_r + iY_i \right), \tag{5.7}$$

where $\omega = 2\pi f$, f is the frequency, k_y is the ground-range wavenumber component in the radar look direction, ρ is a damping factor and Y_r, Y_i are two feedback coefficients representing the long-wave modulation of the wind input to the short waves.

The superposition of the tilt and hydrodynamic modulation transfer function yields the rar (real aperture radar) mtf. This describes the net cross-section modulation, which governs the imaging of ocean waves by a rar.

In addition to these two mechanisms, there exists for the sar an important third modulation mechanism. This is related to the use of phase information by the sar to locate the azimuthal position of a backscattering element. The advection of a backscattering element by the orbital velocity of the long waves induces an additional Doppler shift which is misinterpreted by the sar as an azimuthal offset of the position of the backscattering element. In an oscillating wave field, the oscillating Doppler shift of the return signal produces alternating azimuthal displacements of the backscattering ocean facets, resulting in variations of the apparent facet density in the sar image plane (Larson *et al*, 1976). This 'velocity bunching' process enables waves to be visualized in sar images even when no cross-section modulation is present.

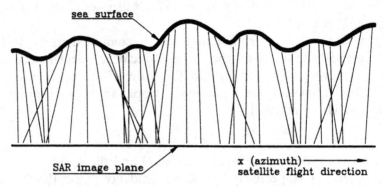

Fig. 5.4. Azimuthal displacement of a backscattering facet in the SAR image plane due to the long wave orbital velocity.

If the azimuthal displacement is significantly smaller than the wavelength of the long waves, the velocity-bunching effect is linear. However, if the facet displacement becomes comparable or large compared with the length of the longer waves, the wave patterns in the SAR image become severely distorted or can even be completely smeared out. The nonlinear distortion generally increases as the waves become shorter and ultimately limits the azimuthal resolution of the SAR at a finite cut-off wavenumber (cf. figure 5.4).

In the linear range, the velocity bunching modulation transfer function is given by (cf. Alpers *et al*, 1981)

$$T^{vb}(\mathbf{k}) = -i\beta k_x T^v(\mathbf{k}), \tag{5.8}$$

where

$$T^v(\mathbf{k}) = -\omega \left(\sin\phi \frac{k_r}{|k|} + i\cos\phi \right) \tag{5.9}$$

denotes the range–velocity transfer function (see also Lamb, 1932), k_x is the wavenumber in the azimuthal direction and β is the ratio of the slant range to the platform velocity. Because of the strong nonlinearity of the motion effects, however, the linear relation has only very limited application for spaceborne SARs, and we must consider generally the fully nonlinear SAR ocean-wave imaging mechanism. (See also Plant *et al*, 1983, Schröter *et al*, 1986, Plant, 1987.)

V.4 Extraction of wave spectra from SAR image spectra

S. Hasselmann, K. Hasselmann and C. Brüning

V.4.1 The forward mapping relation

The velocity bunching mechanism is a purely geometrical, fully determined process. If the RAR transfer function is also known, the mapping of the sea surface into the image plane and the resulting nonlinear transformation of the wave spectrum into a SAR image spectrum can be computed numerically for any given realization of the sea surface. In order to be useful for the operational assimilation of SAR ocean-wave image data in wave models, however, the computation must be accurate, computationally efficient and easy to invert. Two transformation schemes have been developed, both of which have been well validated against data, but of which only the second meets all three requirements for operational data assimilation.

The first method uses a Monte Carlo approach first described by Brüning *et al*, 1988, 1990). For a prescribed spectrum, a series of random realizations of an instantaneous ocean-wave field is created. For each realization, characterized by the instantaneous fields of the surface displacement and velocity, the sea surface is mapped into the SAR image plane pixel by pixel, using the fully nonlinear transformation relation. The images are then Fourier transformed, and the squared Fourier amplitudes averaged over the ensemble of realizations to provide an estimate of the SAR image variance spectrum. The method is relatively costly in computer time, as it typically requires 20 – 50 realizations of the sea surface to achieve acceptable statistical stability, and even then the results are still contaminated by residual statistical sampling errors. Furthermore, since the spectral mapping relation is not specified explicitly, the technique does not provide a straightforward approach for the inversion of the spectral mapping (cf. Brüning *et al*, 1990).

The second scheme makes use of the Gaussian property of the input wave spectrum, which enables all higher order nonlinear properties of the input wave field to be expressed in terms of the wave spectrum, to derive a closed integral transform relation between the ocean-wave spectrum and the SAR image spectrum (Hasselmann and Hasselmann, 1991, Hasselmann *et al*, 1991b, see also Krogstad, 1992). The computation is sufficiently fast to be applied operationally and can be readily inverted.

The Hasselmann and Hasselmann relation for the mapping of the surface wave spectrum $F(\mathbf{k})$ into the SAR image spectrum $P(\mathbf{k})$ (i.e. the symmetrical variance spectrum of the instantaneous SAR surface image) is given by

$$P(\mathbf{k}) = (2\pi)^{-2} \exp(-k_x^2 \xi'^2) \int d\mathbf{r} \Big\{ \exp(-i\mathbf{k} \cdot \mathbf{r}) \exp[k_x^2 \beta^2 f^v(\mathbf{r})]$$

$$\times \Big(1 + f^R(\mathbf{r}) + ik_x \beta [f^{Rv}(\mathbf{r}) - f^{Rv}(-\mathbf{r})]$$

$$+ (k_x \beta)^2 [f^{Rv}(\mathbf{r}) - f^{Rv}(0)][f^{Rv}(-\mathbf{r}) - f^{Rv}(0)] \Big) \Big\}, \qquad (5.10)$$

where

$$\xi'^2 = \beta^2 < v^2 >= \beta^2 \int |T^v(\mathbf{k})|^2 F(\mathbf{k}) \, d\mathbf{k} \qquad (5.11)$$

and $\mathbf{r}$ denotes the position vector, ξ' is the rms azimuthal displacement, $v(x)$ the range component of the local long-wave orbital velocity advecting the small-scale backscattering facet, and f^{Rv}, f^R, and f^v represent the covariance and autocovariance functions of the field v and the frozen surface image I^R. The latter three functions can be expressed as Fourier transforms of the wave spectrum:

$$f^v(\mathbf{r}) = < v(\mathbf{x} + \mathbf{r})v(\mathbf{x}) >= \int F(\mathbf{k}) |T^v(\mathbf{k})|^2 \exp(i\mathbf{k} \cdot \mathbf{r}) \, d\mathbf{k}, \qquad (5.12)$$

$$f^R(\mathbf{r}) = < I^R(\mathbf{x} + \mathbf{r})I^R(\mathbf{x}) >$$

$$= \frac{1}{2} \int [F(\mathbf{k}) |T^R(\mathbf{k})|^2 + F(-\mathbf{k}) |T^R(-\mathbf{k})|^2] \exp(i\mathbf{k} \cdot \mathbf{r}) \, d\mathbf{k}, \qquad (5.13)$$

$$f^{Rv}(\mathbf{r}) = < I^R(\mathbf{x} + \mathbf{r})v(\mathbf{x}) >$$

$$= \frac{1}{2} \int [F(\mathbf{k})T^R(\mathbf{k})T^v(\mathbf{k})^* + F(-\mathbf{k})T^R(-\mathbf{k})^* T^v(-\mathbf{k})] \exp(i\mathbf{k} \cdot \mathbf{r}) \, d\mathbf{k},$$

$$(5.14)$$

where $T^R(\mathbf{k}) = T^{tilt}(\mathbf{k}) + T^{hydr}(\mathbf{k})$ and $T^v(\mathbf{k})$ denote the RAR and range velocity transfer functions, respectively.

The nonlinear transfer integral itself also has the form of a Fourier transform, except for the nonlinear exponential term containing the orbital velocity autocovariance function. However, by expanding this term in a Taylor series one obtains a series of Fourier transforms of the form

$$P(\mathbf{k}) = \exp(-k_x^2 \xi'^2) \sum_{n=1}^{\infty} \sum_{m=2n-2}^{2n} (k_x \beta)^m P_{nm}(\mathbf{k}), \qquad (5.15)$$

where m denotes the nonlinearity order with respect to the velocity bunching parameter β (which always occurs in combination with k_x), n indicates the nonlinearity order with respect to the input wave spectrum and P_{nm} consists of nth order products of the covariance functions $f^R(\mathbf{r}), f^v(\mathbf{r})$ and $f^{Rv}(\mathbf{r})$. The common exponential factor represents the azimuthal cut-off arising from the nonlinear velocity bunching. A good approximation of the series (5.15) can normally be achieved by a truncation after about 12 terms.

It should be noted that the mapping relation (5.10) has been determined from first principles and is only weakly dependent on empirical coefficients. To first order the mapping is governed by the velocity bunching mechanism, a purely kinematic process, and the tilt modulation, which is also largely geometrical (except for the empirical power law for the ripple spectrum). Empirical parameters arise only in the hydrodynamic modulation transfer function (5.7), and affect a relatively narrow band of near-range travelling waves.

An important feature of the SAR imaging of surface waves is that no direct measurement of the mean backscattering cross-section σ_o is required, so that the resultant wave spectra can be calibrated independently of the SAR calibration. This is possible because the cross-section σ_o, which determines the absolute level of the SAR modulation spectrum, can be inferred indirectly from the clutter spectrum, which is measured simultaneously with the wave spectrum (Alpers and Hasselmann, 1982). The calibration method is based on the property that the statistically independent backscattering from individual patches of the sea surface gives rise to an essentially white background clutter spectrum superimposed on the wave modulation spectrum. The energy levels of both the background clutter spectrum and the wave modulation spectrum are proportional to the square of the mean cross-section. Thus the ratio of the two yields an internally calibrated SAR ocean-wave image spectrum, independent of the SAR calibration.

V.4.2 Inversion of the nonlinear mapping relation

The closed integral transform can be readily inverted by minimizing a cost function representing the error between the observed and the computed SAR spectrum (a more general discussion of the use of cost functions for inversion and assimilation problems is given in chapter VI). The inversion is not unique, however, because of the 180° directional ambiguity inherent in the spectrum of a frozen image and the loss of information in the image spectrum beyond the azimuthal cut-off of the SAR. To remove this indeterminacy, an additional regularization term is introduced which makes use of the information from a first-guess wave spectrum $\hat{F}(\mathbf{k})$:

$$ J = \int (P(\mathbf{k}) - \hat{P}(\mathbf{k}))^2 d\mathbf{k} + \mu \int \left[\frac{F(\mathbf{k}) - \hat{F}(\mathbf{k})}{B + \hat{F}(\mathbf{k})} \right]^2 d\mathbf{k}. \qquad (5.16) $$

Here $\hat{P}(\mathbf{k}), P(\mathbf{k})$ denote the observed and fitted SAR spectra, respectively; $\hat{F}(\mathbf{k})$ is the first-guess wave spectrum; $F(\mathbf{k})$ is the optimally fitted wave spectrum from which the optimally fitted SAR spectrum $P(\mathbf{k})$ was derived; and μ is a suitably chosen constant reflecting the confidence in the observed SAR spectrum relative to the first-guess wave spectrum. B is a small constant introduced to prevent the denominator of the second integral from becoming zero. (Formally, however, infinities in normalizing factors are acceptable, as they merely express an infinitely hard side condition.)

The solution of the local minimum condition,

$$\frac{\delta J}{\delta F(k)} = 0, \tag{5.17}$$

is obtained by iteration. The technique is based on alternating applications of the fully nonlinear mapping relation in the forward direction and a simplified mapping relation for the inversion. For the simplified mapping relation, equation (5.15) is truncated after the first term:

$$P(\mathbf{k}) = \exp(-k_x^2 \xi'^2) P_1(\mathbf{k}). \tag{5.18}$$

This 'quasi-linear' approximation (cf. Hasselmann and Hasselmann, 1991) yields a reasonable (for iteration purposes) first order approximation to the fully nonlinear mapping relation which can be inverted explicitly.

The iteration proceeds as follows: Assume that at the iteration level n a wave spectrum F_n and its associated SAR spectrum P_n, computed using the fully nonlinear mapping relation, have been determined. The iteration from the estimates F_n, P_n, to an improved estimate $F_{n+1} = F_n + \Delta F_n$ and $P_{n+1} = P_n + \Delta P_n$ is then constructed by assuming that the change ΔP_n is related to ΔF_n through the quasilinear relation (5.18). The new estimates for F_{n+1} and P_{n+1} are then substituted into (5.16), yielding

$$J = \int \{\Delta P_n(\mathbf{k}) - [\hat{P}(\mathbf{k}) - P_n(\mathbf{k})]\}^2 d\mathbf{k} + \mu \int \left(\frac{\Delta F_n(\mathbf{k}) - [\hat{F}(\mathbf{k}) - F_n(\mathbf{k})]}{B + \hat{F}(\mathbf{k})} \right)^2 d\mathbf{k}$$

$$\tag{5.19}$$

The variational problem with J given by (5.19) can be solved explicitly with respect to ΔF_n, yielding an improved solution F_{n+1}. Application of the full nonlinear mapping relation to F_{n+1} then yields the associated new SAR image spectrum P_{n+1}, and the iteration is repeated.

Normally the inversion method converges within four or five iteration steps. It requires approximately one second single-processor computing time on a CRAY-2 (see also Engen et al, 1994).

V.4.3 Results

Figure 5.5 shows different examples of observed, first-guess and optimally fitted SAR spectra (panels e, b, d, respectively) together with first-guess and optimally fitted wave spectra (panels a, c, respectively). The observed SAR

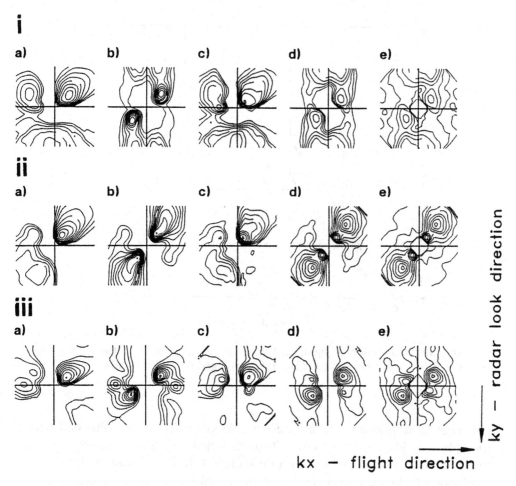

Fig. 5.5. Modelled first-guess wave spectra (panels a), associated computed SAR spectra (panels b), optimally fitted wave spectra (panels c), associated optimally fitted SAR spectra (panels d), and observed SAR spectra (panels e) for three typical cases at 15° S 26.7° W 01/22/1992 11:18 (series i), at 0.8° N 13.5° W 01/22/1992 11:13 (series ii), at 34.6° S 22.2° W 01/22/1992 11:23 (series iii). Isolines are in 3 db for the wave spectra and in 1 db for the SAR image spectra.

image spectra were computed from ERS-1 SAR wave mode data, while the first-guess wave spectra were computed with the operational global WAM model of the European Centre for Medium-Range Weather Forecasts (ECMWF).

The agreement between the observed SAR spectra and the best fit wave spectra is excellent (as it should be) for all cases for which convergence was achieved (about 95%). However, if the first-guess spectrum is very different from the true spectrum (for example if the peak direction differs by more than 50° from the true direction) the minimization routine does not always

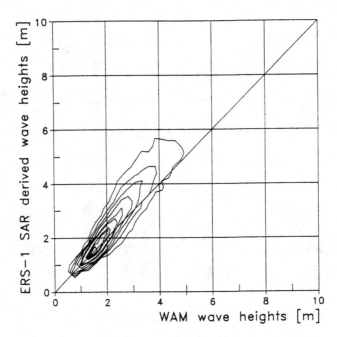

Fig. 5.6. Intercomparison of significant wave heights derived from the ERS-1 SAR with the WAM wave heights for July, 1992 averaged over the global ocean.

converge. The first-guess WAM spectra also generally agree quite well with the optimally fitted wave spectra, although systematic differences can be seen in some cases (usually indicating a deficiency in the wind field analysis).

Figure 5.6 shows a comparison of the mean ERS-1 SAR wave-mode wave heights with the WAM model wave heights for July, 1992, averaged over the global ocean. The level of agreement is comparable with that found in the intercomparison of the altimeter and WAM wave heights (figure 5.2). However, the model tends to underestimate the wave heights relative to the SAR for higher waves. In this respect SAR-derived wave heights seem to be more consistent with the buoy measurements (figure 5.1). The comparable quality of SAR and altimeter wave height data and the overall agreement of both satellite data sources with the WAM model wave heights is borne out also by a comparison of the three panels of figure 5.3.

Intercomparisons of further integral spectral parameters are shown in figures 5.7 (wavenumber), 5.8 (direction) and 5.9 (spectral spread). The mean wavenumber is somewhat overestimated by the WAM model, while the other parameters show reasonably good agreement. The mean parameters are determined in the usual manner as spectrally weighted averages, the mean direction being defined by $\bar{\theta} = \arctan(\overline{\sin\theta}/\overline{\cos\theta})$. The spectral spread is

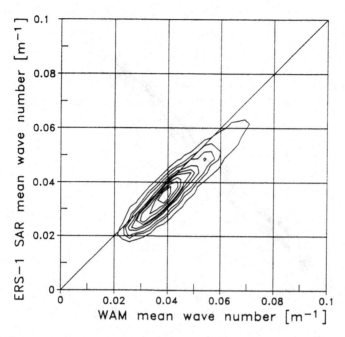

Fig. 5.7. Intercomparison of mean wavenumbers inferred from the ERS-1 SAR-derived wave spectra and from the WAM model wave spectra for the global ocean for July, 1992.

determined according to Yamartino (1984) as

$$\sigma_\theta = \sin^{-1}(\epsilon)[1 + 0.1547\epsilon^3],$$ (5.20)

where $\epsilon = \sqrt{1 - [\overline{\sin\theta}^2 + \overline{\cos\theta}^2]}$, with

$$\overline{\sin\theta} = \frac{\int\int \sin\theta F(f,\theta)\mathrm{d}f\mathrm{d}\theta}{\int\int F(f,\theta)\mathrm{d}f\mathrm{d}\theta}$$ (5.21)

and

$$\overline{\cos\theta} = \frac{\int\int \cos\theta F(f,\theta)\mathrm{d}f\mathrm{d}\theta}{\int\int F(f,\theta)\mathrm{d}f\mathrm{d}\theta}.$$ (5.22)

A comparison of mean spectral parameters alone cannot, however, do justice to the information contained in the full two-dimensional wave spectrum provided by the SAR. A measure of the overall level of agreement between two spectra can be given by the pattern correlation coefficient (cf. Brüning *et al*, 1993, 1994). However, although a high correlation index is useful in confirming good agreement when such agreement exists, a low correlation coefficient establishes only that the agreement is poor, without identifying the origin of the discrepancy. A useful technique for intercomparing the

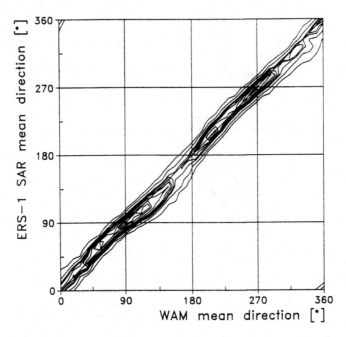

Fig. 5.8. Intercomparison of mean wave directions inferred from the ERS-1 SAR-derived wave spectra and from the WAM model wave spectra for the global ocean for July, 1992.

structure of wave spectra which avoids this difficulty has been proposed by Gerling (1992). Gerling attempts to capture the salient features of real complex wave spectra by representing the spectral distributions by a limited number of spectral parameters characterizing distinct wave systems within the spectrum. In the following applications the original spectral partitioning scheme of Gerling is modified by defining the wave systems as 'inverted catchment areas'.

In hydrology, a topographical domain is decomposed into a set of catchment areas associated with local minima of the topography. The catchment area of a local topographic minimum is defined as the area which drains into the minimum point. We define analogously the spectral *wave system* associated with a given peak of a wave spectrum as the catchment area of the local minimum corresponding to the inverted peak of the inverted spectral topography. Expressed directly, the domain of a *wave system* associated with a given spectral peak consists of all spectral points whose paths of steepest ascent lead to that peak. A path of steepest ascent is defined here on a discretized grid as the directed set of line segments connecting spectral grid points to the highest of the four nearest-neighbour grid points (figure 5.10). Mathematically (and numerically), the domain of a wave sys-

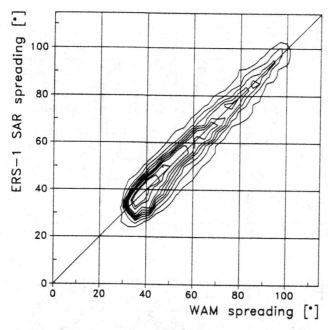

Fig. 5.9. Intercomparison of spectral spreads inferred from the ERS-1 SAR-derived wave spectra and from the WAM model wave spectra for the global ocean for July, 1992.

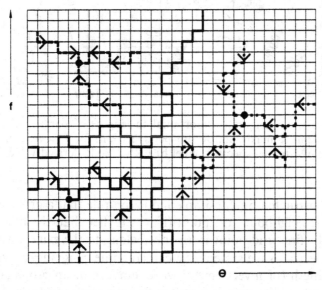

Fig. 5.10. Spectral partitioning scheme. Paths of steepest ascent are defined as the directed set of line segments connecting spectral grid points to the highest of the four nearest-neighbour grid points.

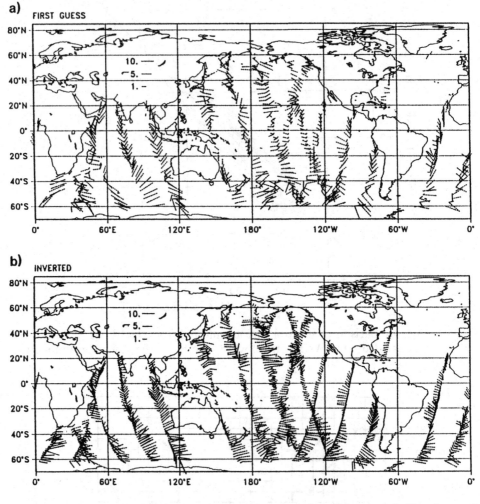

Fig. 5.11. Mean wave heights (proportional to lengths of bars) and directions of major wave systems (a maximum of four per spectrum) for 9 November 1992. Panel a: first-guess WAM model spectra; panel b: ERS-1 SAR-derived wave spectra.

tem can be constructed by the simple induction rule that a grid point and its highest nearest-neighbour grid point (if it is higher than the first grid point – otherwise the first grid point is a peak) belong to the same wave system.

For each spectral wave system one can compute a significant wave height, mean wave length (or inverse mean wavenumber), mean propagation direction and spread in the same way as for the complete spectrum. In practice, auxiliary criteria are needed to merge neighbouring secondary wave systems associated with spurious peaks arising from sampling variability or small-scale variations in the wind field.

Figure 5.11 shows a comparison of the principal wave systems (up to a maximum of four) for a one-day global set of SAR-derived and WAM model wave spectra. The figure demonstrates convincingly the complexity of typical open-ocean wave spectra.

There is a surprisingly good overall agreement between the structures of the SAR-derived and modelled spectra, validating both the WAM model and the ERS-1 wave mode instrument. Nevertheless, systematic differences are seen in individual wave systems, probably indicating errors in the analysed wind fields. These can be located more reliably from the partitioned wave system parameters than when only integral parameters of the complete spectrum are available. The optimal use of the detailed two-dimensional SAR wave mode data for wind and wave data assimilation clearly presents an exciting new challenge for wave modellers.

Chapter VI
Wave data assimilation and inverse modelling
S. Hasselmann *et al*

VI.1 General features of wave data assimilation and inverse modelling

S. Hasselmann and G. J. Komen

VI.1.1 Background

In previous chapters we have discussed the use of a mathematical *model*, the third generation wave prediction model, to compute the state of the sea. We can distinguish between hindcasts, nowcasts and forecasts, the difference being the time for which the sea state is computed relative to the clock time. A forecast field can be computed only by a model. However, the model estimation of hindcast and nowcast fields can be improved using *observations*, which have been considered in the previous chapters of this book only for validation of the underlying physics or for verification of model results. This combination of using model results and observations to create an optimal estimate of the sea state is called *data assimilation* or *analysis*. The word *analysis* originates in early meteorological applications, for which meteorologists would subjectively draw isobaric patterns on the basis of isolated pressure observations. They *analysed* the weather. Later, this work was carried out with numerical models.

Observed data can also be used to validate and improve models. When the model improvement is carried out using numerical automatic model fitting techniques, one speaks of *inverse* modelling: instead of using a given model to compute data, which are then compared with observations, the observations are used in an inverse mode to construct an optimally fitted model. While data assimilation is generally based on a dynamical model, inverse modelling deals with any mathematical model and adjusts free model parameters to achieve an optimal fit to data (Tarantola, 1987). But it can also be used to compute from an initial first-guess field and observations the best estimate initial model state, which is essentially the goal of data assimilation. Inverse modelling methods have many other features in common with data assimilation and will therefore be treated in this chapter together with data assimilation.

Atmospheric data assimilation (see, for example, Hollingsworth, 1986) began in the 1960s with the introduction of numerical weather prediction. Wave modellers, however, did not address this problem until much later. The quality of meteorological forecasts depends to a large extent on the correctness of the initial state from which the forecast starts, which means that the initialization is crucial in weather prediction. In wave modelling, however, this is not the case. In fact, after the model has spun up sufficiently, all memory of the initial conditions has vanished. Also, in early numerical wave models the quality of the wind field forcing was a limiting factor for the quality of the results. With the improvement of wind forecasting and analysis, the need arose for improving the quality of the wave models, together with the wish to combine wave models with more extensive wave observations.

Whereas the model improvement programme was quite successful, data assimilation attempts were for a long time hampered by the lack of high quality wave observations. An exception was the relatively dense observational network of the North Sea, which was the subject of several early wave data assimilation studies. However, on a global scale the data situation remained inadequate: visual observations from ships had large errors and the overall coverage was poor. A similar situation existed in meteorology, where measuring stations were providing sufficient observations over land but were only sparsely distributed over the oceans, with very few stations over the entire southern hemisphere.

The launch of ocean satellites has radically changed the situation and opened new perspectives for atmosphere and ocean modelling and climate research. Until recently, observations were available only at synoptic times (0, 6, 12, 18 hours) and were distributed mainly over land. Modern earth-observing satellites, however, also provide a continuous stream of observations over the oceans (see, for example, Intergovernmental Oceanographic Commission, 1986). The continuous influx of data implies that the present assimilation schemes at fixed time levels need to be extended to full four-dimensional systems including time interpolation. (In the case of wave models it would be more accurate to speak of two-dimensional spatial and three-dimensional space-time schemes, but we will adhere to the commonly used terminology, using the term four-dimensional simply to denote space-time interpolation.)

In the framework of data assimilation the model is also called a system, and both the physical system and the observations are described in terms of statistical variables. Replacing certain model variables by observations is regarded as selecting a particular realization from an ensemble of possible realizations. It is normally assumed that the errors between the data and the model state are normally distributed. Statistical variables were discussed in chapter I. They are quantities characterized by probability distributions rather than by single values. Once the statistical ensemble specifying the system and the probability of measurement outcomes have been specified, statistical methods to construct a best estimate of the sea state can be applied.

VI.1.2 An integrated data assimilation scheme

A major application for wave modelling is the assimilation of scatterometer, altimeter and SAR data in a combined wind and wave data assimilation scheme. Microwave sensors deployed on ocean-observing satellites were discussed in § I.4 and chapter V. A characteristic property of such instruments is that the information they yield is generally incomplete or ambiguous and must be complemented by additional information from other sensors or conventional observing systems:

- *Scatterometers* measure the small scale surface roughness, which is related to near-surface wind and surface stress. These data, together with boundary layer stability data from meteorological models, can be converted into surface winds, which can then be used as input in general atmospheric data assimilation schemes. For optimal inversion scatterometer algorithms require first-guess wind fields and for improved retrievals probably also first-guess wave fields.

- *Altimeters* provide significant wave height and wind speed. The observed wave heights together with first-guess model wave heights can be used to correct the energy of the two-dimensional wave spectrum. To update the complete two-dimensional wave spectrum itself, however, additional information or assumptions are needed (see § VI.5).

- SARS (*Synthetic Aperture Radars*) yield calibrated data on the full two-dimensional wave spectrum. This information is limited, however, by the 180° directional ambiguity inherent in frozen-surface ocean-wave images, and also by a high wavenumber azimuthal cut-off (i.e. in flight direction) arising from the nonlinearity of the imaging mechanism. However, as discussed in chapter V, by using model spectra as first-guess input, useful wave spectra can be extracted from the SAR. This more detailed information can be applied to significantly upgrade the simple data assimilation schemes using wave height information only.

A complete end-to-end wind and wave data assimilation system has been proposed by Hasselmann (1985). An outline of this system is given in figure 6.1. Because of the interrelation of sensor algorithms and first-guess model data, this system includes the sensor algorithms, such as the SAR inversion technique described in chapter V. The data extracted with the sensor algorithms, using first-guess forecasts as input, are then passed into an integrated model-data-assimilation scheme, i.e. a coupled atmosphere and wave model system. There all satellite and conventional data are combined with modelled first-guess fields to provide improved analysed initial fields for wave, wind and general weather forecasts. The improved analysed fields can then be used for various other applications, for example to provide wave climate statistics, or to compute the wind driven global ocean circulation from the surface wind and stress fields. It may still take several more years to realize such a complete system. So far, research has focussed on partial aspects, some of which will be discussed in the remainder of this chapter.

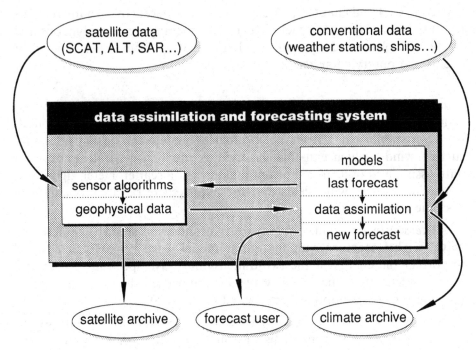

Fig. 6.1. An integrated wind and wave data assimilation system.

VI.1.3 Specific features of wave data assimilation

The assimilation of data into wave models is based on essentially the same technique as used for atmospheric models. However, there are a few differences:

– A characteristic feature of wave data assimilation is the need for a joint analysis of both wind and wave data, as the wave field responds very sensitively to the wind. Thus if the modified wave field consists of swell, the corrected swell would propagate throughout the rest of the forecast; but if the wave field consists mainly of wind sea, and the incorrect winds are not modified with the wave field, the corrected wave field will relax back rapidly into its original incorrect state. In atmospheric data assimilation a similar process is the translation of the wind speed observations into changes in the pressure field, without which the measured information would quickly be erased. Wave measurements provide, in fact, a highly sensitive indicator of the quality of the wind field (Intergovernmental Oceanographic Commission, 1986, Hasselmann *et al*, 1988, Janssen *et al*, 1989b), and it could even be argued that a major motivation for wave data assimilation lies not so much in the improvement of the wave analysis as in the improvement of the wind-field analysis.

- The continuous assimilation of wave data at individual grid points does not cause serious 'dynamical shocks', as in meteorological models, because the correction at a single grid point propagates into the forecast without dynamical coupling to neighbouring grid points (all physical processes are local). Therefore, the problem of dynamically balanced initialization, which requires filtering of gravity wave modes in atmospheric models, has no counterpart in wave data assimilation.
- The available data are typically more incomplete than in meteorological observing systems (containing, for example, only integral or directionally ambiguous information on the wave spectrum). A wind and wave data-assimilation scheme therefore has to address the problem of translating this incomplete information into optimal spectral updates.
- The region of influence of a wave observation cannot be estimated *a priori* but depends strongly on the sea state itself. For example, the spatial correlation scale of swell which has been generated by a distant storm is much larger than that of a wave field generated locally. A wind- and wave-data assimilation scheme needs to compute these dynamical regimes in the data assimilation cycle and to keep track of the 'age' of the spectral wave components, and may need to correct wind fields in the past and at some distance from the measurement position.

An important feature of any data assimilation scheme is that the observed data are inevitably less dense than the model grid, so that the information from data inserted at one grid point has to be distributed over neighbouring points to avoid unrealistic discontinuities in the evolution of the model. The space-time interpolation of data can be achieved using either kinematic or dynamical schemes:

- Kinematic (three-dimensional assimilation) schemes ignore the model constraints and create an analysed field at some time t_o by forming a linear combination of the model first-guess at time t_o and a weighted combination of all data collected in the synoptic time interval $[t_o - t_1, t_o + t_1]$ where t_1 is typically three hours. The weights are empirical functions of the model error, or are computed from the covariances of the model or observational error (optimal interpolation, see §§ VI.2.4 and VI.5).
- Dynamical schemes (four-dimensional assimilation schemes) insert observations under the constraints of the model equations. Examples are the adjoint technique (§§ VI.3, VI.7 – VI.9), Kalman-Bucy filtering (Kalman, 1960, Kalman and Bucy, 1961) and the Green's (impulse) function approach (§ VI.10).

Historically, the first simple wave data assimilation schemes were based on kinematic techniques. They corrected the wave field, while the wind field was updated only for the current time step or not at all (Janssen *et al*,

1987b, 1989b, Hasselmann *et al*, 1988). An example of such a simple scheme, which is running operationally at the European Centre for Medium-Range Weather Forecasting (ECMWF) in Reading, UK, is described in § VI.5. A similar scheme has been installed at the Royal Netherlands Meteorological Institute (KNMI).

Dynamical schemes are relatively complex and require significantly larger computing resources. They are currently being developed, but more experience will be needed before they can be implemented operationally.

General concepts of data assimilation will be introduced in the following section. In § VI.3 the data assimilation or inverse modelling schemes based on a suitably defined cost function will be briefly described. Applications of these schemes will be discussed in §§ VI.4 – 10.

VI.2 The general minimization problem

S. Hasselmann, G. J. Komen and K. Hasselmann

VI.2.1 Basic assumptions and notation

In chapters I and II the dynamical processes governing wave evolution were described. Recognizing that the sea surface elevation $\eta(\mathbf{x}, t)$ could best be treated as a stochastic process, the wave spectrum $F(\mathbf{k}; \mathbf{x}, t)$ was introduced. Data assimilation and inverse modelling involve a 'second stochastization', because the spectrum itself is now also treated as a stochastic process. It can never be computed exactly, due to effects not contained in the model, nor can it be measured exactly, because of measurement errors.

Thus any computed spectrum F should always be regarded as a particular realization of a statistical ensemble of possible spectra classified by a probability distribution $P[F(\mathbf{k}; \mathbf{x}, t)]$. In some approaches the statistical nature of F is explicitly specified, for instance by introducing an additional white noise forcing term in the energy balance equation, or by assuming that the initial conditions are random; in other approaches it is simply left open.

Similarly, measurements are also regarded as random variables, because of errors attributed to the measurement systems or because of imperfections in the algorithms used to translate raw observations into geophysical parameters.

A statistical model for the wave spectrum

The continuous function $F(\mathbf{k}; \mathbf{x}, t)$ can be represented in a discretized form on the model $\mathbf{k}$ and $\mathbf{x}$ grid points as a (random) state vector:

$$\hat{\boldsymbol{\Psi}}(t) = (\hat{\psi}_{ij\alpha\beta}(t)) = \left(F(k_{x_i}, k_{y_j}; x_\alpha, y_\beta, t) \right),$$

$$(i, j, \alpha, \beta,) \in [(1, 1, 1, 1), \cdots (i_{max}, j_{max}, \alpha_{max}, \beta_{max})] . \tag{6.1}$$

For a frequency-directional resolution of, say, 25 by 24, and 5000 active spectral grid points the random vector function $\hat{\boldsymbol{\Psi}}(t)$ already has 3 million components. Expression (6.1) can be simplified by the introduction of one large vector $\hat{\boldsymbol{\Psi}}$ containing all components labelled with a single index. The energy balance equation can then be written as:

$$\frac{d\hat{\boldsymbol{\Psi}}(t)}{dt} = T(\hat{\boldsymbol{\Psi}}(t), \mathbf{U}, t). \tag{6.2}$$

Note that $\hat{\boldsymbol{\Psi}}(t)$ refers to the state vector of the system at time t, T is a nonlinear operator consisting of the advection and source term applied to $\hat{\boldsymbol{\Psi}}$, which depends on the forcing wind fields $\mathbf{U}(t)$. In general these equations represent a nonlinear initial value problem with $\hat{\boldsymbol{\Psi}}(t_o) = \hat{\boldsymbol{\Psi}}_o$ at initial time t_o.

Further simplification of the notation can be achieved by discretizing also in time, and introducing one large vector $\boldsymbol{\Psi}$ of dimension n to describe the trajectory of the state vector of the system:

$$\boldsymbol{\Psi} = (\psi_i) = (\hat{\boldsymbol{\Psi}}(t_1), \hat{\boldsymbol{\Psi}}(t_2), \cdots \hat{\boldsymbol{\Psi}}(t_p)), \quad i = 1, \cdots n. \tag{6.3}$$

In general, n is very large, given by the product of the dimension of $\hat{\boldsymbol{\Psi}}(t)$ (of the order 10^6) and the number p of time levels considered (which is usually of the order of 10^2).

In addition to the system variables $\boldsymbol{\Psi}$, other external variables have to be considered. Some of these can be regarded as deterministic and given, others as random, e.g. the forcing wind field, the initial model state, or free model parameters. We refer to these variables as control variables $\boldsymbol{\Phi}$. In a model fitting or data assimilation application, the control variables are allowed to be modified, for example by optimally fitting the model to data. In equation (6.2), control variables could be the set

$$\boldsymbol{\Phi} = (\phi_i) = (\hat{\boldsymbol{\Psi}}(t_0), \mathbf{U}(t_0), \mathbf{U}(t_1), \cdots \mathbf{U}(t_p)), \quad i = 1, \cdots m. \tag{6.4}$$

With the introduction of discrete time one can rewrite the dynamical model (6.2) in the form:

$$\mathbf{E}(\boldsymbol{\Psi}, \boldsymbol{\Phi}) = (E_i) = 0, \quad i = 1, \cdots n. \tag{6.5}$$

The model equations provide exactly n, in general nonlinear, constraints. It should be noted that the E_i depend on the particular discretization chosen, e.g. the semi-implicit time stepping and first order upwind scheme of the WAM model has a different form for E_i than a higher order explicit advection scheme. The problem of solving equation (6.2) can now be reformulated as follows:

Given m (random) control variables ϕ_i, $i = 1, \cdots m$ solve (6.5) for the trajectory vector $\mathbf{\Psi} = (\psi_i)$, $i = 1, \cdots n$.

In a complete statistical theory, we would like to compute the probability distribution function $P(\mathbf{\Psi})$ or, equivalently, the set of all moments

$$< \psi_{i_1}, \psi_{i_2}, \cdots \psi_{i_l} >=$$
$$\int d\psi_{i_1}, d\psi_{i_2}, \cdots d\psi_{i_l} (\psi_{i_1}, \psi_{i_2}, \cdots \psi_{i_l}) P_{\psi_{i_1}, \psi_{i_2}, \cdots \psi_{i_l}} (\psi_{i_1}, \psi_{i_2}, \cdots \psi_{i_l}). \quad (6.6)$$

Finding $P(\mathbf{\Psi})$ or all moments is, of course, impossible in practice. Instead one usually tries to obtain a reduced set of statistical properties of $\mathbf{\Psi}$, even if only approximately.

Observations as random variables

As indicated above, random variables are also needed to describe observations. Measuring the complete wave spectrum at each model grid point and at all times t_i yields one particular measurement $\mathbf{\Psi}^o$ of the system. Repeated measurements under identical system conditions will yield different $\mathbf{\Psi}^o$s. Of course, $\mathbf{\Psi}^o$ is introduced only for conceptual convenience, because for practical reasons it is impossible to measure all its components completely. Also, often quantities are measured which are only indirectly related to the state vector $\mathbf{\Psi}$, such as values between grid points (which can be obtained by spatial interpolation) or integral spectral parameters, such as the significant wave height. Therefore, in practice, it is not $\mathbf{\Psi}^o$ that is determined but only a relatively small number of quantities, which will be denoted by

$$d_i^o = \mathscr{D}_i(\mathbf{\Psi}^o), \quad i = 1, \cdots n_{obs}, \quad (6.7)$$

with, in general, n_{obs} much smaller than n. Below, these observations will be compared with their model counterparts, which are given by:

$$d_i = \mathscr{D}_i(\mathbf{\Psi}), \quad i = 1, \cdots n_{obs}, \quad (6.8)$$

with $\mathscr{D}_i$ essentially the same operator in (6.7) and (6.8). In a typical example $\mathbf{d}^o = (d^o)_i$ may be the set of all wave heights from conventional buoys on a particular day.

Table 6.1. *Different approaches to data assimilation and inverse modelling, as characterized by the different terms considered in the cost functions (for further explanation see §§ VI.2.2 and VI.2.4).*

	method/application	J^d	J^f	J^c	J^E	J^t	§ VI.n
1)	optimal interpolation/ data assimilation	-	-	-	-	×	5
2)	straightforward least square fitting/ inverse modelling	×	-	(×)	∞	-	6
3)	adjoint model/ inverse modelling	×	-	(×)	∞	-	7
4)	adjoint model/ data assimilation	×	×	×	∞	-	3,8,9
5)	Green's function method/ data assimilation	×	×	×	∞	-	10

VI.2.2 Cost functions

Having established a conceptual framework and notation, the general data assimilation and inverse modelling problem may be formulated as follows:

Find by variation of the control variables Φ the *best* or *optimal* estimate ψ^e for the state vector Ψ of a dynamical model.

We need to define what is meant by *best* or *optimal*. Different methods of data assimilation or inverse modelling differ in their definition of *best*. All methods have in common that the optimal solution is defined as the solution which minimizes some prescribed *cost* function (the terminology is taken from applications in economics). The methods differ in their definition of the cost function. There are also differences in the mathematical technique used to determine the minimal cost solution. We may thus classify the various data assimilation and inverse modelling techniques in a two-dimensional table with respect to the structure of the cost function and the mathematical minimization method (table 6.1).

The cost functions normally represent a trade off between a number of competing requirements. They consist typically of a sum of quadratic forms, each form representing the error incurred in trying to meet a particular requirement. Different weights are assigned to each of the penalty terms in accordance with the relative importance attached to the different requirements. The individual quadratic forms can also vary. Thus the elements of the matrix or *metric* defining the (positive definite) quadratic form can be adjusted to express the relative weights assigned to the different data components appearing in the error expression.

Cost functions are generally constructed from a subset of the following five basic penalty terms:

(i) The misfit between the observed and modelled data $\mathbf{d}^o$ and $\mathbf{d}$, respectively, expressed as an error function

$$J^d = \sum_{i,j}(d_i^o - d_i)M_{ij}^d(d_j^o - d_j), \qquad (6.9)$$

with some suitably defined positive definite metric M_{ij}^d. In matrix notation,

$$J^d = (\mathbf{d}^o - \mathbf{d})^T M^d(\mathbf{d}^o - \mathbf{d}). \qquad (6.10)$$

(ii) The deviation between the model data $\mathbf{\Psi}$ and some first-guess model values $\mathbf{\Psi}^f$, obtained, for example, from a model prediction:

$$J^f = (\mathbf{\Psi} - \mathbf{\Psi}^f)^T M^f(\mathbf{\Psi} - \mathbf{\Psi}^f). \qquad (6.11)$$

(iii) The difference between the optimally fitted control variables $\mathbf{\Phi}$ and some first-guess expectation of the most likely set of control variables $\mathbf{\Phi}^f$, expressed again as an error term

$$J^c = (\mathbf{\Phi} - \mathbf{\Phi}^f)^T M^c(\mathbf{\Phi} - \mathbf{\Phi}^f). \qquad (6.12)$$

(iv) A possible error incurred in satisfying the dynamical model equations (6.5) with the chosen set of control variables $\mathbf{\Phi}$ and model state variables $\mathbf{\Psi}^f$,

$$J^E = \mathbf{E}(\mathbf{\Psi},\mathbf{\Phi})^T M^E \mathbf{E}(\mathbf{\Psi},\mathbf{\Phi}). \qquad (6.13)$$

This error function is meaningful, of course, only if the model used to compute the state of the system is not the true dynamical model, but some approximation, otherwise the control variables always yield a model state for which J^E vanishes identically. However, the error term can be also retained formally in data assimilation or inverse modelling schemes which satisfy the dynamical equations exactly by assigning to it an infinite weighting factor (table 6.1).

(v) The difference between the modelled system state $\mathbf{\Psi}$ and the true state of the system $\mathbf{\Psi}^t$, expressed as an error function

$$J^t = <(\boldsymbol{\Psi} - \boldsymbol{\Psi}^t)^T M^t (\boldsymbol{\Psi} - \boldsymbol{\Psi}^t)>, \qquad (6.14)$$

where the parentheses $< \dots >$ denote statistical ensemble averages. The true state of the system will, of course, generally not be known. However, this is irrelevant, since the error function (6.14), in contrast to the other error functions, is defined here as a statistical expectation value, so that it involves only the first and second statistical moments of the true state of the system, not the true system state itself. These will normally be known, or can at least be estimated.

The error function J^t is normally used only in optimal interpolation schemes (table 6.1 and § VI.2.4), where it is furthermore restricted to the assimilation of data at a given time level. In this case the matrix M^t acts only on data at a given time level (or within a restricted time window), so that the error function can be written

$$J^t = \hat{J}^t = <(\hat{\boldsymbol{\Psi}} - \hat{\boldsymbol{\Psi}}^t)^T M^t (\hat{\boldsymbol{\Psi}} - \hat{\boldsymbol{\Psi}}^t)> . \qquad (6.15)$$

In addition to these five penalty terms, the cost function is sometimes extended to include other terms penalizing properties of the solution which are deemed undesirable, such as spatial or temporal roughness. The cost function is also sometimes augmented by terms which do not correspond to a quadratic form, for example to prohibit the wave spectrum from becoming negative (Long and Hasselmann, 1979). These features are usually restricted to special inverse modelling applications, however, and will not be considered in the following.

VI.2.3 Maximum likelihood derivation of the error matrices

In deciding on the cost function for a data assimilation or inverse modelling problem one needs to define both the relative weights of the separate cost function terms and the error matrices for each of the individual error terms. Whereas the relative weights are usually based on subjective judgements, the error matrices can often be determined from objective criteria. This is the case whenever the variables in question represent random variables with known Gaussian probability distributions. We illustrate the method for the case of the error term J^d.

Consider the probability distribution $P(\mathbf{d} - \mathbf{d}^o; \boldsymbol{\Phi})$ for $\mathbf{d} - \mathbf{d}^o$, given a set of values for the control variables $\boldsymbol{\Phi}$. When regarded as a function of external parameters rather than as a function of its random variables this probability is referred to as a *likelihood* function. In the maximum likelihood approach *best* or *optimal* is defined as the solution for which P, regarded as a function of $\boldsymbol{\Phi}$ for fixed $\mathbf{d}$ and $\mathbf{d}^o$, is maximized. We will show that for the particular case of a Gaussian distribution this translates into a quadratic-form cost function.

The determination of the maximum of P for an arbitrary likelihood function is generally a numerically intractable problem. However, for a linear system with many degrees of freedom it can often be argued by the Central Limit Theorem that P is Gaussian:

$$P(\mathbf{d} - \mathbf{d}^o) \simeq \exp\left(-\frac{1}{2}(\mathbf{d} - \mathbf{d}^o)^T M(\mathbf{d} - \mathbf{d}^o)\right), \tag{6.16}$$

where M is the inverse of the error covariance matrix,

$$M_{ij} = < (d - d^o)_i (d - d^o)_j >^{-1}. \tag{6.17}$$

In order that P is a maximum, the exponent must be a minimum, so that maximizing (6.16) is equivalent to the minimization of the cost function

$$J^d = \frac{1}{2}(\mathbf{d} - \mathbf{d}^o)^T M(\mathbf{d} - \mathbf{d}^o). \tag{6.18}$$

Thus we find $M^d = M$.

We emphasize again that the maximum likelihood justification of the optimal solution as the solution which minimizes J^d with $M^d = M$ is valid only for a Gaussian process. We could also have arrived at the same result by starting from our original definition of *optimal* as the solution which *minimizes* the cost J^d, with an as yet undetermined metric M^d. Although the choice of metric M^d appears at first sight arbitrary, one can argue now that if the variables $d_i - d_i^o$ are transformed linearly to a new set of variables $\tilde{d}_i - \tilde{d}_i^o$ which are statistically orthonormal, the only natural choice of metric $\tilde{M}$ for this new set of variables, all of which have the same variance and are uncorrelated, is the unit matrix $\tilde{M}^d = I$. On transforming back to the original variables, this is seen to be identical to the definition (6.17).

The maximal likelihood definition of the cost function matrix can be applied generally to the cost functions J^d and J^t. The approach is applicable also to the cost functions J^f and J^c, if reasonable estimates of the statistics of the first-guess errors or the probabilities of deviations of the control variables from their first-guess values are available. For the 'cost function' J^E the definition of the cost-function matrix is irrelevant in the cases we shall be concerned with (table 6.1), since in all cases the cost function J^E is either excluded or included with an infinite weight, i.e. the model equations are either ignored or satisfied exactly as a side condition in the minimization problem.

VI.2.4 Cost function classification of methods

(1) Optimal interpolation

One of the simplest and most widely used methods of assimilating data in a model is the *optimal interpolation* method. This is based on a purely statistical interpolation technique without regard to the dynamical model contraints. It is used most frequently in meteorological data assimilation in a single time level mode (Bengtsson *et al*, 1981, Liebelt, 1967). Because of its relative simplicity it was also the first technique to be used in wave data assimilation (see § VI.5).

At each time level, all observations available in a time window (normally 6 hours) centred at the time level are used to construct the 'most likely' estimate of the field at that time level. The basic idea is to compute an analysed or best estimate value $\hat{\boldsymbol{\Psi}}$ of the true state vector $\hat{\boldsymbol{\Psi}}^t$ at time t in the form of a linear combination of a first-guess value and a weighted error between the observations and their model state vector counterparts:

$$\hat{\psi}_i = \hat{\psi}_i^f + \sum_{j=1}^{n_{obs}} W_{ij}(d_j^o - d_j^f), \tag{6.19}$$

where, as before ψ_i^f denotes the first-guess fields, obtained from a previous model run, d_j^o denotes the n_{obs} observations and d_j^f their first-guess model counterparts:

$$\mathbf{d}^f = \mathscr{D}(\hat{\boldsymbol{\Psi}}^f). \tag{6.20}$$

The index i labels the different components of the dynamical state vector. In general the range of i is much larger than the range of j.

The interpolation weights W_{ij} in equation (6.19) can be obtained by requiring that the mean square error between the true state vector $\hat{\boldsymbol{\Psi}}$ and its best estimator $\hat{\boldsymbol{\Psi}}^t$ is minimized (minimum variance of $\hat{\boldsymbol{\Psi}}$), i.e. that the cost function

$$J^t =< (\hat{\boldsymbol{\Psi}} - \hat{\boldsymbol{\Psi}}^t)^T (\hat{\boldsymbol{\Psi}} - \hat{\boldsymbol{\Psi}}^t) >, \tag{6.21}$$

with unit error matrix $M^t = I$ is minimized. Although we apply the minimum condition here to the square of the complete difference vector $\hat{\boldsymbol{\Psi}} - \hat{\boldsymbol{\Psi}}^t$, it should be noted that the minimization problem (6.21) is, in fact, separable into a set of minimization problems for the squares of each of the individual components of the difference vector. This is because the weights W_{ij} are determined independently for each separate index i. The mean square solutions for the individual components i of the difference vector are then seen to be identical, for Gaussian probability distributions, to the maximum likelihood solutions.

To obtain a useful equation for the weights we write both $\mathbf{d}^o$ and $\mathbf{d}^f$ as a mean value (assumed to be the same) and a deviation

$$\mathbf{d}^o = \bar{\mathbf{d}}^o + \tilde{\mathbf{d}}^o, \qquad \mathbf{d}^f = \bar{\mathbf{d}}^f + \tilde{\mathbf{d}}^f, \tag{6.22}$$

where $\tilde{\mathbf{d}}^o$ is the observational error and $\tilde{\mathbf{d}}^f$ the model first-guess error. Inserting (6.22) into (6.19) and the resulting expression for $\hat{\psi}_i$ into (6.21), and differentiating subsequently with respect to W_{ij}, a set of equations for the interpolation weights W_{ij} is obtained:

$$\sum_{j=1}^{n_{obs}} (\sigma_{jk}^o + \sigma_{jk}^f) W_{ij} = \frac{1}{2} \left(<\psi_i^f \tilde{d}_k^f> + <\tilde{d}_i^f \psi_k^f> \right), \quad k = 1, \cdots n_{obs}, \tag{6.23}$$

where

$$\sigma_{ij}^o = <\tilde{d}_i^o \tilde{d}_j^o> \quad \text{and} \quad \sigma_{ij}^f = <\tilde{d}_i^f \tilde{d}_j^f> \tag{6.24}$$

denote the covariance matrix of the observational error and the first-guess field, respectively.

In matrix form (6.23) can be written as

$$\sum_j A_{jk} W_{ij} = B_{ik} \tag{6.25}$$

so that

$$W_{ij} = \sum_k A_{kj}^{-1} B_{ik} \quad . \tag{6.26}$$

(Note that the index i appears in (6.26) only as a dummy index, i.e. the problem is separable with respect to the state vector components, as pointed out above.)

Equation (6.19) is applicable for any form of data, such as buoy data, radar altimeter wave height observations and even two-dimensional spectra observed by a synthetic aperture radar. The distribution of error corrections onto neighbouring grid points, which is dependent on the covariance scales, is automatically taken care of by this technique. However, the determination of the covariance matrix is not straightforward. Long term statistics of the model and the observational errors are needed. If these are not available, simple empirical relations are normally used (see also § VI.5).

(2) Direct least square fitting

The simplest method of tuning a model with respect to a set of free parameters Φ is to define a cost function (cf. table 6.1)

$$J = J^d + J^c \tag{6.27}$$

and then minimize J directly by systematically adjusting the model parameters (control variables), the model data $\mathbf{d}$ being computed explicitly each time the control variables are modified. (Solving the model equations explicitly corresponds formally to including an additional cost function term J^E with infinite weight in J, cf. table 6.1.)

The error metric M^d of the first cost-function term J^d is normally estimated as the inverse of the error covariance matrix (§ VI.2.3), while the term J^c penalizing the deviation of the control variables from some set of first-guess values is optional and usually based on subjective criteria.

The minimization of J can be carried out using a Newton or conjugate gradient method of steepest descent. If the model is relatively complex (which is the case for third generation wave models) the gradient of the cost function can be computed only numerically. The computation is then very expensive if the dimension of the control variable vector $\mathbf{\Phi}$ is large. The direct least square solution method is therefore feasible only if the number of control variables is small. A typical application is the tuning of a wave model with respect to a few empirical coefficients in the source function (§ VI.6).

(3) The adjoint model technique applied to inverse modelling

The adjoint modelling technique was first developed and applied in meteorological and oceanographic problems as a method of initialization in forecasting applications (Lewis and Derber,1985,Talagrand and Courtier, 1987, Courtier and Talagrand, 1987, Thacker and Long, 1988, Long and Thacker, 1989). However, in wave modelling, the first adjoint model applications were in the simpler field of inverse modelling (Lawson, 1992, Snyder *et al*, 1992).

The adjoint model technique avoids the prohibitive computation costs incurred in the determination of the cost-function gradient for a large set of control variables in the direct minimization method. The model equations are not solved explicitly when computing the cost-function gradient, but are included as an additional side condition term with appropriate Lagrange multipliers in the cost function. Once the Lagrange multipliers ('costate variables') have been determined, the gradient of J can be computed as the gradient of the extended cost (Lagrange) function without explicitly imposing the model equations as side condition.

The cost function is thus identical to the form (6.27) (formally with an infinite term J^E to represent the model constraints), but the method of minimization differs from the direct minimization approach. The technique is discussed in more detail in § VI.3.

(4) The adjoint model technique applied to data assimilation

In considering now the application of the adjoint modelling technique to wave data assimilation the only difference with respect to the previous inverse modelling application is that the cost function

$$J = J^d + J^f + J^c \qquad (6.28)$$

must now be augmented by an additional term J^f representing the deviation of the model field from the first-guess field Ψ^f (cf. table 6.1). Furthermore, the error term J^c for the control variables (normally the forcing wind field, but possibly also a current or some other environmental field) is now no longer optional. The error matrices M^f and M^c of both terms can usually be estimated by maximum likelihood methods from the inverse covariance matrices of the model and wind field errors, respectively. Since the first-guess forcing wind fields are typically of higher dimension, an adjoint model or similar technique which avoids solving the model equations explicitly in computing the cost-function gradient is clearly a prerequisite for a successful minimization technique. The model and adjoint model equations nevertheless still need to be integrated many times in successive computations of the cost-function gradient when seeking the cost-function minimum, so the adjoint model technique is still computationally expensive. Applications are discussed in §§ VI.7 – 9.

(5) *The Green's function method of wave data assimilation*

The need to compute the cost-function gradient and to integrate the wave model equations a large number of times in an iterative search for the cost-function minimum is avoided in the Green's function technique. Here the perturbed wave model equations determining the response of the wave spectrum to small perturbations of the forcing wind field are inverted explicitly. The wave spectrum perturbations are thus expressed in terms of the wave-spectrum impulse response or Green's function as a space-time integral over the wind field perturbations.

   The cost function is thereby reduced to a function of the forcing wind field only, and its minimum can be readily computed. Unfortunately, the exact determination of the Green's function of the perturbed wave model equations is, in practice, an insoluble task, so that the approach can be carried through only after several approximations have been introduced. The advantage of the Green's function technique, however, is that the required computation time, in contrast to the adjoint model or the direct minimization approach, is of the same order as integration of the model itself, so that it can be readily implemented operationally. The method is described in more detail and applied in § VI.10.

VI.3 The adjoint model technique

K. Hasselmann and S. Hasselmann

VI.3.1 General principle

In the following we describe the adjoint model method in more detail. The implementation for wave models will be described later in the applications. The method is described for the general data assimilation problem; its application to inverse modelling is identical except that the term J^f penalizing the deviation from the first-guess field is not included.

The problem is to minimize the cost function (cf. table 6.1)

$$
\begin{aligned}
J &= J^d + J^f + J^c \\
&= (\mathbf{d} - \mathbf{d}^o)^T M^d (\mathbf{d} - \mathbf{d}^o) + (\mathbf{\Psi} - \mathbf{\Psi}^f)^T M^f (\mathbf{\Psi} - \mathbf{\Psi}^f) \\
&\quad + (\mathbf{\Phi} - \mathbf{\Phi}^f)^T M^c (\mathbf{\Phi} - \mathbf{\Phi}^f)
\end{aligned}
\tag{6.29}
$$

under the side conditions

$$
E_i(\mathbf{\Psi}, \mathbf{\Phi}) = 0, \quad i = 1, \cdots n.
\tag{6.30}
$$

It is assumed that the equations can be solved for $\mathbf{\Psi}$: $\psi_i = \psi_i(\mathbf{\Phi})$.

For conceptual convenience we assume further that the solution lies close to the initial reference trajectory, which can be defined as $\mathbf{\Psi} = \mathbf{\Psi}^f = 0$, $\mathbf{\Phi} = \mathbf{\Phi}^f = 0$, $\mathbf{d} = 0$, and that the data relations and side conditions can be linearized relative to this reference trajectory.

The linearized model equations then take the form :

$$
P\delta\mathbf{\Psi} + Q\delta\mathbf{\Phi} = 0
\tag{6.31}
$$

or in components:

$$
P_{ij}\delta\psi_j + Q_{ik}\delta\phi_k = 0,
$$

where

$$
P_{ij} = \partial E_i/\partial \psi_j \quad \text{and} \quad Q_{ik} = \partial E_i/\partial \phi_k \ .
$$

The corresponding linearized data relations equation (6.20) are given by:

$$
\delta\mathbf{d} = R\delta\mathbf{\Psi} \quad \text{or in matrix representation:} \quad \delta d_v = R_{vi}\delta\psi_i.
\tag{6.32}
$$

The linearization approximation is not, in fact, a basic restriction, as the minimal solution is normally determined by iteration. The fully nonlinear system equations can then be applied after each iteration step to reinitialize the minimization procedure. As the iterations converge towards the minimal solution, the linearization approximation becomes successively more accurate. To simplify the notation we replace $\delta\mathbf{\Psi}$, $\delta\mathbf{\Phi}$ and $\delta\mathbf{d}$ in the following by $\mathbf{\Psi}, \mathbf{\Phi}$ and $\mathbf{d}$, respectively.

Direct elimination method

An obvious (nonadjoint) method of solving the minimization problem is first to solve the model equations (6.31) for $\boldsymbol{\Psi}$ as a function of $\boldsymbol{\Phi}$,

$$\boldsymbol{\Psi} = -P^{-1}Q\boldsymbol{\Phi} \tag{6.33}$$

and then substitute the solution into (6.29), using (6.32), to express J as a function of $\boldsymbol{\Phi}$ alone:

$$J = \boldsymbol{\Phi}^T A\boldsymbol{\Phi} + \boldsymbol{\Phi}^T B^T \mathbf{d}^o + (\mathbf{d}^o)^T B\boldsymbol{\Phi} + (\mathbf{d}^o)^T M^d \mathbf{d}^o, \tag{6.34}$$

where

$$A = (RP^{-1}Q)^T M^d (RP^{-1}Q) + (P^{-1}Q)^T M^f (P^{-1}Q) + M^c \tag{6.35}$$

and

$$B = M^d (RP^{-1}Q). \tag{6.36}$$

The minimization of (6.34) is a standard least square problem which can be solved by conjugate gradient methods once the matrix P^{-1} has been determined. In this approach P^{-1} is determined once and for all, so that no iterative integrations of the model and adjoint model are needed, in contrast to the adjoint method considered in the following. However, the determination of the $n \times n$ Green's function matrix P^{-1} normally requires much more computation than the determination of one particular solution $\boldsymbol{\Psi}$ for given $\boldsymbol{\Phi}$, which is the approach pursued in the adjoint method. The direct elimination method is therefore feasible only if the inversion of the model equations is significantly simplified through the introduction of appropriate approximations. We discuss this approach in § VI.10.

The adjoint method

From the theory of variations it is known that the minimization of (6.29) under the constraint (6.31) is equivalent to finding the stationary point of the Lagrange function

$$\mathscr{L} = J + \boldsymbol{\lambda}^T \mathbf{E} = J + \lambda_i E_i \tag{6.37}$$

with respect to $\boldsymbol{\Psi}$, and $\boldsymbol{\Phi}$ and the Lagrange multipliers λ_i (Talagrand and Courtier, 1987, Courtier and Talagrand, 1987, Long and Thacker, 1989, Thacker, 1988, and Thacker and Long, 1988). Variation of the Lagrange function (6.37) with respect to λ_i yields the model equations (we use again the linearized versions of the equations introduced above):

$$\frac{\partial \mathscr{L}}{\partial \lambda_i} = 0 \rightarrow \mathbf{E} = P\boldsymbol{\Psi} + Q\boldsymbol{\Phi} = 0, \tag{6.38}$$

while variation with respect to ψ_i yields the adjoint model equations for the Lagrange multipliers,

$$\frac{\partial \mathcal{L}}{\partial \psi_i} = 0 \rightarrow 2R^T M(\mathbf{d} - \mathbf{d}^o) + P^T \lambda = 0. \qquad (6.39)$$

Finally, variation with respect to the control variables yields

$$\frac{\partial \mathcal{L}}{\partial \phi_k} = 0 \rightarrow 2N\mathbf{\Phi} + Q^T \lambda = 0. \qquad (6.40)$$

The set of equations (6.38) – (6.40) can be solved iteratively as follows.

Choose $\mathbf{\Phi} = \mathbf{\Phi}^1$ as a starting point. Compute $\mathbf{\Psi} = \mathbf{\Psi}^1$ as solution of the model equations for given $\mathbf{\Phi}^1$. Compute $\mathbf{d}^1 = \mathbf{d}(\mathbf{\Psi}^1) = R\mathbf{\Psi}^1$ from (6.32). Compute $\lambda^1 = \lambda(\mathbf{\Psi}^1, \mathbf{\Phi}^1)$ as the solution of the adjoint equations (6.39):

$$\lambda^1 = -2P^{-1}R^T M(\mathbf{d}^1 - \mathbf{d}^o). \qquad (6.41)$$

The starting point $(\mathbf{\Phi}^1, \mathbf{\Psi}^1, \lambda^1)$ then satisfies the variational equations (6.38), (6.39), so that $\partial \mathcal{L}/\partial \lambda_i = \partial \mathcal{L}/\partial \psi_i = 0$, but it does not satisfy (6.40) because the first-guess $\mathbf{\Phi}^1$ does not minimize the cost. The residual gradient is given by

$$\frac{\partial \mathcal{L}}{\partial \mathbf{\Phi}^1} = 2N\mathbf{\Phi}^1 + Q^T \lambda^1. \qquad (6.42)$$

We shall show below that the vector of partial derivatives of $\mathcal{L}$ is the same as the total gradient of the cost function with respect to the control variables, and hence can be used in a gradient descent algorithm to move the control variables towards the cost minimum. The appropriate step size in the gradient direction (6.42) can be determined by perturbing $\mathbf{\Phi}^1$ in the down-gradient direction and computing a new value of the cost. Assuming that the cost is approximately quadratic in the controls along this line, one can then fit a parabola to the three values: cost and cost gradient at $\mathbf{\Phi}^1$ and cost at the perturbed position. The minimum point of the parabola then becomes $\mathbf{\Phi}^2$. The entire procedure is then iterated to convergence, using a modified direction algorithm such as the conjugate gradient (e.g. Luenberger, 1969), rather than the straightforward steepest descent, after the first iteration.

Each line minimization step requires two solutions of the model equations and one solution of the adjoint equations. The technique is more efficient than direct elimination if the number of conjugate gradient iterations required for convergence is significantly smaller than n. This is normally the case if n is large. In the wave model case, however, P^{-1} can be written down explicitly under certain approximations, which makes the (approximate) direct inversion method more efficient than the adjoint method (see § VI.10).

Although a known result of variational analysis (Gill *et al*, 1981), we confirm in conclusion that the stationary point of the Lagrange function does indeed correspond to the minimum of the cost function. Specifically, given that (6.37) and (6.38) are satisfied for given values of the controls, (6.40) is identically equal to the total gradient of J with respect to $\mathbf{\Phi}$, whence if the former vanishes, so does the latter.

In the iteration technique used in the adjoint method, all points $(\mathbf{\Phi}, \mathbf{\Psi}, \lambda)$ always satisfy the relations $E_i = \partial \mathcal{L}/\partial \lambda_i = \partial \mathcal{L}/\partial \psi_i = 0$. We wish to show that the total derivative $d_\mathbf{\Phi} J$ of J reduces then to the $\mathbf{\Phi}$ gradient component of $\mathcal{L}$, i.e. to $\partial_\mathbf{\Phi} \mathcal{L}$. For

$$\mathcal{L} = J(\mathbf{\Phi}, \mathbf{\Psi}) + \mathbf{E}^T \lambda, \tag{6.43}$$

the condition

$$\partial_\mathbf{\Psi} \mathcal{L} = 0 = \partial_\mathbf{\Psi} J + P^T \lambda \tag{6.44}$$

yields

$$\lambda = -(P^{-1})^T \partial_\mathbf{\Psi} J, \tag{6.45}$$

so that

$$\begin{aligned} \partial_\mathbf{\Phi} \mathcal{L} &= \partial_\mathbf{\Phi} J + \lambda^T Q \\ &= \partial_\mathbf{\Phi} J - (\partial_\mathbf{\Psi} J)^T P^{-1} Q. \end{aligned} \tag{6.46}$$

Alternatively, we have:

$$\begin{aligned} d_\mathbf{\Phi} J &= \partial_\mathbf{\Phi} J + (\partial_\mathbf{\Psi} J)^T \partial_\mathbf{\Phi} \mathbf{\Psi} \\ &= \partial_\mathbf{\Phi} J - (\partial_\mathbf{\Psi} J)^T P^{-1} Q. \end{aligned} \tag{6.47}$$

Thus

$$\partial_\mathbf{\Phi} \mathcal{L} = d_\mathbf{\Phi} J \tag{6.48}$$

as stated.

In many derivations of the adjoint model equations, the system is discretized in space, but time is retained as a continuous variable. In a numerical implementation time must also be discretized. The translation of the adjoint model formalism into a discretized integration procedure is illuminating, as it demonstrates that the evolution in time is, in fact, not the essential feature underlying the system equations. The fundamental model property is the existence of a numerical algorithm in which new variables are successively computed from previously computed variables, starting from some prescribed set of data, the control variables. These consist, in general, of the state of the system at some initial time $t = 0$, the forcing fields and model parameters. From this point of view the mathematics becomes particularly simple and is readily transformed into computer code.

Stripped down to its essentials, a numerical model thus comprises a sequence of statements assigning values to a sequence of variables. The sequence begins with a prescribed subset of variables, the control variables, which we designate by ψ_n, $n = 1, \cdots M$ (note that we denote the control variables now as ψ_n, rather than as ϕ_n, as before). The remaining variables, by definition, can be computed once the controls are given. We designate these ψ_n, where n runs from $M + 1$ to N and indexes the variables in the same order in which they are calculated in the computer code. The final variable calculated, ψ_{N+1}, is the cost function which is to be minimized.

At its minimal point the gradient of the cost with respect to the control variables vanishes. Computing the gradient of the cost is not trivial because of the possibly long sequence of dependences connecting the value of the cost with the values of the controls; specifically, for all n greater than M,

$$\psi_n = F_n(\psi_{i<n}), \tag{6.49}$$

that is each variable may depend explicitly on any or all of the controls and on any or all of the intermediate variables already defined in the computational sequence. We could, in principle, obtain the needed cost gradient by applying the chain rule to the above relationships, but in practice the algebra would be prohibitive. Instead, we apply the method of Lagrange multipliers, which proceeds in the present case as follows.

We construct the Lagrange function $\mathcal{L}$ by writing the constraint equations (6.30) in homogeneous form (collecting terms on the right), then multiplying each right hand side by a Lagrange multiplier, summing the results, and adding this to the cost:

$$\mathcal{L}(\lambda_{i>M}, \psi_j) = \psi_{N+1} + \sum_{n=M+1}^{N+1} \lambda_n(F_n(\psi_{k<n}) - \psi_n). \tag{6.50}$$

The minimum point of the cost in the space of the control variables is the same as the saddle point of the Lagrange function in the much larger space of all arguments λ_i and ψ_i. We find the saddle point by taking partial derivatives of $\mathcal{L}$ with respect to each of its arguments, treating the remaining arguments as if they were mutually independent, then setting the results equal to zero. Doing so with respect to the multipliers simply returns the model equations as the first conditions that the optimal model solution must satisfy. Doing so with respect to the ψ_i for $i > M$ returns the adjoint equations governing the multipliers, specifically, for $i = N + 1$,

$$\lambda_{N+1} = 1 \tag{6.51}$$

and

$$\lambda_i = \sum_{n=i+1}^{N+1} \lambda_n \frac{\partial F_n}{\partial \psi_i} \quad (M < i < N+1). \tag{6.52}$$

Finally, partial differentiation with respect to the controls yields

$$\frac{\partial \mathscr{L}}{\partial \psi_i} = \sum_{n=M+1}^{N+1} \lambda_n \frac{\partial F_n}{\partial \psi_i} \quad (1 \le i \le M). \tag{6.53}$$

Under the condition that the ψ_i satisfy the model equations and the λ_i satisfy the adjoint equations, these partial derivatives are the same as the components of the gradient of the cost with respect to the controls (as shown in the previous section) and must therefore also vanish at the minimum. For arbitrary values of the control variables, the gradient will be nonzero, but may then be used in an iterative procedure to search out the minimum of the cost as follows: Starting with a first guess for the values of the controls, the model equations are solved, the cost evaluated, and the factors on the right hand side of the adjoint equations, $\partial F_n / \partial \psi_i$, computed. The adjoint equations can then be marched backward from the known value of λ_{N+1} to produce the components of the cost gradient. The control variables may subsequently be updated using an appropriate descent algorithm (such as conjugate gradient) and the procedure repeated to convergence.

The above forms for the adjoint are completely general and therefore appear more complicated than those normally encountered in practice. Typically, only a few terms on the right hand side are nonzero because the functions F_n usually contain only a few of the previously defined variables. Note also that we have placed no constraints on the form of the functions F_n other than that 1) they are differentiable with respect to their arguments in some neighbourhood of the cost minimum, and 2) they represent an algorithm that starts with a given set of control variables and computes a sequence of intermediate variables, terminating with a value for the cost function F_{N+1}. The cost function must also be differentiable with respect to its arguments and must have a minimum. For the cost function J, equation (6.29), this is the case.

VI.4   Applications – Overview

S. Hasselmann

In the previous sections of this chapter we have presented briefly the mathematical and statistical basis for the data assimilation and inverse modelling techniques currently in use or under study. In the following sections, we provide several examples of applications of these procedures to practical problems involving wave data. Except for the assimilation scheme based on optimal interpolation, which is already being applied operationally to ERS-1 altimeter wave height data (§ VI.5), all schemes described in the following are still in the development stage and the results are necessarily preliminary.

This is a general characteristic of this rapidly evolving field. Since the launch of ERS-1 in July, 1991, work on wave data assimilation has received a strong boost through the successful near-real-time retrieval of wave heights and two-dimensional wave spectra from the altimeter and SAR instruments flown on the satellite (cf. chapter V). The availability of a third generation wave model capable of reproducing detailed features of the two-dimensional wave spectrum has furthermore motivated efforts to develop rather more sophisticated adjoint model and Green's function techniques for assimilating these data. The continually expanding global data sets produced by this satellite – together with recent extensive experimental campaigns such as LEWEX and SWADE – have also stimulated the development of adjoint models for application in the inverse modelling mode to optimize the source functions of the WAM model.

The simplest data assimilation scheme is the optimal interpolation method introduced in § VI.2.4. An example will be described in the next section. OI schemes have been implemented operationally at the British Meteorological Office in Bracknell, UK for a second generation wave model (Thomas, 1988), and for the WAM model at the European Centre for Medium-Range Weather Forecasts ECMWF in Reading, UK (cf. Lionello *et al*, 1990, Günther *et al*, 1993, 1994) and at KNMI (Burgers *et al*, 1990). So far only satellite radar altimeter wave heights have been used in these applications.

Since a second generation wave model prescribes the shape of the wind sea spectrum and normally (as in the UK model) contains only one free scale dimension, the adjustment of the energy level of the spectrum to the altimeter wave height data is unique for such a model when the spectrum represents a pure wind sea. If the spectrum also contains swell components, some assumption is needed to distribute the altimeter wave height information between the wind sea and swell.

For a third generation wave model such as WAM , which can have an arbitrary spectral distribution, the translation of a single altimeter wave height value into a set of corrections for the full two-dimensional wave spectrum clearly poses more difficulties. The basic approach of Lionello *et al* (1990) for the WAM model, described in the next section, is nevertheless similar in spirit to that of Thomas (1988) for a second generation model. A significant improvement of the method is to be expected when it has been generalized, as planned, to include two-dimensional wave spectral information from the SAR wave mode. The present problem of distributing the information from a single wave height value over a complete two-dimensional spectrum can then be largely resolved, since the observed data are more closely matched to the modelled data.

In §§ VI.6 and VI.7, we turn to inverse modelling applications. Although it has been shown in previous chapters that the WAM model reproduces rather well our present understanding of wave dynamics and is generally well validated by data, it can none the less be expected that the parametrization of the source terms can be improved by systematically optimizing the model parameters. In § VI.6 we describe an application of the simplest direct cost function minimization technique to tune optimally a small number of WAM source function parameters, using an iterative gradient descent method and a finite difference representation of the cost function derivative. In § VI.7 the same problem is treated using the more powerful adjoint model method, with which a much larger number of free parameters can be optimized simultaneously.

In §§ VI.8 and VI.9, the adjoint model method is applied to the problem of data assimilation. In one example (§ VI.8) the initial wave field and in a second case (§ VI.9) the wind field are optimally adjusted to the observed wave data. Both studies represent first exploratory exercises to identify the problems and potential benefits of the application of the adjoint technique in wave data assimilation. A complete adjoint scheme would, of course, need to correct both the initial wave field and the entire present and past wind field.

Different techniques for the construction of the adjoint model are presented in the three adjoint modelling applications of §§ VI.7 – 9. In the inverse modelling case, § VI.7, the adjoint is constructed for a one-dimensional WAM model (a restricted geometry model allowing fetch or time dependence only) by applying a general adjoint generator line by line to the original model code, as described in § VI.3.2. In the first data assimilation example, § VI.8, the adjoint equations for basically the same single-gridpoint WAM model are derived algebraically and programmed independently of the original model. Finally, in the second data assimilation application, § VI.9, a fully space-time-dependent case is studied, and because of computer limitations, a simpler second generation model is used for both forward (model) and backward (adjoint) integrations. The adjoint is again obtained by algebraic differentiation of the original model and independently coded.

The last application, § VI.10, also concerns data assimilation. In this case a Green's function technique is used to relate variations of the wave spectrum directly to variations in the wind field, thereby expressing the cost (for small perturbations of the wind and wave fields) as a function of the wind field only. The cost can then be minimized directly without the application of expensive iteration descent algorithms, making the method operationally feasible. However, a number of *ad hoc* approximations are required in order to obtain a numerically tractable Green's function.

The longer term goal of wave data assimilation, as outlined in § VI.1.2, is to incorporate one or some combination of these general techniques within an integrated comprehensive system in which all relevant meteorological and ocean wave data are assimilated in a single coupled atmospheric general circulation and global wave model. The results of the exploratory studies summarized in the following sections indicate that considerable work is still required to achieve this goal, but that encouraging progress along this path is at least being made.

VI.5 An optimal interpolation scheme for assimilating altimeter data into the WAM model

P. Lionello, H. Günther, B. Hansen, P. A. E. M. Janssen
and S. Hasselmann

VI.5.1 Introduction

The optimal interpolation method described in this section was developed for the WAM model and is operational at ECMWF (Lionello *et al*, 1992). Similar single-time level data assimilation techniques (cf. § VI.2.4) for satellite altimeter wave heights have been applied by Hasselmann *et al* (1988), Janssen *et al* (1987b, 1989b), Thomas (1988) and Lionello *et al* (1990).

Instead of estimating the full state vector $\hat{\Psi}$, as in equation (6.19), we estimate in this case only the significant wave height field $\mathbf{H}$ (the index S in the notation for the significant wave height is dropped in the following analysis). The data vector $\mathbf{d}^f$ consists then of the first-guess model wave heights, interpolated to the locations of the altimeter observations, while $\mathbf{d}^o$ are the actually observed altimeter wave heights.

The assimilation procedure consists of two steps:

- first a best-guess (in meteorological terminology: analysed) field of significant wave heights is created by optimum interpolation, in accordance with the general OI approach outlined in § VI.2.4 and with appropriate assumptions regarding the error covariances; then
- this field is used to retrieve the full two-dimensional wave spectrum from a first-guess spectrum, introducing additional assumptions to transform the information of a single wave height measurement into separate corrections for the wind sea and swell components of the spectrum.

The problem of using wave height observations for correcting the full two-

dimensional spectrum was first considered by Hasselmann *et al* (1988) and
Bauer *et al* (1992), who assimilated SEASAT altimeter wave heights into the
WAM model by simply applying a constant correction factor, given by the ratio
of altimeter and model wave heights, to the entire spectrum. A shortcoming
of this method was that the wind field was not corrected. Thus although swell
corrections were retained for several days, the corrected wind sea relaxed
back rapidly to the original incorrect state due to the subsequent forcing
by uncorrected winds. Janssen *et al* (1987b) removed this shortcoming by
extending the method to include wind corrections, but nevertheless achieved
only short relaxation times due to the choice of an insufficient correlation
scale (the corrections were essentially limited to a single gridpoint). This was
remedied in later versions of the scheme described below.

A related but somewhat different approach was pursued by Thomas (1988),
who assimilated altimeter wave height data together with alitimeter wind
speed data in the second generation wave model of the UK Meteorological
Office. Using both sources of data and noting that in a second generation
wave model the shape of the wind sea spectrum is specified, it was possible
to derive separate corrections for the wind sea and swell components of
the spectrum with very few additional assumptions (see also Francis and
Stratton, 1990 and Esteva, 1988).

As in most of these schemes, the present method corrects the two-dimensional spectrum by introducing appropriate rescaling factors to the energy
and frequency scales of the wind sea and swell components of the spectrum,
and also updates the local forcing wind speed. The rescaling factors are
computed for two classes of spectra: wind sea spectra, for which the rescaling
factors are derived from fetch and duration growth relations, and swell
spectra, for which it is assumed that the wave steepness is conserved. All
observed spectra are assigned to one of these two classes. This restriction
will be removed in the planned extension of the scheme to include ERS-1 SAR
wave mode data (see chapter V).

VI.5.2 Wave height analysis

First, an estimated best-guess significant wave height field $\mathbf{H}^b = (H_i^b)$ is
created by optimum interpolation (cf. equation (6.19)):

$$H_i^b = H_i^f + \sum_{j=1}^{n_{obs}} W_{ij}(H_j^o - H_j^f), \tag{6.54}$$

where $\mathbf{H}^o$ denotes the significant wave height field observed by the altimeter
and $\mathbf{H}^f$ is the first-guess significant wave height field computed by the WAM
model. Since long-term statistics of the prediction and observational error

covariance matrices equation (6.23) were not available, empirical expressions were taken:

$$\sigma_{ij}^f = \sigma^f \exp(-|x_i - x_j|/L) \quad \text{and} \quad \sigma_{ij}^o = \delta_{ij}\sigma^o. \tag{6.55}$$

Good results were obtained for a correlation length of five gridpoints, or $L = 5 \times 330 = 1650$ km. This is consistent with the optimal scale length found by Bauer *et al* (1992) using a triangular interpolation scheme.

The best-guess wave spectrum

In the next step, the full two-dimensional wave spectrum is retrieved from the best-guess (analysed) significant wave height fields. Two-dimensional wave spectra are regarded either as wind sea spectra, if the direction of the highest spectral peak lies within 15° of the wind direction and the peak frequency is larger than the Pierson-Moskowitz frequency, or, if these two conditions are not satisfied, as swell.

In both cases a best-guess two-dimensional wave spectrum $F^b(f, \theta; \mathbf{x}, t)$ is computed from the first-guess wave spectrum $F^f(f, \theta; \mathbf{x}, t)$ and the optimally interpolated best guess wave heights H_i^b by rescaling the spectrum with two scale parameters A and B:

$$F^b(f, \theta) = AF^f(Bf, \theta). \tag{6.56}$$

Different techniques are applied to compute the parameters A and B for wind sea or swell spectra.

Retrieval of a wind sea spectrum

The parameters A and B in equation (6.56) can be determined from empirical duration-limited growth laws relating, in accordance with Kitaigorodski's (1962) scaling laws, the nondimensional energy $\epsilon_* = u_*\epsilon/g^2$ (where $\epsilon = (H/4)^2$), mean frequency $\bar{f}_* = u_*\bar{f}/g$ and duration $T_* = u_* T/g^2$. Specifically, we take the growth relations proposed by Lionello *et al* (1992):

$$\epsilon_*(t_*) = 955 \tanh(6.02 \times 10^{-5}\, t_*^{0.695}), \tag{6.57}$$

and

$$\epsilon_*(\bar{f}_*) = 1.68 \times 10^{-4}\bar{f}_*^{-3.27}. \tag{6.58}$$

The mean frequency is preferred to the peak frequency because its computation is stabler. Since the first-guess friction velocity was used to generate the waves and the first-guess wave height is known, an estimate of the duration T of the wind sea can be derived from the duration-limited growth laws. Assuming this estimated duration is correct, the best-guess wave height yields from the growth laws, equations (6.57) and (6.58), best estimates of the friction velocity u_*^b and mean frequency $\bar{f}^b$. The best-estimate wave height and mean frequency determine then the two parameters A and B:

$$A = \left(\frac{H^b}{H^f}\right)^2 B \quad \text{and} \quad B = \bar{f}^b/\bar{f}^f. \tag{6.59}$$

The corrected best-estimate winds are then used to drive the model for the rest of the wind time step. In a comprehensive wind and wave assimilation scheme, the corrected winds should be also inserted into the atmospheric data assimilation scheme to provide an improved wind field in the forecast model. This step has not yet been implemented, but is currently being tested at ECMWF.

Retrieval of a swell spectrum

A spectrum is converted to swell and begins to decay at the edge of a storm, before dispersion has separated the swell into spatially distinct frequencies. One can therefore distinguish between a nonlinear swell regime close to the swell source and a more distant linear regime, where dispersion has reduced the swell wave slopes to a level at which nonlinear interactions have become negligible. Because of these complexities, and also because of a lack of adequate data, there exist no empirical swell decay curves comparable to the growth curves in the wind sea case. However, Lionello and Janssen (1990) showed that for the WAM model swell spectra the average wave steepness,

$$s = H\bar{k}/8\pi \tag{6.60}$$

is approximately the same for all spectra at the same decay times, despite the wide range of significant wave heights and mean frequencies of their data set. Assuming that the effective decay time and therefore the wave steepness is not affected by the correction of the wave spectrum, the scale factors are then given by

$$B = (H^B/H^f)^{\frac{1}{2}}, \tag{6.61}$$

$$A = B(H^b/H^f)^2. \tag{6.62}$$

Intuitively, this approach appears reasonable, because a more energetic spectrum will generally also have a lower peak frequency, and increasing the energy without decreasing the peak frequency produces a swell of unrealistic steepness. Since the swell spectrum is not related to the local stress, and only the local wind field is corrected in the assimilation scheme, the wind field is not updated in the case of swell.

The general case

It was shown in Lionello *et al* (1992) that the wind sea and swell retrieval scheme works well for simple cases or pure wind sea or swell. If the spectrum consists of a superposition of wind sea and swell, and the wind sea is well separated from the swell, the wind sea and swell correction methods can, in principle, still be applied separately to the two components of the spectrum. In this case, however, one needs to introduce additional assumptions regarding the partitioning of the total wave height correction between wind sea and swell.

The arbitrariness of the present and similar methods of distributing a single wave height correction over the full two-dimensional wave spectrum could presumably be partially alleviated by using maximum likelihood methods based on a large set of observed data, which is now becoming available through ERS-1. However, a more satisfactory solution is clearly to assimilate additional data, such as two-dimensional SAR spectral retrievals, to overcome the inherently limited information content of altimeter wave height data.

VI.5.3 Results

Although Lionello *et al* (1992) demonstrated the potential benefits of the assimilation of altimeter wave data for wave forecasting, they also found deviations between their assimilated wave heights and buoy observations, which they attributed to discrepancies between altimeter and buoy wave height measurements (cf. § V.2). In preparation for the operational implementation of the assimilation scheme at ECMWF, the scheme of Lionello *et al* (1992) for SEASAT and GEOSAT altimeter data was therefore tested for the ERS-1 altimeter data (Günther *et al*, 1993, 1994). The test was carried out in a quasi-operational mode in near real time. Two global runs were carried out: a reference run without data assimilation, and an assimilation run. Each run was divided into an initial analysis period of one day, during which data were assimilated, and a subsequent forecast period of five days. In both reference and assimilation runs the model was driven by analysed wind fields during the analysis period and forecast winds for the five forecast days. (In the assimilation run, the regular analysed wind field produced by the ECMWF operational analysis scheme was modified in the second half of each 6-hour assimilation window, as mentioned above, by the wind field corrections generated by the wave data assimilation scheme.) The last wave field of an analysis period was used as the initial field for the following integration run. In the following the notation *Aa* refers to the first section of an assimilation run with analysed winds, *An* ($n = 1, \cdots 5$) to the forecast run up to forecast day *n*, and *Ra*, *Rn* to the corresponding reference run without wave data assimilation.

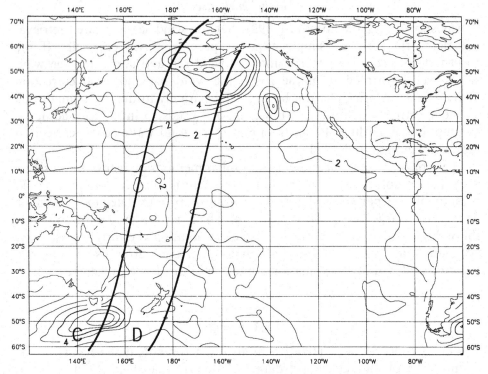

Fig. 6.2. Satellite tracks C and D at 23.05 UTC and 21.25 UTC, respectively, on 18 March 1992, superimposed on the H_S field produced by the Ra run at 24.00 UTC on 18 March 1992. The isoline interval is 2 metres.

Impact on wave analysis

The impact of ERS-1 altimeter data on the wave analysis is illustrated for two swaths, C and D, at 23.05 UTC and 21.25 UTC, respectively, on 18 March 1992, shown in figure 6.2 together with the significant wave height field of run Ra.

The wind seas from two storms are seen, one in the Southern Ocean and one in the North Pacific. The H_S differences between runs Aa and Ra (figure 6.3) indicate that the reference run overestimated the waves in the Gulf of Alaska, while the H_S was underestimated in the intense storm in the Southern Ocean.

Figure 6.4 shows (model and altimeter) wave heights along the satellite track. In general, the agreement is good, particularly in the tropics, where swell dominates the wave field. However, in the storm region of the Southern Ocean (panel a), the observed peak wave height lies further south than indicated by the model. Analysed winds are usually low in this area because of lack of observations. But the discrepancy could also be due to the

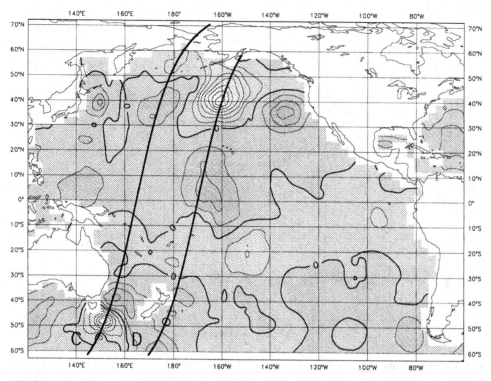

Fig. 6.3. Satellite tracks C and D superimposed on the field of H_S differences between runs *Aa* and *Ra*. The isoline interval is 0.25 metres. Darker shading indicates areas of higher H_S in *Aa*.

bounding of the ECMWF global wave model at 60° South, which ignores waves generated between 60° South and the Antarctic ice boundary (the ECMWF WAM model has since been extended to the weekly updated ice edges in the Antarctic and Arctic).

Figure 6.5 shows the global spatial distribution of the averaged bias between the reference and assimilation runs (using all available altimeter data) for the months of February (panel a) and March (panel b). The bias (altimeter − WAM wave height) is smaller in magnitude in the tropics than in the Southern Oceans (positive bias) and the Northern Pacific (negative bias).

The bias is particularly high in the Indian Ocean (figure 6.5). For this no straightforward explanation was found. There is no evidence in this case of missing swell from the incorrect southern ice boundary of the model. A comparison of the altimeter winds and ECMWF winds shows no large bias in the Indian Ocean (figure 6.6). More investigations of this phenomenon using the extensive ERS-1 altimeter and SAR wave mode data which are now becoming available (cf. chapter V) would be desirable.

The results shown in figures 6.5 and 6.6 appear to be representative of

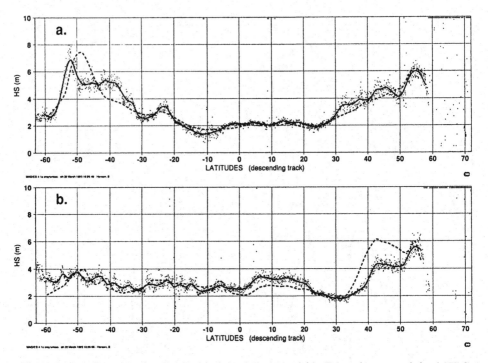

Fig. 6.4. H_S along tracks C (panel a) and D (panel b). Dots denote original ERS-1 altimeter measurements, the continuous line the measurements after quality control and smoothing of the altimeter data along the satellite tracks; the dashed line represents model H_S values interpolated to the altimeter grid points.

most wave situations encountered during a longer analysis period from 1 January – 31 October 1992: waves are overestimated by the model in the northern Pacific and underestimated in the Southern and Indian Oceans.

The data assimilation results have been validated against a set of wave buoys, indicated in figure 6.5, panel a (Günther *et al*, 1993, 1994). Results for buoy 51003 (figure 6.7) off Hawaii are presented here only. Unfortunately, no buoys are available in the Southern Ocean, where the largest differences between model and altimeter wave heights are observed. During February 1992 the significant wave height of the reference run was lower than the buoy observations (panel a). The situation was reversed during March (panel b), when the bias became positive for the greater part of the month. Assimilation of altimeter data improved the agreement. However, extreme situations were still not simulated correctly. This may be due to an underestimation of altimeter wave heights relative to buoy wave heights for high waves (cf. figure 5.3 in § V.2).

In summary, the impact of assimilating ERS-1 altimeter data into the WAM model is strongest in the moderate range of significant wave heights below

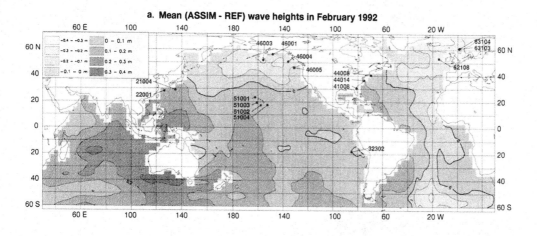

a. Mean (ASSIM - REF) wave heights in February 1992

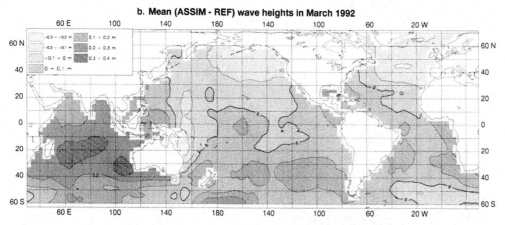

b. Mean (ASSIM - REF) wave heights in March 1992

Fig. 6.5. Monthly averaged bias $< H_A - H_R >$ between wave heights generated by runs *Aa* and *Ra* in February (panel a) and March (panel b). Darker areas denote positive values, lighter areas negative values. Buoy positions are indicated in panel a.

four meters. Although no systematic bias between altimeter, model and buoy data could be found in this range, the assimilation of altimeter data into the WAM model nevertheless improved the agreement between model and buoy wave heights. As to be expected, the assimilation of altimeter wave heights is unable to remove the discrepancies between buoy and model data for higher significant wave heights, where the buoy and altimeter data exhibit a systematic bias.

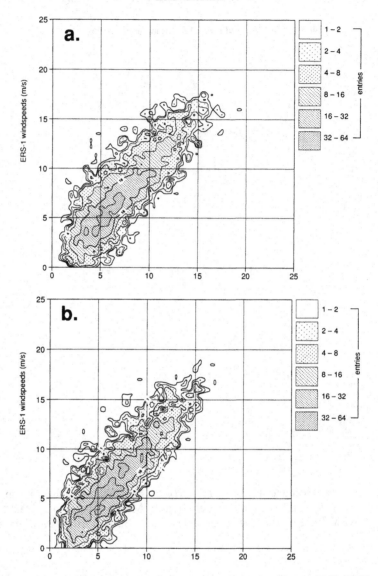

Fig. 6.6. Intercomparison of ERS-1 altimeter wind speeds and ECMWF analysed winds for the Indian Ocean in February (panel a) and March (panel b).

The impact of the assimilation of ERS-1 *altimeter data on wave forecasts*

Figures 6.8 and 6.9 show the impact of data assimilation on model forecasts in the northern and southern hemispheres and in the tropics for March 1992. As expected, the impact of data assimilation on forecast wave heights decays most rapidly in the wind sea regions in the North. The rate of decay of the correction is weaker for the first two days into the forecast in the mixed

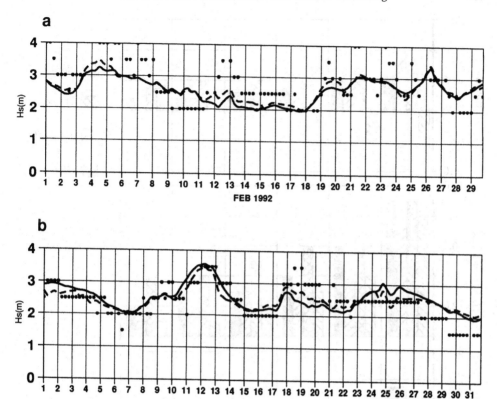

Fig. 6.7. H_S data at buoy 51003 near Hawaii for February (panel a) and March (panel b). Dots represent buoy measurements, the continuous and dashed lines data from runs *Ra* and *Aa*, respectively.

wind sea and swell area of the Southern Hemisphere and for the swell in the tropics. The longest impact, persisting even after five days, is found in the tropics. A similar behaviour is seen in the scatter index.

The ERS-1 altimeter data were not assimilated after the initialization of the fields at the start of the forecast and can therefore be considered as an independent data set for validating the impact of data assimilation on the forecast shown in figure 6.9. As expected, bias and scatter index are lower in the assimilation run than in the reference run on both regional and global scales. The impact on the forecast in the tropics can again be still clearly seen after five days.

VI.5.4 Conclusions

Even a simple data assimilation scheme which uses only the significant wave height to correct the full two-dimensional spectrum improves wave analyses

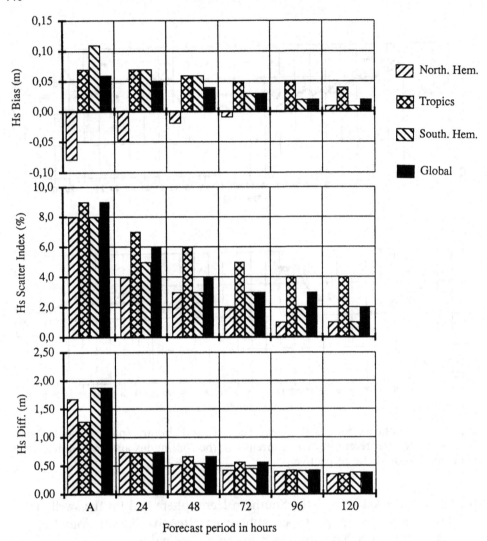

Fig. 6.8. Wave height statistics for reference and assimilation runs for a five day forecast in March 1992. From top to bottom: bias (assimilation – reference), scatter index and maximum absolute difference.

noticeably. The impact is strongest in the high-wind regions of the Southern Ocean, where observations are sparse and analysed wind speeds are often too low.

The impact of the improved analyses is seen well into the wave forecast. In regions dominated by swell the relaxation time for the wave field correction can exceed five days.

The wind field corrections inferred from the corrected wave field represent

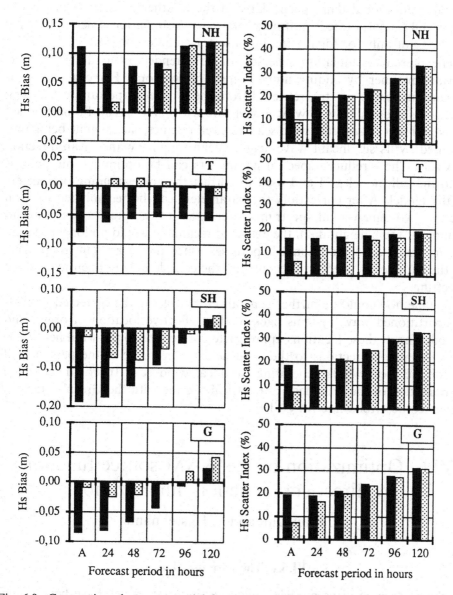

Fig. 6.9. Comparison between model forecast and altimeter observations in March 1992. The panels show the bias (left) and scatter index (right) for the analysis (A) and five forecast ranges. Full bars denote the reference run, dotted bars the assimilation run. From top to bottom: Northern Hemisphere (NH), tropics (T), Southern Hemisphere (SH), and global (G).

valuable additional data, particularly in the Southern Ocean, which could be usefully incorporated in atmospheric data assimilation schemes. This is relevant not only for forecasting but also for climate research.

The present assimilation scheme can be extended in a rather straight-forward manner by adding more degrees of freedom. For example, two-dimensional wave spectra, retrieved from the synthetic aperture radar using first-guess spectra from a WAM model forecast (§ V.4.2), can be partitioned into a number of separate wind sea and swell regimes, each regime being represented by its significant wave height, mean frequency and mean direction (§ V.4.3). These reduced spectral parameters can be optimally interpolated onto the full model grid and used to correct the corresponding parameters of the model. After correction, the parameters can be recombined to form the full two-dimensional spectrum. In contrast to the present limitation to significant wave height data, the extended technique would no longer require the introduction of additional hypotheses, since the measurements provide all the parameters needed to reconstruct the complete two-dimensional wave spectrum.

The method could be further generalized by tracing the corrected individual partitioned wave systems back to their effective location of origin and introducing at these locations appropriate corrections to the wind field. This would enable corrections to the wind fields to be made in the past as well as for the local winds generating the local wind sea. It would require a combination of the extended parametrical scheme with the Green's function approach described later in § VI.10.

VI.6 Optimization of the WAM source functions by direct cost function minimization

J. Monbaliu and S. Hasselmann

VI.6.1 The cost function

In the past, most wave models, including the WAM model, have been tuned to observations more or less by trial and error. A more systematic way to optimize a model, discussed in § VI.2.4, is clearly to minimize a cost function representing the deviations between the model and the observations. If the model contains a large number of free parameters, the most efficient minimization method is the adjoint-model technique. However, if the number of tunable parameters is small, one can minimize the error expression directly using standard methods of steepest descent (see, for example, Scales, 1985). We shall present examples of both techniques, beginning with the direct minimization method in this section and continuing with the more powerful adjoint model method in the following section.

A number of free model parameters will be considered, but in order to apply the direct minimization technique effectively, only two parameters will be varied at a time, the remaining parameters being kept fixed. The cost function is defined as a diagonal quadratic form consisting of the sum of the squared differences between the observed and computed nondimensional energies ϵ_* (nondimensionalized as usual with respect to g and the friction velocity u_*):

$$J = \frac{1}{W} \sum_{i=1}^{n_{obs}} w_i \left(\frac{\epsilon_{*i} - \epsilon_{*i}^o}{\epsilon_{*i}^o} \right)^2, \tag{6.63}$$

where W is the sum over the set of suitably chosen weights w_i (set equal to unity in the present case) assigned to the $n_{obs}(= 18)$ data values for a given set of discrete nondimensional fetch values X_{*i} (spaced logarithmically, with $X_{*i} = 1.2\, X_{*i-1}$) and the superscript 'o' refers to the measured energies. The cost function could be extended to encompass, for example, fetch-dependent data for the peak frequency and Phillips' constant. Both of these are available and will be included in the application in the next section. However, in the present simple example only the energy, as the variable exhibiting the greatest sensitivity to parameter variations, is considered (Monbaliu, 1992a,b). Additional error terms penalizing deviations from possible first-guess parameter estimates (see (6.9) and table 6.1) are not included in the cost function.

The minimization of the cost function was carried out using standard library routines (NAG, 1990), which compute first derivatives with respect to the control variables using finite difference schemes. The minimization is carried out iteratively, computing a new model state and its gradient from a new approximated set of control parameters at each iteration step.

VI.6.2 Source terms in the WAM model

Monbaliu (1992a,b) considered two different types of wind input:
1) Snyder (Komen *et al*, 1984):

$$S_{in}(f,\theta) = \max[0, 0.25\, a_1 \frac{\rho_a}{\rho_w} \left(a_2 \frac{28\, u_*}{c} \cos\theta - 1 \right) \omega\, F(f,\theta)] \tag{6.64}$$

2) Stewart (Stewart, 1974):

$$S_{in}(f,\theta) = 0.04\, a_3 \frac{\rho_a}{\rho_w} \left[\left(a_4 \frac{28\, u_*}{c} \right)^2 - a_5 \left(a_4 \frac{28\, u_*}{c} \right) \right] \cos\theta\, \omega\, F(f,\theta), \tag{6.65}$$

where a_1, a_2, a_3, a_4, and a_5 represent the free parameters. The variation in the parameter a_2 reflects the unknown value for the drag coefficient, which is used to transform Snyder's U_5 wind at 5 m height to friction velocity, which drives the cycle 3 WAM model considered here (in cycle 4 the model is driven by prescribed U_{10} winds and u_* is computed internally, allowing for the sea-state dependence of u_*, cf. chapter III). The parameter a_1 determines the overall wind input level. Both parameters will be treated separately. The parameters a_3 and a_4 in (6.65) are equivalent to the parameters a_1 and a_2 in the Snyder type wind input term (6.64). Setting a_5 equal to zero yields Plant's (1982) expression for the wind input. However, the parameter a_5 was not considered explicitly in Monbaliu (1992a,b) and was set equal to unity. Monbaliu showed furthermore that no satisfactory fit could be found for the Stewart type input, equation (6.65), so that these results are not presented here.

For the dissipation S_{ds} by whitecapping, the expression of Komen *et al* (1984) was used, in accordance with the general whitecapping theory of Hasselmann (1974),

$$S_{ds}(f,\theta) = -c_1\bar{\omega}\left(\frac{\omega}{\bar{\omega}}\right)^n\left(\frac{\hat{\alpha}}{\hat{\alpha}_{PM}}\right)^m F(f,\theta), \qquad (6.66)$$

with free parameters c_1 (overall dissipation level) and m (wave steepness dependence).

The cost function for variations in the dissipation parameter n is rather flat and optimization of this parameter led to only marginal improvements (Monbaliu, 1992a,b). In the present investigations the parameter was kept constant at $n = 2$.

For the nonlinear transfer S_{nl} the discrete interaction approximation (Hasselmann *et al*, 1985) was regarded as sufficiently accurate and was not varied in the present application.

VI.6.3 Fetch laws and optimization routine

The model was fitted to two alternative fetch laws: the original JONSWAP fetch law (Hasselmann *et al*, 1973),

$$\epsilon_* = 1.6 \times 10^{-4} X_* \qquad (6.67)$$

and a fetch law

$$\epsilon_* = 2.4 \times 10^{-3} X_*^{0.78}. \qquad (6.68)$$

derived by Kahma and Calkoen (1992) from a combination of the reanalysed JONSWAP data set and fetch-limited data from Lake Ontario and the Bothnian Sea.

To compute the nondimensional energy, Kahma and Calkoen (1992) used

Wu's (1982) relationship to convert wind speed measured at 10 m height to friction velocity u_*.

The growth curves are valid only in the fetch-limited growth stage in the range $X_* \leq 5 \times 10^6$. For the fully developed state, the energy level was kept constant (cf. Komen *et al*, 1984) :

$$\epsilon_*^{\mathrm{PM}} = 1.1 \times 10^3 \text{ at } X_*^{\mathrm{PM}} = 1.2 \times 10^8, \qquad (6.69)$$

where PM refers to the fully developed state (Pierson and Moskowitz, 1964).

The model integrations were carried out with a deep-water one-dimensional version of the WAM model (limited to one free integration variable, in this case fetch) using a first order forward difference scheme with a dynamically adjusting time step (van Vledder and Weber, 1988, Monbaliu, 1992a,b). The model was initialized at the nondimensional fetch $X_* = 2 \times 10^5$ by a JONSWAP spectrum using the scale parameters v_* (nondimensional peak frequency) = 0.0193 and $\alpha_p = 0.0223$, with shape parameters $\sigma_a = 0.07$, $\sigma_b = 0.09$ and $\gamma = 3.3$ (Hasselmann *et al*, 1973). The model was driven by a constant friction velocity $u_* = 0.5$ m/s (Monbaliu, 1992a,b).

VI.6.4 Results

In each of the numerical experiments shown below, only two parameters were regarded as adjustable control variables, the others being set constant. The following parameter combinations were investigated:

- one of the parameters a_1, a_2, a_3 or a_4 in the wind input term, combined with the overall dissipation level c_1;
- the two parameters in the dissipation term (c_1 and m).

Figure 6.10 shows the nondimensional energy from the model runs for the best four optimal-fit parameter combinations, together with the JONSWAP growth curve, plotted against nondimensional fetch. Reasonably good agreement is achieved in both the growing and the fully developed state of the sea. The fit to the data is still better for the Kahma and Calkoen reanalysed data (figure 6.11). However, for the present parametrizations of the input and dissipation source terms it was not possible to find parameters which reproduced a strictly linear growth (on a log–log plot) in the fetch-limited regime, as required by both growth relations, while also yielding a satisfactory transition to the fully developed state.

Table 6.2 summarizes the optimally fitted parameters. The best-fit parameters for the JONSWAP case are in rather close agreement with the original parameters found by Komen *et al* (1984), which were used in the first three cycles of the WAM model.

For the improved fit of the Kahma and Calkoen growth curves, lower values were obtained for both the wind input parameter a_1 and the dissipation

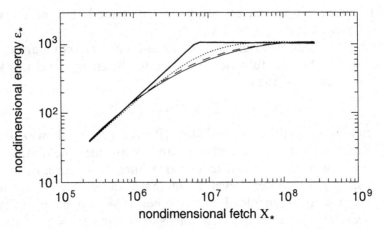

Fig. 6.10. Nondimensional energy as function of nondimensional fetch for JON-SWAP (thick continuous line) and model for best fit parameters: $a_1 = 1.01$, $c_1 = 3.07$ (dashed line); $a_2 = 1.05$, $c_1 = 3.89$ (long-dashed line); $a_2 = 1.17$, $c_1 = 5.66$ (dotted line); and $m = 2.04$, $c_1 = 3.05$ (thin continuous line). The dashed, long-dashed and thin continuous lines nearly coincide.

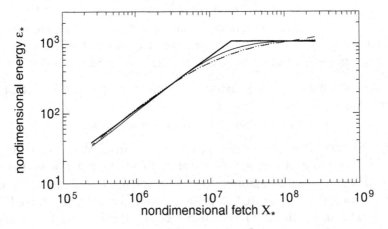

Fig. 6.11. Kahma and Calkoen (thick continuous line) Same as figure 6.10 for best fit parameters $a_1 = 0.61$, $c_1 = 2.13$ (thin continuous line) $a_2 = 1.42$, $c_1 = 12.2$ (dashed line) $m = 2.65$, $c_1 = 3.20$ (dotted line). The dashed and dotted lines nearly coincide.

parameter c_1 than found by Komen *et al* (1984), while the combination of a_2 and c_1 yielded larger values for both parameters.

Although the present definition of the cost function considered only total wave energy, growth curves for peak frequency and Phillips' constant α_p are also in good agreement with the data (Monbaliu, 1992a,b).

Table 6.2. *Summary of parameters in the source terms (Monbaliu, 1992a).*

JONSWAP

parameters	a_1	a_2	$c_1 \ (\times 10^{-5})$	m	cost
(a_1, c_1)	1.01	1	3.07	2	0.015
(a_2, c_1)	1	$1.05 \rightarrow 1.21$	$3.89 \rightarrow 5.91$	2	0.013
(c_1, m)	1	1	3.05	2.04	0.014
Komen	1	1	3.33	2	0.018

Kahma and Calkoen

parameters	a_1	a_2	$c_1 \ (\times 10^{-5})$	m	cost
(a_1, c_1)	0.61	1	2.13	2	0.005
(a_2, c_1)	1	$1.36 \rightarrow 1.42$	$10.7 \rightarrow 12.2$	2	0.003
(c_1, m)	1	1	3.2	2.65	0.004
Komen	1	1	3.33	2	0.06

Comparable minima of the cost function are found for different parameter combinations, suggesting that the problem may be ill-determined for certain parameter combinations. The cost function is not very sensitive in the neighbourhood of the optimal parameter combinations (figure 6.12). The strongest variations in the cost function were introduced by changes in the parameter a_2 (figure 6.13). A better definition of the minimum of the cost function could presumably be obtained by introducing additional constraints into the cost function. The spectral parameter which appears to be most sensitive to the choice of control parameters is Phillips' constant α_p for the fully developed state. However, the question of the dependence of the best-fit solutions on the control parameter choice can be resolved only by a more general optimization method in which all model parameters are optimized simultaneously. This requires a more efficient method of minimization, which is discussed in the following section.

VI.7 Wave model fitting using the adjoint technique

G. Barzel and R. B. Long

In the previous section, parameters of the WAM model atmospheric input and dissipation source terms were optimized by minimizing a cost function using

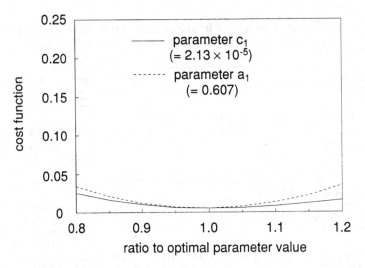

Fig. 6.12. Dependence of the cost function on the parameters c_1 (continuous line) and a_1 (dashed line). The parameters are normalized with respect to their optimally fitted values. The curves were obtained from the path generated by the iterative minimization algorithm and represent a one-dimensional transection through the ellipsoidal cost function surface defined by the two control variables c_1, a_1.

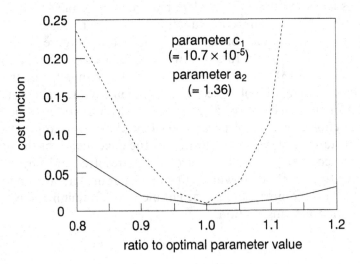

Fig. 6.13. Same as figure 6.12 for the parameter combination c_1. a_2.

a direct gradient descent method, the required gradient being computed by finite differences. Computing the cost gradient this way requires $m + 1$ runs of the numerical wave model over the assimilation interval for each iteration of the descent algorithm, where m is the number of parameters being varied. This represents a severe computational load unless m is quite small.

Table 6.3. *Constants used in the empirical fetch relations for dimensionless energy and peak frequency.*

	energy		frequency	
	A_E	P_E	A_f	P_f
JONSWAP	1.6×10^{-4}	1.00	1.082	−0.33
Kahma	2.4×10^{-3}	0.78	0.358	−0.244

The adjoint method (§ VI.3) provides an alternative means to the same end, yielding an accurate estimate of the cost gradient in the equivalent of two model runs (one model run and one run of the adjoint model) for each iteration of the descent algorithm, regardless of how many parameters are to be varied. Thus the adjoint method is a feasible optimization procedure even when the parameters to be varied number in the hundreds or thousands.

Following the implementation prescription described in § VI.3.2, we have constructed the adjoint of a one-dimensional (fetch-limited) version of the WAM model, similar to the model version used in the previous section, and applied the adjoint to optimize the WAM model source functions, using again the same JONSWAP and Kahma-Calkoen (1992) fetch-limited data sets as in the previous section. However, in contrast to the direct minimization approach, five parameters of the WAM model source terms are now recovered simultaneously, and the cost function is furthermore extended to include the nondimensional peak frequency and Phillips' constant α.

As observations we consider the dimensionless energy, peak frequency and Phillips' 'constant' α, as given by the general fetch-limited growth relations

$$\epsilon_* = A_E X_*^{P_E}, \qquad v_* = A_f X_*^{P_f}, \qquad \alpha = 0.35 X_*^{-0.22} \qquad (6.70)$$

for nondimensional fetch $X_* \leq 5 \times 10^6$, all quantities being made dimensionless, as usual, by scaling with g and the friction velocity u_*. As in the previous section, two growth laws were considered: the original JONSWAP experiment (Hasselmann *et al*, 1973), and a reanalysis of fetch-limited growth data by Kahma and Calkoen (1992). These correspond to the two sets of constants A_E, P_E, A_f and P_f listed in table 6.3. At large dimensionless fetch, the growth relationships are replaced by the limiting values for a fully developed sea (Pierson and Moskowitz, 1964):

$$\epsilon_*^{\text{PM}} = 1.1 \times 10^3, \qquad v_*^{\text{PM}} = 5.6 \times 10^{-3}, \qquad \alpha_{\text{PM}} = 8.1 \times 10^{-3}. \qquad (6.71)$$

The data for the dimensionless energies are the same as used in the previous section.

The shape of the growth curves in the transition region between growth and saturation is uncertain because very few observations are available for this fetch range. Therefore, this region was excluded from the cost function.

The cost function was defined as in equation (6.63), but suitably extended to include data on the nondimensional peak frequency and Phillips' constant:

$$J = \sum_{i=1}^{n_{obs}} \left(\frac{\epsilon_{*i} - \epsilon_{*i}^o}{\epsilon_{*i}^o} \right)^2 + \left(\frac{v_{*i} - v_{*i}^o}{v_{*i}^o} \right)^2 + \left(\frac{\alpha_i - \alpha_i^o}{\alpha_i^o} \right)^2. \qquad (6.72)$$

The minimization of the cost function was carried out simultaneously with respect to five free parameters. Two parameters were used for the WAM wind input term, a global factor W_m and an additional factor W_b for the u_* term:

$$S_{in}(f,\theta) = \max \left(0.0, 0.25 W_m \frac{\rho_a}{\rho_w} \left(W_b \frac{28 u_*}{c} \cos \theta - 1 \right) \right) \omega F(f,\theta). \qquad (6.73)$$

For the dissipation source term, three parameters were introduced, a global factor D_m and two exponents, D_f and D_s, for the mean-frequency and steepness terms, respectively:

$$S_{ds}(f,\theta) = -3.33 \times 10^{-5} D_m \left(\frac{\omega}{\bar{\omega}} \right)^{2D_f} \left(\frac{\hat{\alpha}}{\hat{\alpha}_{PM}} \right)^{2D_s} \bar{\omega} \, F(f,\theta). \qquad (6.74)$$

The third source function, the nonlinear transfer S_{nl}, was retained unchanged as the discrete interaction approximation (Hasselmann et al, 1985). With these definitions, setting the five free parameters to unity reproduces the original WAM cycle 3 model.

The model equivalents of the data were computed as follows: ϵ_* was determined by integration over the model spectrum; v_* was obtained by fitting a parabola to the maximum of the frequency spectrum and the two neighbouring spectral points; and α was computed as the mean of the pre-whitened frequency spectrum (obtained by multiplying by f^{-5}) in the interval $1.35 f_p$ to $2.0 f_p$.

Before fitting the model to the fetch-law data, a series of identical twin experiments was carried out to test the behaviour of the optimization procedure. The model, with all parameters set to unity, was used to create synthetic 'data' in the form of values of ϵ_*, v_* and α at given fetches. The model/adjoint pair were then applied to recover the free parameter values from these 'data' starting from different parameter values in the range 1.0 ± 0.5. Unfortunately, the optimization procedure did not always converge to the known correct values of the parameters. For most starting points in the five-dimensional parameter space, correct parameters were recovered with an error of less than 1%. For other starting points, however, the optimization routine (NAG-routine E04DGE) (NAG, 1990) terminated, signalling convergence of the descent algorithm, even though the solution obtained was not the right solution (all parameters equal to one). Calculating values

Table 6.4. *The five source term parameters (upper table) and associated constants (lower table) for the original WAM model (cycle 3) model and after optimization against the JONSWAP and Kahma fetch relations.*

	dissipation			wind input		cost	
	D_m	D_f	D_s	W_m	W_b	JONSWAP	Kahma
original	1.00	1.00	1.00	1.00	1.00	0.0434	0.0874
JONSWAP	1.77	1.29	1.16	1.31	1.11	0.0230	–
Kahma	0.996	0.950	1.19	0.861	1.06	–	0.0063

	dissipation			wind input	
	general factor	exponents		factors	
		freq.	steepn.	general	directional
original	3.33×10^{-5}	2.00	2.00	0.250	1.00
JONSWAP	5.89×10^{-5}	2.58	2.32	0.328	1.11
Kahma	3.32×10^{-5}	1.90	2.38	0.215	1.06

of the cost at a series of points along the line in parameter space connecting such an erroneous solution with the true solution revealed a 'hill' in the cost surface lying between these points. This suggested (though it certainly does not prove) that the procedure had converged on a secondary minimum. Although it is prohibitively expensive to map out the whole five-dimensional cost surface in detail, experience indicates that these apparent secondary minima occur mostly for exceptionallly low starting values of the dissipation parameters. In spite of these difficulties, we have confidence in the results obtained using real data (described below), since they proved to be not too far removed from the original WAM model values and did indeed result in a reduced value of the cost (i.e. a better fit to the data).

For the real data fits we used 18 points (as in the previous section) in the growing region, spaced logarithmically according to $X_{*i} = 1.2 X_{*i-1}$ and covering the range $X_* = 2.40 \times 10^5$ to $X_* = 5.32 \times 10^6$. In the Pierson-Moskowitz saturation region we used 6 points, starting at $X_* = 1.18 \times 10^8$ and with the same logarithmic spacing. As the initial state at $X_* = 2.4 \times 10^5$ we used a JONSWAP spectrum with $\cos^2$-spreading function and parameters $\alpha = 2.226 \times 10^{-2}$ and $v_* = 1.927 \times 10^{-2}$.

All assimilation runs were carried out with a one-dimensional (fetch-limited) version of the WAM model cycle 3. In addition, forward integration runs were carried out with a model in which the nonlinear transfer was computed exactly using the filtered phase space technique of Hasselmann and Hasselmann (1981, 1985b).

Although exact nonlinear transfer computations are still too time-consuming to be used in iterative adjoint-model optimization procedures, they are useful as a reference. Moreover, it is conceivable that an appropriately truncated approximation to the exact nonlinear transfer algorithm could be implemented that would be sufficiently fast while still providing good accuracy. In fact, the WAM model cycle 3 discrete interaction approximation used here, with only two interacting wave quadruplets, can be viewed as such a truncation carried to its logical extreme. Presumably, less severely truncated approximations to the nonlinear transfer will eventually be incorporated into the source function optimization process, but in the meantime, the wide-spread use of the discrete interaction approximation in the present WAM model formulation is regarded as sufficient justification to treat this source function as given while optimizing the remaining two WAM source functions.

Table 6.4 shows a comparison of the results of the optimizations with the values used in the WAM model cycle 3 model. The corresponding growth curves for the original WAM model, optimized model and empirical data against which the model was fitted are shown in figure 6.14 for the JONSWAP growth relations and figure 6.15 for the corresponding Kahma relations. Results for model integrations using the exact nonlinear transfer expression are not shown, since they are very close to the corresponding WAM results using the discrete interaction approximation, both for the original WAM parameters and for the optimized values.

The curves indicate that the agreement between the data and the model is substantially improved only in the Kahma case in the quasi-linear growth region at short fetches. This is perhaps not surprising, since the data set is not very complex, so that the hand-tuning which was used to derive the original WAM model could be expected to arrive reasonably close to the true optimal solution. A more critical test of the model physics will be provided by the extensive sets of complex two-dimensional wave spectra which are now becoming available on a global scale through the ERS-1 SAR wave mode data (Brüning et al, 1993, 1994, § V.4). It is planned to apply the adjoint technique to these data by extension of the present adjoint model to the full 2+1 dimensions of the WAM model space-time domain.

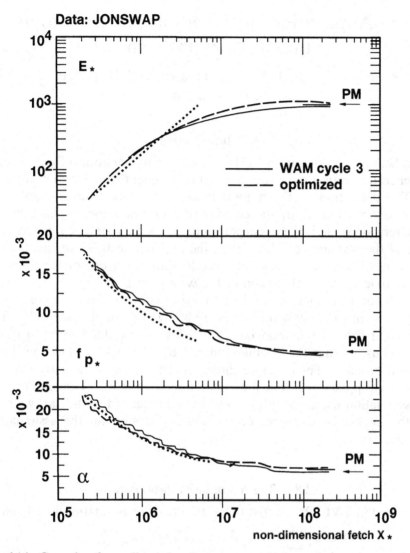

Data: JONSWAP

Fig. 6.14. Growth of nondimensional energy, peak frequency and α as function of nondimensional fetch for original WAM cycle 3 parameters (solid curve) and optimized parameters (dashed curve) fitted to the JONSWAP growth relations (dotted curve)

VI.8 Application of the adjoint method to an initial value problem

M. M. de las Heras, P. A. E. M. Janssen, G. J. H. Burgers and
S. Hasselmann

VI.8.1 Introduction

The application of the adjoint technique to the assimilation of data into a third generation wave model such as WAM is, in general, not a straightforward task. The derivation of the adjoint model equations poses a number of difficulties, such as the computation of the derivatives of the nonlinear wave–wave interactions and the incorporation of the semi-implicit integration scheme of the WAM model. Here, as in the previous section, we simplify the problem by constructing the adjoint model only for a one-dimensional (in this case, time-dependent) version of the WAM model.

The adjoint technique is applied to assimilate significant wave height observations into the WAM model by updating the initial wave field. The forcing wind field is not modified. All fields are assumed to be spatially homogeneous, so that the assimilation reduces to a one-dimensional time-dependent problem. The example chosen is clearly not particularly relevant for practical applications, but illustrates some of the issues encountered in data assimilation using the adjoint model technique. In particular, we shall study the impact of the form of the source function on the assimilation procedure.

VI.8.2 The assimilation procedure

As discussed in § VI.2.2, the cost function for data assimilation is defined as

$$J = w_1 \sum_t (\epsilon^t - \epsilon_o^t)^2 + w_2 \sum_{k,m} (F_{km}^{t=0} - F_{f,km}^{t=0})^2, \tag{6.75}$$

where ϵ_o^t, ϵ^t denote the observed and modelled energy density, $F_f^{t=0}$, $F^{t=0}$ the first-guess and current initial spectra, and t, k and m represent time, direction and frequency indices, respectively. The discrete observation times t are assumed to be identical to the discretized time points used in the numerical integration of the model and its adjoint. Since there are normally less observations than independent variables, the second regularization term must be added in the definition of the cost (§ VI.2.2) to ensure a unique minimal solution.

The adjoint equations of the WAM model were computed (in contrast to the general implementation procedure described in § VI.3.2) by explicitly

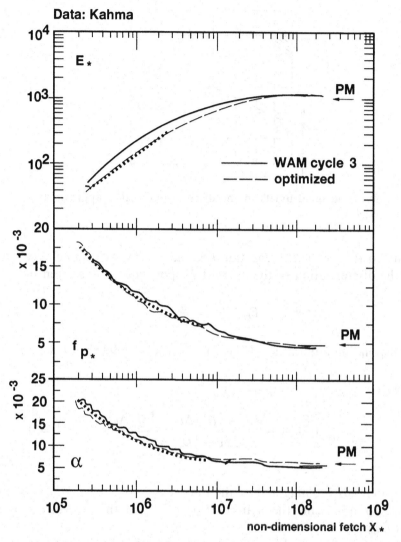

Fig. 6.15. Same as figure 6.14 for the Kahma growth curves

differentiating the corresponding Lagrange function

$$\mathscr{L} = J + \int dt\, df\, d\theta\, \lambda \left(\frac{\partial F}{\partial t} - S(F, u) \right), \qquad (6.76)$$

where λ is the Lagrange multiplier (costate variable), yielding (in discretized notation)

$$\frac{\partial \mathscr{L}}{\partial F^t} = 2w_1 \sum_{t'} (\epsilon^{t'} - \epsilon_o^{t'}) \frac{\partial \epsilon^{t'}}{\partial F^t} - \frac{\Delta}{\Delta t} \lambda^t - \lambda \frac{\partial}{\partial F^t} S^t = 0. \qquad (6.77)$$

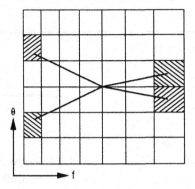

Fig. 6.16. Interacting quadruplets in the discrete interaction approximation (after van Vledder, 1990).

According to the implicit integration scheme of the WAM model (see chapter III), the discretization of the model transport equation is given by:

$$F_{km}^t = F_{km}^{t-1} + \frac{S_{km}^{t-1}\Delta t}{1 - \frac{1}{2}D_{km}^{t-1}\Delta t}. \tag{6.78}$$

The corresponding adjoint equations for the WAM model then become:

$$
\begin{aligned}
&2w_1(\epsilon^t - \epsilon_o^t)\frac{\partial \epsilon^t}{\partial F_{KM}^t} + \lambda_{KM}^t - \lambda_{KM}^{t+1} \\
&- \sum_{km} \lambda_{km}^{t+1} \left(\frac{S_{kmKM}^t \Delta t(1 - \frac{1}{2}D_{km}^t\Delta t) + \frac{1}{2}S_{km}^t \Delta t^2 \Lambda_{kmKM}^t}{(1 - \frac{1}{2}D_{km}^t\Delta t)^2} \right) = 0.
\end{aligned}
$$
$$\tag{6.79}$$

Here D_{km}^t denotes the diagonal of the matrix $S_{kmKM} = \partial S_{km}/\partial F_{KM}$ at time t and Λ_{kmKM} represents the matrix $\partial D_{km}/\partial F_{KM}$ of the derivatives of the diagonal D_{km}.

Equation (6.79) is solved for λ backwards in time t. At the last integration time $t = 0$, the left hand side of the equation is equal to the gradient of J. This yields the direction of steepest descent for a conjugate gradient descent algorithm which computes the minimum of the cost function with respect to the control variables. The rather tedious task of computing the matrices of first and second derivatives of the nonlinear interaction expression is similar to the computations involved in deriving the implicit integration scheme of the WAM model. However, in the case of the adjoint equations, the nondiagonal terms, which were neglected in the implicit scheme, must also be included.

For the nonlinear interactions source term, every point in the frequency-direction plane is the centre of two quadruplets, shown in figure 6.16.

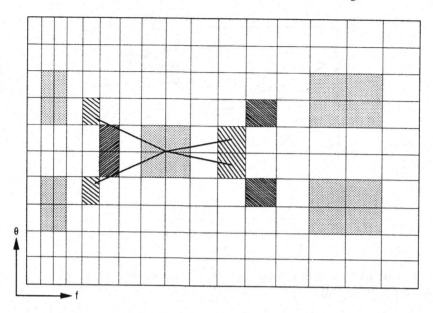

Fig. 6.17. Gridpoints influenced by interactions with the centrepoint. Gridpoints indicated by lightly hatched areas represent the initial interacting main quadruplet and its symmetrical counterpart, with central point (M, K). Heavily hatched areas indicate the centres of all other quadruplets which interact with (M, K). Dotted areas show the remaining points belonging to quadruplets centred in the heavily hatched area.

Each quadruplet involves 9 spectral components. A component of the nonlinear source term receives contributions from 18 quadruplets; through these quadruplets, it depends on 69 spectral components. These 69 points, which must be taken into account in the construction of the adjoint, are shown in figure 6.17.

VI.8.3 Results

An initial 'observed' spectrum was generated by integrating the WAM model for a constant $U_{10} = 18.45$ m/s wind speed for one day to an essentially equilibrium state. An initial first-guess wave spectrum was created similarly by a one-day integration with a constant $U_{10} = 12$ m/s wind speed. The wind was then turned off in both runs, and observations and first-guess wave height fields were generated every 20 minutes for a one day decay period.

Subsequently, the observations were assimilated into the WAM model for various test cases to study the sensitivity of the scheme to the different source terms of the model.

In a first test case the 'observations' were assimilated into the WAM model using the adjoint scheme considering only dissipation in the source term

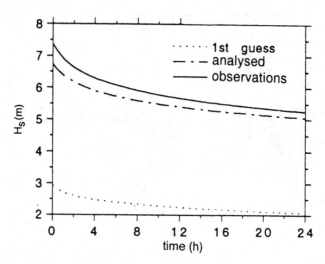

Fig. 6.18. Comparison of observed, first-guess and analysed wave height time series for the case $S = S_{ds}$ only.

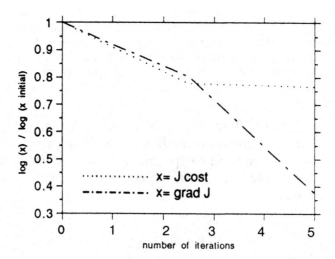

Fig. 6.19. Reduction of the value of the cost function and its gradient for each iteration step, for the case $S = S_{ds}$.

– i.e. $S = S_{ds}$. The resulting analysed significant wave heights are shown in figure 6.18. The reduction of the cost function and its gradient with increasing number of iteration steps are shown in figure 6.19. The cost function is already reduced by a factor of 9 after two iteration steps.

Figures 6.20 and 6.21 show the corresponding results using the full source function $S = S_{ds} + S_{nl}$ (since the wind was turned off after initialization, the wind-input source function is zero).

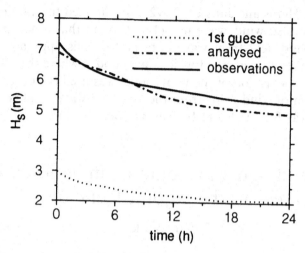

Fig. 6.20. Comparison of observed, first-guess and analysed wave height time series for the case $S = S_{ds} + S_{nl}$.

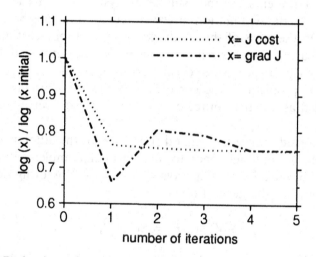

Fig. 6.21. Reduction of the value of the cost function and its gradient for each iteration step, for the case $S = S_{ds} + S_{nl}$.

The results look quite promising. In particular, the rapid modification of the wave field after model initialization and the marked reduction of the cost function after only a few iteration steps suggest that it may be feasible to assimilate wave observations into the WAM wave model operationally using the adjoint technique.

The next step is to assimilate wave height data using winds as the control variable, since especially for growing waves one can expect a stronger impact on the observed wave growth from the wind than from the initial sea state.

In a first attempt, Heras and Janssen (1992) have applied the adjoint method to a simplified coupled wind–wave model and found that it was possible to reconstruct the wind field from wave data alone. This approach has been applied to a more realistic coupled wind–wave model, where the wave model physics of the cycle 4 version of the WAM model was used (Heras *et al*, 1994). The reconstruction of wind fields from the observed wave heights and mean wave directions was found also to be feasible for this more complex system.

VI.9   A wind and wave data assimilation scheme based on the adjoint technique

C. F. de Valk

VI.9.1   General properties of the adjoint of a wave prediction model

This contribution presents another simple analysis of the basic strengths and weaknesses of data assimilation in wave prediction models based on the adjoint model. On the basis of this analysis a general approach towards the realization of a feasible data assimilation scheme will be developed, the ideas being illustrated with some results of numerical experiments. The focus is on the large-scale problem of the correction of the wind forcing, which is generally regarded as a major source of error in numerical wave predictions and analyses.

We begin by considering some general properties of the adjoint of a wave prediction model, without reference to any particular implementation. As discussed in § VI.3, the adjoint of a wave model, which (ignoring depth and current refraction) has the general form

$$\frac{\partial F}{\partial t} + \nabla_x \cdot (\mathbf{c}_g F) = \mathbf{S}(F, \mathbf{u}_*), \tag{6.80}$$

is derived from the tangent model, obtained by linearizing (6.80) around a trajectory:

$$\frac{\partial \delta F}{\partial t} + \nabla_x \cdot (\mathbf{c}_g \delta F) = \mathbf{A}\delta F + \mathbf{B}\delta \mathbf{u}_*, \tag{6.81}$$

where δF denotes the perturbation of the spectrum, and $\mathbf{A}$ and $\mathbf{B}$ represent the tangent components of the source function $\mathbf{S}$, which will depend, in general, on the local sea state F and friction velocity u_*.

It was shown in § VI.3 that the adjoint of a wave model is driven at the gridpoints and time levels at which wave observations are available, the forcing being given by the gradient of the local cost with respect to the spectrum. Thus for a quadratic cost function proportional to $\int (F - F^0)^2 d\mathbf{k}$, for example, the forcing is simply the prediction error $F - F^0$. The error is propagated backwards along the characteristics according to the adjoint transport equation

$$-\frac{\partial \lambda}{\partial t} - \nabla_x \lambda \cdot \mathbf{c}_g = \mathbf{A}^* \lambda + \dots, \tag{6.82}$$

where λ is the local adjoint state, $\mathbf{A}^*$ is the adjoint operator of $\mathbf{A}$ and the dots represent the forcing of the adjoint model. The contributions to $\mathbf{A}^*$ from the wind-input and dissipation source functions can be readily evaluated (dissipation is almost exactly the same as in the forward model, but of opposite sign), but the functional derivative of the nonlinear source function is more difficult to compute (see previous section). Once the adjoint model has been integrated, the gradient of the cost with respect to the local wind vector is readily computed as $\mathbf{B}^* \lambda$.

To investigate the nonlocal aspects of the adjoint, consider the case of waves generated by a storm occupying a small region in an otherwise calm sea. At some distance from the storm, wave interactions vanish and $\mathbf{A}$ becomes diagonal. If at a distant location only the spectral density in a single wavenumber $\mathbf{k}$ is observed, data assimilation based on the adjoint model has the following limitations:

(i) It is impossible to locate the source of observed wave energy in space and time directly from the observational data (only the characteristic is determined, not the location of the source on the characteristic).

(ii) When following a wave package with wavenumber $\mathbf{k}$ generated in a storm field, $\mathbf{B}$ in (6.81) will be large inside the storm field but becomes much smaller or zero after the waves have propagated out of the generating area. This is because $\mathbf{B}$ is nonzero only if the wind component in the direction of wave propagation is greater than about 1.2 times the phase velocity, or expressed in terms of the friction velocity, if $28\mathbf{k}u_* > \omega(\mathbf{k})$ (chapter III and WAMDI, 1988). Thus it is relatively easy to reduce the wind speed within a storm field that was mislocated in the first guess, but it is much more difficult to increase the wind speed at the true location of the storm if the first-guess wind at the true location was small. The adjoint propagates the information from the observations back along the characteristics, but the gradient $\mathbf{B}^* \lambda$ of the cost function with respect to the wind remains small at the location where the correction needs to be applied because $\mathbf{B}$, determined by the first guess, is small. Some improvement can perhaps be achieved through suitable scaling of the Hessian approximation in the descent algorithm of Fletcher (1987).

(iii) The gradient of the cost function with respect to the local wind vector cannot provide independent information about the two components of the wind vector, because the spectral density of the observed wave component is a scalar.

If a time sequence of spectral observations in several frequency bands is available at a given location, problems 2 and 3 above remain, but the first problem can, in principle, be resolved. By making use of wave dispersion, a storm can be located from observations at a single point as the region in which the characteristics of the observed spectral densities intersect (Snodgrass *et al*, 1966), thereby increasing locally the gradient of the cost with respect to the wind. However, a drawback of the use of detailed observations of spectral density is that it may lead to a cost function with multiple local minima: the change of the predicted energy in a spectral bin with increasing wind speed or duration need not be monotonic, because of the shift of the spectral peak. In mathematical terms, the model output manifold is strongly curved in the ambient space, see, for example, Chavent (1991). The problem can be alleviated by formulating the cost in terms of integral parameters of the spectrum which still carry information about the storm location, for example the significant wave height and mean period. This *ad hoc* solution, however, loses information and is not completely satisfactory .

If directional spectra along a swath are gathered in a short time (e.g. by ERS-1), a storm can also be located from the variation of the observed wave direction along the swath, in analogy with the previous case.

A special feature of the adjoint method is that a zero wind vector represents a stationary point of the variational problem because **B** vanishes, but it does not normally represent the optimum. In data assimilation experiments with a single-gridpoint model without advection, the analysis wind vectors converged to zero when the first-guess wind directions were too far off. However, this problem has not been observed in experiments with a full-scale model (see also § VI.9.3).

Apart from this feature, most of the ambiguities listed above are not problems specific to the adjoint method as such but simply reflect inherent limitations of the information content of measurements of individual spectral components with respect to the wind field which generated them. The same difficulties arise, for example, in the Green's function method described in the next section or in simpler single-time-level variational methods. Basically, the problems can be resolved only through a sufficiently large number of independent spectral measurements at different locations and times. These are now becoming available through satellites. With sufficient sampling density, wave measurements potentially provide a sensitive tool for detecting wind field errors in the areas which are often of greatest interest, namely in the regions of highest winds.

VI.9.2 Implementation of the adjoint technique for a second generation wave model

Controls

A traditional difficulty of using wind fields as controls is the huge dimension of a wind field sequence. However, this presents only a minor problem in practice; modern descent methods such as the conjugate gradient method can easily deal with such high-dimensional problems. A second problem is the construction of physically realistic wind fields. We have just discussed the limits to the information on wind fields obtainable from a restricted set of wave observations, and thus we cannot expect that wind field corrections derived from such measurements will necessarily be consistent with the dynamical constraints of the real atmosphere. The ultimate solution is to couple the wave model to an atmospheric model and to develop an assimilation method for the complete system (cf. § VI.1.1 and Hasselmann *et al*, 1988). A provisional solution is to parametrize the wind field sequence in a manner consistent with the dynamical constraints or to incorporate the constraints as additional terms into the cost function. In the current implementation of the adjoint method described below, smoothness of wind field corrections is imposed by a B-spline representation of the vector components. Tests are planned to introduce at least the most important dynamical constraint of quasi-geostrophy by using a simple representation of the surface wind fields in terms of (quasi-)surface pressure fields.

Cost function

The cost function should reflect the statistics of errors in the observations and in the first guesses of model inputs. As discussed in § VI.2.3, this is achieved by the maximum likelihood method. One needs for this the covariance error matrices of the observations and of the model first guesses, which appear in the two separate terms of the cost function (§ VI.2.2). The specification of the statistics of the observational errors is normally relatively straightforward, see, for example, Long (1980), although difficulties can arise when the measurement process is still poorly understood. In most cases different measurements are statistically independent; the cost function expression representing the misfit to the observations reduces then to a diagonal form. Specification of the covariance matrix of the first-guess fields (the controls) is usually more difficult. For wind fields, the relevant statistics can be obtained from meteorological data archives. It is important that the spatial and temporal correlations of errors in first-guess wind fields are properly taken into account, because this determines the relative cost of different wind field error structures. For example, if wind perturbations at different gridpoints are penalized independently, a refinement of the grid would automatically increase the associated cost. The easiest way to avoid such dilemmas is to expand the wind field terms of a set of base patterns and thereby to fix the error correlation scales. The cost of wind field perturbations relative to the cost of a misfit to the observations can be established by regarding the first-guess wind fields formally as 'bogus observations' and treating both wave observations and first-guess wind fields formally as independent components of a common observational data set (Thacker, 1988).

Numerical solution

To solve large-scale minimization problems, several efficient and storage-effective descent methods are available. These are based on iterative procedures and require functional evaluations (integration of the forward model) and gradient evaluations (integration of the adjoint model) at each iteration step. In the present implemention a limited-memory BFGS method was used (Liu and Nocedal, 1988).

There remains nevertheless the computational problem of integrating a full two-dimensional, space- and time-dependent wave model together with its equally time consuming adjoint repeatedly until the minimization has been completed. Some approximation of the model during the minimization procedure appears to be the only feasible solution to this problem. One possibility, generally adopted in meteorological applications, is to use coarser resolution models for the minimization. The technique could be refined by interspersing more accurate higher resolution computations at a lower cycle rate during the iteration sequence.

An alternative is to use a second generation wave model. This is the course pursued in the present implementation. The quality of the resulting wave analysis can be tested by reproducing the observed data subsequently with a third generation model.

The second generation model used in the present implementation employed a standard advection scheme for both wind sea and swell and the WAM dissipation source function for swell. Wind sea growth was modelled by: a) extracting wind sea parameters from the advected spectra (using a sea-swell separation similar to that used by Janssen *et al*, 1989b); b) computing the new wind sea parameters from the local friction velocity, using growth curves tuned to the WAM model; c) computing a new parametrized wind sea spectrum. Weaknesses of the model, common to all second generation models (cf. SWAMP, 1985), are the directional relaxation for turning winds and the adjustment of the spectral wind sea peak after a decrease in wind speed. The parametrization of the wind sea spectrum as currently implemented is similar to the JONSWAP spectrum, but with different coefficients. An alternative approach could be to retain the shape of the first-guess wind sea spectrum and simply modify the frequency and energy scales of the spectrum, as in Lionello *et al*, (1992). This would have the advantage of providing an accurate approximation when the wind fields and thus the wave spectra are close to their first guesses.

VI.9.3 Results of numerical experiments

Numerical data assimilation experiments with this second generation model and its adjoint have so far been carried out only with simulated data. The first tests were made for a North Sea storm which had already been used by de Valk and Calkoen (1989) to test an earlier version of the scheme. Simulated sea truth wave fields were generated by running a 1990 version of the NEDWAM model (the KNMI North Sea implementation of the WAM model, Karsten, 1987) driven by a sequence of true wind fields for January 1976 from KNMI; these wind and wave fields are referred to as *simulations* in the figures. Simulated wave observations of significant wave height, mean wave period and mean wave direction were taken from these wave fields every three hours at the seven locations in the North Sea indicated in figure 6.22. The same wind field sequence was shifted by 6 hours and taken as the first-guess wind field, from which the first-guess wave fields were produced by running the second generation wave model. The data assimilation scheme was then run to test whether the original true wind fields could be recovered from the simulated true wave observations.

Separate terms were introduced into the cost function to penalize the misfit to the simulated observations of significant wave height, mean wave period and mean wave direction. For wave height and period, the squared differences divided by $(0.025 \text{ m})^2$ and $(0.3 \text{ s})^2$, respectively, were used; for wave direction, the corresponding error term was defined as one minus the cosine of the difference in the angular directions. The term in the cost function penalizing perturbations of the first-guess wind fields was of the form described in de Valk and Calkoen (1989), but its contribution was made very small.

The first-guess, analysed and simulated wind fields six hours after the start of the simulation are shown in figure 6.23, the corresponding wave fields nine hours after the start in figure 6.24. The wave analyses were produced by the same second generation wave model that was used in the minimization. The first-guess wind and wave fields are seen to have been corrected only in the area surrounding the gridpoints where observations were simulated. This is to be expected in the vicinity of a storm producing only locally generated waves.

Further tests (not shown) with different wind fields but using the same assimilation technique and wave observables were carried out to investigate whether wind corrections could be inferred for storm fields far away from the locations of wave observations. This is, of course, the distinguishing

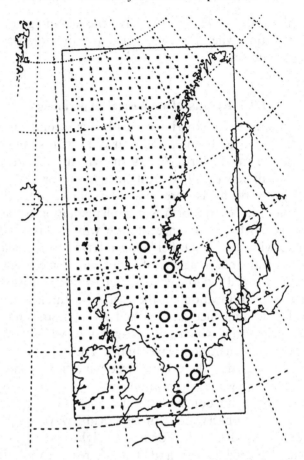

Fig. 6.22. Model grid points (squares) and locations of simulated observations (open circles).

characteristic of a multi-time level method. In the case that predicted swell was completely absent in the simulated observations but was present in the first-guess wave field, the storm in the first-guess wind fields was correctly modified to prevent the generation of waves propagating toward the locations of the observations. However, the same test with first-guess and simulated sea truth interchanged, i.e. with observed swell but no first-guess swell, produced only local corrections of wind fields and sea states: the adjoint technique failed to recognize the simulated wave observations as swell from a distant storm field. This is consistent with the second property of the adjoint discussed in § VI.9.1.

The tests performed so far were designed only to investigate the conditions under which the adjoint method is successful in principle in correcting wind fields and sea states. In applications with real observations, the cost

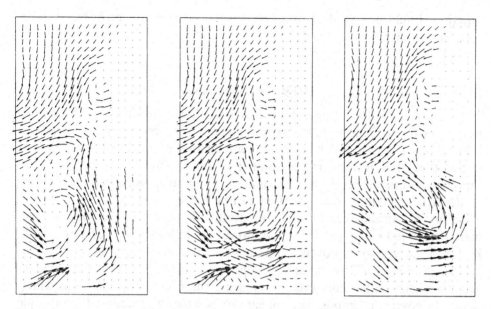

Fig. 6.23. First-guess (left panel), analysed (middle panel) and simulated (right panel) wind fields after six hours.

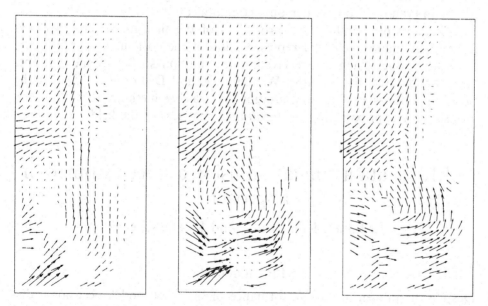

Fig. 6.24. Same as figure 6.23 for wave fields (wave heights and mean directions) after nine hours.

function should be modified to reflect more accurately the error statistics in accordance with § VI.9.2.

VI.9.4 Conclusions

The adjoint modelling approach to data assimilation in wave models can still not be regarded as more than a promising technique, but we now understand better how it works and how the remaining problems can be solved. It is hoped that after introducing a number of improvements in the wind field representations, model approximations and minimization methods, a more conclusive evaluation of the method can be made.

Technical aspects such as the computational speed and memory requirements were not discussed here (see, however, de Valk and Calkoen, 1989), but in this area also improved techniques such as the so-called *decomposition-coordination methods* for large-scale optimal control problems may lead to greater efficiency, possibly even eliminating the need for model approximation. Moreover, decomposition appears to be able to overcome the multiple minima problem which arises when detailed spectral information is used (see discussion above), the wind field estimates converging to the unique global optimum (van der Vooren, personal communication).

Estimation of the statistical properties of the first-guess fields required for the maximum likelihood representation of the cost function is another important issue which has received little attention so far. Some concepts are available (see, for example, Wahba, 1990 and Dee *et al*, 1985), but the problem remains to find computationally feasible methods for estimating and suitably approximating the covariance matrices of the fields.

VI.10 Data assimilation using a Green's function approach

E. Bauer, I. R. Young and K. Hasselmann

VI.10.1 Introduction

Despite the proven overall performance of the WAM model, occasional discrepancies are still observed. There are three possible causes for such problems: inadequate representation of the physical source terms in the model, poor model numerics, and poor specification of the driving wind fields. The data assimilation and inverse modelling techniques described in this chapter address all three sources of error. Under carefully documented wind conditions, for example for fetch-limited wave growth in specially designed field experiments, it is possible to identify and correct for model shortcomings in the physics or numerics (§§ VI.6, VI.7). However, for a well-verified model in normal applications, the errors due to poor specification of the wind field are generally significantly greater than the model errors. To correct these errors the wave data must be assimilated in the wave model by a scheme in which both wind and wave fields are altered in a dynamically mutually consistent manner.

This can be achieved by minimizing a cost function representing the misfit between model and observations under the constraints of the model equations (§ VI.2). The most general technique for doing this is the adjoint method (cf. §§ VI.3, VI.8 and VI.9). Here we discuss the alternative Green's function method (cf. § VI.2.4 and Bauer *et al*, 1995). Although based on a number of physically rather intuitive approximations and therefore lacking the rigour of the adjoint method, it has the advantage of yielding the minimum of the cost directly without going through a computationally expensive iterative descent algorithm requiring multiple integrations of the model and its adjoint. Thus, in contrast to the adjoint method, there is no need to introduce additional approximations, for example in the formulation of the model, in order to reduce the scheme to a computationally feasible level.

VI.10.2 Green's function formulation

The cost function to be minimized is of the general form (§§ VI.2.2, VI.2.4)

$$ J = \sum_r \left\{ \frac{(\Delta d_r - d_r)^2}{(\sigma_r^d)^2} \right\} + C \sum_p \left\{ u_p^2 + v_p^2 \right\}, \qquad (6.83) $$

where $\Delta d_r = d_r^f - d_r^o$ is the difference between the first-guess model equivalent of the data d_r^f and the observed value d_r^o, d_r is the modification of the model data value generated in the course of the optimization relative to the first-guess model value (note that in contrast to the notation of previous sections the model data d_r are defined here as the *deviation* relative to the first-guess model data), σ_r^d represents the standard deviation of the observed data, u_p and v_p are the changes introduced into the x and y components, respectively, of the wind velocity fields (at locations indicated by the index p), and C is a suitably chosen weighting factor. In the examples given below, the WAM model cycle 4 was used. This is driven by the 10 m wind (in contrast to earlier cycles driven by the friction velocity), so that u_p, v_p refer here to the components of the 10 m wind.

In applications to SAR data, which yield the full two-dimensional wave spectrum, each component of the model directional wave spectrum is considered as an independent datum value d_r^f. In practice, however, more reliable corrections are obtained by considering a cluster of spectral components with similar frequencies and directions (see discussion below). The summation index r ranges in this case over the set of clusters. In general, d_r^f can represent any functional of the wave spectrum, such as the total energy (for altimeter data, for example) or the one-dimensional frequency spectrum.

A modification of the wind vectors (u_p, v_p) will cause a modification to the model spectrum, d_r. Thus the task is to minimize (6.83) by choosing an optimal set of wind vectors, with the model data d_r being determined by the wind vectors through the model equations.

For small modifications $\mathbf{u}$ of the velocity field relative to the initial (first-guess) velocity field $\mathbf{U}$, the relation between the modification δF of the wave spectrum F and the wind perturbation $\mathbf{u}$ can be obtained by linearizing the wave model equation (without depth and current refraction)

$$\frac{DF}{Dt} = \frac{\partial F}{\partial t} + \mathbf{c}_g \cdot \nabla F = S_{tot}(F, \mathbf{U}), \qquad (6.84)$$

yielding

$$\frac{D\delta F}{Dt} = \frac{\partial S_{tot}}{\partial F}\delta F + \frac{\partial S_{tot}}{\partial \mathbf{U}}\mathbf{u} \quad \text{or} \quad L\delta F = \frac{\partial S_{tot}}{\partial \mathbf{U}}\mathbf{u}, \qquad (6.85)$$

where $L = (D/Dt - \partial S_{tot}/\partial F)$, $\mathbf{c}_g$ is the wave group velocity and S_{tot} the total source function.

Small changes d_r in the model data can, in turn, be linearly related to the perturbation of the local wave spectrum,

$$d_r = \sum_{\mathbf{k}} A_r(\mathbf{k})\delta F(\mathbf{k}), \qquad (6.86)$$

where $A_r(\mathbf{k})$ is a transfer function which depends on the type of observed data (SAR-derived wave spectrum, significant wave height, etc.). It will be shown below that (6.85) can be integrated explicitly (under suitable approximations), yielding the perturbation of the wave spectrum as a function of the wind field perturbation. The model data changes d_r can then be expressed, through (6.86) and the inverse of (6.85), as linear functions of the changes in the velocity field, and (6.83) becomes a standard quadratic optimization problem for the modification of the velocity field (see also § VI.3.1).

The integration of (6.85) can be written down formally in terms of the Green's (impulse response) function (Roach, 1982). In full generality, however, this expression is not very useful, as the determination of the Green's function requires the inversion of the operator L in (6.85), which in the discretized form represents a very high-dimensional matrix. Moreover, the individual terms of L are highly complex, involving the functional derivative of the total source function, including, for example, the nonlinear transfer integral (cf. § VI.8.2).

Bauer *et al* (1995), however, have attempted to make the problem computationally tractable by assuming that the Green's function is highly localized in space and time, taking the form of a δ-function. Physically, the basis for this assumption is the localized sensitivity of the waves to changes in the wind field. If the history of a given spectral component is traced from its area of generation to the location of its measurement, one can normally distinguish between three regions: 1) a wind-sea region, where the wave component received energy from the wind; 2) a swell region, where the wave component is no longer affected by the wind and is thus insensitive to wind changes; and 3) the transition region between the generation region and the swell region. It will be argued below that the rather narrow transition region is the region of greatest sensitivity to changes in the wind field. If there exists no swell region, i.e. if the wave component belongs to the local wind sea, the most sensitive region is still the region where the wave component last received an input from the wind, i.e the location of the wave component itself.

Under this δ-function approximation, the modification of the wave spectrum can be expressed in the form

$$\delta F(\mathbf{k}, \mathbf{x}_r) = W^u u_p + W^v v_p, \tag{6.87}$$

where $(u_p, v_p) = (u(\mathbf{x}_p, t_p), v(\mathbf{x}_p, t_p))$. The coordinates $\mathbf{x}_p, t_p$ of the 'influence point', the most sensitive point at which the wind vector must be altered to modify the spectral component $\mathbf{k}$ at the point $\mathbf{x}_r$, and the associated (integral) impact factors W^u, W^v will be derived in the next section.

Substitution of (6.86) and (6.87) into (6.83) yields a cost function which is now expressed directly in terms of the minimization variables u_p and v_p,

$$J = \sum_{r,\mathbf{k}} \left\{ \frac{[\Delta d_r - \{B_r^u(\mathbf{k}) u_p + B_r^v(\mathbf{k}) v_p\}]^2}{(\sigma_r^d)^2} + C\left\{ u_p^2 + v_p^2 \right\} \right\}, \tag{6.88}$$

where

$$B_r^u(\mathbf{k}) = A_r(\mathbf{k}) W^u(\mathbf{k}) \tag{6.89}$$

and

$$B_r^v(\mathbf{k}) = A_r(\mathbf{k}) W^v(\mathbf{k}) \tag{6.90}$$

(contrary to the notation of equation (6.83), the sum over p in the last term of (6.88) is now regarded as implied by the sum over the variables r and $\mathbf{k}$, which define the coordinates $\mathbf{x}_p, t_p$ of u_p, v_p). The values of u_p and v_p which minimize (6.88) can be readily evaluated in closed form from the minimization conditions $\partial J / \partial u_p = 0$ and $\partial J / \partial v_p = 0$ (cf. Bauer *et al*, 1995).

In practice, Bauer *et al* (1995) used a more robust and powerful technique by collecting individual spectral observations into spectral clusters and regarding these as the basic data set. This has two advantages. Firstly, it stabilizes the results by smoothing. Secondly, it overcomes the inherent limitation, discussed in § VI.9.2, of any attempt to infer errors in the wind velocity from individual spectral data. Wind errors represent two-dimensional vectors, whereas the energy of a single spectral component is a scalar. It can thus never provide a unique wind velocity correction without additional information. By collecting a number of spectral observations in a cluster it is possible to extract such additional information by considering not only the mean energy of the cluster but also modifications in the local shape of the spectrum.

Bauer *et al* (1995) consider three different weighted averages over the spectral components of a cluster. The weights are defined to yield estimates of the mean energy of the cluster and the two components of the mean local spectral gradient with respect to frequency and direction. A cluster consists of seven spectral points around a variable centre point. The cluster centre point runs through all spectral bins (with the exception of the edge points), so that the number of clusters is essentially the same as the original number of spectral bins. Since cluster data represent linear combinations of the original spectral data, the introduction of cluster data into the cost function does not significantly complicate the minimization problem. Details are given in Bauer *et al* (1995).

VI.10.3 Determination of the impulse response functions

The feasibility of the direct inversion approach relies critically on the validity of the δ-function approximation for the Green's function. It is assumed that if a perturbation to the wind field is applied to the entire model grid, only a given small region will cause any modification of any given component of the wave spectrum at any given observation point. The magnitude of this modification is determined by the integral impact factors W^u and W^v. Hence the given spectral component contains information only about the wind at this point of maximal influence.

The rationale for this assumption is based on the processes governing the transformation of wind sea into swell. Once a given wave component leaves its generation region and enters a region where the wind speed is too small to impart energy to the component, it becomes swell. Hence, the wind speed has no influence on the waves in this region. Within the active growth region, on other hand, the wave component continually receives energy from the wind (if the spectrum has already migrated down to that frequency), but nonlinear interactions at the same time continually exchange this energy with other wave components. Thus the direct effect of the wind input is lost through this scattering process. However, in the narrow transition region between the generation and swell regions, the last impact of the wind is retained and transported into the swell region, where the information can then propagate undisturbed to the measurement location. Thus the spectrum is most sensitive to modifications in the wind field in the narrow region representing the transition from wind sea to swell.

The hypothesis, although intuitively plausible, clearly applies only approximately, since nonlinear interactions within the generation region are known to affect the rate of migration of the wind-sea peak to lower frequencies (cf. §§ II.3, III.3). Thus changes in the wind field in the interior of the generating region will influence, for example, the energy and frequency of the waves corresponding to the wind-sea peak at the point where the peak waves leave the generation region and become swell. The role of this interior wind forcing relative to the 'last wind impact' needs to be studied further, but for the present purposes we will simply assume that it is not significant. We note, however, that its influence is, in fact, implicitly included in the computation of the wave age described below. This is derived from a weighted average of the time of generation of the wave perturbations, the weights in the averaging integral being determined by the generating source functions. Thus contributions from the interior of the generating region are automatically included in the computation of the location of the region of maximum wind influence.

To determine the position of the influence point and the impact factors $W^u(\mathbf{k})$ and $W^v(\mathbf{k})$, we must inspect the structure of the operator $\partial S_{tot}/\partial F$ and the derivative $\partial S_{tot}/\partial \mathbf{U}$ in (6.85). The total source function consists generally of the sum of the input, nonlinear transfer and dissipation source functions, $S_{tot} = S_{in} + S_{nl} + S_{ds}$. In WAM, the dependence on the velocity $\mathbf{U}$ appears only in the input source function S_{in}, which is proportional to the spectrum, so that

$$\frac{\partial S_{tot}}{\partial \mathbf{U}} = \mathbf{a}F, \qquad (6.91)$$

where the vector $\mathbf{a}$ is in general a function of $\mathbf{U}$, $\mathbf{k}$ and (in the cycle 4 WAM model version) also of integral spectral properties. Thus the linearized wave transport equation (6.85) takes the general form

$$\left(\frac{D}{Dt} - \Lambda\right)\delta F = \mathbf{a}\cdot\mathbf{u}\,F, \qquad (6.92)$$

where $\Lambda = \partial S_{tot}/\partial F$.

The operator Λ is complicated. It can be simplified by retaining only the diagonal terms, as in the implicit integration scheme used in WAM (cf. § III.4). This assumption has the shortcoming, however, that the coupling between wave components, which is essential for spectral development and the response to a wind change, is lost. Bauer *et al* (1995) accordingly consider both diagonal and nondiagonal components of the operator, $\Lambda = \lambda + \Lambda_{nd}$. The diagonal term λ is identical to the term computed for the implicit numerical integration scheme of the WAM model, while the nondiagonal term Λ_{nd} is collected together with the wind input term on the right hand side of (6.92) in a new effective net input source term. Decomposing the net input term into components parallel ($\|$) and perpendicular ($\perp$) to the local wind vector **U**, (6.92) then becomes

$$\left(\frac{D}{Dt} + \lambda \right) \delta F = \beta^{\|} u^{\|} + \beta^{\perp} u^{\perp} \qquad (6.93)$$

The problem is then to find appropriate expressions for the net (differential) impact factors $\beta^{\|}$ and $\beta^{\perp}$ without entering into the highly complex computation of the nondiagonal terms of the functional derivative Λ (cf. § VI.8.2). The following relations are based on a combination of physical arguments and preliminary empirical tuning. The expressions could presumably be improved by more systematic model inversion techniques based on a larger set of numerical experiments.

We consider first the effect of a wind change $u^{\|}$ parallel to the wind direction, i.e. a change in the magnitude U of the local wind speed. Physically, the perturbation $u^{\|}$ will produce first a perturbation of the local wind input source term, as expressed by the forcing term in (6.92). This will result in a perturbation in the spectrum similar to the wind input source function S_{in} itself. However, the spectral perturbation is then rapidly redistributed by the nonlinear transfer, which tends to return the wind sea spectrum to a universal quasi-equilibrium form (cf. Hasselmann *et al*, 1973, 1976 and § III.3). The additional spectral perturbation associated with this redistribution will have a form similar to the nonlinear transfer source function S_{nl}, which tends to transfer energy from the spectral peak to lower and higher frequencies. The net effect of both processes is a spectral perturbation which can be expressed approximately as a linear combination of the input and nonlinear transfer source functions. Empirically, the following expression was found to yield satisfactory results:

$$\beta^{\|} u^{\|} = 43.5 \frac{u^{\|}}{c} \left(\frac{S_{in}}{2} + S_{nl} \right) \cos[2(\theta - \theta_w)], \qquad (6.94)$$

where c is the phase velocity, θ the wave propagation (azimuth) angle and θ_w the wind direction.

Similar arguments can be applied to estimate the impact of an orthogonal wind velocity change $u^{\perp}$. Here satisfactory results were obtained using the expression

$$\beta^{\perp} u^{\perp} = 11.3 \frac{u^{\perp}}{c} \frac{\partial}{\partial \theta} \left(S_{in} + \frac{S_{nl}}{2} \right) |\sin[2(\theta - \theta_w)]| \qquad (6.95)$$

The details of the effective source functions (6.94), (6.95) were determined by optimally fitting the spectral perturbations obtained by integrating the perturbation equation (6.93) for given wind field perturbations to the spectral perturbations obtained by integrating the full model. The simplest case for this calibration is a uniform wind field $\mathbf{U} = (0, U)$ with superimposed constant wind field perturbations $u^{\perp}, u^{\|}$ in directions orthogonal or parallel, respectively, to the background wind. The impulse response functions, i.e. the impact factors W^{u}, W^{v}, can then be computed by integrating (6.93). Noting the definition (6.87), the determining equations for the impact factors take the simple form

$$\left(\frac{\mathrm{D}}{\mathrm{D}t} - \lambda\right) W^{u,v} = \beta^{\perp,\|},\tag{6.96}$$

which can be immediately integrated.

The time at which the wind field correction should be applied is specified by the wave age, the time $\tau_p(\mathbf{k})$ since the spectral component last received a significant impact from the wind (Booij and Holthuijsen, 1987). The corresponding location can be determined by tracing the spectral component back along the propagation path at its group velocity. The time of origin t_p corresponding to $\tau_p(\mathbf{k}) = (t - t_p)$ can be evaluated as a time-weighted average, the weights being governed by the differential impact factors $\beta^{\|}, \beta^{\perp}$:

$$t_p(\mathbf{k}, \mathbf{x}) = t'/W',\tag{6.97}$$

where t', W' are determined by integrating the propagation equations

$$\left(\frac{\mathrm{D}}{\mathrm{D}t} - \lambda\right) t'(\mathbf{k}, \mathbf{x}) = \beta t,\tag{6.98}$$

$$\left(\frac{\mathrm{D}}{\mathrm{D}t} - \lambda\right) W' = \beta\tag{6.99}$$

and $\beta = [(\beta^{\|})^2 + (\beta^{\perp})^2]^{\frac{1}{2}}$ represents the isotropic (directionally averaged) differential wind impact factor.

Implementation of this technique involves the integration of four additional transport equations: two equations (6.96) for the integral impact factors W^{u}, W^{v} and equations (6.98), (6.99) for determination of the weighted generation time t' and associated normalization factor W', respectively, which define the wave age τ_p. This requires an increase in the memory storage of the model, but only a minor increase in computation time.

VI.10.4 Test cases

The application of the technique is illustrated for two test cases: the calibration experiment, and the case of a change in the wind field introduced in a localized distant region within an otherwise uniform wind field. The following computations were carried out using the Green's function scheme of Bauer *et al* (1995) based on clustered spectral data.

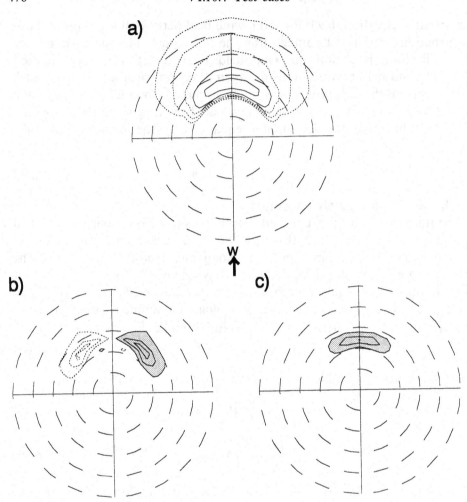

Fig. 6.25. Wave spectrum $F(f, \theta)$ (panel a), impact functions $W^u(f, \theta)$ (panel b) and $W^v(f, \theta)$ (panel c) in the centre of the computational grid for a uniform southerly wind field.

In the calibration experiment, a uniform southerly wind $\mathbf{U} = (0, U)$ with wind speed $U = 10$ m/s is perturbed by a 1 m/s velocity field $(u^\perp, 0)$ or $(0, u^\parallel)$ orthogonal or parallel to $\mathbf{U}$. Figure 6.25 shows the background directional spectrum computed with the WAM model at a point in the centre of the computational grid, together with the impact factors $W^u(\mathbf{k})$ and $W^v(\mathbf{k})$. The impact factors were not computed from the wind field perturbations, but rather by integration of equations (6.96). The integrations were carried out on a $1° \times 1°$ grid in the Southern Hemisphere (figure 6.27) using a standard

WAM cycle 4 model with 15° angular resolution and 10% frequency resolution on a logarithmic frequency scale.

The directional wave spectrum shows the usual symmetrical wind-sea distribution about the wind direction. The fetch was large (1500 km), so that the spectrum is close to fully developed. The impact factors $W^u(\mathbf{k})$ and $W^v(\mathbf{k})$ indicate how the spectrum would be modified according to the Green's function method by unit wind field perturbations in the x (orthogonal or eastward) and y (parallel or northward) directions, respectively.

The impact function W^u is antisymmetric about the local wind direction, with a positive lobe to the east and a negative lobe to the west. This corresponds to a rotation of the spectrum towards the direction of the wind field perturbation. The factor peaks at frequencies slightly below the spectral peak frequency, indicating that these spectral components are most affected by a change in the wind field. In addition, the function falls off rapidly with increasing frequency, in accordance with the concept that the information on direct wind input to these components is rapidly lost through nonlinear interactions.

The W^v function represents the response of the spectrum to a perturbation in the magnitude of the wind velocity. As expected, the function is symmetric about the local wind direction and everywhere positive. The function also peaks at frequencies slightly less than the spectral peak, indicating that the spectrum grows and migrates to lower frequencies if the wind is increased. As before, the impact function falls off rapidly with increasing frequency, the effect of wind-input modifications being again rapidly lost through the redistribution of energy by the nonlinear transfer.

Figure 6.26 shows the corresponding changes in the wave spectrum computed with the WAM model for the prescribed 1 m/s wind velocity perturbations in the x (eastward) and y (northward) directions. The qualitative agreement between the spectral modifications computed by the Green's function method (figure 6.25) and using the full model (figure 6.26) indicates that the calibration of the impact functions was reasonably successful, and that the Green's function method should work at least in this simple case. Similarly good agreement was also found for growing wind seas at smaller fetches. However, the results of the Green's function method and the exact model integration clearly do not agree uniformly well in all spectral bands. This suggests that some form of additional spectral smoothing may be needed for a robust application of the technique.

The calibration exercise used to optimize the impact factors assumed spatial homogeneity for both the background and the perturbation wind fields. It thus provided no test of the locality concept or the wave age computations of the Green's function technique. A second experiment was accordingly devised to test these features explicitly.

The same uniform southerly 10 m/s wind field was again adopted as

a)

b)

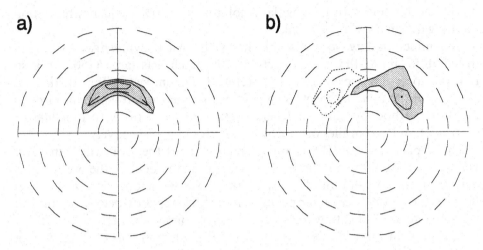

Fig. 6.26. Changes in the wave spectrum computed with the WAM model for unit perturbations in the wind vector in the y (panel a) and x directions (panel b) for the southerly wind field of figure 6.25.

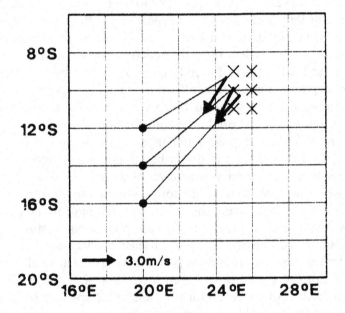

Fig. 6.27. Mean computed wind correction vectors for the idealized swell propagation case obtained from the assimilation of three different observed spectra along 20° E at 16° S, 14° S and 12° S, respectively. The wind corrections show both the expected increase of the wind of about 3 m/s in the direction of 225° and the expected location at which the swell was perturbed. Wind correction vectors and associated observation positions are indicated by connecting great circle paths.

background. The wind velocities at six grid points, shown in figure 6.27, were then tripled in magnitude to 30 m/s and the wind directions changed to 225°. The geometry of the wind field modification was chosen to yield a local wave generation region which would produce swell propagating towards the observation point in the centre of the grid. As the generation region approximates a spatial δ-function, it provides a good test of the ability of the wave age determination scheme to identify this region as the source region of the swell. (The fact that the generating region is not particularly realistic need not concern us for this test of the basic method.)

To test the accuracy of the integral impact factors for this rather extreme configuration, errors were introduced into the wind field by increasing the magnitude of the wind in the generating area by 10% to 33 m/s. The resulting errors in the wave spectra at the observation point were computed with the WAM model, and the Green's function technique was then applied to recover the wind field errors from the differences between the spectra for the original and modified WAM model runs.

Wind correction vectors were computed for all spectral components exhibiting significant spectral energy deviations and were then averaged over the spectrum. Figure 6.27 shows the mean wind corrections obtained from three different observation positions. The Green's function technique correctly indicates that no modification of the wind field is required within the wind sea region, as no errors were found for the high frequency wind sea components whose points of influence lie in this region. For the low frequency spectral bins which originate in the swell source region, however, increases in the wind speed of the correct order $u^{\parallel} \approx 3$ m/s are found. Reliable wind speed and direction retrievals were obtained not only from the most southern observation point which lies directly downwind in the swell propagation direction, but also from observation points off the dominant propagation direction.

VI.10.5 Conclusions

The Green's function assimilation scheme represents an attempt to modify optimally both the wind and wave fields in a nonlocal scheme at acceptable computational cost. In achieving this economy a number of simplifying assumptions have been necessary. The initial test results, however, indicate that little physical information appears to have been lost in these simplifications and that the scheme has promise.

The technique as presented here represents only the first step towards an operational assimilation scheme. Once wind field corrections have been generated by this technique, the next task is to assimilate these corrections into a wind field model and thus to generate an updated wind field. This corrected wind field can then be used to rerun the wave model, thereby generating a corrected wave field in a dynamically consistent manner from the corrected wind field.

It is still open whether the Green's function technique can provide a realistic alternative to the more rigorous but computationally demanding adjoint model techniques described in previous sections of this chapter. To achieve computationally feasible assimilation algorithms, the Green's function method introduces approximations initially at the analytical level, while the adjoint method is forced to make approximations subsequently at the implementation level. Further exploratory studies are needed to pursue the promising aspects of both approaches and develop a feasible operational wave data assimilation scheme.

Ultimately, this will then need to be combined with an atmospheric data assimilation method in a comprehensive data assimilation system for a coupled atmospheric circulation–global wave model. Only when this goal has been achieved will it be possible to realize the full potential of the modern global satellite wind and wave data which motivated most of the recent work on wave data assimilation, thereby opening a new era in ocean wave research and forecasting.

Chapter VII

Summary and outlook

Summary and outlook

The WAM group has realized its objectives: a third generation wave model has been developed; it runs in global and regional modes; it extended our understanding of the underlying physics; and data assimilation schemes have been developed and tested. This book is testimony of what has been achieved. However, careful reading will also reveal open ends, which range from small inconsistencies to major open issues. In this chapter we give an outlook on expected developments.

Compared to the so-called second generation models considerable progress has been made regarding the formulation of the evolution equation for the wave spectrum. The present WAM model is based on an explicit formulation of the physics of generation of waves by wind, nonlinear wave–wave interactions and dissipation due to whitecapping and bottom processes, rather than on the approach of *ad hoc* modelling which was commonplace with second generation models. The latter approach was shown to be inadequate under extreme circumstances such as hurricanes (SWAMP, 1985), while the WAM model gives for rapidly varying wind fields very satisfactory results (see chapter IV). Nevertheless, under 'normal' circumstances both approaches give similar results for the wave height. The reason for this is that although second generation models have inadequate physics they have been tuned to a considerable extent. Thus, the benefits of a third generation model are mainly related to a better representation of the spectrum itself and to a more explicit formulation of the underlying physics of wave evolution.

Despite the progress, we still are not able to make wave predictions that always fall within the error bands of the observations. One may wonder whether it will be possible further to ameliorate modelling of the sea state by introducing 'better' physics, better numerics or higher resolution. In view of the progress that has been made by going from second to third generation models, one should not be too optimistic about the effect of further refinements, possibly with one exception, namely the further improvement of the quality of the driving winds including the effect of gusts. With this in mind we would like to discuss now the three possible sources of error: inadequate input winds, inadequate wave model physics and inadequate numerics and resolution.

The winds

Most studies use winds obtained from the large numerical models that are used for weather prediction. These models assimilate observations. Occasionally a subjective analysis of the observations is still made, but even then the first-guess model fields are an important tool. In general, one may expect better winds from more observations, better assimilation techniques and better (atmospheric) models.

Specific ocean satellites have considerably enlarged the number of observations. It will be a challenge to the atmospheric modelling community to make full use of these extra observations and it is important to realize that wave models can be of help here. If this is successful, one may expect that the present experimental satellites will be replaced by operational ones. At the same time everything possible should be done to continue conventional observations from ships and buoys and – if possible – to increase their quality and reliability.

Better weather (and wind) prediction is not only a matter of improving the atmospheric model, because the initialization is also essential. Therefore, data-assimilation techniques should be improved. There is hope here because the new generation of computers may allow the running of four-dimensional data-assimilation schemes.

A widely accepted strategy for improving atmospheric models is to improve the resolution, its numerics and its physical parametrizations. Although all parametrizations (clouds, for example) are relevant, wave modelling has a particular interest in the parametrization of the turbulent atmospheric boundary layer. In the past, curious inconsistencies have occurred. For instance, the friction that the atmospheric model used to compute the surface winds might be different from the drag computed from these winds, which in turn might be inconsistent with the momentum flux into the waves. This is clearly unacceptable if the differences become too large. In such cases it is essential to introduce a two-way coupling between the atmosphere and the waves. This should take into account also the effects of density stratification caused by air/sea temperature difference. It appears that these stability effects also determine the level of gustiness of the winds, which has an effect on the wave growth. The prediction of gustiness – not done at present – would be highly welcome.

Model physics

The *wind input* term of cycle 4 of the WAM model (§§ II.2 and III.3) is based on the quasi-linear theory, which extends Miles' description of shear flow instability. It is in fair agreement with observations both in the laboratory and in the field, although there is considerable scatter in these observations.

It should be realized that the quasi-linear theory is a semi-analytic approximation to the problem of turbulent air-flow over a given wave profile. The full problem of turbulent flow in the coupled air/sea system has not nearly been solved. Only in such a model could one expect to describe realistically such phenomena as air flow separation and the shear in the top layer of the ocean. It is important to compare the present theories in detail with measurements in the boundary layer over growing waves, to see how accurate they are. At the same time one should try to extend the theory.

A few items deserve special attention. What is the correct scaling velocity: is it u_*, U_{10} or something else? How do stability and density stratification affect wave growth? How should gustiness be described and parametrized? How should the roughness of the very short waves be treated? What is the best turbulence closure in an oscillating boundary layer? What happens in the case of adverse wind? What is the effect of swell on wave growth?

The *wave–wave interaction* is described in the discrete interaction approximation (§ III.3). Wave growth comes out well, but it cannot be denied that the approximation gives transfer rates that differ from the exact ones. It would be useful therefore to search for other economic approximations to the Boltzmann integral.

The section on *deep water dissipation*, § II.4, shows that much work remains to be done. The WAM model has a wave dissipation source term which is quasi-linear in the spectrum, i.e. linear but with proportionality constants depending on integral spectral properties. Such a source term can be justified under quite general conditions. However, the challenge remains to work out the statistics and hydrodynamics of different whitecapping dissipation theories and to find experimental ways of distinguishing between them. In the end it should be possible to determine the constants from first principles. The same applies *mutatis mutandis* to *dissipation at the bottom*.

These 'microscale' approaches to the determination of the source functions are traditionally complemented by comparing model predictions with observations (§§ II.8 – II.11). The usual approach was to compare model behaviour in idealized fetch-limited growth with observed growth curves. It appears that this approach should be abandoned, because we have come to realize that the idealized fetch-limited conditions never occur in the field. Therefore, we should replace the traditional approach by an inverse modelling approach. The results described in §§ VI.6 – VI.7 are a promising first step in this direction but still suffer from the fact that the model is normally fitted to a fit to the data rather than to the data themselves. If inverse modelling techniques are developed sufficiently one could simply use a number of storms to optimize the source terms. For these applications it will be essential to have high quality input wind fields. When shallow water aspects are significant the measurement of the correct bottom parameters (see § II.5) will also be essential.

Numerics and resolution

For propagation several higher order schemes have been considered, but they never seemed to lead to better predictions. One of the reasons is perhaps the fact that source regions of waves (the storms) normally extend over many grid points so that wave modelling is somewhat remote from the usual numerical tests in which one follows the evolution of an initially localized signal. Another reason is perhaps the approximate agreement between the numerical dispersion and the physical dispersion, required by the finite frequency and directional resolution of the spectrum. Yet cases are known in which the propagation accuracy seems to be a limitation, so that one might want to consider higher order or Lagrangian schemes.

An unsolved problem is the parametrization of sub-resolution scale variations of bathymetry for refraction calculations. Further study is also required concerning the interaction of waves and currents.

In chapter IV abundant evidence has been presented of how the resolution limits the quality of the predictions. This explains the natural tendency to strive for higher resolution. The resolution that can be obtained is usually limited for economic reasons by the computer capacity. Grid nesting may sometimes help, but here also more research is needed. The ability to make high resolution models operational will depend to a large extent on the availability and architecture of large computers. It seems likely now that it will be necessary to make a new version of the WAM model (cycle 5) in order to achieve efficient multitasking on the CRAY YMP. Massive parallel processing may make further demands.

Two-way coupling and the use of satellites

The abundance of global surface wind and ocean-wave data which have now become available through ocean satellites such as ERS-1 should have a major impact on ocean-wave and marine forecasting, once appropriate techniques for assimilating these data have been fully developed. Figure 6.1 outlines a system in which a coupled atmosphere–ocean-wave model assimilates all available atmospheric and wave observations in real time. Such a system is not yet complete, although significant progress has been made and parts have already been realized.

The two-way coupling with the atmospheric model needs to be tested further, as should the possible coupling with ocean circulation models.

The presently operational, simple wave data-assimilation schemes should be extended to include more wave parameters. In particular, the use of directional information, such as that obtained with the SAR should be incorporated. In addition the four-dimensional methods should be further developed. As long as a full two-way coupling between atmosphere and waves is not achieved, one may expect the greatest impact of wave data-assimilation to occur for swell. Once the coupling has been realized, one may also expect a beneficial impact on wind sea forecasting. In the coupled system outlined in figure 6.1 waves play their appropriate geophysical role in air/sea interaction. The system will produce atmospheric and wave predictions of high quality. At the same time it will make efficient use of global (satellite) observations for climate research.

Members of the WAM group

G. Barzel
E. Bauer
R. C. Beal
L. Bertotti
C. Brüning
G. J. H. Burgers
C. J. Calkoen
V. Cardone*
J. C. Carretero
L. Cavaleri*
S. Christopoulos
J. J. Conde
J. E. DeLuis
M. Donelan*
D. G. Duffy
J. A. Ewing
R. Flather
S. J. Foreman
P. Francis*
Q. Gao
T. Gerling
H. C. Graber*
A. Greenwood
H. Günther
A. Guillaume*
J. Guddal
B. Hansen
K. Hasselmann
S. Hasselmann
M. de las Heras
M. W. Holt
L. Holthuijsen*
K. P. Hubbert
P. A. E. M. Janssen*
R. E. Jenkins
R. E. Jensen

K. Kahma
G. J. Komen*
M. Khandekar
N. P. Kurian*
A. Laing
J-M. Lefèvre
P. Lionello
R. B. Long
A. K. Magnusson
V. K. Makin
D. Masson
C. Mastenbroek
J. Monbaliu
H. Oliveira-Pires
W. Perrie
M. Reistad
K. Richter
W. Rosenthal*
C. Ross
T. Schilperoort
V. I. Shrira
R. Stratton
J. P. Thomas
H. L. Tolman
B. Toulany
C. F. de Valk
G. van Vledder
W. J. P. de Voogt*
S. L. Weber
T. Wolf
X. M. Wu
I. R. Young
Y. Yuan*
V. E. Zakharov*
L. Zambresky

*Also member of SCOR working group 83

487

Affiliations and addresses of contributors

G. Barzel
Max-Planck-Institut für Meteorologie, Bundesstrasse 55, 20146 Hamburg, Germany

E. Bauer
Institut für Meereskunde, Troplowitzstr. 7, 22529 Hamburg, Germany

L. Bertotti
Istituto per lo Studio della Dinamica delle Grandi Masse (ISDGM), San Polo 1364, 30125 Venice, Italy

C. Brüning
Institut für Meereskunde, Troplowitzstr. 7, 22529 Hamburg, Germany

G. J. H. Burgers
Koninklijk Nederlands Meteorologisch Instituut (KNMI), PO Box 201, 3730 AE De Bilt, The Netherlands

C. J. Calkoen
DELFT HYDRAULICS, PO Box 152, 8300 AD Emmeloord, The Netherlands

V. Cardone
Oceanweather Inc., 5 River Road, Cos Cob CT 06807, USA

L. Cavaleri
Istituto per lo Studio della Dinamica delle Grandi Masse (ISDGM), San Polo 1364, 30125 Venice, Italy

M. Donelan
Canada Centre for Inland Waters (CCIW), PO Box 5050, Burlington, Ontario L7R 4A6, Canada

R. Flather
Proudman Oceanographic Laboratory (POL), Bidston Observatory, Birkenhead, Merseyside L43 7RA, UK

H. C. Graber
Rosenstiel School of Marine and Atmospheric Science, University of Miami, 4600 Rickenbacker Causeway, Miami, Florida, 33149-1098, USA

H. Günther
GKSS Forschungszentrum Geesthacht, Max Planck Strasse, 21502 Geesthacht, Germany

B. Hansen
European Centre for Medium-Range Weather Forecasts (ECMWF), Shinfield Road, Reading, Berkshire 9AX RG2, UK

K. Hasselmann
Max-Planck-Institut für Meteorologie, Bundesstrasse 55, 20146 Hamburg, Germany

S. Hasselmann
Max-Planck-Institut für Meteorologie, Bundesstrasse 55, 20146 Hamburg, Germany

M. de las Heras
Ente Publico Puertos del Estado, Clima Maritimo, Avenida del Partenon 10, 28042 Madrid, Spain

M. W. Holt
Meteorological Office, Bracknell, Berkshire RG12 2SY, UK

L. Holthuijsen
Delft University of Technology, Stevinweg 1, 2628 CN Delft, The Netherlands

P. A. E. M. Janssen
Koninklijk Nederlands Meteorologisch Instituut (KNMI), PO Box 201, 3730 AE De Bilt, The Netherlands and European Centre for Medium-Range Weather Forecasts (ECMWF), Shinfield Park, Reading, Berkshire 9AX RG2, UK

R. E. Jensen
Waterways Experiment Station, Corps of Engineers, 3909 Halls Ferry Road, Vicksburg, MS 39180-6199, USA

K. Kahma
Finnish Institute of Marine Research, PO Box 33, SF-00931 Helsinki, Finland

G. J. Komen
Koninklijk Nederlands Meteorologisch Instituut (KNMI), PO Box 201, 3730 AE De Bilt, The Netherlands

P. Lionello
University of Padua, Dept. of Physics, Via Marzolo 8, 35135 Padova, Italy

R. B. Long
601 NE 2nd Place, Dania, FL 33004, USA

V. K. Makin
Koninklijk Nederlands Meteorologisch Instituut (KNMI), PO Box 201, 3730 AE De Bilt, The Netherlands

D. Masson
Institute of Ocean Sciences, PO Box 6000, Sidney B.C., Canada V8L 4BZ

C. Mastenbroek
Koninklijk Nederlands Meteorologisch Instituut (KNMI), PO Box 201, 3730 AE De Bilt, The Netherlands

J. Monbaliu
Laboratory for Hydraulics, de Croylaan 2, 3001 Heverlee, Belgium

H. L. Tolman
Marine Prediction Branch, Development Division, NOAA/NMC, 5200 Auth Road, Room 206, Camp Springs, MD 20746, USA

C. F. de Valk
DELFT HYDRAULICS, PO Box 152, 8300 AD Emmeloord, The Netherlands

S. L. Weber
Koninklijk Nederlands Meteorologisch Instituut (KNMI), PO Box 201, 3730 AE De Bilt, The Netherlands

X. M. Wu
Proudman Oceanographic Laboratory (POL), Bidston Observatory, Birkenhead, Merseyside L43 7RA, UK

I. R. Young
Australian Defence Force Academy, Northcott Drive, Canberra, ACT 2600, Australia

Y. Yuan
First Institute of Oceanography of the State Oceanographic Administration, Qingdao, PO Box 98, China

L. Zambresky
Fleet Numerical Oceanography Center (FNOC) - 40 Department, Monterey, California 93943-5005, USA

Notation and abbreviations

Frequently used symbols

Roman symbols

a	wave amplitude; (as index) air
$a(\mathbf{x}, t)$	amplitude of WKB train
$a_n, a_{\mathbf{k}}$	Fourier amplitude with rapid oscillation taken out
$a_n(\mathbf{x}, t)$	slowly varying Fourier amplitude
$\tilde{a}$	amplitude which oscillates rapidly
ϕ	small amount of action
A	amplitude of Klein-Gordon solution; constant
A_{tot}	total wave action
$A_{tot,n}$	total wave action in nth mode, packet
$\mathscr{A}$	action per square meter (WKB)
$\mathscr{A}(\mathbf{x}, t)$	mean action per square meter
B	degree of saturation of the spectrum; constant
c	phase velocity
c_p	phase velocity at the peak of the spectrum
c_s	speed of sound
c_g	group velocity
C	proportionality constant
C_{ds}	dissipation constant
C_D	drag coefficient
C_{10}	drag coefficient in the atm boundary layer wrt U_{10}
C_{bot}, C_f, C_p	coefficient in bottom dissipation source term (general, friction, percolation)
$\mathbf{d}^o, (d^o)_i$	set of observations
$\mathbf{d}$	model counterparts of observations
$\mathscr{D}_i$	operator projecting from Ψ on d_i
D_m	sediment grain diameter
D_w	effective diffusion coefficient in the quasi-linear theory of wave generation
E_{tot}	total wave energy
$E_{tot,n}$	total wave energy in mode or wave packet number n
$E(\Psi, \Phi) = 0$	basic model equation in discrete matrix representation
$\mathscr{E}$	wave energy per square metre $= \rho g m_0$
$\bar{\mathscr{E}}$	mean wave energy per square metre
$\bar{\mathscr{E}}(\mathbf{x}, t)$	mean wave energy per square metre (in a WKB approximation)
f	frequency, Coriolis parameter

f_i	fraction of sea covered with ice
$f_A(\mathbf{k}, \mathbf{x}, t)$	action density distribution among wave packets
$f_E(\mathbf{k}, \mathbf{x}, t)$	energy density distribution among wave packets
f_p	peak frequency
$\bar{f}, <f>$	mean frequency
$\bar{f}_*$	dimensionless mean frequency
F_{tot}	total wave variance
$F(\omega), F(f)$	(one-dimensional) (angular) frequency (variance) spectrum
$F_\mathbf{k}, F(\mathbf{k})$	(two-dimensional) wavenumber (variance) spectrum
$\hat{F}_\mathbf{k}, \hat{F}(\mathbf{k})$	(two-dimensional) frozen image spectrum
$F(f, \theta)$	two-dimensional frequency directional (variance) spectrum
$F(\mathbf{x}, t)$	variance per unit area in a WKB train
$F_n(\mathbf{x}, t)$	variance per unit area in the nth mode, packet
$F(\mathbf{k}, \mathbf{x}, t)$	slowly varying variance spectrum
$\mathscr{F}$	covariance function
g	gravitational acceleration
G	Green's function
$h, h(\mathbf{x})$	depth
H	Hamiltonian; Heaviside function
$\mathscr{H}$	Hamiltonian density
$H(\mathbf{x}, t)$	mean sea level
$H_S = H_{m_o}$	significant wave height
i	discrete dummy index
j	discrete dummy index
J	cost function
$\mathbf{k}$	wavenumber vector
$\mathbf{k}(\mathbf{x}, t)$	wavenumber vector (WKB)
$\mathbf{k}_n(\mathbf{x}, t)$	wavenumber vector of nth mode
$\bar{k}$	mean wavenumber
k_B	Bragg wavenumber
k_p	wavenumber at the peak of the spectrum
k_N	bottom roughness
$\mathbf{K}$	dominant wavenumber of wave packet
$\mathbf{K}_n$	dominant wavenumber of nth wave packet
ℓ	mixing length
L	Obukhov length
L	Lagrangian
$\mathscr{L}$	Lagrange density
m	moisture
m_0	zeroth moment of variance spectrum
m_n	nth moment of variance spectrum
M	covariance matrix

$\mathcal{M}$	wave momentum per square metre
$\bar{\mathcal{M}}$	mean wave momentum per square metre
$\mathcal{N}(\mathbf{x}, \mathbf{k}, t)$	density of wave packets in phase space
$\mathcal{N}_{tot}$	total number of wave packets in the ocean
$N(\mathbf{x}, t)$	action density $/(\rho g)$
$N_n(\mathbf{x}, t)$	action density $/(\rho g)$ in nth mode
$N(\mathbf{k})$	wavenumber action spectrum $/(\rho g)$
$N(\mathbf{k}, \mathbf{x}, t)$	slowly varying action spectrum $/(\rho g)$
N_{tot}	$A_{tot}/(\rho g)$
o	(superscript): observed
p	pressure
p_c, p_w	steady and oscillatory part of pressure in the bottom boundary layer
$p(\ ), P(\ )$	probability function
$P(\mathbf{k})$	wavenumber momentum spectrum, SAR spectrum
PM	(subscript) Pierson Moskowitz
p	(Hamiltonian) momentum variable
q	(Hamiltonian) coordinate variable
q	humidity
$\mathbf{r}_a, \mathbf{r}_s$	position vectors
R	gas constant
Ri	Richardson number
$s(\mathbf{x}, t)$	eikonal
$s_n(\mathbf{x}, t)$	eikonal of nth mode
S	generic symbol for source term (in both action balance equation and energy balance equation)
S_{in}	wind input source term
S_{nl}	nonlinear source term
S_{ds}	dissipation (whitecapping) source term
S_{bot}, S_f, S_p	bottom (friction, percolation) source term
S_{ice}	source term describing the interaction with ice floes
SI	scatter index ($\sigma/$ mean observed value)
t	time
T	wave period, duration, temperature, time scale, modulation transfer function, kinetic energy
$\tilde{T}$	dimensionless duration (Tg/U_{10})
$T^\star$	dimensionless duration ($Tg/u_\star$)
T_v	virtual temperature
$\mathbf{u}$	velocity vector, orbital motion (u_1, u_2) or $(u_1, u_2, u_3) = (u_1, u_2, w)$
$u_\star$	friction velocity in the atmospheric boundary layer
$\mathbf{u}_c$	horizontal current in the bottom boundary layer

$\mathbf{u}_w$	horizontal wave-induced velocity in the bottom boundary layer
U	potential energy
$\mathbf{U}(z)$	wind vector, current
$\mathbf{U}_0(z)$	equilibrium wind profile in Miles' theory
$\mathbf{U}_{10}$	wind vector at 10 m
U_c	$\mathbf{U}_{10}$ projected on the mean propagation direction of the waves at the peak of the spectrum
U_λ	wind speed at $z = \lambda$
$\mathbf{U}_c^b$	horizontal current at the top of the bottom boundary layer
$\mathbf{U}_w^b$	horizontal wave-induced velocity at the top of the bottom boundary layer
$(U_w^b)_{rms}$	rms value of the wave velocity at the top of the bottom boundary layer
$\mathbf{v}_D$	Doppler shifted group velocity
w	z-component of velocity vector
w_i	weight function for ith observation
W	$U_0 - c$; also: total action
W^u, W^v	impact factors
$\mathscr{W}$	wronskian
$\mathbf{x}$	(x_1, x_2)
X	fetch
$X_\star$	dimensionless fetch ($X/(gu_\star^2)$)
$\tilde{X}$	dimensionless fetch ($X/(gU_{10}^2)$)
$\mathbf{X}$	position of centre of wave packet
$\mathbf{X}_n$	centre of nth wave packet
y	$y = z - z_c$ (Miles' theory)
z	height (sea surface: $z = 0$), vertical coordinate
z_0	roughness length
z_c	critical height
z_{obs}	height of observation
z_t	roughness length for temperature profile
z_q	roughness length for humidity profile

Greek symbols

α	dimensionless roughness, Charnock parameter
α_p	Phillips' constant
α_{ch}	Charnock's constant
$\hat{\alpha}$	integral steepness parameter, proportionality constant
β	dimensionless growth rate (γ/ω)
β_{ds}	dimensionless dissipation rate
β_M	Miles' parameter
$\beta^{\parallel}, \beta^{\perp}$	differential impact factors

γ	spectral parameter (γ^Γ is the peak enhancement factor)
γ_a	amplitude growth rate (ϵc_1)
γ	relative energy growth rate S_{in}/F or S_{in}/N
Γ_{bot}	$-S_{bot}/F$
Γ_{ds}	$-S_{ds}/F$
δ	Plant's constant; thickness of the (wave) bottom boundary layer
δ_{ij}	Kronecker delta
$\delta()$	Dirac's delta-function: $\int_{-\infty}^{+\infty} dx' f(x')\delta(x-x') = f(x)$, for every function f
Δ	Laplacian, response function
ϵ	expansion parameter
$\tilde{\epsilon}$	dimensionless wave (variance) energy per square metre
ϵ_*	dimensionless wave (variance) energy per square metre
$\varepsilon(x), \varepsilon(k)$	zero except in bin centred on argument, where it equals one
η	elevation of water surface $\eta(\mathbf{x}, \mathbf{t})$
$\hat{\eta}^\pm(\mathbf{k})$	continuous $\pm$ Fourier components of η
$\hat{\eta}(\mathbf{k})$	$\hat{\eta}^+(\mathbf{k})$
$\hat{\eta}(\mathbf{k}, t)$	generalization of $\hat{\eta}(\mathbf{k})$
$\eta_{\mathbf{k}}$	discrete contribution to η from kth component
$\tilde{\eta}$	discrete Fourier coefficient oscillating rapidly
θ	angle, wave direction, incidence angle
θ_p	peak angle
θ_0	mean wave direction
θ_w	wind direction
κ	von Kármán constant
λ	wavelength, latitude, Lagrange multiplier
$\mu_{\mathbf{k}}$	phase angle of kth Fourier mode
v	kinematic viscosity
v_a	molecular viscosity in air
v_e	eddy viscosity
v_{ec}	eddy viscosity in the current bottom boundary layer
v_{ew}	eddy viscosity in the wave bottom boundary layer
$v10$	dimensionless peak frequency
$v_\star$	dimensionless peak frequency
ξ	relative coordinate
π	$3.14 \cdots$
ρ	density, amplitude
ρ_a	density of air
ρ_s	density of sediment
ρ_w	density of water
ρ_e	number of ice flows per unit area

$\sigma, \sigma(\mathbf{k}), \sigma(\mathbf{k}, h)$	intrinsic angular frequency
$\sigma, \sigma_a, \sigma_b$	JONSWAP spectral width parameters
σ	standard deviation
σ_0	radar cross-section
σ_B	cross-section in the Boltzmann integral
Σ	summation
τ, τ_o	time scale
τ_f	frequency-dependent time scale
τ_{ij}	stress tensor
$\tau, \tau(z)$	stress vector (function of height), wind stress, bottom stress
$\tau^b, \tau^b_c, \tau^b_w$	stress in the bottom boundary layer (total, averaged over wave phases and wave supported)
$\tilde{\tau}$	dimensionless time scale; dimensionless total wind stress
ϕ	velocity potential, longitude, solution of Klein-Gordon equation, angle of incidence
ϕ_i, ϕ_s	incident potential, disturbance to the presence of ice floes
Φ	control variables
ψ	$\phi(x, y, z = \eta)$, canonical conjugate of η
$\psi, \psi_m, \psi_H, \psi_E$	profile functions
ψ	(quasi-linear theory) displacement of the mean streamline
ψ^f	first-guess approximation to the state vector
ψ^b	best-guess approximation to the state vector
Ψ	state vector; $(\hat{\eta}, \hat{\mathbf{u}})$
χ	normalized wave induced vertical velocity, $w/w(0)$
ω	angular frequency ($= 2\pi f$)
$\omega(\mathbf{x}, t)$	angular frequency in single WKB train
$\omega_n(\mathbf{x}, t)$	angular frequency in nth WKB train
$\omega_\pm$	$\pm\sigma$
ω_p	angular frequency at the peak
$\bar{\omega}$	mean angular frequency
$\Omega(\mathbf{k}, \mathbf{U})$	$\mathbf{k} \cdot \mathbf{U} + \sigma(\mathbf{k}, h)$, true frequency including Doppler shift
$\Omega(\mathbf{K}, \mathbf{X})$	$\mathbf{K} \cdot \mathbf{U}(\mathbf{X}) + \sigma[\mathbf{K}, h(\mathbf{X})]$, dominant frequency of wave packet
$\Omega(\mathbf{K}_n, \mathbf{X}_n)$	$\mathbf{K}_n \cdot \mathbf{U}(\mathbf{X}_n) + \sigma[\mathbf{K}_n, h(\mathbf{X}_n)]$, dominant frequency of nth wave packet

Other symbols

$\nabla = \nabla_x$	$(\partial_{x_1}, \partial_{x_2}) = (\partial_1, \partial_2) = (\partial/\partial x_1, \partial/\partial x_2)$
∇_k	$(\partial_{k_1}, \partial_{k_2}) = (\partial/\partial k_1, \partial/\partial k_2)$
η_t	$\partial\eta/\partial t$
ϕ_{x_1}	$\partial\phi/\partial x_1$

Acronyms

BMO	British Meteorological Office model
ECMWF	European Centre for Medium-Range Weather Forecasts
ERS-1	European Remote Sensing
ESA	European Space Agency
FNOC	Fleet Numerical Oceanography Center
GONO	Golven Noordzee model
HEXOS	humidity exchange over sea
HYPA	hybrid parametric wave model
JONSWAP	Joint North Sea Wave Project
KNMI	Koninklijk Nederlands Meteorologisch Instituut
LAM	limited area model
LEWEX	Labrador extreme wave height experiment
MIZ	marginal ice zone
MPI(M)	Max-Planck-Institut für Meteorologie
MTF	modulation transfer function
NASA	National Aeronautic and Space Administration
NMC	National Meteorological Center
NOAA	National Oceanographic and Atmospheric Administration
OW/AES	Ocean Weather/Atmospheric Environment Service
SAR	synthetic aperture radar
SCOR	Scientific Committee on Oceanic Research
SWADE	surface wave dynamics experiment
SWAMP	sea wave modelling project
SWIM	shallow water intercomparison of models
TMA	Texel-Marsen-Arsloe
UKMO	UK Meteorological Office
WAM	wave modelling group
WAMDI	wave model development and implementation group
WKB	Wentzel, Kramers, Brillouin

References

Aberson, S.D. and M. De Maria, 1991. A nested barotropic hurricane track forecast model (VICBAR), p81-86 in: Proc. 19th conference on hurricane and tropical meteorology; Amer. Meteor. Soc., Boston.

Allender, J.H., J. Albrecht and G. Hamilton, 1983. Observations of directional relaxation of wind sea spectra. *J. Phys. Oceanogr.* **13**, 1519-1525.

Alpers, W. and K. Hasselmann, 1978. The two-frequency microwave technique for measuring ocean-wave spectra from airplane or satellite. *Boundary Layer Meteorol.* **13**, 215-230.

Alpers, W. and K. Hasselmann, 1982. Spectral signal-to-clutter and thermal noise properties of ocean wave imaging synthetic aperture radars. *Int. J. Remote Sensing* **3**, 423-446.

Alpers, W., D.B. Ross and C.L. Rufenach, 1981. On the detectability of ocean surface waves by real and synthetic aperture radar. *J. Geophys. Res.* **C86**, 6481-6498.

Al-Zanaidi, M.A. and W.H. Hui, 1984. Turbulent air flow water waves. *J. Fluid Mech.* **148**, 225-246.

Amos, C.L., A.J. Bowen, D.A. Huntley and C.F.M. Lewis, 1988. Ripple generation under the combined influence of waves and currents on the Canadian continental shelf. *Contin. Shelf Res.* **8**, 1129-1153.

Anderson, D., A. Hollingsworth, S. Uppala and P. Woiceshyn, 1987. A study on the feasibility of using sea and wind information from the ERS-1 satellite. ESA report part 1, ECMWF, Reading, 121p.

Anthes, R.A., 1982. Tropical cyclones: their evolution, structure and effects. Amer. Meteor. Soc., Boston, 208p.

Atakturk, S.S. and K.B. Katsaros, 1987. Intrinsic frequency spectra of short gravity-capillary waves obtained from temporal measurement of wave height on a lake. *J. Geophys. Res.* **C92**, 5131-5141.

Atlas, R., A.J. Busalachi, M. Ghil, S. Bloom and E. Kalnay, 1987. Global surface wind and flux fields from model assimilation of SEASAT data. *J. Geophys. Res.* **C92**, 6477-6487.

Attema, E., 1992. The ERS-1 geophysical validation program for wind and wave data products, p1-3 in: Proc. of the workshop on ERS-1 Geophysical Validation, Penhors, France, 27-30 April 1992; ESA WPP-36, European Space Agency, Paris.

Banner, M.L., 1990a. The influence of wave breaking on the surface pressure distribution in wind wave interactions. *J. Fluid Mech.* **211**, 463-495.

Banner, M.L., 1990b. Equilibrium spectra of wind waves. *J. Phys. Oceanogr.* **20**, 966-984.

Banner, M.L., I.S.F. Jones and J.C. Trinder, 1989. Wavenumber spectra of short gravity waves. *J. Fluid Mech.* **198**, 321-344.

Barnett, T.P., 1968. On the generation dissipation and prediction of ocean wind waves. *J. Geophys. Res.* **73**, 513-530.

Barrick, D.E., 1968. A review of scattering from surfaces with different roughness scales. *Radio Science* **3**, 865-868.

Barrick, D.E. and J.B. Snider, 1977. The statistics of hf sea-echo Doppler spectra.

IEEE Transactions on Antennas and Propagation **AP-25**, 19-27.

Bascom, W., 1980. Waves and beaches, Anchor Press/Double Day, Garden City, New York, 366p.

Batchelor, G.K., 1967. An introduction to fluid dynamics. Cambridge University Press, Cambridge, 615p.

Bauer, E., S. Hasselmann, K. Hasselmann and H.C. Graber, 1992. Validation and assimilation of Seasat altimeter wave heights using the WAM wave model. *J. Geophys. Res.* **C97**, 12671-12682.

Bauer, E., K. Hasselmann, I.R. Young and S. Hasselman, 1995. Assimilation of wave data into the wave model WAM using an impulse response function method. To appear in *J. Geophys. Res.*

Beal, R.C., 1991. Directional ocean wave spectra. The Johns Hopkins University Press, Baltimore, 218p.

Belcher, S.E. and J.C.R. Hunt, 1993. Turbulent shear flow over slowly moving waves. *J. Fluid Mech.* **251**, 109-148.

Belcher, S.E., T.M.J. Newley and J.C.R. Hunt, 1993. The drag on an undulating surface due to the flow of a turbulent layer. *J. Fluid Mech.* **249**, 557-596.

Bengtsson, L., 1991. Advances and prospects in numerical weather prediction. *Q. J. Royal Meteorol. Soc.* **117**, 855-902.

Bengtsson, L., M. Ghil and E. Kallen, 1981. Dynamic meteorology. Data assimilation methods. Springer, New York, 330p.

Benney, D.J. and P.G. Saffman, 1966. Nonlinear interactions of random waves in a dispersive medium. *Proc. Roy. Soc. London* **A289**, 301-320.

Bergamasco, A., L. Cavaleri and P. Sguazzero, 1986. Physical and mathematical analysis of a wind model. *Il Nuovo Cimento* **9C**, No. 1, 1-16.

Bertotti, L. and L. Cavaleri, 1993. Sensitivity of wave model results to directional resolution, ISDGM, TR 17/93, 18p.

Birch, K.G. and J.A. Ewing, 1986. Observations of wind waves on a reservoir, IOS-rep. No. 234, Wormley, 37p.

Blanc, T.V., 1985. Variation of bulk-derived surface flux, stability and roughness results due to the use of different transfer coefficient schemes. *J. Phys. Oceanogr.* **15**, 650-669.

Booij, N. and L.H. Holthuijsen, 1987. Propagation of ocean waves in discrete spectral wave models. *J. Comp. Phys.* **68**, 307-326.

Borgmann, L.E., 1972. Confidence intervals for ocean wave spectra, p237-250 in: Proc. of the 13th international conference on coastal engineering, Vancouver BC; American Society of Civil Engineers, New York.

Bouws, E., 1986. Provisional results of a wind wave experiment in a shallow lake (Lake Marken, The Netherlands). KNMI Afdeling Oceanografisch Onderzoek memo, OO-86-21, De Bilt, 15p.

Bouws, E. and G.J. Komen, 1983. On the balance between growth and dissipation in an extreme depth-limited wind-sea in the southern North Sea. *J. Phys. Oceanogr.* **13**, 1653-1658.

Bouws, E., H. Günther, W. Rosenthal and C.L. Vincent, 1985. Similarity of the wind wave spectrum in finite depth water, part 1: Spectral form. *J. Geophys. Res.* **C90**, 975-986.

Bretherton, F.P. and C.J.R. Garrett, 1968. Wave trains in inhomogeneous moving media. *Proc. Roy. Soc. London* **A302**, 529-554.

Bretschneider, C.L., 1954. Generation of wind waves over a shallow bottom. Techn. Mem 151, Beach Erosion Board, Corps of Engineers, 24p.

Bretschneider, C.L., 1973. in: Shore protection manual. US Army Coastal
    Engineering Research Center. Corps of Engineers.
Broer, L.J.F., 1974. On the Hamiltonian theory of surface waves. *Appl. Sci. Res.* **30**,
    430-446.
Brooke Benjamin, T. and J.E. Feir, 1967. The disintegration of wave trains on deep
    water. *J. Fluid Mech.* **27**, 417-430.
Brooke Benjamin, T. and P.J. Olver, 1982. Hamiltonian structure, symmetries and
    conservation laws for water waves. *J. Fluid Mech.* **125**, 137-185.
Brüning, C., W. Alpers, L.F. Zambresky and D.G. Tilley, 1988. Validation of a SAR
    ocean wave imaging theory by the shuttle imaging radar-B experiment over
    the North Sea. *J. Geophys. Res.* **93**, 15403-15425.
Brüning, C., W. Alpers and K. Hasselmann, 1990. Monte Carlo simulation studies
    of the nonlinear imaging of a two-dimensional surface wave field by a
    synthetic radar. *Int. J. Remote Sensing* **11**, 1695-1727.
Brüning, C., S. Hasselmann, K. Hasselmann, S. Lehner and T. Gerling, 1993. On
    the extraction of ocean wave height spectra from ERS-1 SAR wave model
    image spectra, in: Proc. of ERS-1 workshop, Cannes, Nov. 1992; ESA SP-359,
    European Space Agency, Paris.
Brüning, C., S. Hasselmann, K. Hasselmann, S. Lehner, T. Gerling, 1994. First
    evaluation of ERS-1 synthetic aperture radar wave mode data. To be
    published in *The Atmosphere Ocean System*.
Burgers, G., 1990. A guide to the Nedwam wave model. KNMI scientific report,
    WR-90-04, De Bilt, 81p.
Burgers, G. and V.K. Makin, 1993. Boundary layer model results for wind-sea
    growth. *J. Phys. Oceanogr.* **23**, 372-385.
Burgers, G., Q.D. Gao and M. de las Heras, 1990. Wave data assimilation in the
    operational North Sea wave model Nedwam, p623-625 in: Proc. of the
    international symposium on assimilation of observations in meteorology and
    oceanography. Clermont-Ferrand, France; World Meteorological
    Organisation, Geneva.
Businger, J.A., J.C. Wyngaard, I. Izumi and E.F. Bradley, 1971. Flux-profile
    relationships in the atmospheric surface layer. *J. Atmospheric Sci.* **28**, 181-189.
Camuffo, D., 1984. Analysis of the series of precipitation at Padova, Italy. *Climatic
    Change* **6**, 57-77.
Cardone, V.J., 1969. Specification of the wind distribution in the marine boundary
    layer for wave forecasting. New York University Geophysical Sciences
    Laboratory, report TR-69-1.
Cardone, V.J., 1992. On the structure of the marine surface wind field in
    extratropical storms, in: Proc. of the third international workshop on wave
    hindcasting and forecasting, Montreal, Quebec, 19-22 May; Environment
    Canada, Ontario.
Cardone, V.J., A.J. Broccoli, C.V. Greenwood and J.A. Greenwood, 1980. Error
    characteristics of extratropical storm wind fields specified from historical data.
    J. Petrol. Techn. **32**, 873-880.
Cardone, V.J., J.G. Greenwood and M.A. Cane, 1990. On trends in historical
    marine wind data. *J. of Climate*, **3**, 113-127.
Cardone, V., H.C. Graber, R. Jensen, S. Hasselmann and M. Caruso, 1994. In
    search of the true surface wind field in SWADE IOP-1: Ocean wave modelling
    perspective. To be submitted to *The Atmosphere Ocean System*.
Carretero Albiach, J.C. and H. Günther, 1992. Wave forecast performed with the

WAM model at ECMWF - Statistical analysis of a one month period (November 1988) PCM Madrid, Publ. no. **49**, 103p.

Cartwright, D.E. and M.S. Longuet-Higgins, 1956. The statistical distribution of the maxima of a random function. *Proc. Roy. Soc. London* **A237**, 212-232.

Cauchy, A.L., 1827. Mémoire sur la théorie des ondes. *Mém. Acad. R. Sci.*, Paris, 1ière série I, 38.

Cavaleri, L., 1979. An instrumental system for detailed wind wave study. *Il Nuovo Cimento* Serie 1, **2C**, 288-304.

Cavaleri, L. and G.J.H. Burgers, 1992. Wind gustiness and wave growth. KNMI Afdeling Oceanografisch Onderzoek memo, OO-92-18, De Bilt, 38p.

Cavaleri, L. and P. Lionello, 1990. Linear and nonlinear approaches to bottom friction in wave motion: a critical intercomparison, *Estuarine, Coastal and Shelf Science* **30**, 355-367.

Cavaleri, L., S. Curiotto, G. Dallaporta and A. Mazzoldi, 1981. Directional wave recording in the Adriatic Sea, *Il Nuovo Cimento* **4C(5)**, 519-534.

Cavaleri, L., L. Bertotti and P. Lionello, 1988. Evaluation of the impact of the availability of wind fields in the Mediterranean Sea - final report, *ESA-ESRIN* study contract 7458/88 HGE-I, European Space Agency, 66p.

Cavaleri, L., L. Bertotti and P. Lionello, 1989. Shallow water application of the third generation WAM wave model. *J. Geophys. Res.* **C94**, 8111-8124.

Cavaleri, L., L. Bertotti and P. Lionello, 1991. Wind wave-cast in the Mediterranean Sea. *J. Geophy. Res.* **C96**, 10739-10764.

Chalikov, D.V., 1976. A mathematical model of wind-induced waves. *Dokl. Akad. Nauk SSR* **229**, 1083-1086.

Chalikov, D.V. and V.K. Makin, 1991. Models of the wave boundary layer. *Boundary Layer Meteorol.* **56**, 83-99.

Charnock, H., 1955. Wind stress on a water surface. *Q. J. Royal Meteorol. Soc.* **81**, 639-640.

Chavent, G., 1991. On the theory and practice of non-linear least squares. *Adv. Water Resources* **14**, 55-63.

Chi Wai Li, 1992. A split operator scheme for ocean wave simulation. *Int. J. for Numerical Methods in Fluids* **15**, 579-593.

Christoffersen, J.B. and I.G. Jonsson, 1985. Bed friction in a combined current and wave motion. *Ocean Engineering* **12**, 387-423.

Collins, J.I., 1972. Prediction of shallow water spectra. *J. Geophys. Res.* **77**, 2693-2707.

Conte, S.D. and J.W. Miles, 1959. On the integration of the Orr-Sommerfeld equation. *J. Soc. Indust. Appl. Math.* **7**, 361-369.

Cook, I, 1974. Advanced nonlinear theory of plasmas in Plasma Physics, p225-242 in: Plasma physics, B.E. Keen (ed); The Institute of Physics, London, 339p.

Courant, R. and D. Hilbert, 1953. Methods of mathematical physics. Interscience, New York.

Courant, R. and D. Hilbert, 1962. Methods of mathematical physics. Vol. II, Interscience, New York - London.

Courtier, P. and O. Talagrand, 1987. Variational assimilation of meteorological observations with the adjoint vorticity equation, part 2: Numerical results. *Q. J. Royal Meteorol. Soc.* **113**, 1329-1347.

Courtier, P., C. Freydier, J.F. Geleyn, F. Rabier and M. Rochas, 1991. The ARPEGE project at Meteo France, p193-231 in: Proc. of ECMWF seminar on numerical methods in atmospheric models, part 2, Reading; ECMWF,

Reading.

Crawford, D.R., B.M. Lake, P.G. Saffman and H.C. Yuen, 1981. Stability of weakly nonlinear deep-water waves in two and three dimensions. *J. Fluid Mech.* **105**, 177-191.

Crombie, D.D., 1955. Doppler spectrum of sea echo at 13.56 Mc/s. *Nature* **175**, 681-682.

Davidan, I.N., 1990. Zakonomernosti phormirovanija spektralnoi struktury vetrovogo volnenija, p48-67 in: Trudy vsesojusnogo nauchno-technicheskoyo obschestva imeni akademika A.N. Krylova; Soudostroenie, Leningrad, 174p.

Davidson, K.L. and A.K. Frank, 1973. Wave-related fluctuations in the airflow above natural waves. *J. Phys. Oceanogr.* **3**, 102-119.

Davidson, R.C., 1972. Methods in nonlinear plasma theory. Academic Press, New York and London, 356p.

Davies, A.G. and A.D. Heathershaw, 1983. Surface wave propagation over sinusoidally varying topography: Theory and observation, *I.O.S. report* No. **159**, part 1, Wormley, 88p.

Davies, A.M. (ed), 1990. Modeling marine systems. Vol. **II**, CRC Press, Boston, 68p.

Dee, D., S.E. Cohn, A. Dalcher and M. Ghil, 1985. An efficient algorithm for estimating noise covariances in distributed system. *IEEE Transactions Autom. Control* **AC-30**, 1057-1085.

Dell'Osso, L., 1990. Some results from the ECMWF spectral limited area model. LAM Newsletter, Deutsche Wetterdienst, Offenbach, 129-143.

Dell'Osso, L., L. Bertotti and L. Cavaleri, 1992. The Gorbush storm in the Mediterranean Sea: atmospheric and wave simulation. *Monthly Weather Review* **120**, 77-90.

Dingler, J.R. and D.L. Inman, 1976. Wave-formed ripples in nearshore sands, p2109-2126 in: Proc. 15th conference coastal engineering, Honolulu; American Society of Civil Engineers, New York.

Dobson, E., F. Monaldo, J. Goldhirsh and J. Wilkerson, 1987. Validation of GEOSAT altimeter-derived wind speeds and significant wave heights using buoy data. *JHU/APL Techn. Dig.* **8**, 222-233.

Dobson, F.W., 1971. Measurements of atmospheric pressure on wind-generated sea waves. *J. Fluid Mech.* **48**, 91.

Donelan, M.A., 1979. On the fraction of wind momentum retained by waves, p141-149 in: Marine forecasting: predictability and modelling in ocean hydrodynamics; Elsevier, Amsterdam.

Donelan, M.A., 1982. The dependence of the aerodynamic drag coefficient on wave parameters, p381-387 in: Proc. of the first international conference on meteorological and air/sea interaction of the coastal zone; Amer. Meteor. Soc., Boston, Mass.

Donelan, M.A., 1983. Attenuation of laboratory swell in an adverse wind. National Water Research Institute, Canada Centre for Inland Waters, 11p.

Donelan, M.A., 1990. Air-sea interaction, p239-292 in: The sea, **9**, Ocean engineering science; Wiley, New York.

Donelan, M.A., 1992. The mechanical coupling between air and sea. An evolution of ideas and observations, p77-94 in: Strategies for future climate research, M. Latif (ed); MPI, Hamburg, 387p.

Donelan, M.A. and W.J. Pierson, 1983. The sampling variability of estimates of spectra of wind generated gravity waves. *J. Geophys. Res.* **C88**, 4381-4392.

Donelan, M.A. and W.J. Pierson, 1987. Radar scattering and equilibrium ranges in wind-generated waves with application to scatterometry. *J. Geophys. Res.* **C92**, 4971-5029.

Donelan, M.A., M.S. Longuet-Higgins and J.S. Turner, 1972. Whitecaps. *Nature* **239**, 449-451.

Donelan, M.A., J. Hamilton and W.H. Hui, 1985. Directional spectra of wind generated waves. *Phil. Trans. Roy. Soc. London* **A315**, 509-562.

Donelan, M.A., M. Skafel, H. Graber, P. Liu, D. Schwab and S. Venkatesh, 1992. On the growth rate of wind-generated waves. *Atmosphere-Ocean* **30**, 457-478.

Dorrestein, R., 1960. Simplified method of determining refraction coefficients for sea waves. *J. Geophys. Res.* **65**, 637-642.

Drake, D.E. and D.A. Cacchione, 1992. Shear stress and bed roughness estimates for combined wave and current flows over a rippled bed. *J. Geophys. Res.* **97**, C2, 2319-2326.

Drazin, P.G. and W.H. Reid, 1981. Hydrodynamic stability. Cambridge University Press, Cambridge, 525p.

Driest, E.R. van, 1951. Turbulent boundary layer in compressible fluids. *J. Aeronautical Science* **18**, 145-160.

Drummond, W.E. and D. Pines, 1962. Nonlinear stability of plasma oscillations. *Nucl. Fusion Suppl.* **3**, 1049-1052.

Duin, C.A. van and P.A.E.M. Janssen, 1992. An analytic model of the generation of surface gravity waves by turbulent air flow. *J. Fluid Mech.* **236**, 197-215.

Duncan, J.H., 1981. An experimental investigation of breaking waves produced by a towed hydrofoil. *Proc. Roy. Soc. London* **337**, 331-348.

Dyer, A.J. and B.B. Hicks, 1970. Flux-gradient relationships in the constant flux layer. *Q. J. Royal Meteorol. Soc.* **96**, 715-721.

Eliasen, E., B. Machenhauer and E. Rasmussen, 1970. On a numerical method for integration of the hydrodynamical equation with a spectral representation of the horizontal fields, report No. 2. Institut for Teoretisk Meteorologi, University of Copenhagen.

Elmore, W.C. and M.A. Heald, 1969. Physics of waves. McGraw-Hill Kogakusha, 477p.

Engen, G., S.F. Barstow, H. Johnson and H.E. Krogstad, 1994. Directional wave spectra by inversion of ERS-1 SAR Imagery. Submitted to *IEEE Transactions on Geoscience and Remote Sensing*.

Essen, H.H., K.W. Gurgel, F. Schirmer and T. Schlick, 1989a. Surface currents in the Norwegian Channel measured by radar in March 1985. *Tellus* **41A(2)**, 162-174.

Essen, H.H., K.W. Gurgel, F. Schirmer and T. Schlick, 1989b. Surface currents during NORCSEX 1988, as measured by land- and a ship-based HF-radar, IGARSS 1989 Proceedings **2**, 730-733.

Esteva, D.C., 1988. Evaluation of preliminary experiments assimilating Seasat significant wave heights into a spectral wave model. *J. Geophys. Res.* **C93**, 14099-14106.

Ewing, J.A., 1971. A numerical wave prediction method for the North Atlantic Ocean. *Dtsch. Hydrogr. Z.* **24**, 241-261.

Ewing, J.A. and E.G. Pitt, 1982. Measurements of the directional wave spectrum off South Uist. Paris, Sept-Oct 1981, in Proc. of wave and wind directionality, applications to the design of structures; Editions Technip, Paris, 573p.

Fabrikant, A.L., 1976. Quasilinear theory of wind-wave generation. *Izv. Atmos. Ocean. Phys.* **12**, 524-526.

Fedor, L.A. and G.S. Brown, 1982. Wave height and wind speed measurements from SEASAT radar altimeter. *J. Geophys. Res.* **C87**, 3254-3260.

Feindt, F., J. Schroeter and W. Alpers, 1986. Measurement of the ocean wave-radar modulation transfer function at 35 GHz from a sea based platform in the North Sea. *J. Geophys. Res.* **91**, 9701-9708.

Fletcher, R., 1987. Practical methods of optimization. Wiley and Sons, Chichester.

Forristall, G.Z., 1981. Measurements of saturated range in ocean wave spectra. *J. Geophys. Res.* **86**, 8075-8089.

Forristall, G.Z. and A.M. Reece, 1985. Measurements of wave attenuation due to a soft bottom: the SWAMP experiment. *J. Geophys. Res.* **C90**, 3376-3380.

Forristall, G.Z., E.H. Doyle, W. Silva and M. Yoshi, 1990. Verification of a soil wave interaction model (SWIM), p41-68 in: Modeling Marine Systems, Vol. **II**, A.M. Davies (ed); CRC Press, Boca Raton, Florida.

Francis, P.E. and R.A. Stratton, 1990. Some experiments to investigate the assimilation of SEASAT altimeter wave height data into a global wave model. *Q. J. Royal Meteorol. Soc.* **116**, 1225-1251.

Garratt, J.R., 1977. Review of drag coefficients over oceans and continents. *Monthly Weather Review* **105**, 915-929.

Gao Q. and G.J. Komen, 1993. The directional response of ocean waves to changing wind direction. *J. Phys. Oceanogr.* **23**, 1561-1566.

Geernaert, G.L. and W.L. Plant, 1990. Surface waves and fluxes, 2 vols.. Kluwer, Dordrecht.

Geernaert, G.L., K.B. Katsaros and K. Richter, 1986. Variation of the drag coefficient and its dependence on sea state. *J. Geophys. Res.* **C91**, 7667-7679.

Gelci, R. and H. Cazalé, 1962. Une équation synthétique de l'évolution de l'état de la mer. *J. Mechan. Phys. Atmosphère* Série 2, **4**, 15-41.

Gelci, R., H. Cazalé and J. Vassal, 1956. *Bull. Inform. Comité Central Océanogr. Etude Côtes* **8**, 170-187.

Gelci, R., H. Cazalé and J. Vassal, 1957. Prévision de la houle. La méthode des densités spectroangulaires. *Bull. Inform. Comité Central Océanogr. Etude Côtes* **9**, 416-435.

Gent, P.R. and P.A. Taylor, 1976. A numerical model of the air flow above water waves. *J. Fluid Mech.* **77**, 105-128.

Gerling, T.W., 1991. A comparative anatomy of the LEWEX wave system, p182-192 in: Directional ocean wave spectra, R.C. Beal (ed); The Johns Hopkins University Press, Baltimore, 218p.

Gerling, T.W., 1992. Partitioning sequences and arrays of directional wave spectra into component wave systems. *J. Atmos. Ocean. Techn.* **9**, 444-458.

Gill, A.E., 1982. Atmosphere-ocean dynamics. Academic Press, New York, 662p.

Gill, P.E., W. Murray and M.H. Wright, 1981. Practical optimization. Academic Press, New York.

Golding, B.W., 1978. A depth-dependent wave model for operational forecasting, p593-606 in: Turbulent fluxes through the sea surface, wave dynamics and prediction, A. Favre and K. Hasselmann (eds); Plenum Press, New York, 677p.

Goodberlet, M.A., C.T. Swift, J.C. Wilkerson, 1992. Validation of the ocean surface wind fields and wave height measurements derived from data of the ERS-1 scatterometer and radar altimeter (early results), p61-64 in: Proc. of the workshop on ERS-1 Geophysical Validation, Penhors, France, 27-30 April 1992; ESA WPP-36, European Space Agency, Paris.

Graber, H.C. and O.S. Madsen, 1988. A finite depth wind wave model, part 1: Model description. *J. Phys. Oceanogr.* **18**, 1465-1483.

Graber, H.C., M. Caruso and R. Jensen, 1991. Surface wave simulation during the October storm in SWADE, p159-164 in: Proc. of MTS 1991 conference, New Orleans, LA; Marine Technology Society, Washington, DC.

Graber, H.C., S. Hasselmann, K. Hasselmann and E. Bauer, 1994. Validation of global wave hindcast with SEASAT altimeter and wave buoys data. Submitted to *J. Geophys. Res.*

Grant, W.D. and O.S. Madsen, 1979. Combined wave and current interaction with a rough bottom. *J. Geophys. Res.* **84**, 1797-1808.

Groen, P. and R. Dorrestein, 1976. Zeegolven. Staatsdrukkerij, 's-Gravenhage. 124p. 3e herz. druk.

Groves, G.W. and J. Melcer, 1961. On the propagation of ocean waves on a sphere. *Geof. Int.* **8**, 77-93.

Guillaume, A., 1990. Statistical tests for the comparison of surface gravity wave spectra with application to model validation. *J. Atm. Ocean. Techn.* **7**, 552-567.

Günther, H., 1981. A parametric surface wave model and the statistics of the prediction parameters. PhD thesis, Univ. Hamburg, 90p.

Günther, H. and M. Holt, 1992. WAM/UKMO wind wave model intercomparison - Summary report. UK Meteorological Office report, Bracknell, 10p.

Günther, H. and W. Rosenthal, 1989. Interpretation of wave model validation statistic, p203-212 in: Proc. of second international workshop on wave hindcasting and forecasting, Vancouver, BC; Environment Canada, Atmospheric Environment Service, Ontario.

Günther, H., W. Rosenthal and M. Dunckel, 1981. The response of surface gravity waves to changing wind direction. *J. Phys. Oceanogr.* **11**, 718-728.

Günther, H., S. Hasselmann and P.A.E.M. Janssen, 1991. Wamodel cycle 4. DKRZ report no. 4, Hamburg.

Günther, H., P. Lionello, P.A.E.M. Janssen, L. Bertotti, C. Brüning, J.C. Carretero, L. Cavaleri, A. Guillaume, B. Hanssen, S. Hasselmann, K. Hasselmann, M. de las Heras, A. Hollingsworth, M. Holt, J.M. Lefevre and R. Portz, 1992. Implementation of a third generation ocean wave model at the European Centre for Medium Range Weather Forecasts, Final report for EC Contract SC1-0013-C(GDF), ECMWF, Reading.

Günther, H., P. Lionello and B. Hanssen, 1993. The impact of the ERS-1 altimeter on the wave analysis and forecast. Report No. GKSS 93/E/44, GKSS Forschungszentrum Geesthacht, Geesthacht, Germany, 56p.

Günther, H., P. Lionello, B. Hanssen, 1994. A global wave forecast experiment assimilating the ERS-1 altimeter data. Submitted to *J. Geophys. Res.*.

Haltiner, G.J. and R.T. Williams, 1980. Numerical prediction and dynamic meteorology. Wiley, New York, 477p.

Harding, J. and A.A. Binding, 1978. Wind fields during gales in the North Sea and the gale of 3 January 1976. *Meteor. Mag.* **107**, 164-181.

Harris, D.L., 1966. The wave-driven wind. *J. Atmos. Sci.* **23**, 688-693.

Hasselmann, D.E. and J. Bösenberg, 1991. Field measurements of wave-induced pressure over wind sea and swell. *J. Fluid Mech.* **230**, 391-428.

Hasselmann, D.E., M. Dunckel and J.A. Ewing, 1980. Directional wave spectra observed during JONSWAP 1973. *J. Phys. Oceanogr.* **10**, 1264-1280.

Hasselmann, D.E., J. Bösenberg, M. Dunckel, K. Richter, M. Grünewald and H. Carlson, 1986. Measurements of wave-induced pressure over surface gravity waves, p353-370 in: Wave dynamics and radio probing of the ocean surface, O.M. Phillips and K. Hasselmann (eds); Plenum, New York, 677p.

Hasselmann, K., 1960. Grundgleichungen der Seegangsvoraussage. *Schiffstechnik* **1**, 191-195.

Hasselmann, K., 1962. On the non-linear energy transfer in a gravity-wave spectrum, part 1: general theory. *J. Fluid Mech.* **12**, 481.

Hasselmann, K., 1966. Feynman diagrams and interaction rules of wave-wave scattering processes. *Rev. Geophys. Space Phys.* **4**, 1-32.

Hasselmann, K., 1967. Nonlinear interactions treated by the methods of theoretical physics (with application to the generation of waves by wind). *Proc. Roy. Soc. London* **A299**, 77-100.

Hasselmann, K., 1968. Weak-interaction theory of ocean waves. *Basic Developments in Fluid Dynamics* **2**, 117 - 182.

Hasselmann, K., 1971a. Determination of ocean wave spectra from Doppler radar return from the sea surface. *Nature* **229**, 16-17.

Hasselmann, K., 1971b. On the mass and momentum transfer between short gravity waves and larger-scale motions. *J. Fluid Mech.* **50**, 189-205.

Hasselmann, K., 1974. On the spectral dissipation of ocean waves due to whitecapping. *Boundary Layer Meteorol.* **6**, 107-127.

Hasselmann, K. 1985. Assimilation of microwave data in atmospheric and wave models and the use of satellite data in climate models, p47-52 in: The use of satellite data in climate models; ESA SP-244, European Space Agency, Paris, 191p.

Hasselmann, K. and J.I. Collins, 1968. Spectral dissipation of finite-depth gravity waves due to turbulent bottom friction. *J. Mar. Res.* **26**, 1-12.

Hasselmann, K. and S. Hasselmann, 1991. On the nonlinear mapping of an ocean wave spectrum into a SAR image spectrum and its inversion, *J. Geophys. Res.* **C96**, 10713-10729.

Hasselmann, K. and O.H. Shemdin, 1982. Remote sensing experiment Marsen. *Int. J. Remote Sensing* **3**, 139-361.

Hasselmann, K., T.P. Barnett, E. Bouws, H. Carlson, D.E. Cartwright, K. Enke, J.A. Ewing, H. Gienapp, D.E. Hasselmann, P. Kruseman, A. Meerburg, P. Müller, D.J. Olbers, K. Richter, W. Sell and H. Walden, 1973. Measurements of wind-wave growth and swell decay during the Joint North Sea Wave Project (JONSWAP), *Dtsch. Hydrogr. Z. Suppl. A* **8(12)**, 95p.

Hasselmann, K., D.B. Ross, P. Müller and W. Sell, 1976. A parametric wave prediction model. *J. Phys. Oceanogr.* **6**, 200-228.

Hasselmann, K., R.K. Raney, W.J. Plant, R.A. Shuchman, D.R. Lyzenga, C.L. Rufenach and M.J. Tucker, 1985. Theory of synthetic aperture radar ocean imaging. A MARSEN view. *J. Geophys. Res.* **C90**, 4659-4686.

Hasselmann, K., S. Hasselmann, E. Bauer, C. Brüning, S. Lehner, H. Graber and P. Lionello, 1988. Development of a satellite SAR image spectra and altimeter wave height data assimilation system for ERS-1. ESA report, Max-Planck-Institute für Meteorologie, Nr. 19, Hamburg, 155p.

Hasselmann, K. S. Hasselmann and K. Bartel, 1990. Use of a wave model as a validation tool for ERS-1 AMI wave products and as an input for the ERS-1 wind retrieval algorithms, report 55, Max-Planck-Institut für Meteorologie, Hamburg, 97p.

Hasselmann, K., S. Hasselmann, C. Brüning and A. Speidel, 1991. Interpretation and application of SAR wave image spectra in wave models, p117-124 in: Directional ocean wave spectra, R.C. Beal (ed); The Johns Hopkins University Press, Baltimore, 218p.

Hasselmann, S. and K. Hasselmann, 1981. A symmetrical method of computing the non-linear transfer in a gravity-wave spectrum. *Hamb. Geophys. Einzelschr. Serie A.* **52**, 138p.

Hasselmann, S. and K. Hasselmann, 1985a. The wave model EXACT-NL. In: Ocean wave modeling, The SWAMP group; Plenum, New York.

Hasselmann, S. and K. Hasselmann, 1985b. Computations and parameterizations of the nonlinear energy transfer in a gravity-wave spectrum, part 1: A new method for efficient computations of the exact nonlinear transfer integral. *J. Phys. Oceangr.* **15**, 1369-1377.

Hasselmann, S., K. Hasselmann, J.H. Allender and T.P. Barnett, 1985. Computations and parameterizations of the nonlinear energy transfer in a gravity wave spectrum, part 2: Parameterizations of the nonlinear energy transfer for application in wave models. *J. Phys. Oceanogr.* **15**, 1378-1391.

Heras, M.M. de las, 1990. WAM hindcast of long period swell. KNMI Afdeling Oceanografisch Onderzoek memo, OO-90-09, De Bilt, 10p.

Heras, M.M. de las and P.A.E.M. Janssen, 1992. Data assimilation with a nonlinear, coupled wind wave model. *J. Geophys. Res.* **C97**, 20261-20270.

Heras, M.M. de las, G.J.H. Burgers and P.A.E.M. Janssen, 1994. Variational data assimilation in a third generation wave model. To appear in *J. Atmos. Ocean. Techn.*.

Herterich, K. and K. Hasselmann, 1980. A similarity relation for the nonlinear energy transfer in a finite depth gravity-wave spectrum. *J. Fluid Mech.* **97**, 215-224.

Holland, J.Z., 1981. Atmospheric boundary layer, in: IFYGL - The international field year for the Great Lakes. E.J. Aubert and T.L. Richards (eds); Pub. NOAA, Ann. Arbor, MI, 410p.

Hollingsworth, A., 1986. Objective analysis for numerical weather prediction, p11-59 in: Short and medium range numerical weather prediction. T. Matsuno (ed); Special volume of the Journal of the Meteorological Society of Japan.

Holt, M.W., 1991. Trials of increased resolution in space and direction for the wave model. UK Meteorological Office, Techn. note no. 58, Bracknell, 20p.

Holthuijsen, L.H., 1980. Methoden voor golfvoorspelling. Technische Adviescommissie voor de Waterkeringen.

Holthuijsen, L.H. and T.H.C. Herbers, 1986. Statistics of breaking waves observed as whitecaps in the open sea. *J. Phys. Oceanogr.* **16**, 290-297.

Holthuijsen, L.H. and H.L. Tolman, 1991. Effects of the Gulf Stream on ocean waves. *J. Geophys. Res.* **C96**, 12755-12771.

Holthuijsen, L.H., A.J. Kuik and E. Mosselman, 1987. The response of wave directions to changing wind directions. *J. Phys. Oceanogr.* **17**, 845-853.

Holton, J.E., 1992. An introduction to dynamic meteorology. 3rd ed. Academic Press, New York, 511p.

Holtslag, A.A.M. and Chin-Hoh Moeng, 1991. Eddy diffusivity and countergradient transport in the convective atmospheric boundary layer. *J. Atmos. Sci.* **48**, 1690-1698.

Hsiao, S.V. and O.H. Shemdin, 1978. Non linear and linear bottom interaction effects in shallow water, p347-372 in: Turbulent fluxes through the sea surface, wave dynamics and prediction, A. Favre and K. Hasselmann (eds); Plenum Press, New York, 677p.

Hsiao, S.V. and O.H. Shemdin, 1980. Interaction of ocean waves with a soft bottom. *J. Phys. Oceanogr.* **10**, 605-610.

Huang, N.E., S.R. Long and L.F. Bliven, 1981. On the importance of significant slope in empirical wind-wave studies. *J. Phys. Oceanogr.* **11**, 569-573.

Huang, N.E., L.F. Bliven, S.R. Long and P.S. DeLeonibus, 1986. A study of the relationship among wind speed, sea state, and the drag coefficient for a developing wave field. *J. Geophys. Res.* **C91**, 7733-7742.

Hubbert, K.P. and J. Wolf, 1991. Numerical investigation of depth and current refraction of waves. *J. Geophys. Res.* **C96**, C2, 2737-2748.

Huntley, D.A. and D.G. Hazen, 1988. Seabed stresses in combined wave and steady flow conditions on the Nova Scotia continental shelf: field measurements and predictions. *J. Phys. Oceanogr.* **19**, 347-362.

Hwang, P.A. and O.H. Shemdin, 1988. The dependence of sea surface slope on atmospheric stability and swell conditions. *J. Geophys. Res.* **C93**, 13903-13912.

Intergovernmental Oceanographic Commission, 1986. Assimilation of satellite wind and wave data in numerical weather and wave prediction models. Report on a workshop at ECMWF, Reading. ECMWF, Reading.

Jackson, F.C., 1987. The radar ocean wave spectrometer, measuring ocean waves from space. *Johns Hopkins APL Techn. Dig.* **8(1)**, 116-127.

Jackson, F.C., W.T. Walton, B.A. Walter, D.E. Hines and C.Y. Peng, 1992. Sea surface mean squared slope from Ku-band radar backscatter data. *J. Geophys. Res.* **C97**, 11411-11427.

Jacobs, S.J. 1987. An asymptotic theory for the turbulent flow over a progressive water wave. *J. Fluid Mech.* **174**, 69-80.

Janssen, P.A.E.M., 1982. Quasilinear approximation for the spectrum of wind-generated water waves. *J. Fluid Mech.* **117**, 493-506.

Janssen, P.A.E.M., 1986. On the effect of gustiness on wave growth. KNMI Afdeling Oceanografisch Onderzoek memo, 00-86-18. De Bilt, 17p.

Janssen, P.A.E.M., 1989a. Nonlinear effects in water waves. Internal report, IC/89/66. ICTP Miramare-Trieste.

Janssen, P.A.E.M., 1989b. Wave-induced stress and the drag of air flow over sea waves. *J. Phys. Oceanogr.* **19**, 745-754.

Janssen, P.A.E.M., 1991a. Quasi-linear theory of wind wave generation applied to wave forecasting. *J. Phys. Oceanogr.* **21**, 1631-1642.

Janssen, P.A.E.M., 1991b. On nonlinear wave groups and consequences for spectral evolution, p46-52 in: Directional ocean wave spectra, R.C. Beal (ed); The Johns Hopkins University Press, Baltimore, 218p.

Janssen, P.A.E.M., 1992. Experimental evidence of the effect of surface waves on the airflow. *J. Phys. Oceanogr.* **22**, 1600-1604.

Janssen, P.A.E.M., 1994. Results with a coupled wind wave model. ECMWF Res. Dept. Techn. report 71. ECMWF, Reading, 58p.

Janssen, P.A.E.M. and G.J. Komen, 1985. Effect of the atmospheric stability on the growth of surface gravity waves. *Boundary Layer Meteorol.* **32**, 85-96.

Janssen, P.A.E.M. and P. Woiceshyn, 1992. Wave age and scatterometer wind retrieval algorithm, p141-145 in: Proc. of the workshop on ERS-1 Geophysical Validation, Penhors, France, 27-30 April 1992; ESA WPP-36, European Space Agency, Paris.

Janssen, P.A.E.M., G.J. Komen and W.J.P. de Voogt, 1984. An operational coupled hybrid wave prediction model. *J. Geophys. Res.* **C89**, 3635-3654.

Janssen, P.A.E.M., G.J. Komen and W.J.P. de Voogt, 1987a. Friction velocity scaling in wind-wave generation. *Boundary Layer Meteorol.* **38**, 29-35.

Janssen, P.A.E.M., P. Lionello, M. Reistad and A. Hollingsworth, 1987b. A study of the feasibility of using sea and wind information from the ERS-1 satellite, part 2: Use of scatterometer and altimeter data in wave modelling and assimilation. ECMWF report to ESA, Reading.

Janssen, P.A.E.M., P. Lionello, L. Zambresky, 1989a. On the interaction of wind and waves. *Philos. Trans. Roy. Soc. London*, **A329**, 289-301.

Janssen, P.A.E.M., P. Lionello, M. Reistad and A. Hollingsworth, 1989b. Hindcasts and data assimilation studies with the WAM model during the Seasat period. *J. Geophys. Res.* **C94**, 973-993.

Janssen, P.A.E.M., A.C.M. Beljaars, A. Simmons and P. Viterbo, 1992. On the determination of surface stresses in an atmospheric model. *Monthly Weather Review* **120**, 2977-2985.

Jeffreys, H., 1924. On the formation of waves by wind. *Proc. Roy. Soc.* **A107**, 189-206.

Jeffreys, H., 1925. On the formation of waves by wind. II. *Proc. Roy. Soc.* **A110**, 341-347.

Jenkins, A.D., 1992. A quasi-linear eddy-viscosity model for the flux of energy and momentum to waves, using conservation law equations in a curvilinear system. *J. Phys. Oceanogr.* **22**, 843-858.

Jensen, B.L., B.M. Sumer and J. Fredsoe, 1989. Turbulent oscillatory boundary layers at high Reynolds numbers. *J. Fluid Mech.* **206**, 265-207.

Jonsson, I.G., 1980. A New approach to oscillatory rough turbulent boundary layers. *Ocean Engineering* **7**, 109-152.

Jonsson, I.G., 1990. Wave-current interactions, p65-120 in: The Sea, **9**, Ocean engineering science; part A, B. Le Mehaute and D.M. Hanes (eds); Wiley, New York.

Kahma, K.K., 1981a. A study of the growth of the wave spectrum with fetch. *J. Phys. Oceanogr.* **11**, 1503-1515.

Kahma, K.K., 1981b. On the wind speed dependence of the saturation range of the wave spectrum, p61-67 in: X Geofysiikan paiavat Helsingissa, 23 - 24 April 1981, M. Lepparanta (ed); Geofysiikan Seura, Helsinki.

Kahma, K.K., 1986. On prediction of the fetch-limited wave spectrum in a steady wind. *Finn. Mar. Res.* **253**, 52-78.

Kahma, K.K. and C.J. Calkoen, 1992. Reconciling discrepancies in the observed growth of wind-generated waves. *J. Phys. Oceanogr.* **22**, 1389-1405.

Kajiura, K., 1968. A model of the bottom boundary layer in water waves. *Bull. Earthquake Res. Inst.* **46**, 75-123.

Kalman, R.E., 1960. A new approach to linear filtering and prediction problems. *Trans ASME, Ser. D. J. Basic Eng.* **82**, 35-45.

Kalman, R.E. and R.S. Bucy, 1961. New results in linear filtering and prediction theory. *Trans ASME, Ser. D. J. Basic Eng.* **83**, 95-108.

Karsten, B., 1987. Implementatie van eerste resultaten met het NEDWAM model, een derde generatie golfverwachtingsmodel voor de Noordzee. KNMI Techn. report, TR-102, De Bilt, 39p.

Katsaros, K.B., S.D. Smith and W.A. Oost, 1987. HEXOS - Humidity Exchange Over the Sea. A program for research on water-vapor and droplet fluxes from sea to air at moderate to high wind speeds. *Bull. Amer. Meteor. Soc.* **68**, 466-476.

Keller, W.C. and J.W. Wright, 1975. Microwave scattering and the straining of wind generated waves. *Radio Science* **10**, 139-147.

Kenney, J.E., E.A. Uliana and E.J. Walsh, 1979. The surface contour radar, a unique remote sensing instrument. *IEEE Transactions on Microwave Theory and Techniques*, **27**, 1980-1992.

Khandekar, M.L., 1989. Operational analysis and prediction of ocean wind waves. Springer, New York, 214p.

Kirwan, A.D., 1985. A review of mixture theory with applications in physical oceanography and meteorology. *J. Geophys. Res.* **C90**, 3265-3283

Kitaigorodskii, S.A., 1962. Application of the theory of similarity to the analysis of wind-generated water waves as a stochastic process, *Bull. Acad. Sci. USSR Geophys.* Ser. no. **1**, 73p.

Kitaigorodskii, S.A., 1970. The physics of air-sea interaction. Israel Program for Scientific Translations, Jerusalem.

Kitaigorodskii, S.A., 1983. On the theory of the equilibrium range in the spectrum of wind-generated gravity waves. *J. Phys. Oceanogr.* **13**, 816-827.

Kitaigorodskii, S.A., V.P. Krasitskii and M.M. Zaslavskii, 1975. On Phillips' theory of equilibrium range in the spectra of wind-generated gravity waves. *J. Phys. Oceanogr.* **5**, 410-417.

Komar, P.D., R.H. Neudeck and L.D. Kulm, 1972. Observations and significance of deep-water oscillatory ripple marks on the Oregon continental shelf, p601-619 in: Proc. of shelf sediment transport: process and pattern, D.G.P. Swift, D.D. Duana and O.H. Pilkey (eds); Bowden, Hutchinson and Ross, Pennsylvania.

Komen, G.J, 1985a. Introduction to wave models and assimilation of satellite data in wave models, p21-25 in: The use of satellite data in climate models; ESA SP-244, European Space Agency, Paris, 191p.

Komen, G.J., 1985b. Activities of the WAM (Wave Modelling) group, p121-127 in: Advances in underwater technology. Ocean science and offshore engineering, Vol **6**, Oceanology; Graham and Trotman.

Komen, G.J., 1986. Wave models, p26-28 in: Report on the workshop on assimilation of wind and wave data in numerical weather and wave prediction models; ICSU-WMO-IOC-SCOR publication, WMO/TD 148-WCP-122. World Meteorological Organisation, Geneva.

Komen, G.J., 1987a. Recent results with a third-generation ocean wave model. *Johns Hopkins APL Techn. Dig.* **8**, 37-41.

Komen, G.J., 1987b. Interactions of wind and waves. *Nature* **328**, 480.

Komen, G.J., 1990. Evolution of ideas and present situation in wave analysis and forecasting, Paris, France, p1-28. WMO/TD 350. World Meteorological Organisation, Geneva.

Komen, G.J., 1991a. Oceans and climate. Presented at the international workshop on Oceans, Climate, Man. Torino, April 1991.

Komen, G.J., 1991b. Modelling wind driven surface waves, p61-76 in: Strategies for future climate research. M. Latif (ed); Max-Planck-Institute für Meteorologie, Hamburg, 387p.

Komen, G.J., K. Hasselmann and S. Hasselmann, 1984. On the existence of a fully developed windsea spectrum. *J. Phys. Oceanogr.* **14**, 1271-1285.

Korevaar, C.G., 1990. North Sea climate based on observations from ships and lightvessels. Kluwer Academic Publishers, Dordrecht, 137p.

Krasitskii, V.P., 1990. Canonical transformation in a theory of weakly nonlinear waves with a nondecay dispersion law. *Sov. Phys. JETP* **71**, 921-927.

Krasitskii, V.P., 1992. Canonical transformation and reduced equations in the Hamiltonian theory of weakly nonlinear surface waves, p66-74 in: Proc. of the Nonlinear Water Waves Workshop, Bristol University, 22-25 October 1991;

University of Bristol, Bristol.

Krauss, W., 1973. Methods and results of theoretical oceanography, vol. I. Dynamics of the homogeneous and the quasihomogeneous ocean. Borntraeger, Berlin, 302p.

Krogstad, H.E., 1992. A simple derivation of Hasselmann's nonlinear ocean-sar transformation. *J. Geophys. Res.* **C97**, 2421-2425.

Krogstad, H.E., R.L. Gordon and M.C. Miller, 1988. High-resolution directional wave spectra from horizontally mounted acoustic Doppler current meters. *J. Atm. Ocean. Techn.* **5**, 340-352.

Krylov, Yu., S.S. Strekalov and V.F. Tsyplukhin, 1976. Vetrovye volny i ich vozdejstvie na sooruzenija. Hydrometeoizdat, Leningrad.

Kuik, A.J. and L.H. Holthuijsen, 1981. Buoy observation of directional wave parameters, p61-70 in: Proc. conference on directional wave spectra applications, University of California, Berkeley, R.L. Wiegel (ed); American Society of Civil Engineers, New York.

Kuik, A.J., G.Ph. van Vledder and L.H. Holthuijsen, 1988. A method for the routine analysis of pitch-and-roll buoy wave data. *J. Phys. Oceanogr.* **18**, 1020-1034.

Lamb, H., 1932. Hydrodynamics. 6th ed. Dover, New York, 738p.

Landau, L., 1946. On the vibrations of the electronic plasma. *J. Phys. (USSR)* **10**, 25.

Landau, L.D. and E.M. Lifshitz, 1960. Mechanics. Pergamon Press Addison-Wesley, Reading, MA.

Large, W.G. and S. Pond, 1982. Sensible and latent heat flux measurements over the ocean. *J. Phys. Oceanogr.* **12**, 464-482.

Larson, T.R., C.I. Moskowitz and J.W. Write, 1976. A note on SAR imagery of the ocean. *IEEE Transactions on Antennas and Propagation* **AP-24**, 393-394.

Lawson, L.M., 1992. Implementation of a parameter optimization procedure for the action balance equation. East Tennessee State University Department of Mathematics, Techn. report MS-92-01, 24p.

Lawson, L.M. and R.B. Long, 1983. Multimodal properties of the surface wave field observed with pitch-roll buoys during GATE. *J. Phys. Oceanogr.* **13**, 474-486.

LeBlond, P.H. and L.A. Mysak, 1978. Waves in the ocean. Elsevier oceanography series, Elsevier, Amsterdam, 602p.

Lewis, J.M. and J.C. Derber, 1985. The use of adjoint equations to solve a variational adjustment problem with advective constraints. *Tellus* **37A**, 309-322.

Liebelt, P.B., 1967. An introduction to optimal estimation. Addison-Wesley, Reading, MA.

Lighthill, M.J., 1962. Physical interpretation of the mathematical theory of wave generation by wind. *J. Fluid Mech.* **14**, 385-398.

Lionello, P. and P.A.E.M. Janssen, 1990. Assimilation of altimeter measurements to update swell spectra in wave models, p241-246 in: Proc. of the international symposium on assimilation of observations in meteorology and oceanography. Clermont-Ferrand, France; World Meteorological Organisation, Geneva.

Lionello, P., H. Günther and P.A.E.M. Janssen, 1990. Assimilation of altimeter wave data in a global ocean wave model. Presented at the Second Conference on Oceans from Space, Venice.

Lionello, P., H. Günther and P.A.E.M. Janssen, 1992. Assimilation of altimeter data in a global third generation wave model. *J. Geophys. Res.* **C97**, 14453-14474.

Liu, D.C. and J. Nocedal, 1988. On the limited memory BFGS method for large scale optimization. Techn. report NAM 03, Northwestern University, Evanston.

Liu, P.C., 1981. Normalized and equilibrium spectra of wind waves on Lake Michigan. *J. Phys. Oceanogr.* **1**, 249-257.

Liu, P.C. and D.B. Ross, 1980. Airborne measurements of wave growth for stable and unstable atmospheres in Lake Michigan. *J. Phys. Oceanogr.* **10**, 1842-1853.

Long, A., 1986. Towards a C-band radar sea echo model for the ERS-1 scatterometer, in: Proc. conference on spectral signatures, Les Arcs, Dec. 1985; ESA SP-247, European Space Agency, Paris.

Long, R.B., 1973. Scattering of surface waves by an irregular bottom. *J. Geophys. Res.* **78**, 7861-7870.

Long, R.B., 1980. The statistical evaluation of directional spectrum estimates derived from pitch/roll buoy data. *J. Phys. Oceanogr.* **10**, 944-952.

Long, R.B. and K. Hasselmann, 1979. A variational technique for extracting directional spectra from multi-component wave data. *J. Phys. Oceanogr.* **9**, 373-381.

Long, R.B. and W.C. Thacker, 1989. Data assimilation into a numerical equatorial ocean model, part 1: The model and the assimilation algorithm. *Dynam. Atmos. Ocean* **13**, 379-412.

Longuet-Higgins, M.S., 1969. On wave breaking and the equilibrium spectrum of wind-generated waves. *Proc. Roy. Soc. London* **A310**, 151-159.

Longuet-Higgins, M.S., 1978. The instabilities of gravity waves of finite amplitude in deep-water. II. Subharmonics. *Proc. Roy. Soc. London* **A360**, 489-505.

Longuet-Higgins, M.S., 1988. Mechanisms of wave breaking in deep water, p1-30 in: Sea surface sound, B.R. Kerman (ed); Kluwer Press, Boston MA.

Longuet-Higgins, M.S. and E.D. Cokelet, 1978. The deformation of steep surface waves on water. II. Growth of normal mode instabilities. *Proc. Roy. Soc. London* **A364**, 1-28.

Longuet-Higgins, M.S. and R.W. Stewart, 1961. Radiation stress and mass transport in gravity waves, with application to 'surf-beats'. *J. Fluid Mech.* **10**, 529-549.

Longuet-Higgins, M.S., D.E. Cartwright and N.D. Smith, 1963. Observations of the directional spectrum of sea waves using the motions of a floating buoy, p111-136 in: Ocean Wave Spectra; Englewood Cliffs, New York.

Lorenz, E.N., 1987. Deterministic and stochastic aspects of atmospheric dynamics, p159-179 in: Irreversible phenomena and dynamical systems analysis in geoscience, C. Nicolis and G. Nicolis (eds); Reidel, Dordrecht.

Luenberger, D., 1969. Optimization by vector space methods. Wiley, New York.

Luke, J.C., 1967. A variational principle for a fluid with a free surface. *J. Fluid Mech.* **27**, 395-397.

Lundgren, T.S., 1967. Distribution function in the statistical theory of turbulence. *J. Phys. Fluids* **10**, 969-975.

Maat, N., C. Kraan and W.A. Oost, 1991. The roughness of wind waves. *Boundary Layer Meteorol.* **54**, 89-103.

Madsen, O.S., Y.-K. Poon and H.C. Graber, 1989. Spectral wave attenuation by bottom friction: theory, p492-504 in: Proc. 21st conference coastal engineering, Malaga; American Society of Civil Engineers, New York.

Makin, V.K. and D.V. Chalikov, 1979. Numerical modeling of air structure above waves. Izv. Akad. Nauk. SSSR. *Atmos. Ocean Phys.* **15**, 292-299. (Engl. transl.: *Izv. Atm. Oc. Phys.* **15**, 199-204).

Mardia, K.V., 1972. Statistics of directional data. Academic Press, London.

Masson, D., 1990. Observations of the response of sea waves to veering winds. *J. Phys. Oceanogr.* **20**, 1876-1885.

Masson, D. and P.H. LeBlond, 1989. Spectral evolution of wind-generated surface gravity waves in a disperced ice field. *J. Fluid Mech.* **202**, 43-81.

Mastenbroek, C., G.J.H. Burgers and P.A.E.M. Janssen, 1993. The dynamical coupling of a wave model and a storm surge model through the atmospheric boundary layer. *J. Phys. Oceanogr.* **23**, 1856-1866.

Mastenbroek, C., V.K. Makin, A.C. Voorrips and G.J. Komen, 1994. Validation of ERS-1 altimeter wave height measurements in a North Sea wave model. To appear in *The Atmosphere Ocean System.*

McWilliams, J.C., W. Brechner Owens and B. Lien Hua, 1986. An objective analysis of the Polymode local dynamics experiment, part 1: general formalism and statistical model selection. *J. Phys. Oceanogr.* **16**, 483-522.

Melville, W.K. and R.J. Rapp, 1985. Momentum flux in breaking waves. *Nature* **317**, 514-516.

Miles, J.W., 1957. On the generation of surface waves by shear flows. *J. Fluid Mech.* **3**, 185-204.

Miles, J.W., 1959a. On the generation of surface waves by shear flows, part 2. *J. Fluid Mech.* **6**, 568-582.

Miles, J.W., 1959b. On the generation of surface waves by shear flows, part 3. *J. Fluid Mech.* **6**, 583-598.

Miles, J.W., 1961. On the stability of heterogeneous shear flows. *J. Fluid Mech.* **10**, 496.

Miles, J.W., 1977. On Hamilton's principle for surface waves. *J. Fluid Mech.* **83**, 153.

Mises, R. von, 1952. Wahrscheinlichkeit, Statistik und Wahrheit, 3rd ed. Vienna, Austria. English translation: Probability, statistics and truth. New York, 1957.

Mitsuyasu, H., 1968. On the growth of the spectrum of wind-generated waves. 1. *Rep. Res. Inst. Appl. Mech., Kyushu Univ.* **16**, 251-264.

Mitsuyasu, H., 1969. On the growth of the spectrum of wind-generated waves. 2. *Rep. Res. Inst. Appl. Mech., Kyushu Univ.* **17**, 235-243.

Mitsuyasu, H., R. Nakayama and T. Komori, 1971. Observations of the wind and waves in Hakata Bay. *Rep. Res. Inst. Appl. Mech., Kyushu Univ.* **19**, 37-74.

Mitsuyasu, H., F. Tasai, T. Suhara, S. Mizuno, M. Ohkusu, T. Honda and K. Rikiishi, 1975. Observation of the directional spectrum of ocean waves using a cloverleaf buoy. *J. Phys. Oceanogr.* **5**, 750-760.

Mizuno, S., 1976. Pressure measurements above mechanically generated water waves. 1. *Rep. Res. Inst. Appl. Mech., Kyushu Univ.* **23**, 113-129.

Moerkerken, R.A. van, 1991. LAM and NEDWAM statistics over the period October 1990 - April 1991. KNMI Techn. report, TR-137. De Bilt.

Molteni, F. and T.N. Palmer, 1991. A real-time scheme for the prediction of forecast skill. *Monthly Weather Review* **119**, 1088-1097.

Monahan, H.H. and P. Armendariz, 1971. Gust factor variations with height and atmospheric stability. *J. Geophy. Res.* **76**, 5807-5818.

Monaldo, F.M., 1988. Expected differences between bouy and radar altimeter estimates of wind speed and significant wave height and their implications on buoy-altimeter comparison. *J. Geophys. Res.* **C93**, 2285-2302.

Monaldo, F.M. and D.R. Lyzenga, 1986. On the estimation of wave slope and height variance spectra from SAR imagery. *IEEE Transactions on Geoscience and Remote Sensing* **GE-24**, 543-551.

Monbaliu, J., 1992a. Wind driven seas: Optimal parameter choice for the wind input term. *J. Wind Engineering and Industrial Aerodynamics* **41-44**, 2499-2510.

Monbaliu, J., 1992b. Wind and waves. Investigation of an optimalization approach to parameter estimation. PhD thesis, Department of Civil Engineering, Katholieke Universiteit, Leuven.

Monin, A.S. and A.M. Yaglom, 1971. Statistical fluid mechanics 1. The MIT Press, Cambridge, MA, 769p.

Monin, A.S. and A.M. Yaglom, 1975. Statistical fluid mechanics 2. The MIT Press, Cambridge, MA, 874p.

Moor, G. de, J. Lanckneus, F. van Overmeire, P. van den Broeck and E. Martens, 1989. Volumetric analysis of residual sediment migrations on continental shelf sand banks in the southern bight (North Sea). Progress in Belgian Oceanographic Research, G. Pichot (ed); 129-146. Science Policy Office, Brussels.

Morris, V.F., 1991. The Bonner Bridge storm. *Mariners Weather Log* **35**, No. 2, 4-9.

Morse, P.M. and H. Feshbach, 1953. Methods of theoretical physics. Vol. I and II. McGraw-Hill, New York, 1978p.

Müller, P., 1976. Parameterization of one-dimensional wind wave spectra and their dependence on the state of development. Hamburger Geophysikalische Einzelschriften, Heft 31, 177p.

NAG (Numerical algorithm group), 1990. The NAG Fortran Library Introductory Guide, Mark 14. NAG, Oxford, UK.

Newell, A.L. and V.E. Zakharov, 1992. Rough sea foam. *Phys. Rev. Letters* **69**, 1149-1151.

Nikolayeva, Y.I. and L.S. Tsimring, 1986. Kinetic model of the wind generation of waves by a turbulent wind. *Izv. Acad. Sci. USSR, Atmos. Ocean Phys.* **22**, 102-107.

Nordeng, T.E., 1991. On the wave age dependent drag coefficient and roughness length at sea. *J. Geophys. Res.* **C96**, 7167-7174.

Offiler, D., 1987. Final report of a study on the development of ERS-1 scatterometer wind retrieval algorithms. UK Meteorological Office, Bracknell, EP/JVD/14.

Orzag, S.A., 1970. Transform method for calculation of vector coupled sums: application to the spectral form of the vorticity equations. *J. Atmos. Sci.* **27**, 890-895.

Ou, S.H. and F.L.W. Tang, 1974. Wave characteristics in the Taiwan Straits, p139-158 in: Proc. int. symp. on ocean wave measurement and analysis, New Orleans, Louisiana, vol 2; American Society of Civil Engineers, New York.

Peierls, R.E., 1929. Zur kinetischen Theorie der Wärmeleitungen in Kristallen. *Ann. Phys.* **3**, 1055-1101.

Peierls, R.E., 1955. The quantum theory of solids. Clarendon Press, Oxford.

Perrie, W. and B. Toulany, 1991. Directional spectra from the B10 hindcast during LEWEX: the wave-ice interaction, p173-176 in: Directional ocean wave spectra, R.C. Beal (ed); The Johns Hopkins University Press, Baltimore, 218p.

Phillips, O.M., 1957. On the generation of waves by turbulent wind. *J. Fluid Mech.* **2**, 417-445.

Phillips, O.M., 1958. The equilibrium range in the spectrum of wind-generated water waves. *J. Fluid Mech.* **4**, 426-434.

Phillips, O.M., 1960. The dynamics of unsteady gravity waves of finite amplitude, part 1. *J. Fluid Mech.* **9**, 193-217.

Phillips, O.M., 1977. The dynamics of the upper ocean, Cambridge University Press, Cambridge, 336p.

Phillips, O.M., 1985. Spectral and statistical properties of the equilibrium range in wind-generated gravity waves. *J. Fluid Mech.* **156**, 505-531.

Pierson, W.J., 1990. Examples of, reasons for, and consequences of the poor quality of wind data from ships for the marine boundary layer: Implications for remote sensing. *J. Geophys. Res.* **C95**, 13313-13340.

Pierson, W.J.,Jr. and L. Moskowitz, 1964. A proposed spectral form for fully developed wind seas based on the similarity theory of S.A. Kitaigorodskii. *J. Geophys. Res.* **69**, 5181.

Pierson, W.J., G. Neumann and R.W. James, 1955. Practical methods for observing and forecasting ocean waves by means of wave spectra and statistics. H.O. Pub 603, US Navy Hydrographic Office.

Pierson, W.J., M.A. Donelan and W.H. Hui, 1992. Linear and nonlinear propagation of water wave groups. *J. Geophys. Res.* **C97**, 5607-5621.

Plant, W.J., 1982. A relation between wind stress and wave slope. *J. Geophys. Res.* **C87**, 1961-1967.

Plant, W.J., 1987. The microwave measurement of ocean-wave directional spectra. *Johns Hopkins APL Techn. Dig.* **1**, 55p.

Plant, W.J. and J.W. Wright, 1977. Growth and equilibrium of short gravity waves in a wind-wave tank. *J. Fluid Mech.* **82**, 767-793.

Plant, W.J., W.C. Keller and A. Cross, 1983. Parametric dependence of the ocean wave radar modulation transfer function. *J. Geophys. Res.* **C88**, 9747-9756.

Poisson, S.D., 1816. Mémoire sur la théorie des ondes. *Mém. Acad. R. Sci.* **1e** série, i.

Prigogine, I., 1962. Non-equilibrium statistical mechanics. Wiley Interscience, New York.

Putnam, J.A. and J.W. Johson, 1949. The dissipation of wave energy by bottom friction. *Trans. Am. Geophys. Union* **30**, 67-74.

Rapp, R.J. and W.K. Melville, 1990. Laboratory measurements of deep water breaking waves. *Philos. Trans. R. Soc. London*, **A331**, 735-780.

Reid, R.O. and K. Kajiura, 1957. On the damping of gravity waves over a permeable sea bed. *Trans. Am. Geophys. Union*, **38**, 662.

Resio, D.T., 1987, Shallow-water waves 1: Theory. *J. Waterway, Port. Coastal and Ocean Engineering* **113**, 264-281.

Riepma, H.W. and E. Bouws, 1989. Preliminary results of the NEDWAM wave model, p248-256 in: Proc. of the second international workshop on wave hindcasting and forecasting; Environment Canada, Ontario.

Riley, D.S., M.A. Donelan and W.H. Hui, 1982. An extended Miles' theory for wave generation by wind. *Boundary Layer Meteorol.* **22**, 209-225.

Roach, G.F., 1982. Green's functions. Cambridge University Press, Cambridge, 325p.

Robert, A.J., J. Henderson and C. Turnbull, 1972. An implicit time integration scheme for baroclinic models of the atmosphere. *Monthly Weather Review* **100** 329-335.

Robinson, A.R. and L.J. Walstad, 1987. The Harvard Ocean Model: Calibration and applications to dynamical process, forecasting and data assimilation studies. *J. Appl. Num. Math.* **3**, 89-131.

Roeckner, E., L. Dumenil, E. Kirk, F. Lunkeit, M. Ponater, B. Rockel, R. Sansen and U. Schlese, 1989. The Hamburg version of the ECMWF model (ECHAM), GARP report 13, WMO/TD, World Meteorological Organisation, Geneva, 332p.

Roeckner, E., K. Arpe, L. Bengtsson, S. Brinkop, L. Dümenil, M. Esch, E. Kirk, F. Lunkeit, M. Ponater, B. Rockel, R. Sausen, U. Schlese, S. Schubert and M.

Windelband, 1992. Simulation of the present-day climate with the ECHAM model: impact of model physics and resolution. Max-Planck-Institut für Meteorologie report 93. Hamburg, 193p.

Roest, P.W., 1960. Wave recording on the IJsselmeer, p53-58 in: Proc. seventh coastal engineering conference, Den Haag; Council Wave Res, Berkeley.

Romeiser, R., 1993. Global validation of the wave model WAM over a one year period using GEOSAT wave height data. *J. Geophys. Res.* **C98**, 4713-4726.

Rosenthal, W., 1978. Energy exchange between surface waves and motion of sediments. *J. Geophys. Res.* **83**, 1980-1982.

Ross, C.M., 1988. The operational wave models, UK Meteorological Office, documentation paper 5.1 of Operat. Num. Weath. Pred. System. Bracknell.

Saffman, P.G. and D.C. Wilcox, 1974. Turbulence-model predictions for turbulent boundary layers. *AIAA J.* **12**, 541-546.

Scales, L.E., 1985. Introduction to non-linear optimization. Springer Verlag, New York.

Schröter, J., F. Feindt, W. Alpers and W.C. Keller, 1986. Measurements of the ocean wave-radar modulation transfer function at 4.3 GHz. *J. Geophys. Res.* **C91**, 932-946.

Seliger, R.L. and G.B. Whitham, 1968. Variational principles in continuum mechanics. *Proc. Roy. Soc. London* **A305**, 1-25.

Sethuraman, S., 1979. Atmospheric turbulence and storm surge due to hurricane Belle (1976). *Monthly Weather Review* **107**, 314-321.

Shapiro, M.A., E.G. Donall, P.J. Neiman, L.S. Fedor and N. Gonzales, 1991. Recent refinements in the conceptual models of extratropical cyclones, p6-14 in: First international winter storms symposium, New Orleans, LA, January 13-18; Amer. Meteor. Soc., Boston.

Shearman, E.D.R., 1986. A review of methods of remote sensing of sea surface conditions by HF radar and design conditions for narrow-beam stystems. *IEEE J. of Oceanic Engineering*, OE-11 **2**, 150-157.

Shemdin, O.H. and P.A. Hwang, 1988. Comparison of measured and predicted sea surface spectra of short waves. *J. Geophys. Res.* **C93**, 13883-13890.

Shemdin, P., K. Hasselmann, S.V. Hsiao and K. Herterich, 1978. Non-linear and linear bottom interaction effects in shallow water, p347-372 in: Turbulent fluxes through the sea surface, wave dynamics and prediction, A. Favre and K. Hasselmann (eds); Plenum, New York, 677p.

Shyu, J.H. and O.M. Phillips, 1990. The blockage of gravity and capillary waves by longer waves and currents. *J. Fluid Mech.* **217**, 115-141.

Simmons, A., 1991. Development of the operational 31-level T213 version of the ECMWF forecast model. *ECMWF Newsletter* **56**, 3-13.

Sleath, J.F.A., 1984. Sea bed mechanics. Wiley Interscience, New York, 334p.

Sleath, J.F.A., 1987. Turbulent oscillatory flow over rough beds. *J. Fluid Mech.* **182**, 369-409.

Smith, S.A. and E.G. Banke, 1975. Variation of the sea surface drag coefficient with wind speed. *Q. J. Royal Meteorol. Soc.* **101**, 665-673.

Smith, S.D., 1980. Wind stress and heat flux over the ocean in gale force winds. *J. Phys. Oceanogr.* **10**, 709-726.

Smith, S.D., 1988. Coefficients for sea surface wind stress, heat flux and wind profiles as a function of wind speed and temperature. *J. Geophys. Res.* **C93**, 15467-15472.

Smith, S.D., K.B. Katsaros, W.A. Oost and P.G. Mestager, 1990. Two major experiment in the humidity exchange over the sea (HEXOS) program. *Bull. Amer. Meteor. Soc.* **71**, 161-172.

Smith, S.D., R.J. Anderson, W.A. Oost, C. Kraan, N. Maat, J. DeCosmo, K.B. Katsaros, K.L. Davidson, K. Bumke, L. Hasse and H.M. Chadwick, 1992. Sea surface wind stress and drag coefficients: the HEXOS results. *Boundary Layer Meteorol.* **60**, 109-142.

Snodgrass, F.E., G.W. Groves, K. Hasselmann, G.R. Miller, W.H. Munk and W.H. Powers, 1966. Propagation of ocean swell across the Pacific. *Philos. Trans. Roy. Soc. London* **A249**, 431.

Snyder, R.L., 1974. A field study of wave-induced pressure fluctuation above surface gravity waves. *J. Marine Res.* **32**, 497-531.

Snyder, R.L., F.W. Dobson, J.A. Elliott and R.B. Long, 1981. Array measurements of atmospheric pressure fluctuations above surface gravity waves. *J. Fluid Mech.* **102**, 1-59.

Snyder, R.L., L.M. Lawson and R.B. Long, 1992. Inverse modelling of the action-balance equation, part 1: Source expansion and adjoint-model equations. *J. Phys. Oceanogr.* **22**, 1540-1555.

Snyder, R.L., W.C. Thacker, K. Hasselmann, S. Hasselmann and G. Barzel, 1993. Implementation of an efficient scheme for calculating nonlinear transfer from wave-wave interactions. *J. Geophys. Res.* **C98** 14507-14525.

SPM, 1973, 1977 and 1984. Shore Protection Manual, vol I. US Army Coastal Engineering Research Center, Washington.

Staub, C., I.G. Jonsson and I.A. Svendsen, 1985. Variation of sediment suspension in oscillatory flow, p2310-2321 in: Proc. 19th conference coastal engineering, Houston 1984, Vol. III; American Society of Civil Engineers, New York.

Stewart, R.H. and C. Teague, 1980. Dekameter radar observations of ocean wave growth and decay. *J. Phys. Oceanogr.* **10**, 128-143.

Stewart, R.W., 1967. Mechanics of the air-sea interface. *Phys. Fluids Suppl.* **10**, S47-S55.

Stewart, R.W., 1974. The air-sea momentum exchange. *Boundary Layer Meteorol.* **6**, 151-167.

Stoffelen, A. and D.L.T. Anderson, 1992. ERS-1 scatterometer calibration and validation activities at ECMWF: a) the quality and the characteristics of the radar backscatter measurements, in: Proc. central symposium of the international space year conference, Munich, Germany, 30 March - 4 April; ESA SP-341, European Space Agency, Paris.

Stokes, G.G., 1847. On the theory of oscillatory waves. *Trans. Camb. Phil. Soc.* **8**, 441-455.

Stull, R.B., 1988. An introduction to boundary layer meteorology. Atmospheric sciences library, Kluwer Academic Publishers, Dordrecht, 666p.

SWAMP group: J.H. Allender, T.P. Barnett, L. Bertotti, J. Bruinsma, V.J. Cardone, L. Cavaleri, J. Ephraums, B. Golding, A. Greenwood, J. Guddal, H. Günther, K. Hasselmann, S. Hasselmann, P. Joseph, S. Kawai, G.J. Komen, L. Lawson, H. Linné, R.B. Long, M. Lybanon, E. Maeland, W. Rosenthal, Y. Toba, T. Uji and W.J.P. de Voogt, 1985. Sea wave modeling project (SWAMP). An intercomparison study of wind wave predictions models, part 1: Principal results and conclusions, in: Ocean wave modeling; Plenum, New York, 256p.

SWIM group: E. Bouws, J.A. Ewing, J. Ephraums, P. Francis, H. Günther, P.A.E.M. Janssen, G.J. Komen, W. Rosenthal and W.J.P. de Voogt, 1985. Shallow water intercomparison of wave prediction models (SWIM). *Q. J.*

Royal Meteorol. Soc. **111**, 1087-1113.

Sverdrup, H.U. and W.H. Munk, 1947. Wind sea and swell: Theory of relations for forecasting. H.O. Pub. 601, US Navy Hydrographic Office, Washington, DC, 44p.

Talagrand, O. and P. Courtier, 1987. Variational assimilation of meteorological observations with the adjoint vorticity equation, part 1: Theory. *Q. J. Royal Meteorol. Soc.* **113**, 1311-1328.

Tarantola, A., 1987. Inverse problem theory, methods for data fitting and model parameter estimation. Elsevier, Amsterdam, 613p.

Taylor, P.A. and R.J. Lee, 1984. Simple guidelines for estimating wind speed variations due to small-scale topographic features. *Climatol. Bull.* **18**, 3-32.

Tennekes, H., 1973. Similarity laws and scale relations in planetary boundary layers, p177-232 in: Proc. of the workshop on micrometeorology, D.A. Haugen (ed); Amer. Meteor. Soc., Boston. MA.

Tennekes, H., 1981. Similarity relations, scaling laws and spectral dynamics, p37-69 in: Atmospheric turbulence and air pollution modelling, F.T.M. Nieuwstadt and H. van Dop (eds); Reidel, Dordrecht.

Thacker, W.C., 1988. Three lectures on fitting numerical models to observations. GKSS 87/E/65, Geesthacht, 64p.

Thacker, W.C. and R.B. Long, 1988. Fitting dynamics to data. *J. Geophys. Res.* **C93**, 1227-1240.

Thomas, J., 1988. Retrieval of energy spectra from measured data for assimilation into a wave model, *Q. J. Royal Meteorol. Soc.* **114**, 781-800.

Thijsse, J.Th., 1949. Dimensions of wind-generated waves. p80-81 in: Association d'Océanographie Physique, Procès-Verbaux no. 4, General Assembly at Oslo; Secrétariat de l'Association: Geofysisk Institutt, Bergen, Norway.

Toba, Y., 1973. Local balance in the air-sea boundary process, 3. On the spectrum of wind waves. *J. Oceanogr. Soc., Japan* **29**, 209-220.

Toba, Y. and M. Koga, 1986. A parameter describing overall conditions of wave breaking, whitecapping, sea-spray production and wind stress, p37-47 in: Oceanic whitecaps, E.C. Monahan and G. Mac Niocaill (eds); Reidel, Dordrecht, 294p.

Tolman, H.L., 1990a. Wind wave propagation in tidel seas. Doctoral thesis. Delft Univ. of Techn., also Comm. Hydr. Geotechn. Eng., Delft Univ. of Techn. report 90-1, 135p.

Tolman, H.L., 1990b. The influence of unsteady depths and currents of tides on wind wave propagation in shelf seas, *J. Phys. Oceanogr.* **20**, 1166-1174.

Tolman, H.L., 1991a. Effects of tides and storm surges on North Sea wind waves. *J. Phys. Oceanogr.* **21**, 766-781.

Tolman, H.L., 1991b. A third-generation model for wind on slowly varying, unsteady and inhomogeneous depths and currents. *J. Phys. Oceanogr.* **21**, 782-797.

Tolman, H.L., 1992. Effects of numerics on the physics in a third-generation wind-wave model. *J. Phys. Oceanogr.* **22**, 1095-1111.

Trowbridge, J. and O.S. Madsen, 1984. Turbulent wave boundary layers: 1. Model formulation and first order solution. *J. Geophys. Res.* **C89**, 7989-7999.

Ulbrich, U., G. Bürger, D. Schriever, H. von Storch, S.L. Weber and G. Schmitz, 1993. The effect of a regional increase in ocean surface roughness on the tropospheric circulation: a GCM experiment. *Climate Dynamics* **8**, 277-285.

Valenzuela, G.R., 1978. Theories for the interaction of electromagnetic and ocean waves - A review. *Boundary Layer Meteorol.* **13**, 61-85.

Valk, C.F. de, and C.J. Calkoen, 1989. Wave data assimilation in a third generation wave prediction model for the North Sea - An optimal control approach. Delft Hydraulics Laboratory, report X38, Delft, 123p.

Vedenov, A.A., E.P. Velikhov and R.Z. Sagdeev, 1961. Nonlinear oscillations of a rarefied plasma. *Nucl. Fusion* **1**, 182.

Vledder, G.Ph. van, 1990. Directional response of wind waves to turning winds. Doctoral thesis. Delft Univ. of Techn., also Comm. Hydr. Geotechn. Eng., Delft Univ. of Techn. report 90-2, 255 p.

Vledder, G.Ph. van, and L.H. Holthuijsen, 1993. The directional response of ocean waves to turning winds. *J. Phys. Oceanogr.* **23**, 177-192.

Vlugt, A.J.M. van der, 1984. Experiences with the WAVEC buoy, in: Proc. symposium on description and modelling of directional seas. June 18-20, Technical University, Denmark, Danish Hydraulics Institute and Danish Maritime Institute, paper A3.

Voorrips, A.C., V.K. Makin and G.J. Komen, 1992. On the coupling between LAM and NEDWAM in the APL. KNMI Afdeling Oceanografisch Onderzoek memo, OO-92-15, 10p.

Voorrips, A.C., V.K. Makin and G.J. Komen, 1994a. Atmospheric stratification: consequences for operational wave modelling. To appear in: The Air-Sea Interface, M.A. Donelan, W.H. Hui and W.J. Plant (eds); The University of Toronto Press, Toronto.

Voorrips, A.C., V.K. Makin and G.J. Komen, 1994b. The influence of stratification on the growth of water waves. Submitted for publication in *Boundary Layer Meteorol.*

Wahba, G., 1990. Spline models for observational data. Society for Industrial and Applied Mathematics, Philadelphia.

Wakimoto, R.M., W. Blier and C. Liu, 1991. A radar perspective of IOP4 during ERICA, First international winter storms symposium, New Orleans, LA, January 13-18, p124-129, Amer. Meteor. Soc. Boston, p124-129.

Walsh, E.J., D.W. Hancock III, D.E. Hines, R.N. Swift and J.F. Scott, 1985. Directional wave spectra measured with the surface contour radar. *J. Phys. Oceanogr.* **15**, 566-592.

WAMDI group: S. Hasselmann, K. Hasselmann, E. Bauer, P.A.E.M. Janssen, G.J. Komen, L. Bertotti, P. Lionello, A. Guillaume, V.C. Cardone, J.A. Greenwood, M. Reistad, L. Zambresky and J.A. Ewing, 1988. The WAM model - a third generation ocean wave prediction model. *J. Phys. Oceanogr.* **18**, 1775-1810.

Weber, S.L., 1988. The energy balance of finite depth gravity waves. *J. Geophys. Res.* **C93**, 3601-3607.

Weber, S.L., 1989. Surface gravity waves and turbulent bottom friction. PhD thesis, Univ. of Utrecht, The Netherlands, 128p.

Weber, S.L., 1991a. Bottom friction for wind sea and swell in extreme depth-limited situations. *J. Phys. Oceanogr.* **21**, 149-172.

Weber, S.L., 1991b. Eddy-viscosity and drag-law models for random ocean wave dissipation. *J. Fluid Mech.* **232**, 73-98.

Weber, S.L., H. von Storch, P. Viterbo and L. Zambresky, 1993. Coupling an ocean wave model to an atmospheric general circulation model. *Climate Dynamics* **9**, 63-69.

Weber, S.L., 1994. Statistics of the air-sea fluxes of momentum and mechanical energy in a coupled wave-atmosphere model. To appear in *J. Phys. Oceanogr.*

Weller, R.A., M.A. Donelan, M.G. Briscoe and N.E. Huang, 1991. Riding the crest: a tale of two wave experiments. *Bull. Amer. Meteor. Soc.* **72**, 163-183.

Whitham, G.B., 1965. A general approach to linear and non-linear dispersive waves using a Lagrangian. *J. Fluid Mech.* **22**, 273-283.

Whitham, G.B., 1974. Linear and nonlinear waves. Wiley, New York, 636p.

Willebrand, J., 1975. Energy transport in a nonlinear and inhomogeneous random gravity wave field. *J. Fluid Mech.* **70**, 113-126.

Wilson, K.C., 1989. Friction of wave-induced sheet flow. *Coastal Engineering* **12**, 371-379.

Woiceshyn, P. and P.A.E.M. Janssen, 1992. Sensitivity study-scatterometer retrievals with wave age parameter, p133-139 in: Proc. of the workshop on ERS-1 Geophysical Validation, Penhors, France, 27-30 April 1992; ESA WPP-36, European Space Agency, Paris.

Wolf, J., K.P. Hubbert and R.A. Flather, 1988. A feasibility study for the development of a joint surge and wave model. Report No. 1 Proudman Oceanographic Laboratory, Birkenhead, 109p.

Wright, J.W., 1968. A new model for sea clutter. *IEEE Transactions on Antennas and Propagation*, **AP-16**, 217-223.

Wu, J., 1980. Wind stress coefficients over the sea surface near neutral conditions - a revisit. *J. Phys. Oceanogr.* **10**, 727-740.

Wu, J., 1982. Wind-stress coefficients over sea surface from breeze to hurricane. *J. Geophy. Res.* **C87**, 9704-9706.

Wu, X. and R.A. Flather, 1992. Hindcasting waves using a coupled wave-tide-surge model, p159-170 in: Third international workshop on wave hindcasting and forecasting, Montreal, Quebec, 19-22 May; Environment Canada, Ontario.

Yan, L. and E. Bouws, 1987. Possible causes of the attenuation of swell in the southern North Sea. KNMI Afdeling Oceanografisch Onderzoek memo, OO-87-21. 16p.

Yamamoto, T. and T. Tori, 1986. Seabed shear modulus profile inversion using surface (gravity) wave-induced bottom motion. *Geophys. J. Roy. Astr. Soc.* **85**, 413-431.

Yamartino, R.J., 1984. A comparison of several single-pass estimators of the standard deviation of wind direction. *J. Climate Appl. Meteor.* **23**, 1362-1366.

Young, I.R. and M.L. Banner, 1992. Numerical experiments on the evolution of fetch limited waves, p267-275 in: Breaking waves, M.L. Banner and R.H.J. Grimshaw (eds); Springer, Berlin.

Young, I.R. and R.H. Sobey, 1985. Measurements of the wind-wave energy flux in an opposing wind. *J. Fluid Mech.* **151**, 427-442.

Yuan Yeli, C.C. Tung and Norden E. Huang, 1986. Statistical characteristics of breaking waves, p265-272 in: Wave dynamics and radio probing of the ocean surface, O.M. Phillips and K. Hasselmann (eds); Plenum, New York, 694p.

Yuen, H.C. and B.M. Lake, 1982. Nonlinear dynamics of deep water gravity waves. *Adv. Appl. Mech.* **22**, 67-229.

Zakharov, V.E., 1968. Stability of periodic waves of finite amplitude on the surface of a deep fluid. *J. Appl. Mech. Techn. Phys.* **9**, 190-194.

Zambresky, L., 1989. A verification study of the global WAM model. December 1987 - November 1988. ECMWF Techn. report 63, ECMWF, Reading.

Zambresky, L., 1991. An evaluation of two WAM hindcasts for LEWEX, p167-172 in: Directional ocean wave spectra, R.C. Beal (ed); The Johns Hopkins University Press, Baltimore, 218p.

Index

and directional relaxation 191(f), 193
  dependence of drag 74, 102, **104**, 110(f), 219, 248(f), **373**
  dependence of Phillips' constant **104**, 111
  for wind/wave data assimilation 409, 473, 475, 477
wave–bottom interactions (*see also* shallow water waves) 156
wave breaking **143**
  probability model **149**
  quasi-saturated model **145**
  whitecap model **144**
wave climatology
  in the North Sea 195
wave–current interaction 17, 43, 213, **349**
  current ring **353**
  Gulf Stream 357
  large-scale tidal flows **350**
  open ocean **355**
  quasi-stationary approach 352
  swell trapping 358
wave diffraction 263
  in sea ice 172
wave diffusion coefficient **99**
wave direction (mean) 33, **190**, 396
wave dissipation/attenuation
  bottom elasticity **169**, 341
  bottom friction **156**, 164, 335
  percolation **156**, 157, 201, 335
  surface processes (*see also* wave breaking) **143**
  swell 93, 148, 165, 167, **200**, 222, **251**, 252(f)
wave energy: *see* energy
wave field
  analysis (*see also* analysed fields) **300**
  comparisons 290(f), 296(f), 298(f), 313(f)
  corrected 408, 409, 432, 439, 470
wave forecast 301, **307**
  prediction of forecast skill 308, 311
wave generation/growth **71**
  dimensionless growth vs. dimensionless frequency 85(f), 219(f)
  effect of atmospheric stability 86(f), 87(f)
  effect of gustiness 88
  Miles' theory 72, **75**, 218
  Phillips' theory 43, 72
  quasi-linear theory 74, **95**, 111, 220
  wind input source function 47, 88, 92, **217**, **314**
wave-induced stress **52**, 73, 103, 217, 220
  linear theory of wave generation 78
  quasi-linear theory of wave generation 96, 99
  reduction of, due to quasi-linear theory 109(f)
  time dependence 247(f)
  vortex forces produced by 81
  wave age dependence 104, 105(f)
  wave momentum spectrum and 103
wave maxima, distribution of 34
wave models
  characteristics **265**
  first generation models 4, 206, 381
  large-scale (coarse) models **264**
  limited-area (fine) models (LAM) 263, **264**

NEDWAM 303, 304(t), 333, 465
  scale **263**
  second generation wave models 4, 197, 206, 381
  spatial/spectral resolution: *see* resolution
  third generation models 4, 199, 206, 381
  validation/refutation 261
  WAM: *see* WAM model
  WAVEWATCH 357
wave momentum spectrum **103**
wave–mean flow interactions 81, 96
wave packets **20**
  in the particle picture **37**, **44**, 135
  number of 37, 46, 139
wave ray equations: *see* Hamilton's equations
wave spectrum: *see* spectrum
wave steepness
  overall (integral) 145
  as small parameter 13, 121
  sensitivity of wave breaking to 154
wave system 398
wave tank experiments on bottom scattering 167
wave vorticity 15
wavenumber scaling 199
WAVEWATCH 357
wave–wave interactions (*see also* nonlinear energy transfer) **113**
  short wave–long wave interactions 68, **131**, 388
waves: *see* ocean surface waves
weak-in-the-mean processes **41**, 144
weak turbulence description 39
whitecapping dissipation 47, **144**, **268**, **221**
whitecaps **143**, 144(f)
  geometric similarity 145
wind **48**, **483**
  altimeter measurements **66**, 67(f)
  critical height of wind-wave interaction 43
  distortion by waves **99**
  drag: *see* drag coefficient
  effective neutral wind 269, 270(t)
  errors **283**
  geostrophic 48, **50**, 57, 463
  gustiness 73, 88, **271**, **322**
  interpolation **299**
  logarithmic profile 54, 83, 216
  measurements 62
  mesocale variability 90
  model validation statistics 289(t), 302(t), 303(t)
  modelled vs observed 288(f), 305(f), 330(f)
  orographic effects **293**, 295(f), 297(f)
  resolution **273**, **294**
  scatterometer measurement **68**
  stress: *see* stress
  turning **188**, 248, 250(f)
  variability **320**
  wave-driven 95
wind fields **268**, 287(f), 294(f), 312(f)
  accuracy **279**
  corrected 408, 432, 440, 460, 470, 472, 478(f)
  modelled 287(f), 294(f), 297(f), 312(f)
  scales of motion **268**
wind profile